D1153964

Advances in Geophysical and Environmental Mechanics and Mathematics

Series Editor

Holger Steeb, Institute of Applied Mechanics (CE), University of Stuttgart, Stuttgart, Germany

The series *Advances in Geophysical and Environmental Mechanics and Mathematics* (AGEM2) reports new developments on all topics of geophysical, environmental and astrophysical mechanics and mathematical research and teaching - quickly and informally, but with high quality and the explicit aim to summarize and communicate current knowledge in an accessible way. Books published in this series are conceived as bridging material between upper level textbooks and the forefront of research to serve the following purposes:

- to be a compact and up-to-date source of reference on a well defined topic;
- to serve as an accessible introduction to a specialized field to advanced students and non-specialized researchers from related areas;
- to be a source of advanced teaching material for specialized seminars, courses and schools.

Both monographs and multi-author volumes will be considered for publication. Edited volumes should, however, consist of a very limited number of contributions only. Proceedings will not be considered for the series.

Volumes published in AGEM2 are disseminated both in print and in electronic formats, the electronic archive being available at springerlink.com. The series content is indexed, abstracted and referenced in many abstracting and information services, bibliographic networks, subscription agencies, library networks, and consortia.

Proposals should be sent to the Series Editor or directly to the Responsible Editor at Springer.

Jonathan Kirby

Spectral Methods for the Estimation of the Effective Elastic Thickness of the Lithosphere

 Springer

Jonathan Kirby 🆔
School of Earth and Planetary Sciences
Curtin University
Perth, WA, Australia

ISSN 1866-8348   ISSN 1866-8356   (electronic)
Advances in Geophysical and Environmental Mechanics and Mathematics
ISBN 978-3-031-10860-0   ISBN 978-3-031-10861-7   (eBook)
https://doi.org/10.1007/978-3-031-10861-7

This Springer imprint is published by the registered company Springer Nature Switzerland AG
The registered company address is: Gewerbestrasse 11, 6330 Cham, Switzerland

For Ruth

Preface

The effective elastic thickness—commonly known by its mathematical symbol, T_e—describes an important property of the lithosphere, namely its flexural rigidity, or resistance to flexure by loading. But besides describing this particular behaviour of the Earth, T_e has been shown to play a vital role in tectonic evolution, with zones of low T_e deforming much more readily than regions with high elastic thickness, such as the ancient cratons that core the continents. However, its estimation has proven difficult and even controversial. One of the main reasons for this is that there are three approaches one can take to find its value, and while these methods do not yield wildly different answers, often they do not agree. One approach is *thermo-rheological modelling*, which combines knowledge of the properties of rocks under various conditions with information about the regional temperature and state of stress, and from these determines the integrated strength of the lithosphere and thence T_e. The other two approaches both use the topography and gravity field of the Earth. The method known simply as *forward modelling* fits predictions of the gravity or topography from thin, elastic plate models to actual observations of these data sets, with T_e being one of the primary controlling factors. *Inverse modelling* finds the statistical relationship between gravity and topography in the frequency domain—embodied in the coherence and admittance—and estimates T_e by inversion of those spectral measures against a plate model. It is the third method that is the subject of this book.

To date, while the topics of T_e and plate flexure have formed a small part of broad-content geophysics textbooks, there has been only one book dedicated to them: 2001's *Isostasy and Flexure of the Lithosphere* by Tony Watts. This excellent volume contains almost everything one needs to know about the effective elastic thickness, but while it has a whole chapter on spectral methods of T_e estimation, it does not inform the reader on how to calculate the coherence and admittance. That is where my book comes in. Correct estimation of the gravity and topography spectra is crucial in order to avoid bias in the resulting T_e estimates, and there has been much debate in the literature over the past two decades as to how this should be done most accurately.

I first came across the subject of T_e estimation during my Ph.D. at Edinburgh University, and although my thesis was primarily geodesy, I couldn't resist including a short chapter on some T_e estimates I made, essentially because I found the subject so captivating. I loved the way it brought together such diverse topics as spectral estimation, geology, algebra and calculus, differential equations and finite difference approximations, civil engineering, signal processing, fractal geometry, plate tectonics, mechanics and elasticity theory, and all this just to find out what happens when you pop a volcano on the crust. But therein lay the problem. In order to get up to speed on all these different subjects, one is thrown into a blizzard of journals and textbooks in physics, geophysics, geology, engineering and mathematics. It is both daunting and time-consuming, and the notes I made were all over the place. So one day, I set about tidying them up, and in the process wrote a book, the kind of book I would have wanted when I was a Ph.D. student.

My book is intended to contain everything one might need when embarking on studies in T_e estimation using spectral methods, so that one doesn't always need to consult myriad other sources. I try to derive equations from first principles wherever possible, to give a fuller understanding of the subject. I also focus on the practical application of the theory, which is often half the battle. The book is divided into three parts. Part I is short, non-mathematical and sets the scene, briefly describing what we mean by the lithosphere's effective elastic thickness and why it is important. This is the only part where further reading, while not essential, is probably a good idea, especially if one is new to the subject.

Part II concerns spectral estimation. Many geophysics textbooks and journal articles gloss over this subject, saying '... then take the Fourier transform and ...', but it is actually the most important part of T_e estimation: get the spectrum wrong and you'll get T_e wrong. So I spend a considerable number of pages describing how Fourier transforms in theory are very different to Fourier transforms in practice. Part II then continues with a chapter each on multitaper spectral estimation and the continuous wavelet transform, now the two most popular techniques of T_e estimation. There is a chapter devoted to the admittance and coherence and their computation, and to map projections, as this is also an important—and oft-neglected—aspect of planar spectral estimation of ellipsoidally coordinated data. Importantly, Part II is stand-alone, written with hardly any mention of T_e. For this reason, it will be useful to any scientist who needs the Fourier transform but isn't terribly well-versed in its use (and abuse).

Part III gets to the core of the subject, which is really T_e estimation using Donald Forsyth's load deconvolution method. But since the observables are gravity and topography, each gets a chapter to itself. The 'topography chapter' concerns elasticity and flexure; the 'gravity chapter' deals with potential theory and physical geodesy. Then there is the capstone chapter on load deconvolution, supplemented by a chapter on its practical application to real-world data. But real-world data are not the only kind, so I have included a chapter on generating synthetic gravity and topography models from the flexure of a thin, elastic plate—models that can be used to test the accuracy of the spectral analysis methods.

The book is aimed primarily at geologists and geophysicists of advanced undergraduate or postgraduate level and higher, and is targeted specifically at those wishing to estimate T_e. However, because most of its content is derived from first principles, and I have tried not to assume too much prior knowledge, it may also benefit those who wish to learn about multitaper spectral estimation or the wavelet transform, or elasticity or the finite difference method, for example, and have no interest in the Earth sciences at all. For best results, the reader is encouraged to read Tony Watts's book, and also a review article I wrote in 2014 for *Tectonophysics*, which summarises and charts the history of T_e estimation using spectral methods. The difference between this book and that article is that the former introduces the subject from scratch while the latter was intended more as a useful reference so that established researchers could find who did what and when without wasting time.

While writing this book has been a solo effort, it would not exist were it not for the collaborations and conversations I have had with colleagues around the world. I thank them sincerely. Foremost amongst these is Chris Swain, whom one grant reviewer once described as a 'sage geophysicist'. He is that indeed, and I have enjoyed our work together over the past 20 years immensely. I have also been lucky enough to spend time with some of the other leading researchers in the field, notably Tony Lowry, Marta Pérez-Gussinyé, Frederik Simons and Mark Wieczorek; I thank them for their wisdom, knowledge and company. And of others who have influenced my thinking and thus had an indirect input into this book, I would like to thank Alberto Jiménez-Díaz, my Ph.D. supervisor Roger Hipkin and his colleague Roger Banks at Edinburgh, and the unfortunately late Simon Holmes, who—one memorable day in Vesuvio—strongly encouraged me to read Percival and Walden (1993). And for their more direct input, I thank Tony Lowry for providing his yield strength envelope code, which I used to generate Figs. 1.12–1.15, Tony Watts for supplying Fig. 1.16 and Frederik Simons for assistance with a tricky derivation in Sect. 3.3.2. I would also like to acknowledge the Generic Mapping Tools (GMT) software [Wessel et al. (2019) Geochem Geophys Geosyst vol. 20], which I used to plot most of the figures. And finally, I thank my wife, Ruth, whose patience is no less than godlike.

Perth, Australia   Jonathan Kirby
December 2021

Contents

Abbreviations

1D	One-dimensional
2D	Two-dimensional
3D	Three-dimensional
BCE	Before Common Era
BDT	Brittle-ductile transition
CBA	Complete Bouguer anomaly
CFT	Continuous Fourier transform
CL	Confidence level
CoI	Cone of influence
CWT	Continuous wavelet transform
DEM	Digital elevation model
DFT	Discrete Fourier transform
DoG	Derivative of Gaussian (wavelet)
DPSS	Discrete prolate spheroidal sequence
DPSWF	Discrete prolate spheroidal wave function
DTFT	Discrete-time Fourier transform
DTM	Digital terrain model
fBm	Fractional Brownian motion
FDE	Finite difference equation
FT	Fourier transform
FFT	Fast Fourier transform
GIA	Glacial isostatic adjustment
GNSS	Global navigation satellite systems
InSAR	Interferometric synthetic aperture radar
LAB	Lithosphere-asthenosphere boundary
LiDAR	Light detection and ranging
LTI	Linear time-invariant
MSL	Mean sea level
NSIC	Normalised squared imaginary coherency
NSRC	Normalised squared real coherency
PDE	Partial differential equation

RAPS	Radially averaged power spectrum
RMS	Root mean square
RTM	Residual terrain model
SAR	Synthetic aperture radar
SBA	Simple Bouguer anomaly
SIC	Squared imaginary coherency
SRC	Squared real coherency
SRTM	Shuttle Radar Topography Mission
STFT	Short-time Fourier transform
WFT	Windowed Fourier transform
WT	Wavelet transform
YSE	Yield strength envelope

s_κ	autocovariance sequence
T	period
T_c	crustal thickness
T_e	effective elastic thickness
T_M	mechanical thickness
t	time; thickness
t_0	Airy-Heiskanen isostatic compensation depth
t_c	cone of influence
U_k	1D discrete prolate spheroidal wave function
u	distance
V	volume; deflection (wavenumber domain); gravitational potential
v	deflection (space domain)
$\mathbf{v}_k$	1D discrete prolate spheroidal sequence (Slepian taper)
$\mathsf{var}\{\cdot\}$	variance of a sequence or signal
W	weight; bandwidth; spectrum of a window function; Moho relief (wavenumber domain); gravity potential
w	deflection
$\mathbf{w}_{jk}$	2D discrete prolate spheroidal sequence (Slepian taper)
w_τ	window function
x	distance; Cartesian coordinate
$\mathbf{x}$	2D space-domain position vector
Y_n	surface spherical harmonic function of harmonic degree n
y	Cartesian coordinate
z	Cartesian coordinate; depth
z_m	depth to Moho
α	azimuth; flexural constant
α^2	time concentration measure
β	spectral exponent
β_g	multiplication factor for gravity anomaly type
β^2	frequency concentration measure
Γ	coherency
γ	shear strain; normal gravity
γ^2	coherence
Δ	deconvolution equations matrix determinant
Δg_B	Bouguer gravity anomaly
Δg_F	free-air gravity anomaly
Δx	grid spacing in x-direction
$\Delta \rho_{ab}$	$= \rho_a - \rho_b$
δ	Dirac delta function; phase difference
δg_B	Bouguer correction
δg_F	free-air correction
δg_T	terrain correction
δ_{jk}	Kronecker delta
ε	normal strain; error

η	discrete distance index; density ratio (equivalent topography)
η_0	zero topography transfer function
θ	polar coordinate; angle; geocentric colatitude
κ	summation index; wavenumber
κ_T, κ_B	flexural deconvolution coefficients
λ	wavelength; geodetic longitude; Lamé parameter (bulk modulus)
λ_e	equivalent Fourier wavelength
λ_F	flexural wavelength
λ_k	eigenvalue of the concentration problem
λ_N	Nyquist wavelength
λ_t	coherence transition wavelength
μ	point scale factor; Lamé parameter (shear modulus)
μ_T, μ_B	gravity deconvolution coefficients
ν	Poisson's ratio ($= 0.25$)
ν_T, ν_B	load deconvolution coefficients
ξ	meridional deflection of the vertical; Forsyth's parameter; discrete distance index
Π	rectangle function
ρ	density
ρ_0	Bouguer reduction density ($= 2670$ kg m^{-3})
ρ_c	mean crust density ($= 2800$ kg m^{-3})
ρ_i	ice density ($= 917$ kg m^{-3})
ρ_m	mean mantle density ($= 3300$ kg m^{-3})
ρ_s	sediment or infill density
ρ_w	seawater density ($= 1030$ kg m^{-3})
σ	standard deviation; normal stress
τ	dummy time variable; discrete time index; shear stress
Φ	astronomic latitude; flexural parameter
ϕ	geodetic latitude; Forsyth's parameter
φ	phase
χ^2	chi-squared statistic
Ψ	wavelet, frequency or wavenumber domain
ψ	wavelet, time or space domain
Ω	rotation matrix
$\lfloor x \rfloor$	floor: rounds x to the greatest integer less than or equal to x
∇^2	Laplacian operator
III	Dirac comb function
$\tilde{G}$	wavelet transform of a signal g
g^*	complex conjugate of a signal g
$*$	convolution operator
$\mathbf{a}^\mathsf{T}$	transpose of vector $\mathbf{a}$

Part I
Context

While the bulk of this book concerns the details of effective elastic thickness (T_e) estimation using spectral methods, the first part sets the scene. What is T_e? How did its use originate in the Earth sciences? Why is it important? Part I—comprising only one chapter—answers these questions in a largely qualitative manner, leaving the mathematics and physics to Parts II and III. It provides a brief history of the development of isostasy, from its beginnings in the eighteenth century up to the mid-twentieth century, when elastic plate models began to replace those of hydrostatic equilibrium. It provides context by explaining the relationship between the elastic plate and the Earth's crust, mantle and lithosphere—the tectonic plates—and concludes with a short discussion about the significance of the effective elastic thickness.

Chapter 1
Isostasy, Flexure and Strength

1.1 Isostasy

Isostasy describes the principle of hydrostatic equilibrium when it is applied to the Earth's lithosphere—the crust and uppermost, rigid part of the mantle associated with tectonic plates. That is, the mass excesses of mountains, plateaus and other topographic features are compensated by mass deficiencies deeper in the Earth, much like an iceberg floating on the ocean. This section leads up to a formulation of two isostatic models with a brief history of isostasy.

1.1.1 Beginnings

In the late 1730s Pierre Bouguer (1698–1758) was part of a team taking measurements of the shape of the Earth in Ecuador. By then, the Earth was known to have an ellipsoidal—rather than spherical—shape, and they sought to determine its flattening by observing the length of an arc of a meridian, a procedure that required observations of the astronomic latitude. This was achieved by determining the angle between the equatorial plane and the local vertical as indicated by a plumb line (Fig. 1.1). Aware of Isaac Newton's relatively recent theory of gravitation, Bouguer was concerned that the team's measurements would be distorted by the proximity of large mountains, reasoning that the mass of the mountains would cause the direction of the local gravity vector—which defines the local vertical—to point towards them. However, it turned out that the measurements were not as affected as Bouguer's model suggested, so he concluded that the gravitational attraction of the mountains was not as large as he had thought (Watts 2001).

A similar scenario unfolded over a hundred years later during the Great Trigonometrical Survey of India. Initially tasked with the measurement of a meridian arc, George Everest (1790–1866)—using different methods than Bouguer's team— essentially found that there was a difference between the astronomic and geodetic latitudes at stations close to the Himalayan mountains, when they were thought to be identical (Fig. 1.1). While Everest put these differences down to both instrument

© Springer Nature Switzerland AG 2022   3
J. Kirby, *Spectral Methods for the Estimation of the Effective Elastic Thickness of the Lithosphere*, Advances in Geophysical and Environmental Mechanics and Mathematics, https://doi.org/10.1007/978-3-031-10861-7_1

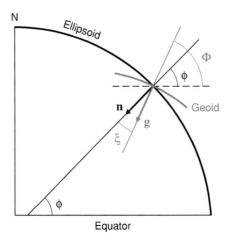

Fig. 1.1 The relationship between geodetic (ϕ) and astronomic (Φ) latitudes. The vector **n** lies along the normal to the surface of the ellipsoid, while the gravity vector **g** (which defines vertical) lies along the normal to the geoid. The meridional deflection of the vertical, ξ, is the angle between these two vectors, meaning that $\Phi = \phi + \xi$. The dashed line is parallel to the equator, and N is the north pole

and geodetic model errors, John Henry Pratt (1809–1871), like Bouguer before him, proposed that the gravitational attraction of the Himalayas was responsible (Pratt 1855), though, again like Bouguer, he overestimated the effect of the attraction, and so dropped the matter.

The discrepancy between model and observation found by Bouguer and Pratt was finally explained by George Biddell Airy (1801–1892) soon after (Airy 1855). Airy proposed that the gravitational attraction of the visible mass of the mountains was being reduced by a mass deficit underneath them, making the net effect smaller. He invoked the same principle of hydrostatic equilibrium, or buoyancy, that keeps an iceberg afloat: if the Earth's crust was a low-density solid floating on a high-density fluid substratum, then the visible mass of a mountain must be compensated by the displacement of the fluid underneath it; this replacement of high-density material by low-density material gives rise to a negative gravity anomaly that almost (but not quite) cancels out the positive anomaly from the mountain. This phenomenon—of the crust existing in a state of hydrostatic equilibrium—would later be termed *isostasy* by Clarence Edward Dutton (1841–1912) (Dutton 1889).

While Fig. 1.2 summarises the problem and Airy's solution, we should first refresh our knowledge of *geodesy*, the study of the shape of the Earth. It is now well known that the size and shape of the Earth can be determined not only from geometric measurements using optical surveying or satellite instruments (geometric geodesy), but also from gravity observations (physical geodesy). As will be explored in Chap. 8, the gravity field can be described in two, complementary ways: either by the direction and magnitude of the vector representing the gravity acceleration at any given location, or by the potential energy in the gravity field surrounding a massive body. The

(a)

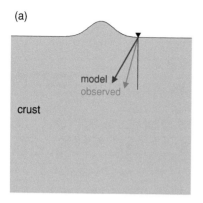

(b)

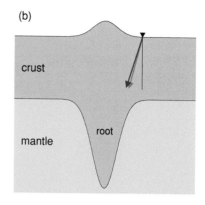

Fig. 1.2 Observed (red) and model-predicted (blue) gravity vectors at a station close to a mountain. The left-hand model **a** has homogeneous, solid crust, while the right-hand (Airy) model **b** specifies a low-density solid crust sitting on a high-density fluid mantle, and incorporates the gravity effect of the compensating root. The vertical black line is the normal to the ellipsoid at the station; the deflection of the vertical is the angle between this normal vector and the gravity vector

two are complementary in that, at any point, the (3D) gradient of the gravity potential defines the gravity vector. Furthermore, the gravity vector is always perpendicular to an *equipotential surface*, being a surface surrounding a body on which the gravity potential is constant. There are, of course, an infinite number of such surfaces. For instance, the *geoid*, mentioned in Fig. 1.1, is that particular equipotential surface that most closely corresponds with mean sea level. Importantly—and here is the link between geometric and physical geodesy—the gravity vector at a point defines the local vertical, while the tangent plane to the equipotential surface passing through that point defines the local horizontal. And as gravity depends upon mass density, any spatial variation in density will generate a spatial variation in gravity, which in turn generates spatial variations in what we understand to be 'horizontal' and 'vertical'. So, when surveying equipment is levelled prior to use, it is actually measuring the direction of the local gravity field vector and orientation of the local equipotential surface.

Hence, we can explain Bouguer's and Everest's observations—and Pratt's and Airy's interpretations—using geodesy. Consider Fig. 1.1. The ellipsoid is the simplest mathematical figure describing the geometric size and shape of the Earth; it is used as a basis for the definition of geodetic latitudes and longitudes. The geoid is a surface defining the physical (i.e. including gravity effects) size and shape of the Earth; it is used as a datum for elevations, and used in the definition of astronomic latitudes and longitudes, among other things. The vector **n** is perpendicular, or 'normal', to the ellipsoidal surface; the vector **g** is normal to the geoid, as noted above. The angle between these two vectors is called the *deflection of the vertical*[1]. If the Earth were a smooth, rotating, homogeneous ellipsoid, then the gravity vector would always

---

[1] In addition to the meridional (north-south) deflection of the vertical shown in Fig. 1.1, there is also an east-west component, the latitudinal deflection of the vertical.

point along the ellipsoidal normal and the geoid would coincide with the surface of the ellipsoid; the deflection of the vertical would be zero everywhere and geodetic and astronomic latitude would always be equal. Instead, the presence of topography and internal density contrasts causes the geoid to undulate and the deflection of the vertical to vary from place to place.

So, the reason that Bouguer's and Everest's surveying equipment detected a slight angular 'error' when measuring latitude was actually the result of the local gravity vector being oriented away from the ellipsoidal normal, giving a non-zero deflection of the vertical. However, while the mountains (Andes and Himalayas) swung the gravity vector towards them, it was not by as much as Bouguer and Pratt thought. Consider Fig. 1.2. In the presence of a mountain, but no internal density contrasts, the deflection of the vertical is reasonably large and directed towards the mountain (the blue vector in Fig. 1.2a); this represents Bouguer's and Pratt's modelling of the scenario, which unfortunately did not agree with the observations (the red vector in Fig. 1.2a). However, Airy's model of a compensating root did predict Everest's observations, shown by the agreement between the red and blue vectors in Fig. 1.2b. The negative gravity effect of the root (the displacement of higher-density mantle by lower-density crust) weakens the overall gravity in the region, causing the model gravity vector to swing back towards the ellipsoidal normal and align with the observed vector, resulting in a smaller deflection of the vertical.

Not to be outdone, Pratt updated his model so that it, too, would predict the observed deflection of the vertical (Pratt 1859, 1864, 1871). Instead of reducing gravity by the replacement of high-density mantle material with a low-density crustal root, he proposed that the density of the entire crust (and perhaps uppermost mantle) under a mountain would be reduced in proportion to the height of the mountain, thus weakening gravity. As discussed in Watts (2001), Pratt's model came to be adopted in North America, while Airy's model gained favour in Europe. There seems to be no particular scientific reason for this difference in preference, because although the models describe vastly different Earth structures, they make very similar predictions of gravity and deflections of the vertical, and as far as geodesists are concerned, that is all that matters.

It is instructive to note at this point that, in addition to the deflection of the vertical predictions, the case for isostasy can also be made by a glance at a Bouguer gravity anomaly map of the Earth (Fig. 1.3). The Bouguer anomaly is explained in Sect. 8.2.5, but briefly, it shows the gravity field without the attractions of the topography above the geoid and the gross planetary mass; it therefore shows the residual gravity field (or 'anomaly') due to density variations in the deeper crust, mantle and core. The most striking feature of Fig. 1.3 is the low gravity over the continents, which is explained equally well by both Airy's and Pratt's models: a low-density crustal root underneath mountains for Airy, and low-density mountains and crust for Pratt.

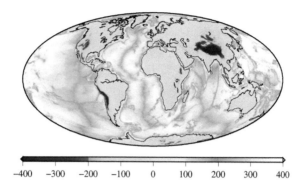

Fig. 1.3 The Earth's Bouguer gravity anomaly from the EGM2008 model of Pavlis et al. (2012). Mollweide projection; units in mGal

1.1.2 Pressure

Before exploring the Airy and Pratt models of isostasy more formally, we must introduce the concept of *hydrostatic pressure*. Pressure is defined as force per unit area, when the force, F, is applied perpendicular to an area A of the surface of a body. Thus, we can write for the pressure, p,

$$p = \frac{F}{A}. \tag{1.1}$$

Hydrostatic pressure is the pressure experienced by a body immersed in a fluid at depth h below the fluid's surface. Consider Fig. 1.4a. The body B is at depth h in a fluid of density ρ. If we imagine a cylinder (or prism) of fluid of height h and cross-

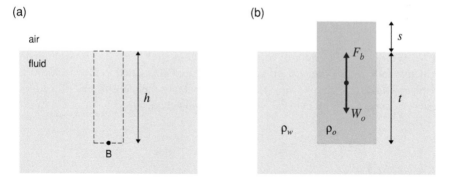

Fig. 1.4 a A body B at depth h in a fluid of density ρ. The dashed box represents a cylindrical column of fluid of height h and cross-sectional area A. The pressure at the depth of the body is $p = \rho g h$ (Eq. 1.3). **b** If we now place a block of material of density ρ_o in the fluid (whose density is now written as ρ_w), if $\rho_o < \rho_w$ then the block floats, and an amount s protrudes above the fluid's surface. If the block is in equilibrium, then the weight of the object (W_o) must equal the buoyancy force (F_b) which in turn must equal the weight of the displaced fluid

sectional area A, then from Newton's second law the weight of the fluid column—and thus the force pressing down on the body—is mg, where m is the column's mass and g is the gravity acceleration. Equation 1.1 can then be written as

$$p = \frac{mg}{A} \,. \tag{1.2}$$

If we now write the mass of the column as $m = \rho V$, where its volume is $V = Ah$, then Eq. 1.2 becomes

$$p = \rho g h \,. \tag{1.3}$$

Strictly speaking, we should also include the weight of the air above the fluid. While air has a very low density ($1.225 \, \text{kg} \, \text{m}^{-3}$ at sea level), atmospheric pressure is approximately 101.3 kPa at sea level, meaning that the pressure at 10 m below the surface of the ocean is not merely $1030 \times 9.79 \times 10 = 100.8$ kPa (from Eq. 1.3), but $100.8 + 101.3 = 202.1$ kPa, or approximately two atmospheres, as any scuba diver knows. However, when considering pressures deep within continents—where Eq. 1.3 predicts a pressure of $\sim$1 GPa at 37 km depth, for example—atmospheric pressure can be neglected.

If we now place an object of density ρ_o in a fluid of density ρ_w, then if $\rho_o < \rho_w$ the object will float. We can calculate how much of the object will be visible above the fluid's surface, and how much will be submerged, using *Archimedes' principle*. This states that if an object is in equilibrium, then the upward buoyancy force on the object (F_b) must be equal to the weight of fluid displaced by the object (W_w), or

$$F_b = W_w \,. \tag{1.4}$$

Figure 1.4b shows the scenario. A height s of the object protrudes above the surface of the fluid, while the rest of the object is submerged to depth t. If the object has cross-sectional area A, then its volume is $V_o = A(s + t)$ and its weight is $W_o = \rho_o g A(s + t)$. The volume of fluid displaced by the submerged part of the object is $V_w = At$, and the weight of this displaced fluid is $W_w = \rho_w g A t$. Archimedes' principle tells us that the upward buoyancy force (F_b) is equal to the weight of the displaced fluid, or $F_b = W_w$ (Eq. 1.4). A force balance then tells us that, if the block is at rest (in equilibrium) then the weight of the object must be equal to the buoyancy force, or

$$W_o = F_b \,,$$

or, using Eq. 1.4:

$$W_o = W_w \,.$$

That is, the weight of the object must be equal to the weight of the displaced fluid. Using the expressions for W_o and W_w given above, we can write

$$\rho_o g A(s + t) = \rho_w g A t \,,$$

which gives the depth of the submerged portion of the object as

$$t = s\left(\frac{\rho_o}{\rho_w - \rho_o}\right) . \tag{1.5}$$

For example, if the object is an iceberg of density $917\,\text{kg m}^{-3}$ and the fluid is seawater of density $1030\,\text{kg m}^{-3}$, then we find that $t = 8.12\,s$. Alternatively, if the object is continental crust of density $2800\,\text{kg m}^{-3}$ and the fluid is mantle of density $3300\,\text{kg m}^{-3}$, then $t = 5.6\,s$.

1.1.3 Airy-Heiskanen Isostasy

In the 1930s, the Finnish geodesist Weikko Aleksanteri Heiskanen (1895–1971) formalised Airy's model of isostasy, representing the topography and seafloor bathymetry by blocks of crustal material floating on a fluid substratum he called a 'magmatic layer' (Heiskanen 1931). We will call it the 'mantle' for now, and stress that it behaves like a zero-viscosity ('*inviscid*') fluid, offering no resistance to movement through it. We will refer to the crust-mantle interface as the 'Moho', in keeping with current terminology, although neither Airy nor Heiskanen used this term. We also need to stress that the blocks representing the crust are free to move vertically with no frictional (or other) resistance to movement from adjacent blocks; that is, a load placed on one block will cause that block, and only that block, to sink further into the mantle. Finally, all crustal blocks have the same density, ρ_c, while the mantle has uniform density ρ_m.

Figure 1.5 shows the Airy-Heiskanen model, and our goal is to find an expression relating the thickness of a continental mountain root (w) to the height of the topography (h). Rather than considering forces, as was done in the derivation of Eq. 1.5, we will consider the pressure at some constant *compensation depth*. The choice of compensation depth is arbitrary, but most interpretations choose the depth of the thickest crustal column. We also need to specify a reference column, and for this we choose a coastal column whose upper surface is at sea level; the thickness of this column is t_0. Using Eq. 1.3, we can write the pressure at point A in Fig. 1.5 as the combined effect of the pressure due to the crustal column above it and the pressure of the intervening mantle, or

$$p_A = \rho_c g t_0 + \rho_m g w .$$

Similarly, the total pressure at point B is just the pressure due to a column of the crust of thickness $h + t_0 + w$, or

$$p_B = \rho_c g (h + t_0 + w) .$$

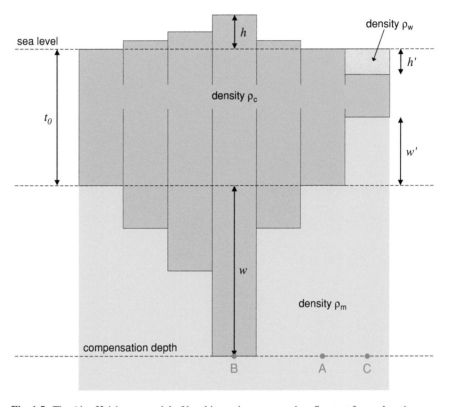

Fig. 1.5 The Airy-Heiskanen model of local isostatic compensation. See text for explanation

If the system is in equilibrium, then the pressure at the compensation depth is constant. So, if we equate p_A and p_B we find

$$w = h \left(\frac{\rho_c}{\rho_m - \rho_c} \right) , \tag{1.6}$$

the same as Eq. 1.5.

Next, we need to apply the same treatment to the oceanic crust. As shown in Fig. 1.5, the top of an oceanic crustal column is below sea level, and its bottom is at a higher level than the base of a coastal reference column. The encroaching mantle (of thickness w') is known as an 'anti-root'. So the pressure at point C in Fig. 1.5 is the sum of the pressures due to the overlying ocean, crust and mantle, or

$$p_C = \rho_w g h' + \rho_c g \left(t_0 - h' - w' \right) + \rho_m g \left(w' + w \right) ,$$

where $h' > 0$. Since the pressure at points B and C is identical, equating p_A and p_C gives

$$w' = h' \left(\frac{\rho_c - \rho_w}{\rho_m - \rho_c} \right) . \tag{1.7}$$

It is interesting to note that Eqs. 1.6 and 1.7 are independent of the reference crustal thickness, t_0. This parameter only becomes important when one needs to calculate the gravity effect of the deflected Moho (Chap. 8). In general, values of t in a range of 20–40 km give predicted gravity values that agree with observations, which suggest that t_0 corresponds to the mean crustal thickness of the Earth.

1.1.4 Pratt-Hayford Isostasy

Just as Heiskanen formalised Airy's isostatic model, Pratt's model was given mathematical rigour by John Fillmore Hayford (1868–1925) (Hayford 1909). The Pratt-Hayford model is similar to the Airy-Heiskanen model in that frictionless crustal columns float upon an inviscid mantle. However, the densities of the columns are variable and are chosen such that their bases are all at the same depth, the depth of compensation (Fig. 1.6). Such a scenario can be achieved when the column that has the highest topography has the lowest density, the column with the next highest topography has the next lowest density, and so on, with the lowest-topography column (at sea in Fig. 1.6) having the highest density. While the goal of Airy isostasy is to find the thickness of the crustal root, the aim of Pratt isostasy is to find the densities of the crustal columns.

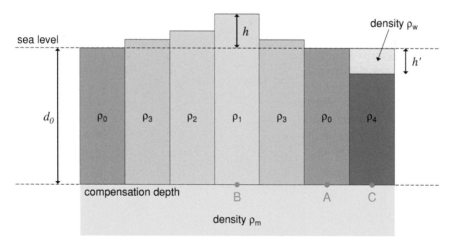

Fig. 1.6 The Pratt-Hayford model of local isostatic compensation, where $\rho_1 < \rho_2 < \rho_3 < \rho_0 < \rho_4 < \rho_m$. The densities ρ_0 through ρ_3 describe continental crust, the density ρ_4 is that of oceanic crust and ρ_w is the density of seawater. See text for further explanation

Proceeding as for the Airy case, we equate the pressures at the bases of the various columns, at the compensation depth. At point A in Fig. 1.6, at the base of a coastal, reference column of density ρ_0, the pressure is

$$p_A = \rho_0 g d_0 .$$

At point B, underneath a continental column of density ρ_1, the pressure is

$$p_B = \rho_1 g(d_0 + h) .$$

Equating these two pressures gives the density of the continental crust column as

$$\rho_1 = \rho_0 \left(\frac{d_0}{d_0 + h} \right) .$$

In the oceans, we equate the pressures at the bases of columns C and A. The pressure at the base of column C is

$$p_C = \rho_w g h' + \rho_4 g(d_0 - h') ,$$

where $h' > 0$. Setting $p_A = p_C$ gives the density of the oceanic crust column as

$$\rho_4 = \frac{\rho_0 d_0 - \rho_w h'}{d_0 - h'} .$$

Unlike the Airy model—which is independent of t_0, the thickness of a reference, coastal column—the Pratt model has a dependence on this thickness, the compensation depth, d_0. The value of this parameter is typically found by computing a predicted gravity field for many values of d_0, and choosing that whose model best fits the observed gravity. Typically, values of the order of 100–200 km are required. While the Pratt model does not depend upon the mantle density, unlike the Airy model, modern interpretations posit that Pratt-Hayford isostasy is the mechanism that provides isostatic support at mid-ocean ridges. But rather than the lateral density variations existing within the crust—which is very thin at mid-ocean ridges—they are modelled as residing in the subcrustal mantle.

1.2   Flexural Isostasy

Isostatic theory became more sophisticated when it began to accept contemporary ideas of Earth structure from geologists, one of which being the lithosphere-asthenosphere duality which complemented the already-known crust-mantle duality. Thus, we begin this section by exploring the idea that the lithosphere has an elastic strength that supports loads, and conclude it with a brief discussion about our current understanding of the lithosphere.

1.2.1 Regional Support

The problem with both Airy and Pratt models is the lack of resistance between columns as they move up and down in response to loading. Even the smallest and narrowest load will cause a response, and then only directly underneath the load, a phenomenon termed *local compensation*. Back around the turn of the twentieth century, this conflicted with contemporary geologists' understanding of the Earth's crust, which they believed possessed a strength—or more accurately a *rigidity*—that enabled it to support some loads elastically. The geologists were also puzzled as to why they should care about isostasy at all: because the predictions from both models were largely identical, gravity or deflection of the vertical measurements could not be used to distinguish between them and provide insight into the regional geology. For their part, the geodesists of the day were really only concerned that as long as their models were fit for purpose (correcting surveying observations, for example), it did not matter whether or not these models described the Earth's structure accurately[2].

So over a period lasting some decades, geologists pursued the concept of *regional support* of loads by the mechanical strength of the crust. Building on work by several—mainly North American—geologists, Joseph Barrell (1869–1919) proposed that the crust behaved like an elastic solid, such that it was able to absorb and transmit stresses outwards from an applied load, over a greater area than that occupied by the load itself (Barrell 1914a) (Fig. 1.7a). Thus, depending on the size of the load and the rigidity of the crust, loads could be either locally compensated by isostatic forces, or regionally supported by shear stresses. Large loads, or loads on a weak crust, would be locally compensated, as in the isostatic models of Airy and Pratt; small loads, or loads on a rigid crust, would be regionally supported by its mechanical strength; for intermediate cases, there would be a proportion of both mechanisms in play.

The problem was formalised by Harold Jeffreys (1891–1989), Felix Andries Vening Meinesz (1887–1966) and Ross Gunn (1897–1966), who chose the equations governing the flexure of a *thin, elastic plate* to model deflections and stresses of the crust. Although the theory of elastic plate flexure was well-known at the time (Hertz 1884), these authors were the first to apply it to isostasy, with Jeffreys (1926) and Gunn (1937) choosing a one-dimensional (1D) beam model, and Vening Meinesz

---

[2] A view that, unfortunately, persists to this day. As Hofmann-Wellenhof and Moritz (2006, p. 145) write: "Although Vening Meinesz' refinement of Airy's theory is more realistic, it is more complicated and is, therefore, seldom used by geodesists because, as we will see, any isostatic system, if consistently applied, serves for geodetic purposes as well." For example, in a recent textbook on the geodesists' take on isostasy and the gravity field in general, Sjöberg and Bagherbandi (2017) perpetuate the practice of approximating a straightforward solution to the differential equation of flexure by an arbitrary Gaussian function.

(a) T_e = 100 km   (b) T_e = 0 km (Airy)

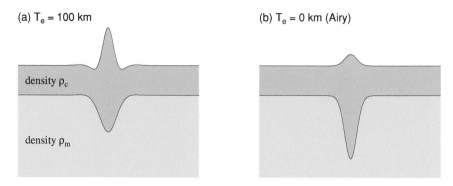

density ρ_c

density ρ_m

Fig. 1.7 a Regional support, and **b** Airy isostatic compensation of a Gaussian load on a 'crust' of density ρ_c overlying a 'mantle' of density ρ_m. The deflections of the crust-mantle boundary were calculated using the theory in Chap. 7

(1931) a two-dimensional (2D) plate. And while not estimating it directly, these authors placed importance upon the thickness of the elastic beam or plate that they used to represent the crust, a parameter that we know today as the *effective elastic thickness*, T_e.

In Chap. 7, we will derive the equations of thin, elastic plate flexure, but for now they will be stated without proof. First, let us rewrite the equation of Airy isostasy, Eq. 1.6, more rigorously as

$$(\rho_m - \rho_c)\, gw \;=\; -\rho_c gh \,, \tag{1.8}$$

where the inclusion of the minus sign on the right-hand side now reflects the fact that the deflection w and topography h are measured in opposite directions (a positive topographic load induces a crustal root in the opposite direction; Fig. 1.5). If we just consider the flexure of a 1D beam rather than a 2D plate, then elasticity theory gives us the fourth-order differential equation

$$D\frac{d^4 w}{dx^4} \;+\; (\rho_m - \rho_c)\, gw \;=\; -\rho_c gh \,. \tag{1.9}$$

Equation 1.9 is known as the *biharmonic equation*, given here in one dimension (cf. Eqs. 7.36 and 7.80). The first term on the left-hand side represents the restoring force (bending stress) within the beam that acts to try to return it to its undeflected position; the other two terms are identical to Eq. 1.8. Thus, the downward-directed weight of the load is balanced by the upward-directed buoyancy force from the displaced mantle and by the upward-directed restoring force.

The parameter D in Eq. 1.9 is the *flexural rigidity*, a measure of the stiffness of an elastic beam or plate[3]. Flexural rigidity is measured in newton metres (the same units as moments), and values of the order of 10^{22}–10^{25} Nm are common on the Earth. A plate with no rigidity (strength) would have $D = 0$, and if we set this restriction in Eq. 1.9 then it reverts to Eq. 1.8, that is, Airy isostasy. Hence, Airy isostasy is a limiting case of regional support. As will be proved in Chap. 7 (Eq. 7.37), the flexural rigidity is proportional to the cube of the effective elastic thickness, or

$$D = \frac{E\,T_e^3}{12\,(1 - \nu^2)}\,,$$

where E and ν are the elastic constants *Young's modulus* and *Poisson's ratio*, respectively.

Regional support is demonstrated in Fig. 1.7, which shows the post-flexure geometry of a synthetic Gaussian mountain after its emplacement on a 'crust' of density ρ_c that overlies an inviscid, fluid 'mantle' of density ρ_m. In Fig. 1.7a, the crust has an effective elastic thickness of $T_e = 100\,\text{km}$ ($D = 8.9 \times 10^{24}$ Nm), while in Fig. 1.7b the elastic thickness is zero and the load is locally compensated by an Airy mechanism. Two observations stand out. First, compared to Airy isostasy, the deflected 'Moho' ('crust'-'mantle' boundary) under the mountain is much broader when the support is regional, as noted above. The crustal root extends well beyond the base of the load with the strong crust, while it is restricted to being directly underneath the load in the Airy case. Second, the strong plate does not flex as much as the weak plate: the Moho is much shallower when regional support plays a role. This has implications for studies that seek to estimate the location of the Moho by inverting gravity data, for example (Kirby 2019).

1.2.2 Crust, Mantle, Lithosphere and Asthenosphere

In the discussion so far, we have spoken about columns of crustal material floating on a fluid mantle, and the flexure of an elastic plate, but how do all these terms— crust, mantle and plate—relate to one another? In the isostatic and flexural models considered here, the material forming the topography needs to be solid, while the layer upon which the solid sits must be fluid, offering no resistance to deformation or movement of the solid. Furthermore, the density of the solid must be less than that of the fluid, so that it floats. Finally, the solid layer must potentially be able to behave elastically.

---

[3] Many articles and books in the literature (including this one) loosely use the term 'strength' instead of 'rigidity', as in 'a strong plate' or 'a weak plate'. Strictly speaking, the terms 'a rigid plate' and 'a flexible plate' should be used, as *strength* is the limit of elastic yielding beyond which failure (brittle fracture or ductile flow) occurs (see Sect. 1.3.1).

Both Airy (1855) and Pratt (1859) referred to the solid layer as the 'crust', as did Heiskanen (1931), Vening Meinesz (1931) and Gunn (1937). And while none of these authors used the term 'mantle', the idea of a solid crust and a flowing mantle had been around since the eighteenth century (Schubert et al. 2001). Airy (1855) refers to the fluid layer as 'lava', while Heiskanen (1931, p. 110) writes

> there is beneath the earth's crust a layer of denser magma supporting the continents and mountains, which are floating on it like icebergs on the sea. This layer is not liquid in the ordinary sense since it behaves like an elastic solid under the action of forces of short period ...; this layer behaves like a liquid only under the uninterrupted pressure of the mountains. We shall use the expression *magmatic layer* in that sense.

This concept of a substance that behaves both like a fluid and a solid was gaining traction around that time, as a resolution to the long-standing issue of whether the mantle was either solid or liquid; it would later develop into the study of mantle convection, and thence the theory of plate tectonics (Schubert et al. 2001). But another transformative development that also fed into plate tectonics later (Dietz 1961) was the concept of the *asthenosphere*, a term coined by Barrell (1914b) meaning 'sphere of weakness'. Rather than flexure of the crust above the mantle, Barrell proposed that it was the *lithosphere*—a term already in use for the rocky, rigid outer layer of the Earth—that could behave elastically in response to loading, while the asthenosphere flowed viscously underneath it.

Today, since the postulation and proof of plate tectonics in the 1960s, we view the upper few hundred kilometres of the Earth in two, complementary ways: a crust-mantle duality, and a lithosphere-asthenosphere duality. The distinction between crust and mantle is primarily *compositional*: the rocks and minerals found in the crust are quite different from those in the mantle, and this imparts a large difference in density at their interface. The distinction between lithosphere and asthenosphere is *rheological*[4]: while the lithosphere can behave as a brittle, elastic, plastic or viscous solid—depending upon the local temperature and pressure—the asthenosphere behaves as a low-viscosity fluid[5].

The lithosphere comprises the crust and the uppermost part of the mantle (the *lithospheric mantle*); the asthenosphere is that part of the mantle directly below the lithosphere (Fig. 1.8). In plate tectonic theory, the lithosphere is divided up into several tectonic plates (Fig. 1.9), and the movement, collision and subduction of the lithospheric plates, as they travel over, and into, the asthenosphere, involve the cohesive movement of the entire crust and lithospheric mantle. Thus, isostasy is more than just a model of blocks of crust floating on a magma: we must instead consider the elastic bending of the entire lithosphere. However, the thickness, T_e, of the elastic plate is not simply the thickness of the lithosphere, which, as we will see, has several definitions.

---

[4] *Rheology* is the study of the way materials flow under various pressure and temperature conditions (see Sect. 1.3.1).

[5] For the purposes of flexural isostasy, the asthenosphere is an *inviscid* (zero-viscosity) fluid, offering no resistance to movement through it.

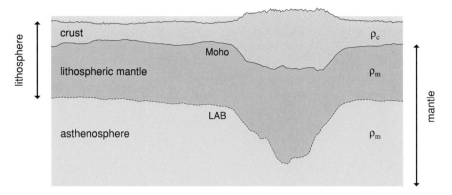

Fig. 1.8 Schematic cross-section (not to scale) through the lithosphere and asthenosphere, showing how the lithosphere comprises the crust and lithospheric mantle. The LAB is the lithosphere-asthenosphere boundary

Fig. 1.9 Tectonic plate boundaries (Bird 2003). Mollweide projection

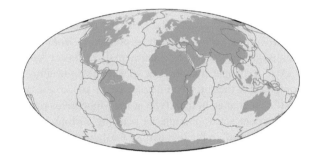

    Continental and oceanic crusts generally have dissimilar compositions. Crudely speaking, the most common type of rocks in the continental crust are granites and rhyolites (called 'felsic' because they are rich in the elements that form the minerals quartz and feldspar: silicon and aluminium), while oceanic crust is primarily composed of basalt and gabbro (called 'mafic' because they are rich in magnesium and iron). Continental crust rocks are less dense than those in the oceanic crust—a reason why the continents 'float' higher than the seafloor—and continental crust is much thicker than oceanic crust. Globally, the average thickness of the continental crust is around 39 km, though its range is approximately 20–80 km, being thinner in rift zones and continental margins, and thicker under younger mountain ranges. The oceanic crust is somewhat more uniform; its global mean thickness is approximately 6–7 km beneath the ocean basins, but it is much thinner at some spreading centres and transform faults (3–4 km), and much thicker under oceanic plateaus and volcanic provinces and above mantle plumes (up to 30 km). The crust is often modelled as having three layers whose density increases with depth (Fig. 1.10). The rheology of the layers also changes, with the upper and middle crust being brittle while the lower crust is ductile.

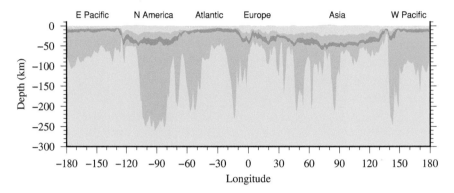

Fig. 1.10 Cross-section through the upper, middle and lower crusts (brown shadings), the litho-spheric mantle (orange) and asthenosphere (yellow), as given by the LITHO1.0 model (Pasyanos et al., 2014). The transect is at a constant latitude of 43°N

The interface between the crust and the mantle, at the base of the crust, is called the *Mohorovičić discontinuity*, shortened to *Moho*. While crustal rocks are predom-inantly granites/rhyolites (continental) or basalts/gabbros (oceanic), the dominant rock of the upper mantle is peridotite, a magnesium-rich, ultramafic, igneous rock composed of the silicate minerals olivine and, to a lesser extent, pyroxene. This com-positional change in rock type at the Moho gives rise to strong seismic velocity and density contrasts (although the petrologic Moho does not always coincide with the seismic Moho). In the majority of isostatic and flexural studies, a density contrast of $500 \, \mathrm{kg \, m^{-3}}$ is used, though the actual range is of the order of 200–$300 \, \mathrm{kg \, m^{-3}}$ in the oceans (except at mid-ocean ridges, where it is much lower), and 350–$600 \, \mathrm{kg \, m^{-3}}$ in the continents.

The layer below the Moho is the lithospheric mantle, which has a broadly similar rheology to the crust even though its rocks are different. But while the lithospheric mantle and asthenosphere both contain peridotite, the peridotite in the cooler litho-spheric mantle is capable of great elastic strength, whereas the temperature and pres-sure conditions in the asthenospheric mantle make peridotite much less viscous and more ductile. This encourages convective flow and enables the overlying lithospheric plates to move without much resistance.

The base of the lithosphere is known as the *lithosphere-asthenosphere boundary* (LAB). While one can say that the depth of the LAB defines the thickness of the lithosphere, the situation is complex. First, there are several ways of defining the boundary: changes in rheology, temperature, heat transport, seismic velocity, seismic anisotropy, composition, electrical resistivity, hydration and mechanical properties. Second, the boundary is not necessarily sharp or distinct. And third, there is often a large uncertainty in the measurements or models of the LAB depth.

Kinematically, the lithosphere is that layer of crust and mantle that moves coher-ently with the plate. This suggests a rheological distinction, whereby asthenospheric materials are more easily deformed than those in the lithosphere, identified by a sharp increase in strain rate as one descends. But the rheological LAB can also be defined by a decrease in viscosity when crossing the boundary. The mechanical LAB, in contrast, is much shallower and can be defined either by the temperature or stress associated with the change from brittle to ductile behaviour of mantle olivine. The elastic thickness of the lithosphere is very much related to its mechanical thickness (Sect. 1.3.1). But while these thicknesses—rheological, mechanical and elastic—can all be quite different in value, they are all connected and very much dependent upon temperature, with hotter materials flowing much more easily than cooler ones. The thermal lithosphere is defined as being bounded by the depth to the 1300°C isotherm, being the temperature at which mantle peridotite starts to melt and moves from an elastic to a plastic rheology (though the presence of water and carbon dioxide in the mantle can lower this temperature). The LAB can also be viewed as the boundary between modes of heat transport: in the lithosphere heat dissipates via conduction, while in the asthenosphere the mechanism is convection. One can also define the LAB seismically, since shear waves travel faster in the colder, denser lithosphere (though water is known to reduce seismic velocities in ultramafic rocks), and the top of the asthenosphere is characterised by a noticeable low-velocity zone. Shear waves also travel faster in one preferential direction, governed primarily by the orientation of olivine crystals in the mantle. A change in such seismic anisotropy fast directions has been observed at depths around the LAB, where the anisotropy in the asthenosphere reflects the present-day plate motion, while lithospheric anisotropy was 'frozen-in' when the plate was formed and reflects the motion at that time. Finally, the lithosphere is electrically resistive, while the asthenosphere is conductive, providing another definition of the LAB.

But while these definitions might seem clear, each metric, above, gives different values for the thickness of the lithosphere. Furthermore, the boundary itself is often more of a gradual transition than a sharp change. While the LAB is more well-defined in oceanic lithosphere and younger parts of the continents, under old, continental cratons it is too diffuse and heterogeneous to resolve clearly, and can itself be up to 50 km thick. For our purposes—the estimation of effective elastic thickness—density is one of the most important parameters, with perturbations in the density contrast between interfaces providing buoyant compensation of loads above. However, the lack of a sharp and clear seismic velocity boundary makes the density change at the LAB hard to quantify, compounded by the generally poor vertical resolution of seismic tomography data. While most lithospheric models specify a lithosphere that is slightly denser than the asthenosphere—so that a subducted oceanic plate has negative buoyancy and will tumble into the mantle—importantly for flexural studies the density contrast at the LAB is small enough to not be involved in isostatic compensation and can be neglected. Many studies adopt 40–50 kg m^{-3} as a density contrast at the LAB, though the actual value probably depends on age and setting, and could have a range of 0–100 kg m^{-3} (Afonso et al. 2007; Boonma et al. 2019; McClain 2021).

Taking all the above into account, the continental lithosphere is generally 50–250 km thick, while the oceanic lithosphere is thinner, with a thickness of approximately 20–100 km (Fig. 1.10). Above all, though, the specific nature of the LAB is generally poorly understood, especially under old, cratonic, continental lithosphere, and is the subject of extensive, ongoing research (Rychert et al. 2020).

The above discussion is a simplification of the sometimes very complex reality, but for the purposes of flexural modelling, it is adequate. We can safely assume that the primary density contrast that takes part in isostatic compensation lies at the Moho, and that the asthenosphere behaves as an inviscid fluid. When it comes to modelling the lithosphere as an elastic plate of thickness T_e, though, we have to be a little more careful, and need to consider the relationship between the rheology of the lithosphere and its mechanical and elastic thicknesses.

1.3   The Significance of T_e

What does the effective elastic thickness mean, in real, physical terms? Why is it important? Here, we attempt to answer these questions, beginning with a discussion on the relationship between flexural rigidity—measured by the effective elastic thickness—and strength.

1.3.1   Plate Strength

We have spoken so far about the lithosphere being represented by an elastic plate or beam, but how true is this in reality? The problem with a purely elastic material is that it can absorb an unlimited amount of stress without yielding. That is, no matter how far or fast one bends an elastic beam or plate, it will never break. Furthermore, once the applied force is released, an elastic material will return to its pre-flexed shape and state exactly. Obviously, no rocks on Earth behave in such a way. Indeed, data from experimental rock mechanics—where small quantities of rock are subjected to stresses and strains in a laboratory—show that, in general, crustal and mantle rocks will deform permanently after a certain applied stress is reached, called the *yield strength*.

When the lithosphere is loaded and bends in response, stresses are induced in the fibres of its rocks. If the plate is flexed downwards (Fig. 1.11), then the upper part of the plate will experience compressive stresses, while the lower half will undergo extension; the *neutral surface* will therefore experience no stress. If the plate is flexed upwards the stresses are reversed. If these induced stresses are not large enough to exceed the yield strength of the local rocks, then the plate behaves elastically, and will return to its original shape if the load is removed. But if these induced stresses exceed the local yield strength, then permanent deformation will occur. The type of deformation depends upon the rock type, its depth within the lithosphere and the amount of water within the rock (the presence of water reduces the yield strength).

Generally speaking, the deformation pattern exhibited by crustal rocks is that of *brittle fracture*, the mechanism responsible for faulting that produces earthquakes.

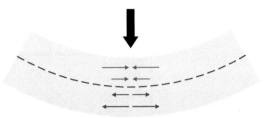

Fig. 1.11 Cross-section through a plate deflected by a load (black arrow). The flexure induces compressive stresses (red arrows) in the upper half of the plate, and extensional stresses (blue arrows) in the lower half. The dashed black line is the *neutral surface* separating the two regimes, which is unstressed

Near the surface, crustal rocks are relatively weak, but as one descends through the crust the increase in pressure due to the overburden (the *lithostatic* or *confining pressure*) suppresses crack propagation and makes the rocks stronger; the yield strength increases linearly. A convenient way of representing the yield strength of lithospheric rocks is through a *yield strength envelope* (YSE) (Goetze and Evans 1979), shown in Fig. 1.12. The YSE is divided into four sectors: compressional stresses to the left (negative), extensional stresses to the right (positive); yielding through brittle fracture at the top, and through a mechanism called ductile flow at the bottom. Point A in the figure shows the yield strength at a depth of 20 km in an extensional setting (approximately 127 MPa). If the rocks here are subject to extensional stress of less than 127 MPa, then they will behave elastically; but once the stress exceeds this value, the rocks will deform brittly. Note the asymmetry of the YSE about the zero stress line. This reflects the fact that brittle solids have less strength under extension than under compression.

As depth increases and the confining pressure increases, rocks get stronger. However, temperature also increases with depth, and there comes a point where the heat overcomes the confining pressure and rocks begin to deform plastically through *ductile flow*. There are at least two mechanisms of plastic deformation by ductile flow (such as dislocation and diffusion creep), but in both the yield strength decreases exponentially as temperature (i.e. depth) increases. Point B in Fig. 1.12 shows the yield strength at a depth of 50 km in an extensional setting (approximately 68 MPa). If the rocks at this depth are subject to extensional stress of less than 68 MPa, then they will behave elastically; but once the stress exceeds this value, the rocks will deform by ductile flow.

For a given type of rock, the depth corresponding to the change in the deformation mode—from brittle fracture to ductile flow—is called the *brittle-ductile transition* (BDT). The BDT is generally not a sharp boundary (as Fig. 1.12 suggests) and can sometimes extend over several kilometres, but it does represent the depth where the lithosphere is strongest. The depth of the BDT depends upon many different properties of both the rock type and the lithosphere in the region. For example, in continental crust rich in quartz and feldspars, the BDT lies at depths corresponding to

Fig. 1.12 A schematic yield strength envelope (YSE) (solid black line) showing the brittle regime as red shading—where rocks fracture brittly—and the ductile regime as green shading—where rocks flow ductilely. The magenta circle marked BDT shows the brittle-ductile transition (for compression). The area shaded in grey (i.e. enclosed by the YSE) equals the integrated strength of the lithosphere. Points A and B on the YSE are discussed in the text

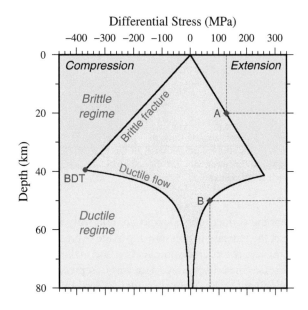

the ~250–450°C isotherms, while in the mantle—dominated by olivine—the BDT is at temperatures of ~600–750°C.

Oceanic lithosphere is relatively homogeneous and comprises basalts that are as strong as olivine in mantle peridotites. It therefore tends to behave as a single mechanical layer with a YSE resembling that in Fig. 1.12. Continental lithosphere, in contrast, is much more heterogeneous in its composition, and its crustal rocks—especially those in the lower crust—have lower strength than those in the lithospheric mantle. Its YSEs reflect this. In Fig. 1.13, the model consists of a continental crust dominated by quartz with a high water content (making it weaker than when it is dry), while the lithospheric mantle is rich in dry olivine. Starting at the surface—in the quartz brittle regime—the yield strength increases with depth until the wet quartz BDT is reached, where the ambient temperature becomes hot enough for the quartz to begin ductile flow. The yield strength of the quartz then decreases until the Moho is reached. Here, in the lithospheric mantle, the lithology changes to peridotite, dominated by olivine which is generally much stronger than any crustal mineral. At these depths, the ambient temperature is still cool enough for olivine to behave brittly under stress, and the yield strength increases once again due to the effects of the confining pressure. Eventually, the olivine BDT will be reached, whereupon it will begin ductile flow, reducing the yield strength.

So how does this discussion relate to the effective elastic thickness? Consider a perfectly elastic plate of *mechanical thickness T_M* subjected to a downward load at its surface, as in Fig. 1.11. The strength-depth profile for this scenario is shown in Fig. 1.14a, reflecting the fact that the upper and lower surfaces of the plate experience the most stress (compressive and extensional, respectively), while the neutral surface is unstressed. The slope of the perfectly-elastic strength profile is inversely propor-

Fig. 1.13 A yield strength envelope for typical continental lithosphere. The crust has the rheology of wet quartz, the lithospheric mantle that of dry olivine, with the Moho at 30 km depth. Points A and B are the BDTs of wet quartz and dry olivine, respectively. The shape of YSEs in continents gives them the informal moniker 'Christmas tree diagrams'

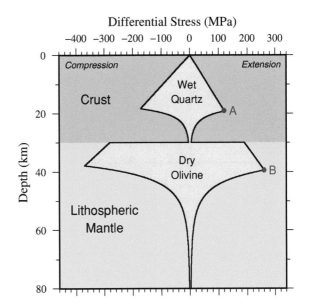

tional to the curvature of the plate: the greater the load, the greater the deflection and thus the greater the plate's curvature there; since greater curvatures produce greater stresses, the slope of the strength profile must decrease (to make the fibre stress arrows longer in Fig. 1.14a). Importantly, because the plate is made of a perfectly elastic material, it can absorb an unlimited amount of stress without failing.

Now consider the flexure of a realistic plate, which can fail brittly and ductilely. At the top of the plate, the fibre stresses generated by the flexure exceed the plate's yield strength, and the rocks undergo brittle fracture; at the base of the plate, the large fibre stresses exceed the yield strength there and the rocks undergo ductile flow. Importantly, only a smaller, central portion of the plate continues to behave elastically, referred to as the *elastic core*. Figure 1.14b shows the scenario in terms of strength-depth profiles. The perfectly-elastic strength profile of Fig. 1.14a is modified by the yield strength envelope, appearing as the solid red line in Fig. 1.14b. The intersection of the perfectly-elastic strength profile with the YSE defines the upper and lower depths of the elastic core: the thickness of this core is T_e, the *effective elastic thickness*, which is now less than the mechanical thickness ($T_e < T_M$). If the curvature increases, the perfectly-elastic strength profile has a shallower slope and the effective elastic thickness becomes even smaller.

The above example of strength reduction by bending is directly applicable to oceanic lithosphere due to the shape of the YSE in Fig. 1.14b, which resembles that of most oceanic YSEs. As noted above, the oceanic lithosphere tends to behave as a single mechanical layer due to the strength similarities between basalt and peridotite, with the crust welded, or *coupled*, to the lithospheric mantle. The elastic and mechanical thicknesses of oceanic lithosphere are therefore of similar value (except in regions of high plate curvature such as oceanic trenches and some seamounts).

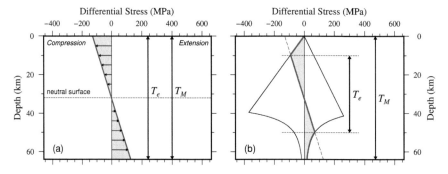

Fig. 1.14 a Strength profile (red line) of a purely elastic plate of mechanical thickness T_M, under downward flexure as in Fig. 1.11. The small, horizontal arrows represent the stresses on the fibres within the plate, which are under compression in the upper half of the plate, under tension in the lower half and are unstressed on the neutral surface. The elastic thickness (T_e) and mechanical thickness (T_M) are equal. **b** Flexure of a realistic plate. The strength profile from (a) (the dashed red line) is now limited by the YSE for the system, and the thickness of the elastic core is now less than the actual thickness of the plate ($T_e < T_M$). Note that the T_e indicated here is the minimum possible value; in reality, it will extend slightly into the brittle and ductile failure zones

YSEs in continental lithosphere are more complex, as suggested by Fig. 1.13. The weak, ductile and quartz-rich lower crust of continental lithosphere may sometimes *decouple* the upper crust from the mantle, meaning that the mechanical strength of the lithosphere is governed by the strength of its strongest layer, rather than being the combined strength of all layers in coupled lithosphere. Using a 'leaf spring' engineering model, applied to a lithosphere comprising n elastically competent layers (elastic cores), Burov and Diament (1995) proposed that T_e of decoupled lithosphere was given by[6]

$$T_e = \sqrt[3]{\sum_{i=1}^{n} h_i^3} , \tag{1.10}$$

where h_i is the elastic thickness of the ith competent layer; any layer that has no strength would have zero elastic thickness (Fig. 1.15). For coupled lithosphere, the relation is instead given by

$$T_e = \sum_{i=1}^{n} h_i . \tag{1.11}$$

For example, consider a lithosphere comprising two elastically competent layers, the crust and mantle, where the elastic thickness of the competent crust is h and that of the competent mantle is $2h$. If the lithosphere is coupled, then by Eq. 1.11 the total

---

[6] Eq. 1.10 is a good approximation to the exact solution, which is obtained by depth-integrating the YSE, since the area enclosed by the YSE is a measure of the integrated strength of the plate (Burov and Diament 1995).

Fig. 1.15 As Fig. 1.14b, but for continental lithosphere where the crust is decoupled from the mantle by a ductile lower crust. From Eq. 1.10, the elastic thickness of the plate is $T_e = \sqrt[3]{h_c^3 + h_m^3}$, where h_c is the minimum elastic thickness of the crustal elastic core, and h_m is that of the mantle elastic core

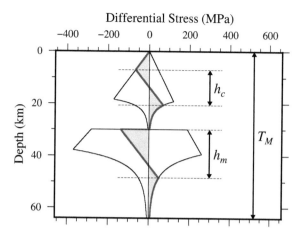

elastic thickness will be $T_e = h + 2h = 3h$. If, however, the crust and mantle are decoupled by a weak lower crust, the total elastic thickness will be $T_e = h\sqrt[3]{9} \approx 2h$, from Eq. 1.10.

The model developed by Burov and Diament (1995) predicted that continental lithosphere with thicker crust would be decoupled and have lower T_e, while lithosphere with thinner crust would be coupled and have a large T_e. We can write the mechanical thickness of the lithosphere as

$$T_M = T_c + h_m ,$$

where T_c is the crustal thickness, and h_m is the thickness of the elastically competent mantle[7] (Burov and Diament 1996). Hence, for a given T_M, if the crust is thick then h_m must be small, and most of the strength will reside in the crust; furthermore, the Moho will be deeper and hotter, and the lower crust is likely to be more ductile, decoupling the plate. But, for the same T_M, if the crust is thin then h_m must be large, and most of the strength resides in the mantle; the Moho will be shallower and cooler, and the lower crust less—or not at all—ductile, resulting in a coupled plate. As olivine is generally stronger than quartz or feldspar, regions with thinner crust tend to have a high T_e. From their modelling, Burov and Diament (1995) proposed a critical crustal thickness—separating the coupled and decoupled regimes—of 35–40 km for old (>750 Ma) continental lithosphere.

Regarding the mechanical thickness, T_M, this is not well defined and has several interpretations (as do most of the other LAB definitions, as we saw in Sect. 1.2.2). The most common seems to be the depth where the yield strength is lower than 10–20 MPa, or where the yield strength is less than 1–5% of the lithostatic pressure. Such depths are suggested to correspond to isotherms of 500–600°C in oceanic

---

[7] In coupled lithosphere, $h_m = T_e - T_c$; in decoupled lithosphere, $h_m = \sqrt[3]{T_e^3 - h_c^3}$, where h_c is the thickness of the elastically competent upper crust.

lithosphere, and 700–900°C in the continents (Burov 2015). When compared to the 1300°C isotherm marking the thermal LAB, this makes the mechanical lithosphere almost half as thick as the seismic or thermal lithosphere.

1.3.2 T_e—Causes and Effects

We have just seen how, in general, the effective elastic thickness does not correspond to any particular depth within the lithosphere, and is only equal to the mechanical thickness when the plate is coupled and unflexed by loads. In the oceans and other coupled lithospheric plates, the elastic core is a single layer (Fig. 1.14b), and here it is easy to identify T_e as the thickness of that core. However, in the decoupled lithosphere of continents (Fig. 1.15) there may exist a number of cores separated by weak zones; here, the geometric conception of a depth or thickness breaks down and T_e is best thought of as a measure of the *integrated strength* of the lithosphere, including the contributions from all brittle, ductile and elastic layers. So perhaps the best way to conceive T_e is as a geometric analogue of the flexural rigidity, for it is often defined as the thickness of a perfectly elastic plate that, when loaded, gives the same deflections as the actual lithosphere.

But what geological or tectonic factors affect the effective elastic thickness? We saw in Sect. 1.3.1 how T_e depends strongly on plate curvature and crustal thickness. Curvature—caused by loading or subduction, for example—weakens the plate as its layers fail by brittle fracture or ductile flow. Thick crust implies a deeper and hotter Moho, with a ductile lower crust decoupling and weakening the plate. Thin crust suggests a greater contribution from the mantle lithosphere, whose olivine is stronger than crustal quartz.

T_e is also strongly dependent upon age and temperature, though while clear in the oceans, the relationship in the continental lithosphere is more obscure. Oceanic T_e values have been shown to be bounded by the depths to the 300–600°C isotherms. Since oceanic geotherms are strongly dependent on the age of the lithosphere, T_e in the oceans also has an age dependence. But rather than being proportional to the age of the lithosphere or to the age of the load, Watts (1978) found that T_e was proportional to the square root of the age of the lithosphere at the time the loads were emplaced. That is, around loads that were emplaced on young oceanic lithosphere (on or near a mid-ocean ridge, for example) T_e is low, while similar age loads that were emplaced on old lithosphere (far from the ridge) are associated with high T_e values. Oceanic T_e values lie in the approximate range 0–40 km, so given that the oceanic crust averages 7 km in its thickness, this implies that the upper part of the oceanic lithospheric mantle contributes considerably to the long-term support of loads in many cases.

In contrast, no such clear relationships have been found for continental lithosphere, due, no doubt, to their more complex geological and tectonic histories producing a very heterogeneous structure and composition. Generally speaking, T_e has been found to correlate reasonably well with the age since the last major thermo-tectonic event, meaning the older parts of the continents tend to have higher T_e values than the younger parts. The most rigid continental lithosphere can be found in the Archaean and Proterozoic cratons—where T_e exceeds 100 km—while Phanerozoic mountain belts and rifted margins generally have lower values (20–60 km); rifts created from mantle plumes have lower values still. Other factors that reduce T_e are the presence of fluids and melts, faulting and tectonic stress, and thick sediments, which, in addition to weakening through loading, can thermally insulate the lower crust and encourage decoupling. Composition also plays a significant role, as seen in our discussion of yield strength envelopes, with quartz in particular being more susceptible to ductile deformation. Interestingly, though, no consistent correlation has been found between continental T_e and a particular isotherm, and while several studies have identified a strong relationship between T_e and other geophysical observables such as surface heat flow, shear wave velocities and seismicity, many others have determined only weak correlation or even none at all.

Nevertheless, the effective elastic thickness is an important property of the lithosphere. The strength and rigidity of the cratonic cores are the reason why they have persisted for billions of years, preserving their general shape and structure while the surrounding weaker parts of the lithosphere—margins, rifts and mobile belts—absorb the stresses and strains of tectonic upheaval and the supercontinent cycle (Audet and Bürgmann 2011). In the oceans too, modelling has shown how subduction could not occur if the lithospheric mantle were uniformly weak, as some would have us believe (Jackson 2002; Burov 2010). Indeed, the formation of rifts, passive margins, foreland basins and mountain ranges, and the landscape itself, all depend upon the mechanical strength of the lithosphere.

In short, the magnitude of T_e in a region both influences and is influenced by the tectonic evolution of that region. However, the variation of continental lithospheric strength over time is not completely understood. Over very long timescales (>1 Myr), the lithosphere does indeed behave like an elastic plate overlying an inviscid asthenosphere, but at shorter timescales this model needs adjustment (Fig. 1.16). At the timescales of co-seismic deformation (seconds to minutes) the entire crust and mantle behave elastically, enabling the propagation of seismic waves. Then, over the following 10 years or so, post-seismic relaxation takes place, in which only the upper crust behaves elastically, while the rest of the lithosphere and mantle deform as a viscoelastic material. Over the much longer timescales of glacial isostatic adjustment, thousands to tens of thousands of years, the response to loading and unloading is similar to the post-seismic model, but the elastic layer is much thicker, spanning the whole depth of the lithosphere. Finally, over geodynamic timescales, millions to tens of millions of years, while there is some evidence that the mechanical lithosphere is initially elastic and then enters a viscoelastic relaxation phase, it eventually settles into its long-term elastic thickness, which becomes 'frozen-in'.

The deformation history and current state of the lithosphere are reflected in its gravity field and topographic surface, providing the data for the T_e-estimation method that is the subject of this book. While the topography signal supplies some of the loads acting on the lithosphere, the gravity reveals how the internal structure and strength of the plate respond to and accommodate the loads. Hence, a comparison

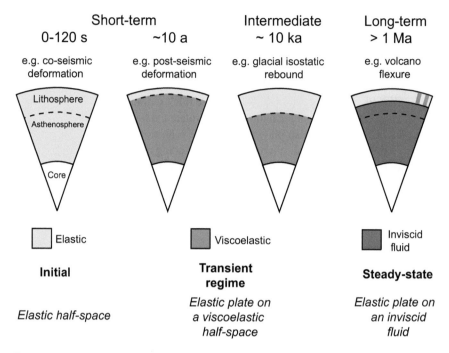

Fig. 1.16 Schematic diagram illustrating how the Earth's outermost layers respond to loads of different timescales. (From Watts (2015), reproduced by permission of Elsevier.)

of the two—in this case performed in the spatial frequency, or spectral, domain— yields the effective elastic thickness, specifically the T_e that was frozen-in after the last major thermo-tectonic event. Alternatively, if one performs the modelling in the space domain (called the 'forward modelling' approach, as opposed to the 'inverse modelling' of the spectral method), then it is possible to estimate the T_e that was active at some event in the past. Forward modelling, however, is more time-consuming and limited in its geographic area of analysis.

1.4   What This Book Does Not Cover

First, what this book does cover. It concerns the estimation of effective elastic thickness by inverting the *coherence* or *admittance* between gravity and topography data, the so-called 'inverse spectral method'. The coherence is a frequency-domain version of the square of the Pearson correlation coefficient, while the admittance is a transfer function, or filter, from topography to gravity. The method can be summarised as follows.

1. Project observed gravity and topography data onto a plane using a map projection.
2. Using some method of planar spectral estimation, form the coherence and admittance between the gravity and topography data as functions of 2D Cartesian spatial frequency, or 'wavenumber'. The planar spectral estimation methods used here are the continuous wavelet transform (CWT) and the multitapered, windowed Fourier transform (the 'multitaper method').
3. Using a model of the flexure of a thin, elastic plate, calculate the predicted coherence and admittance for a range of T_e values. The T_e value that gives a predicted coherence (or admittance) that best fits the observed coherence is selected as representative of the plate at that location.

The book covers everything needed to implement these steps. Part II deals with spectral estimation: the Fourier transform, the multitaper method and the continuous wavelet transform. Part III contains enough elasticity theory to derive the biharmonic equation of thin plate flexure from first principles. It also introduces the relevant portions of gravitational potential theory, and then explains in detail the *load deconvolution* method, the favoured procedure of T_e estimation. The book also discusses the generation of synthetic flexural models by which T_e-estimation methods may be tested, and finishes with some data sources and practical examples of T_e estimation using real data.

Topics this book does not cover are spectral estimation on the surface of a sphere, and thin (spherical) shell flexure. Although such approaches have been developed, they do not seem to have caught on in the literature, at least on planet Earth. For the interested reader, estimation of the coherence and admittance using the spherical multitaper method was developed by Wieczorek and Simons (2005), while the spherical CWT approach was pioneered by Audet (2011). Equations for the flexure of a thin, elastic shell have been presented by Brotchie and Silvester (1969) in the space domain, and Turcotte et al. (1981), Lowry and Zhong (2003), Beuthe (2008) and Audet (2014) in the spherical harmonic domain.

Furthermore, the book presents only the inverse spectral method of T_e estimation, so does not cover forward modelling, whereby predicted profiles of topography, Moho deflection and gravity anomaly are fitted to observed data in order to estimate T_e (although a space-domain solution to the 1D biharmonic equation is briefly presented in Sect. 7.7).

Finally, we do not cover mechanical anisotropy, being the estimation of anisotropic, directional elastic thickness. Although there may be some merit in these studies, Kirby and Swain (2014) and Kalnins et al. (2015) showed that results of the so-called 'weak direction' of mechanical anisotropy might be biased by anisotropy in the topography and/or gravity field.

1.5   Conventions

The mathematical conventions used in this book will be noted as they are introduced. It is worth mentioning now, though, that I always use a Cartesian reference frame where the z-axis is positive up, away from the centre of the Earth. This is not uncommon, but is by no means standard practice, and many textbooks and papers have z positive down, especially when dealing with elasticity theory and some aspects of gravity potential theory (e.g. Parker (1972)'s formula). My motivation is consistency, and a fundamental wish to be increasing in altitude as I climb a mountain.

The other convention I stick to concerns dimensionality. Unless we are dealing with relativistic physics, our world is three-dimensional (3D), so quantities such as gravity potential are functions of three orthogonal coordinates ('dimensions'), as in $V(x, y, z)$ or $V(r, \phi, \lambda)$. Sometimes, we can ignore the third dimension, in which case we might be modelling a field given on a 2D surface. This is where things can get confusing. Topography, for example, is a 2D quantity as it is dependent on latitude and longitude, or eastings and northings, only: $h(\phi, \lambda)$ or $h(x, y)$. Even though topography extends into a third dimension, we do not speak of the topography at 5 km below the Empire State Building, or the height of the topography 6 km above Mount Everest; we say that the value of the topography at (N27°59′9.8″, E86°55′21.4″) is $h = 8848.86$ m above sea level. The same goes for flexure: a flat plate might be flexed into a third dimension, but—when concerned with incompressible materials—the value of the deflection is depth-independent and is thus 2D, $w(x, y)$. Similarly, a 1D quantity is a function of one coordinate only, be it distance, time, depth, longitude, or latitude. I stress this now because many books and articles refer to the flexure of a beam as a 2D problem, when it is actually 1D: the magnitude of the deflection, $w(x)$, is a function of distance along the beam from some reference point only, and of no other coordinate.

1.6   Summary

The concept of isostasy was originally borne out of geodesy and the need to explain perplexing surveying observations. Deflections of the vertical that were not as large as expected could be explained by mass deficits compensating the mass excesses of the topography. In the late nineteenth and early twentieth centuries, two models were postulated to describe these mechanisms, the Airy-Heiskanen and Pratt-Hayford models of local isostatic compensation. However, these models could not explain many geological phenomena and a model of regional support of loads by an elastic plate was proposed by Barrell, formulated later by Vening Meinesz. This work developed into the modern study of flexural isostasy, whereby the lithosphere is modelled as a thin, elastic plate floating on a fluid asthenosphere. The resistance of the plate to vertical deformation by loading is measured by the flexural rigidity, commonly represented by the plate's effective elastic thickness, T_e, a geometric analogue for the rigidity. In general, T_e does not correspond to a particular depth within the lithosphere, but is a proxy for the integrated strength of all elastic, brittle and ductile layers within the crust and uppermost mantle. While rigidity and strength are related, they should not be confused, with the latter being the stress required for a material to fail by brittle fracture or ductile flow, as represented by yield strength envelopes. The lithosphere's effective elastic thickness has geodynamic importance, and controls many tectonic movements and landscape features.

1.7 Further Reading

It is said that one of the best ways to understand a subject is to understand its historical development; why was it felt at the time that the subject and its paradigm shifts were needed? So, essential reading for anyone seriously studying these topics is the book by Watts (2001), which contains two whole chapters on the origins and development of isostasy and flexure from the early eighteenth century through to the 1970s.

The reader will have perhaps noted that this chapter is light in citations. This is deliberate as much of the content is elementary, well-known and uncontroversial, at least for geologists and geophysicists. Fortunately for the reader unversed in these disciplines, there are many textbooks and review articles covering the material. Even more fortunately, two encyclopaedias of geophysics have recently been published, and the following articles in Elsevier's *Treatise on Geophysics* (Schubert 2015) are well worth reading: Burov (2015), Fischer (2015), Mooney (2015), Watts (2015) and Wessel and Müller (2015). Published not long after was Springer's *Encyclopedia of Solid Earth Geophysics* (Gupta 2021), and the reader should investigate the articles by Burov (2021), Davis and Chapman (2021), James (2021), Jaupart and Mareschal (2021), McClain (2021) and Watts (2021). These review articles cover our current understanding of the lithosphere and flexural isostasy, and could be joined by the reviews of the lithosphere-asthenosphere boundary by Eaton et al. (2009) and Rychert et al. (2020). Alternatively, for a broader background, the textbooks by Turcotte and Schubert (2002) and Fowler (2005) can be consulted, while that by Artemieva (2011) deals specifically with all matters concerning the lithosphere.

The discussion on yield strength here is very elementary and qualitative, and merely provides a context to give meaning to T_e. For a more thorough, quantitative introduction to yield strength and rheology in general, turn to the books by Ranalli (1995), Turcotte and Schubert (2002), Karato (2008) or Artemieva (2011). For a discussion on the particular relationship between YSEs and T_e the book by Watts (2001) is a good starting point, together with the journal papers by Lowry and Smith (1995), Burov and Diament (1995, 1996), Watts and Burov (2003), Burov and Watts (2006) and Burov (2010), plus those encyclopaedia articles cited above, particularly Burov (2015) or Burov (2021).

Finally, for material on the importance of T_e and its relationship to tectonics and other matters, Watts (2001), Watts (2015) and Watts (2021) are recommended.

References

Afonso JC, Ranalli G, Fernàndez M (2007) Density structure and buoyancy of the oceanic lithosphere revisited. Geophys Res Lett 34:L10302. https://doi.org/10.1029/2007GL029515

Airy GB (1855) On the computation of the effect of the attraction of mountain-masses, as disturbing the apparent astronomical latitude of stations in geodetic surveys. Philos Trans R Soc Lond 145:101–104

Artemieva IM (2011) The lithosphere: an interdisciplinary approach. Cambridge University Press, Cambridge

Audet P (2011) Directional wavelet analysis on the sphere: application to gravity and topography of the terrestrial planets. J Geophys Res 116:E01003. https://doi.org/10.1029/2010JE003710

Audet P (2014) Toward mapping the effective elastic thickness of planetary lithospheres from a spherical wavelet analysis of gravity and topography. Phys Earth Planet Inter 226:48–82

Audet P, Bürgmann R (2011) Dominant role of tectonic inheritance in supercontinent cycles. Nature Geosci 4:184–187

Barrell J (1914a) The strength of the Earth's crust: Part IV. Heterogeneity and rigidity of the crust as measured by departures from isostasy. J Geol 22:289–314

Barrell J (1914b) The strength of the Earth's crust: Part VI. Relations of isostatic movements to a sphere of weakness–the asthenosphere. J Geol 22:655–683

Beuthe M (2008) Thin elastic shells with variable thickness for lithospheric flexure of one-plate planets. Geophys J Int 172:817–841

Bird P (2003) An updated digital model of plate boundaries. Geochem Geophys Geosyst 4:1027. https://doi.org/10.1029/2001GC000252

Boonma K, Kumar A, Garcia-Castellanos D, Jiménez-Munt I, Fernández M (2019) Lithospheric mantle buoyancy: the role of tectonic convergence and mantle composition. Sci Rep 9:17953. https://doi.org/10.1038/s41598-019-54374-w

Brotchie JF, Silvester R (1969) On crustal flexure. J Geophys Res 74:5240–5252

Burov EB (2010) The equivalent elastic thickness (T_e), seismicity and the long-term rheology of continental lithosphere: time to burn-out "crème brûlée"? Insights from large-scale geodynamic modeling. Tectonophys 484:4–26

Burov EB (2015) Plate rheology and mechanics. In: Schubert G (ed) Treatise on geophysics, vol 6, 2nd edn. Elsevier, Amsterdam, pp 95–152

Burov E (2021) Lithosphere, mechanical properties. In: Gupta HK (ed) Encyclopedia of solid earth geophysics, 2nd edn. Springer, Cham, pp 884–893

Burov EB, Diament M (1995) The effective elastic thickness (T_e) of continental lithosphere: what does it really mean? J Geophys Res 100(B3):3905–3927

Burov EB, Diament M (1996) Isostasy, equivalent elastic thickness, and inelastic rheology of continents and oceans. Geology 24:419–422

Burov EB, Watts AB (2006) The long-term strength of continental lithosphere: "jelly sandwich" or "crème brûlée"? GSA Today 16:4–11

Davis EE, Chapman DS (2021) Lithosphere, oceanic: thermal structure. In: Gupta HK (ed) Encyclopedia of solid earth geophysics, 2nd edn. Springer, Cham, pp 906–915

Dietz RS (1961) Continent and ocean basin evolution by spreading of the sea floor. Nature 190:854–857

Dutton CE (1889) On some of the greater problems of physical geology. Bull Philos Soc Wash 11:51–64 [reprint J Wash Acad Sci 15:359–369, 1925]

Eaton DW, Darbyshire F, Evans RL, Grütter H, Jones AG, Yuan X (2009) The elusive lithosphere-asthenosphere boundary (LAB) beneath cratons. Lithos 109:1–22

Fischer KM (2015) Crust and lithospheric structure—seismological constraints on the lithosphere-asthenosphere boundary. In: Schubert G (ed) Treatise on geophysics, vol 1, 2nd edn. Elsevier, Amsterdam, pp 587–612

Fowler CMR (2005) The solid earth, 2nd edn. Cambridge University Press, Cambridge

Goetze C, Evans B (1979) Stress and temperature in the bending lithosphere as constrained by experimental rock mechanics. Geophys J R Astron Soc 59:463–478

Gunn R (1937) A quantitative study of mountain building on an unsymmetrical Earth. J Franklin Inst 224:19–53

Gupta HK (ed) (2021) Encyclopedia of solid earth geophysics, 2nd edn. Springer, Cham

Hayford JF (1909) The figure of the Earth and isostasy from measurements in the United States. Government Printing Office, Washington DC

Heiskanen WA (1931) Isostatic tables for the reduction of gravimetric observations calculated on the basis of Airy's hypothesis. Bull Géod 30:110–153

Hertz H (1884) Ueber das Gleichgewicht schwimmender elastischer Platten. Ann Phys 258:449–455

Hofmann-Wellenhof B, Moritz H (2006) Physical geodesy, 2nd edn. Springer, Vienna

Jackson J (2002) Strength of the continental lithosphere: time to abandon the jelly sandwich? GSA Today 12:4–10

James DE (2021) Lithosphere, continental. In: Gupta HK (ed) Encyclopedia of solid earth geophysics, 2nd edn. Springer, Cham, pp 866–872

Jaupart C, Mareschal J-C (2021) Lithosphere, continental: thermal structure. In: Gupta HK (ed) Encyclopedia of solid earth geophysics, 2nd edn. Springer, Cham, pp 872–884

Jeffreys H (1926) On the nature of isostasy. Beitr Geophys 15:153–174

Kalnins LM, Simons FJ, Kirby JF, Wang DV, Olhede SC (2015) On the robustness of estimates of mechanical anisotropy in the continental lithosphere: a North American case study and global reanalysis. Earth Planet Sci Lett 419:43–51

Karato S (2008) Deformation of earth materials: an introduction to the rheology of solid earth. Cambridge University Press, Cambridge

Kirby JF (2019) On the pitfalls of Airy isostasy and the isostatic gravity anomaly in general. Geophys J Int 216:103–122

Kirby JF, Swain CJ (2014) On the robustness of spectral methods that measure anisotropy in the effective elastic thickness. Geophys J Int 199:391–401

Lowry AR, Smith RB (1995) Strength and rheology of the western US Cordillera. J Geophys Res 100(B9):17,947–17,963

Lowry AR, Zhong S (2003) Surface versus internal loading of the Tharsis rise. Mars. J Geophys Res 108(E9):5099. https://doi.org/10.1029/2003JE002111

McClain JS (2021) Lithosphere, oceanic. In: Gupta HK (ed) Encyclopedia of solid earth geophysics, 2nd edn. Springer, Cham, pp 893–906

Mooney WD (2015) Crust and lithospheric structure—global crustal structure. In: Schubert G (ed) Treatise on geophysics, vol 1, 2nd edn. Elsevier, Amsterdam, pp 339–390

Parker RL (1972) The rapid calculation of potential anomalies. Geophys J R Astron Soc 31:447–455

Pasyanos ME, Masters TG, Laske G, Ma Z (2014) LITHO1.0: an updated crust and lithospheric model of the Earth. J Geophys Res Solid Earth 119:2153–2173

Pavlis NK, Holmes SA, Kenyon SC, Factor JK (2012) The development and evaluation of the Earth Gravitational Model 2008 (EGM2008). J Geophys Res 117:B04406. https://doi.org/10.1029/2011JB008916, (Correction, J Geophys Res Solid Earth 118, 2633. https://doi.org/10.1002/jgrb.50167, 2013)

Pratt JH (1855) On the attraction of the Himalaya mountains, and of the elevated regions beyond them, upon the plumb-line in India. Philos Trans R Soc Lond 145:53–100

Pratt JH (1859) On the deflection of the plumb-line in India, caused by the attraction of the Himmalaya mountains and of the elevated regions beyond; and its modification by the compensating effect of a deficiency of matter below the mountain mass. Philos Trans R Soc Lond 149:745–778

Pratt JH (1864) On the degree of uncertainty which local attraction, if not allowed for, occasions in the map of a country, and in the mean figure of the Earth as determined by geodesy; a method of obtaining the mean figure free from ambiguity by a comparison of the Anglo-Gallic, Russian, and Indian arcs; and speculations on the constitution of the Earth's crust. Proc R Soc Lond 13:253–276

Pratt JH (1871) On the constitution of the solid crust of the Earth. Philos Trans R Soc Lond 161:335–357

Ranalli G (1995) Rheology of the earth, 2nd edn. Chapman and Hall, London

Rychert CA, Harmon N, Constable S, Wang S (2020) The nature of the lithosphere-asthenosphere boundary. J Geophys Res Solid Earth 125: e2018JB016463. https://doi.org/10.1029/2018JB016463

Schubert G (ed) (2015) Treatise on geophysics, 2nd edn. Elsevier, Amsterdam

Schubert G, Turcotte DL, Olson P (2001) Mantle convection in the earth and planets. Cambridge University Press, Cambridge

Sjöberg LE, Bagherbandi M (2017) Gravity inversion and integration: theory and applications in geodesy and geophysics. Springer, Cham

Turcotte DL, Schubert G (2002) Geodynamics, 2nd edn. Cambridge University Press, Cambridge

Turcotte DL, Willemann RJ, Haxby WF, Norberry J (1981) Role of membrane stresses in support of planetary topography. J Geophys Res 86:3951–3959

Vening Meinesz FA (1931) Une nouvelle méthode pour la réduction isostatique régionale de l'intensité de la pésanteur. Bull Géod 29:33–51

Watts AB (1978) An analysis of isostasy in the world's oceans: 1 Hawaiian-Emperor seamount chain. J Geophys Res 83(B12):5989–6004

Watts AB (2001) Isostasy and flexure of the lithosphere. Cambridge University Press, Cambridge

Watts AB (2015) Crustal and lithosphere dynamics: an introduction and overview. In: Schubert G (ed) Treatise on geophysics, vol 6, 2nd edn. Elsevier, Amsterdam, pp 1–44

Watts AB (2021) Isostasy. In: Gupta HK (ed) Encyclopedia of solid earth geophysics, 2nd edn. Springer, Cham, pp 831–847

Watts AB, Burov EB (2003) Lithospheric strength and its relationship to the elastic and seismogenic layer thickness. Earth Planet Sci Lett 213:113–131

Wessel P, Müller RD (2015) Plate tectonics. In: Schubert G (ed) Treatise on geophysics, vol 6, 2nd edn. Elsevier, Amsterdam, pp 45–93

Wieczorek MA, Simons FJ (2005) Localized spectral analysis on the sphere. Geophys J Int 162:655–675

Part II
Spectra

Part II of the book describes the spectral estimation methods used to determine the admittance, coherency and coherence. The most commonly used methods in contemporary T_e estimation are the windowed (or short-time) Fourier transform with Slepian tapers, and the continuous wavelet transform using the fan wavelet. Part II begins with Chap. 2 on the Fourier transform, the basis for most spectral estimation techniques, describing some of the relevant theory before moving on to its practical application. The theory and practice of Fourier transformation are different: while the theory concerns infinite and continuous signals, the practice acknowledges that real-world data are of finite length and discretely sampled. These two features generate artefacts that must be mitigated.

Chapter 3 describes the discrete prolate spheroidal sequences, the tapers most commonly used in implementations of the multitaper method. It then shows how the tapers are applied to moving windows, enabling a time- or space-varying estimation of the spectrum. Chapter 4 deals with the continuous wavelet transform, specifically the Morlet wavelet, which, in its 2D form, is used to build the fan wavelets used in T_e estimation. In Chaps. 2–4, the theory is introduced in one dimension—as in time series analysis—and then extended to 2D signals. Chapter 5 then turns to the admittance, coherency and coherence, the spectral-domain quantities that describe the relationship between two signals, in our case gravity and topography. The chapter also addresses their errors. The final chapter of Part II does not concern spectral estimation, but describes map projections. Because the spectral estimation methods presented and used in this book are planar rather than spherical, the geodetically coordinated gravity and topography data must be projected onto a plane, incurring minimal distortion. Chapter 6 describes that process and recommends some projections.

Chapter 2
The Fourier Transform

2.1 Introduction

The Fourier transform forms the foundation of spectral estimation. First developed by Joseph Fourier in the early nineteenth century for the purpose of solving a differential equation (the heat equation), it has since been embraced by almost all facets of modern study where the analysis of numerical data is important. The reason for this is that Fourier analysis identifies periodic, or repeating, cycles in the data and categorises them by amplitude and frequency. This very useful feature lends itself to the detection of patterns in data, patterns that might not be visible to the human eye. And once a pattern is identified, predictions can be made.

In this section, the concept of Fourier transforms will be introduced in terms of one independent variable only, where, for a function or signal $y = g(x)$, x is the independent variable, and y is the dependent variable. Even though much of the data used in modern elastic thickness estimation are functions of two spatial variables (latitude and longitude, or grid easting and northing), we will choose time as our independent variable, for simplicity and alignment with other texts on the Fourier transform (see Sect. 2.8).

2.1.1 Dimensionality

It is worthwhile stressing here the meaning of the *dimensionality* of the data. A signal or function $g(x)$ is a one-dimensional (1D) signal because it is a function of one independent variable (coordinate) only: the x. A signal or function $g(x, y)$ is a two-dimensional (2D) signal because it is a function of two independent variables: x and y. Unfortunately, some authors refer to signals such as $g(x)$ as 2D signals, possibly because they mistake the dependent variable (the g) as another independent variable (the x). They might also talk about topography as a 3D signal; true, it penetrates into

© Springer Nature Switzerland AG 2022
J. Kirby, *Spectral Methods for the Estimation of the Effective Elastic Thickness
of the Lithosphere*, Advances in Geophysical and Environmental Mechanics
and Mathematics, https://doi.org/10.1007/978-3-031-10861-7_2

a third dimension ('up'), but that penetration is but the value of the surface elevation at a given x and y: $h(x, y)$. Eventually, their convention breaks down, with gravity and magnetism, for example, which, presumably, they would call 4D signals because these potential fields are defined throughout 3D space: $g(x, y, z)$. However, there can be no doubt that gravity is in fact a 3D signal[1] and does not extend into some sort of 4D hyperspace[2].

2.1.2 Harmonics

As mentioned, Fourier analysis identifies cycles of different frequencies within data. The cycles are represented by sinusoids—sine and cosine waves—which are referred to as the *harmonics* of the signal. The relative strength of a harmonic within a signal is measured by the *amplitude* of the corresponding sinusoid (Fig. 2.1). The *frequency* of the sinusoid—how many oscillations or cycles there are in a given interval—is the reciprocal of its *period* (for temporal data) or *wavelength* (for spatial data), thus,

$$f \ = \ \frac{1}{T} \quad \text{or} \quad f \ = \ \frac{1}{\lambda},$$

where T is the period (in units of time) (Fig. 2.1a), and λ is the wavelength (in units of distance) (Fig. 2.1b). The wavelength (or period) is the distance between adjacent positive peaks of the sinusoid, so that high-frequency harmonics have a short wavelength/period, and low-frequency harmonics have a long wavelength/period.

As will be seen, the frequency can take negative values, a somewhat peculiar concept. One way to imagine them would be when the independent variable is a spatial position, rather than time: positive frequencies give the wavelengths of harmonics one encounters when moving in an easterly direction; negative frequencies give those when moving to the west. This directionality to frequency will become important when we consider 2D spatially distributed data, later in the book.

2.1.3 Continuous Signals Versus Discrete Sequences

Many natural phenomena display properties that can be measured at any point in time or space. The data from these phenomena are called *continuous signals*: they are potentially infinite in extent or duration and are potentially observable in infinite

---

[1] Though the *gravity anomaly* is a 2D signal, as we shall see later, because it is defined on a particular surface, and is thus independent of the z-coordinate.

[2] Though according to general relativity, gravity *is* a 4D field because it is a function of 4D spacetime, $g(t, x, y, z)$. Indeed, gravity is a consequence of the curvature of spacetime, though this need not concern us here as we can ignore its time-dependence and our gravity fields are weak, relatively speaking.

Fig. 2.1 Two sinusoids. **a** A
sine wave in the time
domain, with amplitude
$A = 2$ units, period
$T = 0.5$ s and frequency
$f = 1/T = 2$ Hz. **b** A sine
wave in the space domain,
with amplitude $A = 1$ unit,
wavelength $\lambda = 0.25$ m and
frequency $f = 1/\lambda = 4 \, \text{m}^{-1}$

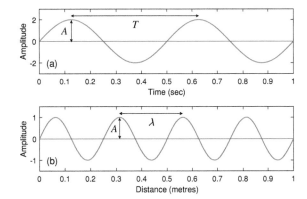

detail at an infinitesimal observation spacing. Examples of continuous signals from
the natural world could be atmospheric temperature, topographic altitude or ground
motion due to earthquakes; they all have observable values at any given time or
location.

Yet more often than not the signal must be sampled at defined moments in time
or locations in space, so that the data can be stored or manipulated in an efficient
manner, for example, on a computer. In addition, the sampled data sets must be of
finite duration or extent, being truncated versions of the signal. Thus, the atmospheric
temperature can be recorded every hour at a single location over five years or simultaneously at 100 stations globally; topographic altitude can be measured using a
precise level every 10 m along a 2 km traverse. Such sampled data are called *discrete
sequences*.

These two restrictions—sampling and truncation—mean that signals and
sequences are treated very differently when their Fourier transforms are desired.
For many studies—including this one—the data are sampled and so demand the use
of the discrete, rather than continuous, Fourier transform. However, the concept of
Fourier theory is best explained using continuous signals, so the practicalities of
transforming discrete sequences will be deferred to Sects. 2.3, 2.4 and 2.5.

2.2 Fourier Theory

It is often useful to be able to write down the Fourier transform of a mathematical
equation as another mathematical equation. When this is possible, such an ability
can tell us much about how certain signals are expected to behave in theory, if those
signals obey some underlying physical process. For instance, if a phenomenon g
obeys the rule $g(t) = e^{-|t|}$, then knowing that its Fourier transform looks like $G(f) = 2/(1 + 4\pi^2 f^2)$ informs us how its harmonic constituents are meant to behave, and
predictions based on observed data can be made.

In order to calculate the Fourier transform of an analytic expression (such as $g(t) = e^{-|t|}$), one uses the *continuous Fourier transform* (CFT) because the independent variable (t) is continuous. One would also use the CFT on continuous data such as the electric current flowing through a circuit. Another option (for the electrical signal) would be to construct a *Fourier series*, and we shall use this as our entry to the world of Fourier transforms.

2.2.1 Fourier Series

Although the representation of a signal by a Fourier series does not feature in the method of elastic thickness estimation, it is a useful tool by which to introduce the concept of Fourier analysis to a beginner. In a nutshell, the Fourier method seeks to represent a signal as a sum of sine and cosine functions of varying amplitude and frequency. However, because sine waves are periodic (with period 2π), the signal must also be periodic, and repeat itself every T seconds. The Fourier series representation (or *Fourier synthesis*) of a real-valued, continuous signal $g(t)$ is

$$g(t) = \frac{A_0}{2} + \sum_{j=1}^{\infty} A_j \cos(2\pi f_j t) + B_j \sin(2\pi f_j t) , \tag{2.1}$$

where the summation, in practice, must be terminated at some integer N. The sinusoids of differing frequencies are the harmonics, and the frequency of the jth harmonic is given by

$$f_j = \frac{j}{T} ,$$

where T is the period of the signal, and $j = 1, 2, \ldots$. The frequency of the first harmonic is referred to as the *fundamental frequency*. The amplitudes of the harmonics are given by the Fourier coefficients A_j and B_j. Figure 2.2 shows a crude example of Fourier synthesis (Eq. 2.1) in action: the sum of the constant term and the four sine waves gives the signal in the top panel.

The inverse procedure, *Fourier analysis*, determines the Fourier coefficients from the signal. This can be done using the equations

$$A_0 = \frac{2}{T} \int_0^T g(t) \, dt$$

$$A_j = \frac{2}{T} \int_0^T g(t) \cos(2\pi f_j t) \, dt$$

$$B_j = \frac{2}{T} \int_0^T g(t) \sin(2\pi f_j t) \, dt$$

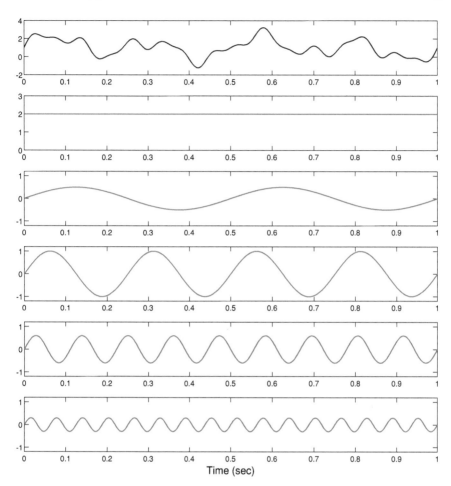

Fig. 2.2 A signal with period 1 s (blue), being the sum of its harmonic constituents (red). These comprise a constant (amplitude 2) and then four sine waves of amplitudes 0.5, 1, 0.6 and 0.3, at frequencies of 2, 4, 9 and 16 Hz, respectively. The process of Fourier analysis determines the harmonics from the signal, while Fourier synthesis constructs the signal from its harmonics

($j = 1, 2, \ldots$). Hence $A_0/2$ in Eq. 2.1 is the mean value of the signal over the interval $[0, T]$, and the coefficient $B_0 = 0$ which is why it is not used.

Equation 2.1 can be written in a more compact form with complex variables. Using Euler's identity[3], we can write Eq. 2.1 as

$$g(t) = \sum_{j=-\infty}^{\infty} C_j \, e^{2\pi i f_j t} \, , \tag{2.2}$$

---

[3] Euler's identity is $e^{i\theta} = \cos \theta + i \sin \theta$.

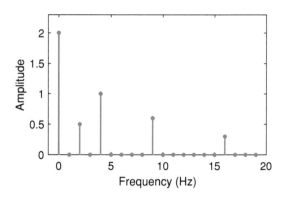

Fig. 2.3 The (one-sided) amplitude spectrum of the signal in the top panel of Fig. 2.2. The values of the amplitudes and frequencies of the non-zero harmonics here correspond to the amplitudes and frequencies of the harmonic constituents in Fig. 2.2

where the Fourier coefficients C_j are now complex numbers, related to the Fourier coefficients A_j and B_j by

$$C_{\pm j} = \frac{A_j \mp i B_j}{2} = \frac{1}{T} \int_0^T g(t)\, e^{-2\pi i f_j t}\, dt \ . \tag{2.3}$$

C_j is referred to as the *spectrum* of the signal and is an alternative way to represent a signal, categorising it by the amplitudes of its harmonics instead of variations in amplitude over time. It is important to note that here, in the Fourier series representation, while the signal is a function of a continuous time-coordinate, its spectrum is a discrete sequence, dependent upon the discrete integers j.

The C_j are complex numbers, but they are often portrayed as real numbers by conversion to a *power spectrum*, S_j:

$$S_j = |C_j|^2 \equiv C_j^* C_j = \frac{A_j^2 + B_j^2}{4} \ , \tag{2.4}$$

where the superscript * indicates the complex conjugate[4] of a complex number. The square root of the power spectrum is often called the *amplitude spectrum*, shown in Fig. 2.3, and gives the amplitudes of the various harmonics of the signal.

2.2.2 The Continuous Fourier Transform

Recall the statement in Sect. 2.2.1 that a signal must be periodic in order to be synthesised from sines and cosines. As most real-world signals are not periodic (or *'aperiodic'*), their Fourier representation must be determined using another approach. This approach is called the *continuous Fourier transform*, or often just 'Fourier

---

[4] For a complex number $z = x + iy$, its complex conjugate is $z^* = x - iy$. If the complex number is given in polar form as $z = e^{i\theta}$, its complex conjugate is $z^* = e^{-i\theta}$.

transform', and is based on the fact that an aperiodic signal can indeed be completely constructed from sines and cosines, but only as long as all frequencies are present. In that sense, an aperiodic function may be thought of as a limiting case of a periodic function where the period tends to infinity and the frequency separation tends to zero (James 2011). This leads us to rewrite the Fourier series of $g(t)$ in Eq. 2.2 as its *Fourier integral* representation,

$$g(t) = \int_{-\infty}^{\infty} G(f) e^{2\pi i f t} \, df , \qquad (2.5)$$

for $|t| < \infty$. This is also called the *inverse Fourier transform* of the spectrum $G(f)$, and it can be seen that the discrete frequencies, f_j, of Eq. 2.2 have become continuous. The *Fourier transform* of $g(t)$ is

$$G(f) = \int_{-\infty}^{\infty} g(t) e^{-2\pi i f t} \, dt , \qquad (2.6)$$

for $|f| < \infty$, analogous to Eq. 2.3 for Fourier series. Equation 2.6 is the equation used to find analytic expressions for the Fourier transforms of analytic expressions.

To introduce some notation used in this book, we can write the Fourier transform as

$$G(f) \equiv \mathsf{F}\{g(t)\} ,$$

and the inverse Fourier transform as

$$g(t) \equiv \mathsf{F}^{-1}\{G(f)\} .$$

Alternatively, if $G(f)$ is the Fourier transform of $g(t)$, and $g(t)$ is the inverse Fourier transform of $G(f)$, then

$$g(t) \longleftrightarrow G(f)$$

denotes a 'Fourier transform pair'.

2.2.3 Amplitude, Power and Phase Spectra

In general, the Fourier transform, $G(f)$, is a complex variable and can be written as

$$G(f) = G_R(f) + i \, G_I(f) ,$$

where the real part is $G_R(f) = \mathsf{Re}[G(f)]$, and the imaginary part is $G_I(f) = \mathsf{Im}[G(f)]$. The Fourier transform can also be written in polar form as

$$G(f) = |G(f)| \, e^{i\phi(f)} ,$$

where the *amplitude spectrum* of G is

$$|G(f)| = \sqrt{G_R^2(f) + G_I^2(f)},$$

and the *phase spectrum* is

$$\varphi(f) = \tan^{-1}\left(\frac{G_I(f)}{G_R(f)}\right), \tag{2.7}$$

both being real-valued. One can also define a *power spectrum* as the square of the amplitude spectrum, superficially as

$$G^*(f)\,G(f) = |G(f)|^2 = G_R^2(f) + G_I^2(f), \tag{2.8}$$

where the * indicates the complex conjugate. Note, though, that power spectral estimation forms a subject in its own right and equations such as Eqs. 2.4 and 2.8 should be used with caution (see Sect. 3.2).

In many disciplines, certainly the Earth sciences, the time (or space)-domain signal will be real-valued. But it will also generally be 'asymmetric', in the mathematical sense of being neither an even (symmetric about the ordinate (y) axis) nor odd (antisymmetric about the ordinate) function. Asymmetric real-valued functions have the property of having Fourier transforms that are called *Hermitian* (Bracewell 1986).

Hermitian functions are complex variables whose real part is even and whose imaginary part is odd. This implies that half of the information contained in the spectrum is redundant as the negative-frequency harmonics are just reflections of the positive-frequency harmonics: $G_R(-f) = G_R(f)$ and $G_I(-f) = -G_I(f)$, as shown in Fig. 2.4. However, even though half of the information is redundant, there is no data loss because the frequency-domain spectrum is described by twice the number of pieces of information (real and imaginary) than the time-domain signal (real only).

Fig. 2.4 A simple Hermitian function in the frequency domain, with an even real part (red) and an odd imaginary part (blue)

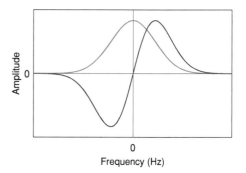

2.2.4 Signal Translation

If the time-domain signal is *shifted*, or *translated*, then its Fourier transform undergoes a phase change but not an amplitude change. This is easily proven. Consider a signal $g(t)$ with Fourier transform $G(f)$ that is shifted by an amount t_0 to $g(t - t_0)$. The Fourier transform of the shifted signal is given by

$$\mathsf{F}\{g(t - t_0)\} = \int_{-\infty}^{\infty} g(t - t_0) \, e^{-2\pi i f t} \, dt \ .$$

Being clever, we can rewrite this as

$$\mathsf{F}\{g(t - t_0)\} = \int_{-\infty}^{\infty} g(t - t_0) \, e^{-2\pi i f t} \, e^{2\pi i f t_0} \, e^{-2\pi i f t_0} \, d(t - t_0)$$

$$= \int_{-\infty}^{\infty} g(t - t_0) \, e^{-2\pi i f (t - t_0)} \, e^{-2\pi i f t_0} \, d(t - t_0) \ ,$$

where we can write $dt = d(t - t_0)$ because t_0 is a constant and its derivative with respect to t is zero, or

$$\frac{d(t - t_0)}{dt} = \frac{dt}{dt} - \frac{dt_0}{dt} = 1 \ .$$

If we let the dummy variable $\tau = t - t_0$, then we find

$$\mathsf{F}\{g(t - t_0)\} = \int_{-\infty}^{\infty} g(\tau) \, e^{-2\pi i f \tau} \, e^{-2\pi i f t_0} \, d\tau$$

$$= e^{-2\pi i f t_0} \int_{-\infty}^{\infty} g(\tau) \, e^{-2\pi i f \tau} \, d\tau \ ,$$

or, since τ is just a dummy variable, this can be written as

$$\mathsf{F}\{g(t - t_0)\} = e^{-2\pi i f t_0} \, G(f) \ . \tag{2.9}$$

If we now write $G(f)$ in polar form, we see

$$\mathsf{F}\{g(t - t_0)\} = |G(f)| \, e^{i \, [\varphi(f) - 2\pi f t_0]} \ ,$$

showing that the translated signal has experienced a phase change of $2\pi f t_0$. Equation 2.9 is referred to as the *shift theorem*.

2.2.5 Energy Conservation

Since the signal, $g(t)$, and the spectrum, $G(f)$, are merely different ways of expressing the same quantity, it stands to reason that they should possess equal energies.
This is embodied in a relationship known as *Parseval's theorem*, which says that the
total energy (E) in a signal is given by

$$E = \int_{-\infty}^{\infty} |g(t)|^2 \, dt = \int_{-\infty}^{\infty} |G(f)|^2 \, df \; . \tag{2.10}$$

A proof of Eq. 2.10 can be found in Percival and Walden (1993). Note that various texts also refer to this theorem as *Plancherel's theorem* or *Rayleigh's identity*,
sometimes applied to discrete time or frequency data, sometimes to continuous: there
appears to be no consistency.

2.2.6 Resolution and the Uncertainty Relationship

There is a law of quantum mechanics—Heisenberg's uncertainty principle—which
states that if one were able to measure the position of a sub-atomic particle to a very
high precision, then if one were also to attempt to measure its momentum (the product
of its mass and velocity) at the same time, the precision on that measurement would
be very poor. Or, the better you know where it is, the worse you know where it's
going. The reverse also holds. This principle is embodied in the formula $\sigma_x \sigma_p \geq \hbar/2$,
where σ_x is the standard deviation of the position, σ_p is the standard deviation of the
momentum and $\hbar$ is Planck's constant divided by 2π. Hence, if the position precision
improves (decreasing σ_x), then the momentum precision must worsen (increasing σ_p)
proportionately. Heisenberg's uncertainty principle is a fundamental law of physics
that cannot be overcome, even with super-precise instruments.

A similar relationship holds between the time and frequency-domain representations of a signal: the more localised a signal is in the time domain, the broader its
Fourier transform; the more localised the spectrum, the longer is the duration of the
signal (Fig. 2.5). Thus, for the Fourier transform pair, $g(t)$ and $G(f)$, their standard
deviations are related by

$$\sigma_g \sigma_G \geq \frac{1}{4\pi} \; , \tag{2.11}$$

a proof of which can be found in Percival and Walden (1993). This *uncertainty
relationship* demonstrates that a signal cannot be both band-limited and time-limited.

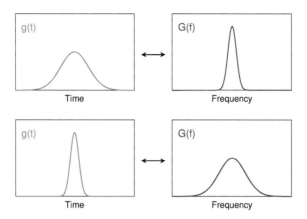

Fig. 2.5 Two Fourier transform pairs, illustrating resolution in the two domains. Top: a function $g(t)$ with a poor resolution (large standard deviation, σ_g) in the time domain has a good resolution (small σ_G) in the frequency domain. Bottom: a function $g(t)$ with a good resolution (small σ_g) in the time domain has a poor resolution (large σ_G) in the frequency domain

2.2.7 Differentiation

It was mentioned at the start of this chapter that the Fourier transform was originally developed for the purpose of solving a differential equation. Its utility in the topic of this book, estimating elastic thickness, also arises from this property, so we will investigate it further.

Take the Fourier integral representation of a signal $g(t)$, Eq. 2.5,

$$g(t) = \int_{-\infty}^{\infty} G(f)\, e^{2\pi i f t}\, df \, ,$$

and differentiate it with respect to its independent variable, t, giving

$$\frac{\partial g(t)}{\partial t} = \int_{-\infty}^{\infty} G(f)\, \frac{\partial e^{2\pi i f t}}{\partial t}\, df$$

$$= \int_{-\infty}^{\infty} G(f)\, (2\pi i f)\, e^{2\pi i f t}\, df$$

$$= \mathsf{F}^{-1}\{2\pi i f\, G(f)\} \, .$$

Thus, the Fourier transform of a derivative of a signal is a function of the signal's spectrum. A similar proof holds for higher derivatives, such that, in general,

$$\mathsf{F}\left\{ \frac{\partial^n g(t)}{\partial t^n} \right\} = (2\pi i f)^n\, G(f) \, . \tag{2.12}$$

This very useful property enables the ready solution of some differential equations. For example, take the differential equation in $y(t)$ given by

$$\frac{\partial^4 y}{\partial t^4} + \alpha^4 y = q,$$

where $q(t)$ is a known function, and α is a known constant. This equation can be solved by taking the Fourier transform of both sides, giving

$$(2\pi i f)^4 Y(f) + \alpha^4 Y(f) = Q(f),$$

where $Y(f)$ is the Fourier transform of $y(t)$, and $Q(f)$ is the Fourier transform of $q(t)$. Rearrangement gives the solution

$$Y(f) = \frac{Q(f)}{16\pi^4 f^4 + \alpha^4},$$

which can be inverse Fourier transformed for the time-domain solution.

2.2.8   Convolution

Convolution is a mathematical operation that takes two functions and, from these, generates a third function by taking the integral of the product of the two functions after one is reversed and shifted. While this definition might not make much sense to a novice, convolution crops up in many fields of science and engineering.

The convolution between two functions $p(t)$ and $q(t)$ produces a third function, say $g(t)$, and is written as

$$g(t) = (p * q)(t),$$

where the asterisk symbol, $*$, indicates convolution, defined by

$$(p * q)(t) \equiv \int_{-\infty}^{\infty} p(\tau) q(t - \tau) d\tau \tag{2.13}$$

for $|t| < \infty$, where τ is a dummy variable of integration. Convolution is commutative, associative and distributive over addition:

$$p * q = q * p$$
$$r * (p * q) = (r * p) * q$$
$$r * (p + q) = (r * p) + (r * q).$$

Figure 2.6a shows an example of convolution, where a finite-duration polynomial function, $p(t)$, is convolved with a Gaussian function, $q(t)$, producing a third

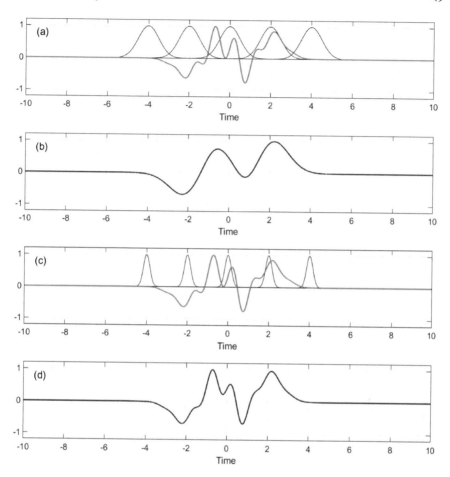

Fig. 2.6 a Five stages of the convolution of a finite-duration polynomial function, $p(t)$ (red), with a Gaussian function, $q(t)$ (black). As the Gaussian moves in stages across the polynomial, the two functions are multiplied and the result (blue), $(p * q)(t)$, is plotted in (**b**). **c** and **d** As (**a**) and (**b**), but for a narrower Gaussian function

function, $g(t)$, shown in Fig. 2.6b. According to Eq. 2.13, a value of t is chosen, say t_1, which defines the position of $q(t)$, for example, the first Gaussian on the left in Fig. 2.6a; the two functions are multiplied together and these products integrated (or summed); the sum—a single number—is stored at $g(t_1)$. The Gaussian is then translated or shifted across the polynomial, to time t_2, and the same procedure is performed. Figure 2.6a shows five such translations. The result, $g(t)$ shown in Fig. 2.6b, is a smoothed version of the polynomial. If the width of the Gaussian is decreased, as in Fig. 2.6c, then the convolution result (Fig. 2.6d) is less smooth and more similar to the polynomial. Hence, the procedure of convolution with a symmetric function like a Gaussian is often called *smoothing* and is equivalent to low-pass filtering (see Sect. 2.3.5).

Convolution is computationally time-consuming, involving repeated integrations. Fortunately, it can be shown that the Fourier transform of a convolution of two functions is equal to the product of their spectra, or

$$(g * h)(t) \longleftrightarrow G(f) H(f) .$$

This is the *convolution theorem*, and it also applies to the convolution of two spectra:

$$g(t) h(t) \longleftrightarrow (G * H)(f) .$$

2.3   Sampling a Continuous Function

We now move on to the real world of practical Fourier transforms, where the signals are not continuous functions of infinite extent, but are finite, sampled versions of such. First, though, we need to understand how sampling is performed, and then explore its consequences. In order to do this, we introduce two very useful functions.

2.3.1   Delta and Comb Functions

2.3.1.1   Dirac Delta Function

The Dirac delta function crops up everywhere. Heuristically, it is defined through

$$\delta(t - t_0) = \begin{cases} +\infty, & t = t_0 \\ 0, & t \neq t_0 \end{cases}$$

as shown in Fig. 2.7, but it has unit area, or

$$\int_{-\infty}^{\infty} \delta(t) \, dt = 1 .$$

Its proper mathematical definition, though, is through an integral formula:

$$\int_{-\infty}^{\infty} g(t) \, \delta(t - t_0) \, dt = g(t_0) . \tag{2.14}$$

That is, the delta function takes a signal and returns its value at a specified time (t_0 here).

Fig. 2.7 A Dirac delta
function with a delay time of
t_0: $\delta(t - t_0)$

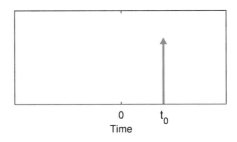

Convolution with a delta function returns the time-delayed function. From Eq. 2.13, we see that

$$g(t) * \delta(t - t_0) = \int_{-\infty}^{\infty} g(\tau)\,\delta(t - t_0 - \tau)\,d\tau = g(t - t_0) ,$$

where we have used the property that the delta function is an even function, $\delta(-t) = \delta(t)$, to reach the final equality. Thus, when convolved with a delta function, the signal $g(t)$ is shifted along the t-axis by an amount t_0.

The Fourier transform of a delta function is a complex exponential,

$$F\{\delta(t - t_0)\} = e^{-2\pi i f t_0} ,$$

which is readily demonstrated from Eqs. 2.6 and 2.14:

$$\int_{-\infty}^{\infty} \delta(t - t_0)\, e^{-2\pi i f t}\, dt = e^{-2\pi i f t_0} .$$

So if $t_0 = 0$, the Fourier transform is just 1. Similarly, the inverse Fourier transform of a frequency-domain delta function is a complex exponential of period $1/f_0$, or

$$F\{e^{2\pi i f_0 t}\} = \delta(f - f_0) ,$$

which the reader can prove for themselves.

2.3.1.2 Dirac Comb

A Dirac comb of period T is a sum of a potentially infinite number of Dirac delta functions, each shifted by an integer multiple of T, or

$$\text{III}_T(t) = \sum_{k=-\infty}^{\infty} \delta(t - kT) ,$$

as shown in Fig. 2.8a. It is typically given the symbol III (pronounced 'shah').

Fig. 2.8 A Dirac comb in **a** the time domain and **b** its Fourier transform in the frequency domain. The period of the comb is T (seconds) in the time domain and $1/T$ (Hertz) in the frequency domain

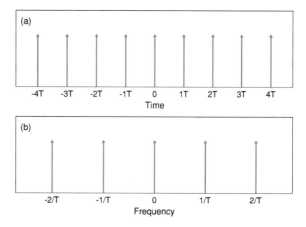

The Fourier transform of a Dirac comb is also a Dirac comb, with the period of the comb in the frequency domain being the reciprocal of that in the time domain, or

$$\mathsf{F}\{\mathrm{III}_T(t)\} \;=\; \frac{1}{T}\,\mathrm{III}_{1/T}(f)\,, \tag{2.15}$$

where

$$\mathrm{III}_{1/T}(f) \;=\; \sum_{k=-\infty}^{\infty} \delta\!\left(f - \frac{k}{T}\right)$$

as shown in Fig. 2.8b.

2.3.2   Sampling

Formally, the sampling action is represented by multiplication of the continuous function, $g(t)$, by a Dirac comb. The spacing of the delta functions in the comb is the sampling interval, Δt (and therefore the period, T, of the comb, i.e. $T = \Delta t$). Thus, if $g(t)$ is the continuous function, then its sampled version is

$$g(t)\,\mathrm{III}_{\Delta t}(t) \;=\; g(t) \sum_{k=-\infty}^{\infty} \delta(t - k\Delta t)\,.$$

We will represent the sequence thus sampled as g_τ or $g(\tau\Delta t)$, where the index τ ($\tau = 0, \pm 1, \pm 2, \ldots, \pm\infty$) denotes the τth sample.

From the convolution theorem and Eq. 2.15, the Fourier transform of the product $g\,\mathrm{III}$ is the convolution of their Fourier transforms, or

$$\mathsf{F}\{g(t)\,\mathrm{III}_{\Delta t}(t)\} \;=\; \frac{1}{\Delta t}\,(\mathrm{III}_{1/\Delta t} * G)(f)\,,$$

where we can also write

$$(\mathrm{III}_{1/\Delta t} * G)(f) \;=\; \sum_{k=-\infty}^{\infty} G\!\left(f - \frac{k}{\Delta t}\right)$$

(Bracewell 1986), which shows that the spectrum is periodic with period $1/\Delta t$. Hence, sampling a function results in an infinite, periodic repetition of the function's spectrum, as shown in Fig. 2.9b.

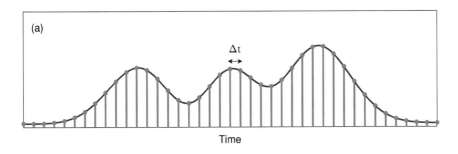

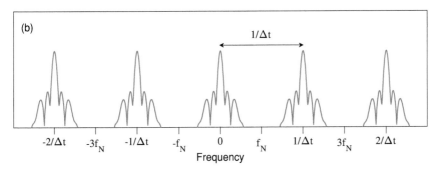

Fig. 2.9 **a** A continuous time-domain signal (blue curve) is sampled (red circles) at a sampling interval of Δt. **b** The Fourier transform of the sampled signal is a continuous, periodic, band-limited repetition of the spectrum to $f \to \pm\infty$, referred to as the discrete-time Fourier transform (DTFT), with a period of $1/\Delta t$. The frequency labelled f_N is the Nyquist frequency, discussed in Sect. 2.3.4

2.3.3 Aliasing

When the signals from natural phenomena are represented as Fourier series, we can identify their harmonic content—which frequencies are present, and in what amounts. When dealt with in such a way, the signal is often categorised by the highest frequency (or shortest period/wavelength) it contains. If the highest frequency present in a signal $g(t)$ is W, then W is said to be the *bandwidth* of the signal, and its Fourier transform is such that

$$G(f) = 0, \quad |f| > W .$$

Since both period and wavelength are equal to 1/frequency, the lowest period (or shortest wavelength) present in the signal is therefore

$$T_{\min} = \frac{1}{W} . \tag{2.16}$$

Figure 2.10 shows a signal belonging to a natural phenomenon that has only one harmonic: the sine wave with a period $T = 2\pi$ s (shown as the blue lines). It is then sampled three times, each with a constant time interval, Δt. In Fig. 2.10a, the sampling interval is $\Delta t = \pi$ s; fitting a model sine wave to these samples would show that the signal can be faithfully reconstructed from the samples. In fact, whenever the sampling interval is less than or equal to half the minimum period of the signal, or

$$\Delta t \leq \frac{T_{\min}}{2} , \tag{2.17}$$

then accurate reconstruction occurs. However, when the sampling interval becomes greater than half the minimum period ($\Delta t > T_{\min}/2$), then a sine wave with a longer period than that of the original signal can be fitted to these samples. In Fig. 2.10b, the sampling interval is now longer than in Fig. 2.10a ($\Delta t = 9\pi/7$ s) and the modelled sine wave (red line) has a period of approximately 3.6π s. In Fig. 2.10c, the sampling interval is longer still ($\Delta t = 2\pi$ s), equal to the period of the signal, and now a straight line can be fit to the samples, effectively a sine wave with an infinite period.

This phenomenon—of longer-period harmonics spuriously appearing in the sampled signal—is called *aliasing*, and it is an artefact caused by undersampling a signal.

Typically, analysts use frequency rather than period or wavelength to categorise harmonic content. So, phrasing the above discussion in terms of frequency, aliasing results in the high-frequency harmonics of the signal appearing as lower-frequency harmonics in the sampled model. From Eqs. 2.16 and 2.17, we see that in order to prevent aliasing, the sampling frequency ($f_s = 1/\Delta t$) must be greater than or equal to twice the bandwidth of the signal being sampled, or

$$f_s \geq 2W . \tag{2.18}$$

Fig. 2.10 A signal,
$g(t) = \sin t$ (blue line in all
graphs), has period $T = 2\pi$ s
and is sampled at various
intervals (black circles),
$\Delta t = \mathbf{a}\,\pi$ s, $\mathbf{b}\,9\pi/7$ s and $\mathbf{c}$
2π s. In (**b**) and (**c**), the
sampling frequencies do not
meet the Nyquist criterion
and aliases are generated (red
lines) with periods of (**b**)
3.6π s and (**c**) infinite period

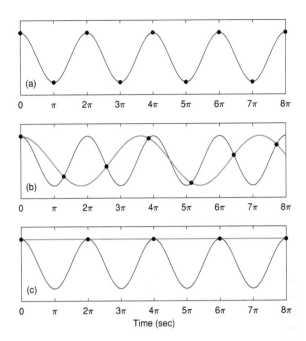

This condition is called the *Nyquist criterion*, and the quantity $2W$ is the *Nyquist rate*. Expressed in terms of the sampling interval, it becomes

$$\Delta t \ \leq \ \frac{1}{2W}\,, \tag{2.19}$$

giving a maximum value of the sampling interval to avoid aliasing. The case of infinite bandwidth will be discussed in Sect. 2.3.5.

So, returning to Fig. 2.10, we calculate that the bandwidth of the continuous signal is $W = 1/(2\pi) = 0.159$ Hz, and its Nyquist rate is $2W = 0.318$ Hz. Now, when the sampling interval is $\Delta t = \pi$ s (Fig. 2.10a), the sampling frequency is $f_s = 1/\pi$ Hz or 0.318 Hz; Eq. 2.18 tells us that the Nyquist criterion is met and aliasing will not occur. For the case in Fig. 2.10b though, the sampling interval is $\Delta t = 9\pi/7$ s, giving a sampling frequency of $f_s = 0.248$ Hz, which is less than the Nyquist rate of 0.318 Hz; the Nyquist criterion is not met and aliasing will occur. And in Fig. 2.10c, the sampling interval is $\Delta t = 2\pi$ s, giving a sampling frequency of $f_s = 0.159$ Hz, which is also less than the Nyquist rate, generating aliasing.

2.3.4   The Nyquist Frequency

A quantity related to the Nyquist rate, but subtly different, is the *Nyquist frequency.*
It applies to the sampled (discrete) sequence, rather than the signal the samples were
taken from, and this distinction is often not made clear, leading to confusion. For a
sampled sequence, if the sampling interval is Δt, then the Nyquist frequency of the
sampled data is given by

$$f_N \;=\; \frac{1}{2\Delta t} \;=\; \frac{f_s}{2}\,, \tag{2.20}$$

where f_s is the sampling frequency. That is, the Nyquist frequency is a function of
the sampling rate, Δt, that created the sequence. It does not depend on the harmonic
content of the signal being sampled.

Whether or not a sampled sequence contains aliased harmonics depends upon the
relationship between the Nyquist rate and the Nyquist frequency. Using Eqs. 2.19
and 2.20, the Nyquist criterion may be written in terms of the continuous signal's
bandwidth (W) and the Nyquist frequency of the sampled sequence, as

$$f_N \;\geq\; W\,, \tag{2.21}$$

or as

$$f_N \;\geq\; \frac{2W}{2}\,,$$

showing that the Nyquist frequency of the sample needs to be at least half the Nyquist
rate ($2W$) of the signal to prevent aliasing.

Again, it is important to remember that the Nyquist rate is a property of the signal
being sampled, while the Nyquist frequency is a property of the sampled sequence.
For example, consider a continuous signal with a bandwidth of $W = 1$ Hz; that is, it
contains no harmonics with higher frequencies. Its Nyquist rate is thus $2W = 2$ Hz.
If one needs to sample this signal, then, in order to prevent aliasing, sampling must
occur with sampling frequencies $f_s \geq 2$ Hz, as prescribed by the Nyquist criterion,
Eq. 2.18. Or we could use Eq. 2.19 to express this as $\Delta t \leq 0.5$ s.

1. Suppose the signal is sampled at a rate of $\Delta t = 2$ s. From Eq. 2.20, the resulting
   sequence has a Nyquist frequency of $f_N = 0.25$ Hz. Consulting Eq. 2.21, we
   see that this is not greater than W ($0.25 < 1$), so the sequence contains aliased
   harmonics.
2. Now suppose the signal is sampled at a rate of $\Delta t = 0.05$ s; this gives a sequence
   with a Nyquist frequency of $f_N = 10$ Hz. Since f_N is greater than W ($10 > 1$),
   the sequence is not aliased. However, just because the sequence has a Nyquist
   frequency 10 Hz, this doesn't mean that it contains harmonics of 10 Hz: remember,
   the original signal only contains harmonics with a maximum frequency 1 Hz. So
   harmonics with frequencies 1–10 Hz do not exist in the sequence.

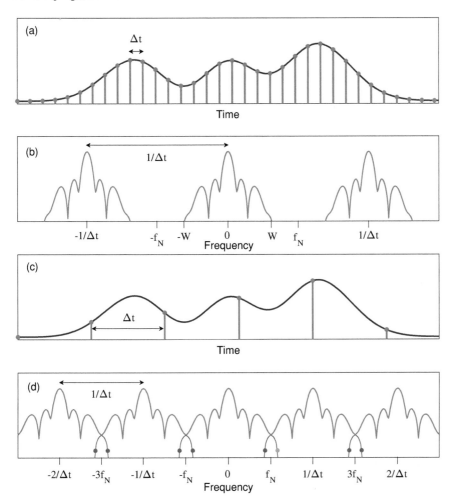

Fig. 2.11 When a signal is sampled such that the sampling frequency $f_s = 1/\Delta t > 2W$ (**a**), where W is the bandwidth, then the periodic repetitions of the signal spectrum are distinct and do not overlap, as shown in (**b**). In contrast, when the sampling interval is increased, shown in (**c**), so that $1/\Delta t < 2W$, the spectra overlap, and the harmonics shown as magenta lines in (**d**) are aliased. The magenta dots in these aliased harmonics are aliases of the green dot, which is an undersampled sequence

Figure 2.11 shows the relationship between f_N and W, and how aliasing arises. In Fig. 2.11a, the sampling interval has been chosen small enough so that $f_N > W$ (Fig. 2.11b), and the Nyquist criterion (Eq. 2.21) is met; the periodic repetitions of the signal spectrum are distinct (recall from Sect. 2.3.2 how sampling a continuous signal generates periodic repetitions of the spectrum). However, in Fig. 2.11c, the sampling interval has been increased too much and the sequence no longer accurately captures and reproduces the higher-frequency harmonics—the signal is aliased. This is shown in Fig. 2.11d where the periodic repetitions of the signal spectrum overlap, and $f_N < W$. The high-frequency harmonics of the signal spectrum are 'folded' back into the spectra at lower frequencies, and for this reason, the Nyquist frequency is sometimes referred to as the *folding frequency*.

The frequencies of the aliased harmonics are readily calculated. A harmonic of frequency f_0 at a frequency higher than f_N by an amount δf (the green dot in Fig. 2.11d), such that $f_0 = f_N + \delta f$ and $f_N < f_0 \leq W$, has an alias at frequency $f_N - \delta f$ (the magenta dot just to its left). But there are also aliased frequencies at all the magenta dots in Fig. 2.11d, occurring on either side of odd multiples of f_N. In general, one can show that the aliases occur at frequencies

$$f_{k\pm} = (2k - 1)f_N \pm \delta f$$

for $k = 1, 2, \ldots, \infty$, where $\delta f = f_0 - f_N$ for $f_0 > f_N$. For the signal being (under)sampled, $f_0 = f_{1+}$. Note that each $f_{k\pm}$ has a negative-frequency image, $-f_{k\pm}$.

We can now calculate the periods of the aliased sine waves in Fig. 2.10. Recall the signal had a period of 2π s. In Fig. 2.10b, the sampling interval is $\Delta t = 9\pi/7$ s, giving a Nyquist frequency of $f_N = 7/(18\pi)$ Hz. If we take the harmonic of interest to be the signal itself, with frequency $f_0 = 1/(2\pi)$ Hz, then

$$\delta f = f_0 - f_N = \frac{1}{2\pi} - \frac{7}{18\pi} = \frac{1}{9\pi} \text{ Hz} .$$

The aliased harmonic shown in Fig. 2.10b will be the first one at a slightly lower frequency

$$f_{1-} = f_N - \delta f = \frac{7}{18\pi} - \frac{1}{9\pi} = \frac{5}{18\pi} \text{ Hz} ,$$

with a period of $T_1 = 18\pi/5$ s $= 3.6\pi$ s, as shown in the figure. Performing a similar calculation for Fig. 2.10c, where the sampling interval is $\Delta t = 2\pi$ s, we find that

$$\delta f = f_0 - f_N = \frac{1}{2\pi} - \frac{1}{4\pi} = \frac{1}{4\pi} \text{ Hz} .$$

The aliased harmonic shown in Fig. 2.10c is thus

$$f_{1-} = f_N - \delta f = \frac{1}{4\pi} - \frac{1}{4\pi} = 0 \text{ Hz} ,$$

with an infinite period, as shown.

2.3.5 Anti-Aliasing (Frequency) Filter

When continuous signals with a defined bandwidth are sampled, it is straightforward to adjust the sampling frequency so that aliasing doesn't occur, as in Eq. 2.18. However, many natural phenomena have a potentially infinite bandwidth, and even the highest technologically possible sampling frequency might not capture all the high-frequency harmonics, resulting in aliasing. In these cases, another approach is needed, and this involves the application of an anti-aliasing filter to the recorded signal before it is sampled, if that is possible.

A low-pass *frequency filter* is a suitable tool and will remove those harmonics with frequencies high enough to cause aliasing, generating a band-limited signal that can then be sampled (e.g. Fig. 2.12). Frequency filters work by frequency-domain multiplication of the Fourier transform of a signal by a filter function. For a signal $g(t)$ with Fourier transform $G(f)$, the filtered signal, $\breve{g}(t)$, is calculated by

$$\breve{g}(t) = \mathsf{F}^{-1}\{\Theta(f)\,G(f)\}\,, \tag{2.22}$$

where $\Theta(f)$ is the frequency filter function. Filters that suppress or remove the high-frequency harmonics are called *low-pass filters*, filters that suppress or remove the low-frequency harmonics are called *high-pass filters* and those that suppress or remove both low- and high-frequency harmonics are called *band-pass filters*. Figure 2.12b shows an example of a low-pass filter with a cut-off frequency (f_c) of 0.3 Hz; that is, it removes those harmonics with frequencies higher than 0.3 Hz. Note how the filter does not have a sharp boundary at 0.3 Hz, but has a smooth transition given by

$$\Theta(f) = \begin{cases} 1\,, & f \leq f_1 \\ \cos^2\left[\frac{\pi}{2}\frac{(f-f_1)}{(f_c-f_1)}\right]\,, & f_1 < f < f_c \\ 0\,, & f \geq f_c \end{cases}$$

where $f_1 = 0.2$ Hz and $f_c = 0.3$ Hz. This smooth roll-off is designed to reduce the *Gibbs phenomenon*, which will introduce artefacts into the filtered signal if the transition from 1 to 0 in the filter is sharp (Sect. 2.5.4).

Specifically, the cut-off frequency of the filter (f_c) must be less than or equal to the desired Nyquist frequency, or

$$f_c \leq f_N\,.$$

So if the signal is to be sampled at an interval of Δt_0 s, then it must first be filtered with a low-pass filter having a cut-off frequency of $1/(2\Delta t_0)$ Hz, from Eq. 2.20. Then the highest frequency present in the continuous signal will be the same as the Nyquist frequency of the to-be-sampled data. For example, if the sampling interval is 0.05 s, then the cut-off frequency of the filter must be at 10 Hz; if the sampling

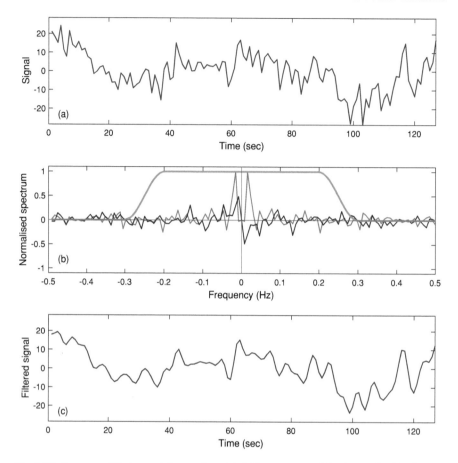

Fig. 2.12 Frequency filtering. **a** A time series, $g(t)$. **b** The spectrum of the time series [$G(f)$, with real (red) and imaginary (blue) parts], and a low-pass filter, $\Theta(f)$ (aqua curve), with a cosine rollover and a cut-off frequency of $f_c = 0.3$ Hz. **c** The filtered time series with suppressed high-frequency harmonics, $\breve{g}(t)$

interval is longer, say 0.1 s, then the cut-off frequency must be reduced, 5 Hz at most. Evidently, the more frequent the sampling, the more frequency content the sampled version will contain.

When used in this fashion, frequency filters are also known as *transfer functions*, converting one signal, $g(t)$, into another, $\breve{g}(t)$ in this case. Using the convolution theorem (Sect. 2.2.8), we can write Eq. 2.22 as

$$\breve{g}(t) \;=\; \theta(t) * g(t) \,, \tag{2.23}$$

where $\theta(t)$ is called the *impulse response function*, so-called because if g were a Dirac delta function (an impulse), then $\breve{g}$ would look like θ, since convolution of a function with a delta function yields the function (Sect. 2.3.1). Note that the impulse response is the inverse Fourier transform of the filter, or $\theta(t) = \mathsf{F}^{-1}\{\Theta(f)\}$. We will return to this topic in Sect. 5.2.1.

2.4 Fourier Transforms of Discrete Data

When one needs to find the Fourier transform of some sampled data, invariably one will use the *fast Fourier transform* (FFT), an algorithm that implements the *discrete Fourier transform* (DFT). But how do the DFT and FFT relate to the continuous Fourier transform (CFT), and to each other for that matter?

The CFT tells us both the theory of Fourier transforms and how to calculate the Fourier transform of an analytic equation, using Eq. 2.6. Its assumptions and requirements are listed in Table 2.1, but importantly, it assumes that the data are continuous, aperiodic and of infinite length. The reality, though, is different. As we saw in Sect. 2.3, while a natural phenomenon may be described by or emit a continuous signal, data representing it are most often collected as samples that create a sequence of finite length. As we cannot use the CFT on sampled data, we must use the DFT, which makes very different assumptions (Table 2.1).

At first glance, the information in the table seems uncontroversial. But a closer look reveals an apparent contradiction, where the DFT is said to have finite support in both domains ($T_1 \leq t \leq T_2$ and $-W \leq f \leq W$). Recall that a signal or sequence cannot be both band-limited and time-limited, embodied, for example, in the uncertainty

Table 2.1 Summary of some of the requirements and assumptions of the continuous and discrete Fourier transforms (CFT and DFT). The first three rows concern the time domain, and the last three concern the frequency domain

CFT	DFT
The signal is continuous, with an observation at every moment in time, no matter how closely separated	The sequence is discrete, with samples separated by constant time intervals, Δt
The signal has infinite support. That is, it has values for all $|t| < \infty$	The sequence has finite support: it has defined beginning and end times, $T_1 \leq t \leq T_2$, and duration $T = T_2 - T_1$
The signal is aperiodic, never repeating itself	The sequence is periodic, repeating itself every T s
The spectrum is continuous, with a harmonic of every frequency	The spectrum is discrete, with harmonics separated by constant frequency intervals, $\Delta f = 1/T$
The spectrum has infinite support. That is, it has values for all $|f| < \infty$	The spectrum has finite support, being band-limited by the bandwidth $|W| \leq f_N$
The spectrum is aperiodic, never repeating itself	The spectrum is periodic, repeating itself every $1/\Delta t$ Hz

relation (Sect. 2.2.6). It appears that the sequences of the DFT are breaking the uncertainty relation. So how is this apparent contradiction resolved?

The key lies in the periodicity of the signals and sequences, but can only be fully understood by exploring what actually happens when you take the FFT of some data. The first thing to appreciate is that the discrete version of the Fourier transform comes in two 'flavours': the discrete-time Fourier transform (DTFT), where the time domain is discrete but the frequency domain is continuous; and the discrete Fourier transform (DFT) where both domains are discrete.

2.4.1 The Discrete-Time Fourier Transform

We have met the *discrete-time Fourier transform* (DTFT) before, in Fig. 2.9, where we saw how sampling a signal in the time domain results in a periodic, continuous, band-limited spectrum. The DTFT spectrum is periodic and continuous because multiplication by a Dirac comb in one domain is equivalent to convolution with a comb in the other domain, and the continuous CFT spectrum is replicated with period $1/\Delta t$ out to $f \to \pm\infty$. And the DTFT spectrum is band-limited because aliasing would occur if it wasn't. Note that, while the spectrum has infinite support, it is still band-limited.

Formally, consider a continuous, band-unlimited, aperiodic signal $g(t)$, with infinite support ($|t| < \infty$), and spectrum (CFT) $G(f)$, also with infinite support ($|f| < \infty$). The signal is sampled at an interval of Δt, giving a sequence g_τ where $\tau = 0, \pm 1, \pm 2, \ldots, \pm\infty$. The discrete-time Fourier transform of g_τ is given by

$$G_{1/\Delta t}(f) = \sum_{\tau=-\infty}^{\infty} g_\tau \, e^{-2\pi i f \tau \Delta t} \,, \tag{2.24}$$

analogous to Eq. 2.6, the CFT. As shown in Sect. 2.3.2, as a consequence of the sampling, $G_{1/\Delta t}(f)$ is band-limited to $|f| \leq f_N$ and is periodic with period $1/\Delta t$, such that $G_{1/\Delta t}(f + p/\Delta t) = G_{1/\Delta t}(f)$, $p = \pm 1, \pm 2, \ldots, \pm\infty$. The DTFT of the sequence shown in Fig. 2.13a is shown in Fig. 2.13b.

The inverse DTFT gives the original sampled data sequence

$$g_\tau = \Delta t \int_{-f_N}^{f_N} G_{1/\Delta t}(f) \, e^{2\pi i f \tau \Delta t} \, df \,, \tag{2.25}$$

where $\tau = 0, \pm 1, \pm 2, \ldots, \pm\infty$. If the original signal, $g(t)$, is band-limited to $|f| \leq W$, and if the Nyquist criterion is met ($W \leq f_N$), then $g(t)$ can be perfectly reconstructed from its DTFT.

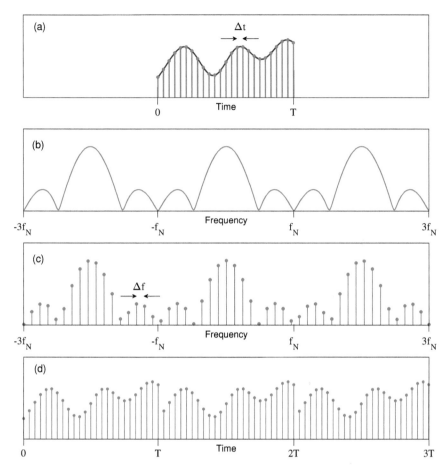

Fig. 2.13 **a** A continuous signal (blue) is sampled in the time domain so that the resulting sequence (red circles) is aperiodic and of finite duration T. **b** The DTFT of the sequence. It is continuous, periodic with period $2f_N = 1/\Delta t$ and band-limited to $\pm f_N$. **c** The DFT of the sequence, being the DTFT sampled at an interval of $\Delta f = 1/T$. **d** The sampled spectrum in (**c**) is inverse DFT'd, giving a periodic repetition of the original sequence in (**a**), with period T

Finally, Parseval's theorem (Sect. 2.2.5) in the context of the DTFT becomes

$$\sum_{\tau=-\infty}^{\infty} |g_\tau|^2 = \Delta t \int_{-f_N}^{f_N} |G_{1/\Delta t}(f)|^2 \, df \qquad (2.26)$$

(cf. Eq. 2.10).

2.4.2 The Discrete Fourier Transform

The DTFT spectrum is continuous, but in order to manipulate our actual data using the FFT and inverse FFT, we need a discrete sequence of Fourier coefficients. As the reader may suspect, discretisation is achieved by multiplying the DTFT by a Dirac comb in the frequency domain; the result is the *discrete Fourier transform* (DFT) and is shown in Fig. 2.13c.

The following points of note concerning the DFT are important to bear in mind when one computes an FFT. (1) The spacing of the delta functions in the comb is usually chosen to be the reciprocal of the length of the time-domain sequence, T, giving the sampling interval of the DFT as

$$\Delta f = \frac{1}{T} . \tag{2.27}$$

(2) Since the DTFT is periodic with period $1/\Delta t$, so is the DFT. (3) The periodic repetition of the DFT extends to $f \to \pm\infty$.

Furthermore, sampling of the DTFT causes the inverse transform to become periodic. The multiplication of the DTFT by the comb in the frequency domain is equivalent to the convolution of the sequence with a comb in the time domain. Thus, the sampled sequence, g_τ, is made periodic (Fig. 2.13d).

In other words, the DFT considers the time-domain sequence to be infinitely long and periodic. This has advantages, discussed below, and disadvantages (Sects. 2.5.5 and 2.5.6).

The assumption of periodicity in the time-domain sequence means that, rather than using an equation such as the DTFT (Eq. 2.24) to compute the Fourier transform, with its infinite summation limits, one can restrict the summation to finite limits because one cycle of the time-domain sequence contains all the available information about the signal. Thus, the discrete Fourier transform can be computed using

$$G_n = \sum_{\tau=0}^{N-1} g_\tau \, e^{-2\pi i n\tau/N} , \tag{2.28}$$

where $n = 0, 1, \cdots, N - 1$, and there are N observations in one cycle. The DFT spectrum G_n is associated with the frequency

$$f_n = n\Delta f = \frac{n}{N\Delta t} \tag{2.29}$$

from Eq. 2.27 with $T = N\Delta t$, the duration of the sequence in the time domain. The *zero frequency* ($f_0 = 0$ Hz) describes a harmonic with an infinite period or wavelength, and its Fourier coefficient, G_0, is often called the *DC value* (from the electronics term 'direct current', indicating an unchanging or constant value). The frequency of the first harmonic (f_1) is the *fundamental frequency* (see also Sect. 2.2.1) and is also equal to the frequency spacing, Δf.

The inverse discrete Fourier transform is a periodic summation of the original sequence, but, again, only needs to be computed over one cycle of the frequency-domain spectrum because that too contains all the available information about the signal. Thus, the inverse DFT is

$$g_\tau = \frac{1}{N} \sum_{n=0}^{N-1} G_n \, e^{2\pi i n \tau / N} \, , \tag{2.30}$$

where $\tau = 0, 1, \ldots, N - 1$. Both g_τ and G_n are periodic sequences with period N: $g_{(\tau + pN)} = g_\tau$, and $G_{(n+pN)} = G_n$, where $p = \pm 1, \pm 2, \ldots, \pm \infty$.

For completeness, the DFT version of Parseval's theorem (Sect. 2.2.5) is written as

$$\sum_{\tau=0}^{N-1} |g_\tau|^2 = \frac{1}{N} \sum_{n=0}^{N-1} |G_n|^2$$

(cf. Eqs. 2.10 and 2.26).

2.4.3 The Fast Fourier Transform

The *fast Fourier transform* (FFT) is not a transform per se, but rather an algorithm for rapid computation of the discrete Fourier transform. The forward FFT computes one cycle of the forward DFT (Eq. 2.28), while the inverse FFT computes one cycle of the inverse DFT (Eq. 2.30). The sequences in both domains must have the same number of elements (which in this section we will call M to distinguish it from N which will be used as a subscript representing properties of the Nyquist frequency).

The FFT is computationally very efficient when compared to the DFT. As shown by Eq. 2.28, the DFT takes $O(M^2)$ operations to compute a Fourier transform: the summation over τ takes M operations just to compute G_n at a single value of n, and there are M elements of G_n to determine. In contrast, the FFT is coded so that it only takes $O(M \log M)$ operations to fully determine the elements of G_n. The details of the FFT algorithm are beyond the scope of this book, but it is worth noting that the FFT is most efficient when M is an integer power of two (e.g. $2048 = 2^{11}$). If one's data set does not meet this restrictive condition, then one can either pad the data with zeros (Sect. 2.5.6.1), or use another FFT algorithm such as a mixed-radix FFT (Singleton 1968) where M only need be a product of prime numbers less than a specified value (e.g. $43{,}560 = 2^3 \times 3^2 \times 5 \times 11^2$). Note though that the larger the prime factors, the slower the FFT.

Unfortunately, the FFT requires the input data sequence to have a fixed and unchanging spacing between its values, i.e. constant Δt. This requirement is one of the features that enables the FFT to be executed much more quickly than the DFT, which does not enforce such a restriction on the data. The need for a constant sampling interval thus prohibits the existence of gaps in the data. If data sequences have such 'drop-outs', then the gaps must be assigned a value, usually arrived at by some interpolation scheme.

A feature of most FFT algorithms is their less-than-obvious indexing of the observations in the frequency domain. Approximately the first half of the elements of the FFT output sequence are assigned to positive frequencies, with the second half assigned to negative frequencies. The exact split, though, depends upon whether the sequence has an odd or even number of observations.

In order to manipulate and plot the spectrum computed by an FFT, we need to understand its indexing structure. Let the FFT output sequence be H_m ($m = 1, 2, \ldots, M$). The first element of H_m (H_1) is always the DC value (at the zero frequency) and is the sum of all the values of the sequence elements ($\sum g_\tau$, from Eq. 2.28 with $n = 0$). Subsequent frequencies increment by Δf, where

$$\Delta f = \frac{1}{M \Delta t}$$

(see Eqs. 2.27 and 2.29). The incrementation is suspended at an index m_N, which I will call a 'Nyquist index', defined by[5]

$$m_N = \left\lfloor \frac{M}{2} \right\rfloor + 1 . \tag{2.31}$$

The Nyquist index identifies the last positive frequency in the FFT sequence; after m_N, all the frequencies are negative (see Table 2.2). It does not, in general, identify the Nyquist frequency (such that H_{m_N} always occurs at f_N): this is only true when M is even; for odd M the Nyquist frequency does not have a designated element in the sequence.

The rule describing the assignation of frequencies $f_{H,m}$ to harmonics of the FFT output sequence, H_m, is

$$f_{H,m} = \begin{cases} (m-1)\,\Delta f, & 1 \le m \le m_N \\ (m-M-1)\,\Delta f, & m_N + 1 \le m \le M \end{cases} . \tag{2.32}$$

The rule in Eq. 2.32 is tabulated in the first two columns of Table 2.2 and displayed in Fig. 2.14a and b, which shows the amplitude spectra of an even (a) and odd (b) sequence of data, as output from the FFT. Of note is that the even sequence spectrum contains the Nyquist frequency as one of its elements, while the odd sequence spectrum does not (it lies 'off-grid').

When displayed graphically, amplitude spectra typically have the origin (zero frequency) at their centre with negative frequencies to the left and positive to the right. To achieve this look, we need to rearrange the FFT spectra (H_m). However, this is not as simple as it sounds owing to the variable transition index from negative to

---

[5] The notation $\lfloor A \rfloor$ in Eq. 2.31 rounds the real number A to the greatest integer less than or equal to A. Thus, for even M, $\lfloor M/2 \rfloor$ will have the same value as $\lfloor (M+1)/2 \rfloor$. In computer speak, it is called 'floor'.

Table 2.2 FFT and DFT indexing. The table only shows the key indices important for understanding the relationship between the FFT output (H_m) and the DFT (G_k). The index m denotes elements of the spectrum H_m outputted from the FFT algorithm and corresponds to frequency values $f_{H,m}$ which are calculated from Eq. 2.32. The index k denotes elements of the spectrum G_k and corresponds to frequency values $f_{G,k}$ which are calculated from Eq. 2.34; note that the indexing used in the DFT equation, Eq. 2.28, is $n = [0, N-1]$ whereas here I use $k = [1, M]$ where $M = N$ (hence $n = k - 1$). The index m_n is the 'Nyquist index' (Eq. 2.31), while the index m_s is the 'separator index' (Eq. 2.33)

m	$f_{H,m}$	k	$f_{G,k}$
1	0	1	$-m_s \, \Delta f$ [b]
...	...	...	...
...	...	$m_s + 1$ [c]	0
m_N	$(m_N - 1) \, \Delta f$ [a]	...	...
$m_N + 1$	$-m_s \, \Delta f$ [b]	...	...
...	...	...	...
M	$-\Delta f$	M	$(m_N - 1) \, \Delta f$

[a] When M is even, this frequency is the Nyquist frequency (f_N); this is not so for odd M, when the frequency value here is $f_N - \Delta f/2$

[b] For even M, these frequency values will equal $-(m_N - 2) \, \Delta f$; for odd M, they will equal $-(m_N - 1) \, \Delta f$

[c] For even M, this frequency index (identifying the zero frequency in f_G) will equal $m_N - 1$; for odd M, it will equal m_N

positive frequencies, depending on whether the sequence has an even or odd number of elements. Once reordered, the spectrum will resemble the DFT spectra shown in Fig. 2.13. Let the DFT sequence be G_k ($k = 1, 2, \ldots, M$). My usage here is slightly different to the standard usage, shown, for example, in Eq. 2.30, the inverse DFT, where the index n has values $n = 0, 1, \ldots, N - 1$ (remember, $M = N$). This choice is made (i) to maintain ready comparison with the index m, and (ii) to facilitate coding of the equations because arrays in computer programs default to indexing beginning at 1, not 0. Therefore, the relationship between n and k is $n = k - 1$.

It is useful to specify the index, m_s, which gives the value of the first negative frequency, as in $-m_s \, \Delta f$. I call this the 'separator index' for reasons given below, with

$$m_s = M - m_N . \tag{2.33}$$

While m_N has the same value for some even M and odd $M + 1$, the corresponding values of m_s differ. For example, in Fig. 2.14c (even M), the lowest frequency is $-9\Delta f \, (= -f_N + \Delta f)$, while in Fig. 2.14d (odd M), it is $-10\Delta f \, (= -f_N + \Delta f/2)$. Now we are in a position to define a formula to compute the frequencies assigned to G_k, being

$$f_{G,k} = (k - m_s - 1) \, \Delta f , \qquad k = 1, \ldots, M . \tag{2.34}$$

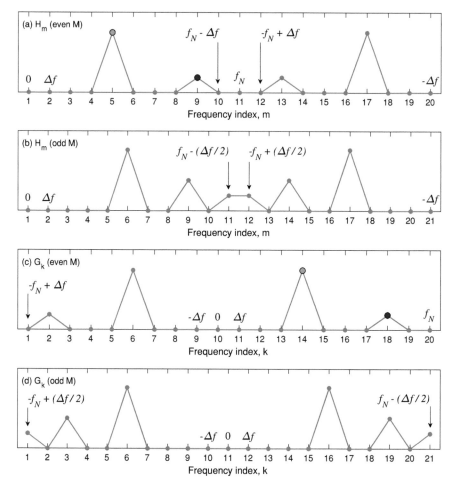

Fig. 2.14 Amplitude spectra of sequences with an even (**a** and **c**) and odd (**b** and **d**) number of elements. Panels (**a**) and (**b**) show H_m, the direct output of an FFT algorithm; panels (**c**) and (**d**) show G_k, the reordered DFT sequences. The even sequence has $M = 20$, Nyquist index $m_N = 11$, separator index $m_s = 9$, and Nyquist frequency $f_N = 10$ Hz. The odd sequence has $M = 21$, $m_N = 11$, $m_s = 10$, and $f_N = 10.5$ Hz. Both sequences have frequency interval $\Delta f = 1$ Hz. The green dots in (**a**) and (**c**) show identical harmonics, and likewise for the blue dots

Equation 2.34 also shows that the zero frequency in the G_k sequence occurs at f_{G,m_s+1}, and that the minimum value of $f_{G,k}$ will be $-m_s \Delta f$, and its maximum will be $(m_N - 1)\Delta f$ (see Table 2.2).

Finally, we can deduce the relationship between the G_k and H_m sequences, being

$$G_k = \begin{cases} H_{k+m_N}, & 1 \le k \le m_s \\ H_{k-m_s}, & m_s + 1 \le k \le M \end{cases} \tag{2.35}$$

Thus, the elements $[1 : m_s]$ of G come from the elements $[m_N + 1 : M]$ of H, and the elements $[m_s + 1 : M]$ of G come from the elements $[1 : m_N]$ of H. The naming of 'separator index' for m_s now becomes apparent. Figure 2.14 shows graphically the mapping in Eq. 2.35, with the green and blue peaks mapped from Figs. 2.14a to 2.14c.

2.5 Artefacts and How to Avoid Them

We have already seen, in Sect. 2.3, how the action of sampling can introduce artefacts into the Fourier transform if not performed correctly, in this case in the form of aliased harmonics. But sampling is only one part of the discretisation process that converts a signal into a sequence. The other part is the necessity to sample a finite segment of the signal, or, in other words, perform *truncation* of the signal.

2.5.1 Signal Truncation

The effect that signal or sequence truncation has upon the Fourier transform can be understood by determining the relationship between the discrete-time Fourier transform, $G_{1/\Delta t}(f)$, and the discrete Fourier transform, G_n. Begin with the DFT, Eq. 2.28, but evaluated between the indices $\pm m$ instead of 0 and $N - 1$, chosen so that one cycle of N points is still evaluated, or

$$G_n = \sum_{\tau=-m}^{m} g_\tau\, e^{-2\pi i n\tau/N} , \tag{2.36}$$

and substitute g_τ from the inverse DTFT (Eq. 2.25), giving

$$
\begin{aligned}
G_n &= \sum_{\tau=-m}^{m} \left(\Delta t \int_{-f_N}^{f_N} G_{1/\Delta t}(f)\, e^{2\pi i f\tau\Delta t}\, df \right) e^{-2\pi i n\tau/N} \\
&= \Delta t \int_{-f_N}^{f_N} G_{1/\Delta t}(f) \left(\sum_{\tau=-m}^{m} e^{2\pi i \tau(f\Delta t - n/N)} \right) df \\
&= \Delta t \int_{-f_N}^{f_N} G_{1/\Delta t}(f) \left(\sum_{\tau=-m}^{m} e^{2\pi i (f - f_n)\tau\Delta t} \right) df ,
\end{aligned}
$$

where we have used Eq. 2.29 for f_n. We now use the identity

$$\sum_{k=-m}^{m} e^{ikx} = \frac{\sin\left[\left(m + \frac{1}{2}\right)x\right]}{\sin\left(\frac{x}{2}\right)} \equiv D_m(x) \tag{2.37}$$

for any non-integer x, where $D_m(x)$ is *Dirichlet's kernel*, two examples of which are shown in Fig. 2.15a. (A proof of Eq. 2.37 is provided in the Appendix to this chapter.) Therefore,

$$\sum_{\tau=-m}^{m} e^{2\pi i(f - f_n)\tau \Delta t} = D_m(2\pi(f - f_n)\Delta t) \,.$$

We now have that

$$G_n = \Delta t \int_{-f_N}^{f_N} G_{1/\Delta t}(f)\, D_m(2\pi(f - f_n)\Delta t)\, df$$

$$= \Delta t \int_{-f_N}^{f_N} G_{1/\Delta t}(f)\, D_m(2\pi(f_n - f)\Delta t)\, df \,,$$

since Dirichlet's kernel is an even function. This is now a convolution integral, showing that the DFT is the convolution of the DTFT spectrum with the Dirichlet kernel, or

$$G_n = \Delta t\, (G_{1/\Delta t} * D_m)(f_n) \,. \tag{2.38}$$

Hence, if we are calculating G_n to obtain an approximation to $G_{1/\Delta t}(f_n)$, we will obtain a distorted spectrum. The nature of the distortion is discussed below.

For those readers concerned over the absence of the sinc function, consider the above discussion but using the continuous Fourier transform rather than the discrete versions. If the continuous signal, $g(t)$, is time-limited rather than of infinite duration, its time-limited CFT is, from Eq. 2.6,

$$G_T(f) = \int_{-T/2}^{T/2} g(t)\, e^{-2\pi i f t}\, dt \,.$$

Then, substituting the inverse CFT for $g(t)$ from Eq. 2.5, we have

$$G_T(f) = \int_{-T/2}^{T/2} \left(\int_{-\infty}^{\infty} G(f')\, e^{2\pi i f' t}\, df' \right) e^{-2\pi i f t}\, dt$$

$$= \int_{-\infty}^{\infty} G(f') \left(\int_{-T/2}^{T/2} e^{2\pi i (f' - f)t}\, dt \right) df' \,.$$

At this point, we use the identity

$$\int_{-X}^{X} e^{iax}\, dx = \frac{2\sin(aX)}{a} \equiv 2X\, \text{sinc}(aX) \,, \tag{2.39}$$

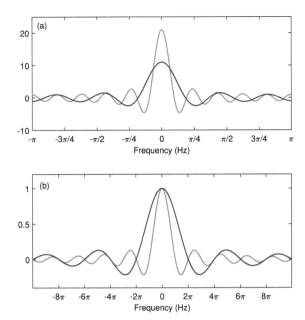

Fig. 2.15 **a** Two Dirichlet kernels, $D_m(f)$ from Eq. 2.37, for $m = 10$ (red line) and $m = 5$ (blue line). **b** Two sinc functions, sinc f (red line) and sinc $\frac{1}{2}f$ (blue line), from Eq. 2.39

(Percival and Walden 1993) where sinc $x = \sin x / x$ is the *sinc function*, shown in Fig. 2.15b. Therefore,

$$\int_{-T/2}^{T/2} e^{2\pi i (f'-f)t}\, dt \;=\; T\, \text{sinc}(\pi(f'-f)T)\,,$$

and we now have

$$G_T(f) = T \int_{-\infty}^{\infty} G(f')\, \text{sinc}(\pi(f'-f)T)\, df'$$

$$= T \int_{-\infty}^{\infty} G(f')\, \text{sinc}(\pi(f-f')T)\, df'\,,$$

because the sinc function is even. This convolution integral shows that the spectrum of a time-limited function is the convolution of the true spectrum with a sinc function, or

$$G_T(f) \;=\; T\, G(f) * \text{sinc}(\pi f T)\,. \tag{2.40}$$

Figure 2.15 shows that the Dirichlet and sinc functions are very similar in appearance. Indeed, they generate similar artefacts in the Fourier transforms when signals or sequences are truncated, being

1. loss of resolution;
2. spectral leakage;
3. the Gibbs phenomenon.

2.5.2   Loss of Resolution

As m in Eq. 2.36 decreases, less of the sequence is sampled in the truncated DFT. In addition, the width of the central lobe of the Dirichlet kernel $D_m(f)$ increases (Fig. 2.15a). This broadening of the central lobe causes higher frequencies in the sequence to be suppressed during convolution—recall the discussion in Sect. 2.2.8. Specifically, those features in $G_{1/\Delta t}(f)$ that are narrower than the width of the central lobe of $D_m(f)$ are smoothed out and will not appear in the time-domain sequence if $G_{1/\Delta t}(f)$ is inverse Fourier transformed.

Conversely, as m increases the central lobe of $D_m(f)$ becomes narrower (indeed, as $m \to \infty$ the Dirichlet kernel becomes a delta function). Hence, the convolution, $G_{1/\Delta t} * D_m$, sees more features of the sequence's spectrum being preserved in G_n, the DFT.

2.5.3   Spectral Leakage

The sidelobe oscillations of both the Dirichlet and sinc functions cause energy at one harmonic in the spectrum to *leak* into adjacent harmonics. This is seen in Fig. 2.16, where a sine wave of a single frequency (1/64 Hz) is truncated to varying degrees. In the top row, while not infinite, the signal is long and the spectrum comprises two spikes (at $\pm 1/64$ Hz) that are almost delta functions, though their bases are extended, flagging the onset of leakage.

As the truncation becomes more pronounced (m decreasing), the Dirichlet kernel reveals itself as each spike becomes a broadening central lobe with sidelobes that become both taller and wider. These sidelobes constitute the leakage of the energy that should be concentrated at $\pm 1/64$ Hz into surrounding frequencies.

2.5.4   The Gibbs Phenomenon

The third artefact is known as the *Gibbs phenomenon*, *Gibbs effect* or *ringing* and is caused by the inability of a finite sum of sinusoids to accurately reproduce discontinuities in a signal or its derivatives. It can be illustrated by looking at how a square wave is represented by a superposition of sine waves (Fig. 2.17), via the formula

$$\sum_{k=1}^{\infty} \frac{\sin((2k-1)t)}{(2k-1)} .$$

As the number of sine waves (k) increases, their sum tends to the square wave except at the discontinuities. Here, the superposition retains an oscillation on either side of the discontinuity no matter how many sine waves comprise it, and the amplitude of the

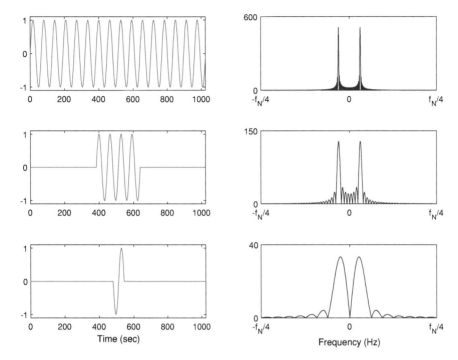

Fig. 2.16 Showing how truncation of a signal increases leakage. The left-hand panels show a sine wave that is truncated to an increasing extent in the time domain (multiplied by a rectangle function of decreasing duration). The right-hand panels show the amplitude spectra of the truncated sine waves (plotted between $\pm f_N/4$ only, to enhance detail, where f_N is the Nyquist frequency)

first oscillation stabilises at approximately 9% of the magnitude of the discontinuity (Stone and Goldbart 2009). However, the width of the oscillations does decrease and will eventually reach zero width after an infinite sum of harmonics.

Discontinuities can occur at data set edges where data truncation takes place, but can also occur within a sequence, as shown in Fig. 2.18a. The sequence shown in this figure is the sum of two sine waves ($g_\tau = g_{1,\tau} + g_{2,\tau}$), where g_1 has frequency 0.017 Hz, and g_2 has frequency 0.063 Hz, but g_τ has a discontinuity at 256 s. If one were to take the DFT of the sequence ($G_n = \mathsf{F}\{g_\tau\}$), immediately followed by the inverse DFT ($g'_\tau = \mathsf{F}^{-1}\{G_n\}$), then one would recover exactly the original truncated sequence (i.e. $g_\tau - g'_\tau = 0$, plotted in Fig. 2.18b), even though there exists a discontinuity in the sequence.

However, if one were to take the DFT as before ($G_n = \mathsf{F}\{g_\tau\}$), but then apply a low-pass filter to remove the high-frequency harmonic (with cut-off frequency at 0.045 Hz, shown in Fig. 2.18c) giving $\check{G}_n$, and then take the inverse DFT of the result then one obtains the sequence $\check{g}_\tau = \mathsf{F}^{-1}\{\check{G}_n\}$. The plot in (d) shows the difference $g_{1,\tau} - \check{g}_\tau$, i.e. the difference between the low-frequency component of the original sequence and the filtered sequence. The oscillations on either side of the discontinuity

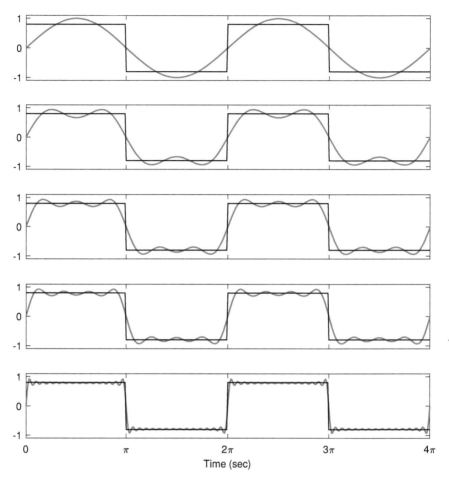

Fig. 2.17 Showing how a square wave (black line) is built up from a sum of sine waves of increasing frequency (red lines), to illustrate the Gibbs phenomenon. From top to bottom, $k = 1, 2, 3, 4, 15$ (see Sect. 2.5.4)

location constitute the Gibbs phenomenon. It can be seen that they are also generated at the data set edges.

The oscillations, or ringing, are easily explained. When the spectrum is low-pass filtered, it is multiplied in the frequency domain by a rectangle function, Π_n (Fig. 2.18c). This is equivalent to the convolution of the time-domain signal with the inverse (discrete) Fourier transform of the rectangle function, being the Dirichlet kernel,[6] or

$$G_n \Pi_n \quad \longleftrightarrow \quad (g * D_m)_\tau \, .$$

---

[6] We have already seen how the DFT of a rectangle function is a Dirichlet kernel, while its CFT is a sinc function—consider Eqs. 2.37 and 2.39.

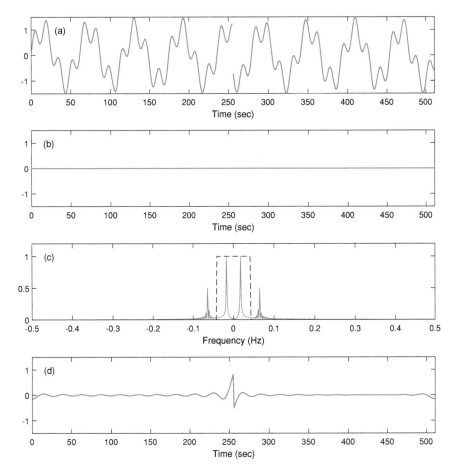

Fig. 2.18 Ringing in the time domain. **a** The sequence g_τ is the sum of two sine waves and has a discontinuity at 256 s. **b** If the DFT is taken of the sequence in (**a**) and the resulting spectrum immediately inverse transformed, the difference between that and the original signal is zero. **c** The spectrum of the sequence in (**a**) (red line), and a rectangle function Π_n acting as a low-pass filter (dashed black line). **d** The difference between the low-frequency harmonic of the original sequence and the filtered sequence. See Sect. 2.5.4

If the sequence is smooth and slowly changing, the convolution has little effect. (Indeed, the sinc function is smooth enough to be frequently used as an interpolating function.) But at a discontinuity, the oscillations of the Dirichlet kernel are amplified, creating the ringing.

2.5.5 Cyclic, Discrete Convolution

Similarly to the continuous case (Sect. 2.2.8), the convolution of two discretely sampled sequences is a translated multiplication of the sequences, with one of them time-reversed, or

$$(g * h)_\tau \equiv \sum_{k=-\infty}^{\infty} g_k \, h_{\tau-k} \, .$$

And as for the continuous convolution theorem, the following relationships hold:

$$(g * h)_\tau \longleftrightarrow G_n \, H_n$$

and

$$g_\tau \, h_\tau \longleftrightarrow (G * H)_n \, .$$

In practice, however, the sequences will be of finite length, so the infinite summation is not achievable. And furthermore, as explained in Sect. 2.4.2, the DFT considers the time-domain sequence to be periodic (see Figs. 2.13d and 2.19a). So if h_τ is periodic with period N, such that $h_{\tau \pm N} = h_\tau$, the sequence 'wraps-around' like a snake biting its own tail, and $h_N = h_0$, $h_{N+1} = h_1$, etc. or $h_{-1} = h_{N-1}$, $h_{-2} = h_{N-2}$, etc. If g and h have the same number of elements (N), the discrete convolution is given by

$$(g * h)_\tau = \sum_{k=0}^{N-1} g_k \, h_{\tau-k} \, , \tag{2.41}$$

where $\tau = 0, 1, \cdots, N - 1$. For example, the first element of the convolution requires wrap-around,

$$(g * h)_0 = g_0 h_0 + g_1 h_{-1} + g_2 h_{-2} + \cdots + g_{N-2} h_{-(N-2)} + g_{N-1} h_{-(N-1)}$$
$$= g_0 h_0 + g_1 h_{N-1} + g_2 h_{N-2} + \cdots + g_{N-2} h_2 + g_{N-1} h_1$$

though the last element does not,

$$(g * h)_{N-1} = g_0 h_{N-1} + g_1 h_{N-2} + g_2 h_{N-3} + \cdots + g_{N-2} h_1 + g_{N-1} h_0 \, .$$

Obviously, if the first and last elements of the sequence (h_0 and h_{N-1}) have very different values, there will be a significant discontinuity between them when wrap-around occurs (e.g. Fig. 2.19a) and this will affect the convolution in ways already described.

2.5.6 Mitigation Methods

There are several techniques one can employ to mitigate the adverse effects of sequence (or signal) truncation, and a few of these will be discussed here. However, not all of them work all of the time, and in certain circumstances, some of them can even make things worse. The best approach is trial and error, and intuition borne of experience.

2.5.6.1 Zero Padding

Appending a sequence of elements, all of value zero, to a data sequence is called *zero padding* (Fig. 2.19b). While not strictly a mitigation technique, it can be used to avoid adverse artefacts generated during wrap-around in cyclic convolution. Recall (Sect. 2.5.5) that wrap-around means that, for a sequence h_τ ($\tau = 0, 1, \ldots, N - 1$), the elements at times before and after the interval $[0, N - 1]$ have the values $h_N = h_0$, $h_{N+1} = h_1$, etc. and $h_{-1} = h_{N-1}$, $h_{-2} = h_{N-2}$, etc. (Fig. 2.19a). However, if the sequence is padded with M zeros, then $h_N = h_{N+1} = h_{N+2} = \cdots = h_{N+M-1} = 0$. Furthermore, the DFT considers the new, padded sequence to be periodic with period $N + M$, which means the elements to the left would also be zero: $h_{-1} = h_{-2} = h_{-3} = \cdots = h_{-M} = 0$ (Fig. 2.19b or c). Thus, the first few elements of the discrete convolution of Eq. 2.41 would be

$$(g * h)_0 = g_0 h_0$$
$$(g * h)_1 = g_0 h_1 + g_1 h_0$$
$$(g * h)_2 = g_0 h_2 + g_1 h_1 + g_2 h_0 .$$

In such a fashion, any large discontinuities between the first and last elements of the sequence (h_0 and h_{N-1}) might be reduced in magnitude if the difference $|h_{N-1} - 0|$ or $|h_0 - 0| \ll |h_0 - h_{N-1}|$, potentially reducing the severity of the Gibbs effect.

The number of zeros to be appended to the sequence is a choice of the user, though it is customary to pad to an integer power of 2 to make the FFT run faster (Sect. 2.4.3). Alternatively, if the length of the data sequence is N, then N zeros are added. Importantly, padding with zeros does not alter the spectrum per se, but it does refine it by decreasing the sampling interval, Δf, of the sequence's DFT, hence providing more estimates of the spectrum. Recall Eq. 2.29

$$\Delta f = \frac{1}{N \Delta t} .$$

So if the length of a sequence is doubled from N to $2N$ by padding it with zeros, then the number of spectral estimates doubles and the spacing between harmonics halves, i.e. it is sampled at a finer spacing. This is seen in Fig. 2.20a–c, where the

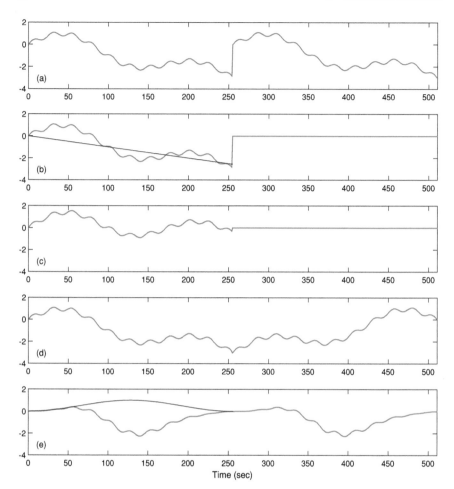

Fig. 2.19 Methods employed to reduce errors caused by sequence truncation in the DFT. The sequence has length 256 s. **a** How the sequence looks to the DFT (i.e. periodic). **b** Zero padding: the length of the sequence is extended by adding zeros. The black line is a best-fitting straight line. **c** Detrending: the best-fitting straight line in (**b**) is subtracted from the sequence; the plot here also shows zero padding applied. **d** Mirroring: the sequence is doubled in length by reflection about its edges. **e** Windowing: the sequence is multiplied by a window function (shown as the black line). Note that the periodicity requirement of the DFT implies that all these sequences are periodic, with (**a**) and (**e**) having period 256 s, while (**b**), (**c**) and (**d**) have period 512 s

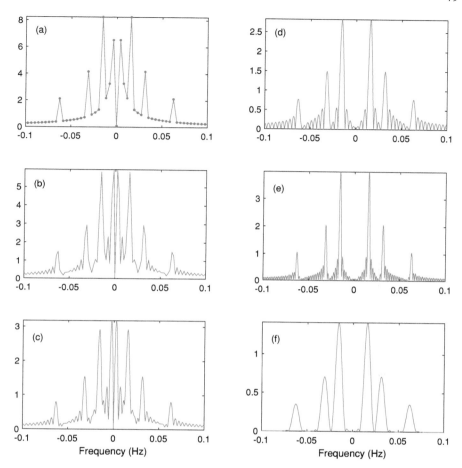

Fig. 2.20 Amplitude spectra of a sequence similar to that in Fig. 2.19, estimated using different mitigation techniques. The sequence has 256 samples, with a sampling interval of 1 s, and is the sum of three sine waves of frequencies 4/256 Hz, 8/256 Hz and 16/256 Hz. **a** Spectrum of the sequence, with the dots showing the actual estimates. **b** Spectrum of the sequence padded with 256 zeros, giving a new sequence of length 512 s (dots now removed for clarity). **c** Spectrum of the sequence padded with 1792 zeros, giving a new sequence of length 2048 s. **d** Spectrum of the detrended sequence. **e** Spectrum of the detrended and mirrored sequence. **f** Spectrum of the detrended and windowed sequence, using a Hann window. (**d**), (**e**) and (**f**) are all zero padded to length 2048 elements to show the detail in the spectra

spectra show more detail as more and more zeros are appended, but the important aspects of the spectrum (the peaks of the dominant harmonics) are unaltered.

Note that this refinement of the spectrum does not represent an increase in spectral resolution, despite some texts claiming so. The resolution is given by the highest frequency of the signal that can be reproduced in the sampled sequence, and this is determined by the sampling interval and thus the Nyquist frequency (Sect. 2.3.4). As Eq. 2.20 tells us, the Nyquist frequency is unaltered by zero padding since it does not depend on the length of the sequence:

$$ f_N \; = \; \frac{1}{2\Delta t} \; . $$

The refinement can reveal some extra information, though. A comparison of Figs. 2.20a and 2.20c shows that the oscillations of the Dirichlet kernel are revealed in the zero-padded sequence, whereas in the unpadded sequence, they are manifested as a higher average spectral amplitude in the tails of the spectrum.

2.5.6.2 Detrending

Examination of Figs. 2.20a–c shows four spectral peaks (ignoring those at negative frequencies), whereas the sequence was only generated with three harmonics. The extra peak (shown in Fig. 2.20a as the first positive-frequency peak) is due to the linear slope on the data (Fig. 2.19b), which is treated by the Fourier transform as a part of a long-wavelength sine wave. This extra, distracting peak can be removed by finding the best-fitting straight line through the data sequence (Fig. 2.19b) and removing it (Fig. 2.19c), which may also reduce the magnitude of any discontinuity at the ends of the sequence. This procedure is called *detrending* or *bias and tilt removal* as it also removes a mean value (the bias) from the data. The resulting spectrum no longer has the extra peak (Fig. 2.20d).

2.5.6.3 Mirroring

Discontinuities at the ends of the sequence can be removed by *mirroring* or reflecting the sequence about its ends (Fig. 2.19d). However, this method invariably generates a discontinuity in the first derivative of the sequence at its end. And just as for discontinuities in the sequence values, the Fourier transform cannot accurately represent sharp changes in the sequence derivatives, and the Gibbs phenomenon may arise, shown in Fig. 2.20e as the low-amplitude oscillations from the Dirichlet kernel. Note, however, that mirroring has narrowed the peaks of the spectra.

2.5.6.4 Windowing or Tapering

A technique frequently employed in spectral estimation is the multiplication of the time-domain sequence by a *window function* or *taper* (Fig. 2.19e). The main purposes of windowing are to reduce spectral leakage and the Gibbs phenomenon. We saw in Sects. 2.5.1 and 2.5.3 that truncation of a sequence (or signal) is equivalent to convolution of the spectrum with a Dirichlet kernel (or sinc function) (Eqs. 2.38 and 2.40). In general, if a sequence g_τ is multiplied by a window function w_τ in the time domain, the 'observed' or estimated spectrum will be a convolution of the sequence's DTFT, $G(f)$, with the spectrum of the window, $W(f)$, or

$$g_\tau w_\tau \quad \longleftrightarrow \quad (G * W)(f_n) \tag{2.42}$$

(cf. Eq. 2.38). For example, if w_τ is a time-domain rectangle function (Π), then W is a frequency-domain Dirichlet kernel, as discussed. The aim of window design is to find a window w whose spectrum W does not cause too many unwanted artefacts in the estimated spectrum, $(G * W)(f_n)$.

There are many windows in common usage, each having different properties suited to the data and desired outcomes. For example, certain windows are designed to reduce the Gibbs phenomenon by having both a zero value and zero first derivative at their ends. Other windows are designed to have a very narrow central lobe in their spectrum to increase spectral resolution, or to have very low-amplitude sidelobes to reduce spectral leakage. Harris (1978) provides a good summary of many of them.

However, it is not possible to design windows with all of these properties together. In the field that is the subject of this book—the estimation of effective elastic thickness using spectral methods—it has been customary to use the family of windows called Slepian tapers, and these will be discussed in Chap. 3.

The window applied to the data in Fig. 2.19e is a Hann window (Harris 1978), yielding the spectrum shown in Fig. 2.20f. When compared with the spectra of the detrended, and detrended and mirrored sequences (Figs. 2.20d and 2.20e), it can be seen that the low-amplitude oscillations have been reduced so as to almost disappear, though this comes at the expense of broadened spectral peaks.

2.6 The 2D Fourier Transform

Much of the data dealt with in this book will be 2D (see Sect. 2.1.1), for example, gravity anomaly or topography, which are given on the sphere (or ellipsoid) as functions of the curvilinear coordinates latitude and longitude. However, while the Fourier transform can be defined in any number of dimensions, these spaces must be Euclidean (or 'flat'): the Fourier transform is not defined on a sphere. Thus, the spherically coordinated data must be treated as existing on (or transformed to) a plane with a spatial Cartesian coordinate system, (x, y). Details about such transformations ('map projections') are dealt with in Chap. 6.

2.6.1 Spatial Frequency—Wavenumber

When using the Fourier transform with spatially coordinated data, the corresponding *spatial frequency* is known as *wavenumber*. This is usually chosen to be an *angular frequency*, with one cycle being 2π radians. So instead of using the units of Hertz (cycles per second), we use radians per second (in the time sense) or radians per kilometre (rad/km) in the spatial sense. The 1D wavenumber (k) is thus related to frequency (f) by

$$k = 2\pi f = \frac{2\pi}{\lambda},$$

where λ is wavelength (see Sect. 2.1.2).

2.6.2 Sampling Theory in the Space Domain

Here, we will put the sampling theory of Sects. 2.3.3 and 2.3.4 in the context of space-domain coordinates, for use in subsequent chapters. Consider a 1D, continuous, band-limited signal, $g(x)$, with Fourier transform $G(k)$ such that

$$G(k) = 0, \quad |k| > W,$$

where W is its bandwidth, calculated from the shortest wavelength present in the signal $(\lambda_{\min})$ by

$$W = \frac{2\pi}{\lambda_{\min}}$$

(cf. Eq. 2.16). The *Nyquist criterion* states that if the signal is sampled at a spacing of Δx metres (or kilometres, miles, etc.) to form a sequence g_ξ, then $g(x)$ can be perfectly reconstructed from g_ξ if and only if the sample spacing is less than or equal to half the minimum wavelength present in the signal, or

$$\Delta x \leq \frac{\lambda_{\min}}{2} = \frac{\pi}{W} \tag{2.43}$$

(cf. Eqs. 2.17 and 2.19). When expressed in terms of wavenumber, the Nyquist criterion states that the signal can be perfectly reconstructed if and only if the sampling wavenumber $(k_s = 2\pi/\Delta x)$ is greater than or equal to twice the bandwidth of the signal being sampled, or

$$k_s \geq 2W \tag{2.44}$$

(cf. Eq. 2.18). The quantity $2W$ is the *Nyquist rate*. If the Nyquist criterion is not met, then aliasing occurs, whereby the high-wavenumber (short-wavelength) harmonics of the signal appear as spurious lower-wavenumber (longer-wavelength) harmonics in the sampled sequence.

The *Nyquist wavenumber* of the sampled sequence gives the wavenumber of the highest-wavenumber harmonic contained in the sequence and is given by

$$k_N = \frac{2\pi}{2\Delta x} = \frac{k_s}{2} \tag{2.45}$$

(cf. Eq. 2.20). The Nyquist wavenumber does not need to have the same value as the Nyquist rate, because the former depends upon the sampling interval (grid spacing) of the sampled sequence, while the latter is a property of the bandwidth of the continuous signal. However, if one wishes to avoid aliasing, then one must satisfy the Nyquist criterion (Eq. 2.44), whereby

$$k_N \geq \frac{2W}{2}$$

(cf. Eq. 2.21); that is, the Nyquist wavenumber must be at least half the Nyquist rate. Equivalently, in the space domain, one can define a *Nyquist wavelength* (λ_N) as 2π divided by the Nyquist wavenumber, or

$$\lambda_N = 2\Delta x , \tag{2.46}$$

from Eq. 2.45. The Nyquist wavelength of the sampled sequence gives the wavelength of the smallest-wavelength harmonic contained in the sequence. The Nyquist criterion is then given by

$$\lambda_N \leq \lambda_{\min} ,$$

or

$$\Delta x \leq \frac{\lambda_{\min}}{2}$$

which is also Eq. 2.43.

2.6.3 The Non-Unitary 1D Fourier Transform

When angular frequency is used, the Fourier transform equations look slightly different. For 1D signals, the (continuous) Fourier transform is

$$G(k) = \int_{-\infty}^{\infty} g(x) e^{-ikx} dx ,$$

while the inverse transform is

$$g(x) = \frac{1}{2\pi} \int_{-\infty}^{\infty} G(k)\, e^{ikx}\, dk \; .$$

The asymmetry in these two equations (the extra factor of $1/2\pi$) leads this convention to be called *non-unitary*. In the *unitary* convention, the single $1/2\pi$ factor in
the inverse transform is split into two $1/\sqrt{2\pi}$ factors in both forward and inverse
transforms.

2.6.4   The 2D Continuous Fourier Transform

Here, we will use the position vector $\mathbf{x} = (x, y)$ to coordinate the data. In the Fourier
domain, the wavenumber vector (or sometimes '*wavevector*') is $\mathbf{k} = (k_x, k_y)$, where
k_x is the wavenumber in the x-direction, and k_y is the wavenumber in the y-direction.
   In the non-unitary convention, the 2D Fourier transform of a signal $g(\mathbf{x})$ is

$$G(\mathbf{k}) = \int_{\mathbb{R}^2} g(\mathbf{x})\, e^{-i\mathbf{k}\cdot\mathbf{x}}\, d^2\mathbf{x} \tag{2.47}$$

for $|\mathbf{k}| < \infty$. The inverse Fourier transform of $G(\mathbf{k})$ is

$$g(\mathbf{x}) = \frac{1}{4\pi^2} \int_{\mathbb{R}^2} G(\mathbf{k})\, e^{i\mathbf{k}\cdot\mathbf{x}}\, d^2\mathbf{k}$$

for $|\mathbf{x}| < \infty$. We define the notation by

$$\int_{\mathbb{R}^2} d^2\mathbf{x} \equiv \int_{-\infty}^{\infty} \int_{-\infty}^{\infty} dx\, dy$$

where $\mathbb{R}$ is the real number line (1D), and $\mathbb{R}^2$ is the real number plane (2D).
   The shift theorem (Sect. 2.2.4) in two dimensions is

$$\mathsf{F}\{g(\mathbf{x} - \mathbf{x}_0)\} = e^{-i\mathbf{k}\cdot\mathbf{x}_0}\, G(\mathbf{k}) \; , \tag{2.48}$$

where the function is translated in both x- and y-directions by an amount $\mathbf{x}_0 = (x_0, y_0)$.
   It will also be useful to define the 2D derivative theorem:

$$\mathsf{F}\left\{ \frac{\partial^m}{\partial x^m} \frac{\partial^n}{\partial y^n} g(x, y) \right\} = (ik_x)^m (ik_y)^n\, G(k_x, k_y) \tag{2.49}$$

similarly to Eq. 2.12, the 1D equivalent in Sect. 2.2.7.

2.6.5 The Hankel Transform

Some 2D functions have circular symmetry about a point in the xy-plane, or—if one prefers to think about them as solids extended into a third dimension—'cylindrical symmetry' about an axis perpendicular to the xy-plane (Fig. 2.21a). A function $g(x, y)$ is said to have cylindrical symmetry if it can be written in polar coordinates, (r, θ), as $g(r)$, with no θ-dependence. The relationship between rectangular Cartesian coordinates and polar coordinates is

$$
\begin{aligned}
x &= r \cos \theta \\
y &= r \sin \theta \\
r &= \sqrt{x^2 + y^2} .
\end{aligned}
\tag{2.50}
$$

The 2D Fourier transform of a cylindrically symmetric function is a special case and becomes a *Hankel transform*. The Hankel transform of a function $g(r)$ is given by

$$
\mathsf{H}\{g(r)\} \equiv \mathcal{G}(k) = \int_0^\infty g(r) \, J_0(kr) \, r \, dr
\tag{2.51}
$$

where k is the radial wavenumber, and in polar coordinates, (k, ϕ), we have

$$
\begin{aligned}
k_x &= k \cos \phi \\
k_y &= k \sin \phi \\
k &= \sqrt{k_x^2 + k_y^2} .
\end{aligned}
\tag{2.52}
$$

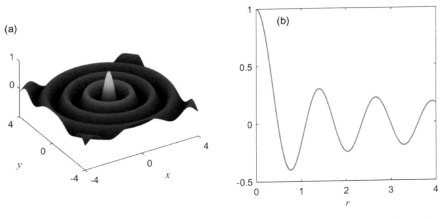

Fig. 2.21 a A A 2D visualisation of a zeroth-order Bessel function of the first kind, $J_0(kr)$, where $r = \sqrt{x^2 + y^2}$ and $k = 5$, exhibiting cylindrical symmetry. **b** The same function in 1D form, used in Eq. 2.51

The function $J_0(x)$ is a *zeroth-order Bessel function of the first kind*, defined by

$$J_0(x) = \frac{1}{2\pi} \int_0^{2\pi} e^{\pm ix\cos\theta} \, d\theta \, , \tag{2.53}$$

and shown in Fig. 2.21b. The inverse Hankel transform is given by

$$\mathsf{H}^{-1}\{\mathcal{G}(k)\} \equiv g(r) = \int_0^\infty \mathcal{G}(k) \, J_0(kr) \, k \, dk \, .$$

As an example of how to evaluate a Hankel transform, we shall use a function that commonly appears in geophysics and geodesy: the cube of the reciprocal distance, l^{-3}, which we will come across again in Sect. 8.3. For a point in 3D Euclidean space, P at (x, y, z), its distance from the origin is given by Pythagoras's theorem as

$$l_0 = \sqrt{x^2 + y^2 + z^2} \, . \tag{2.54}$$

Our task is to find the 2D Fourier transform of l_0^{-3}, which is

$$\mathsf{F}\left\{\frac{1}{l_0^3}\right\} = \int_{-\infty}^\infty \int_{-\infty}^\infty \frac{e^{-i(k_x x + k_y y)}}{\left(x^2 + y^2 + z^2\right)^{3/2}} \, dx \, dy \, , \tag{2.55}$$

from Eq. 2.47. Note, we are not transforming with respect to the z-coordinate, which is taken to be constant. Equation 2.55 is most easily solved by converting to polar coordinates, (r, θ) and (k, ϕ), using Eqs. 2.50 and 2.52. In doing so, we must also convert the Cartesian integrals to polar integrals, which is achieved with the replacements

$$\int_{-\infty}^\infty \int_{-\infty}^\infty dx \, dy \longrightarrow \int_0^{2\pi} \int_0^\infty r \, dr \, d\theta \, .$$

If we let $r = \sqrt{x^2 + y^2}$ from Eq. 2.50, then Eq. 2.54 becomes

$$l_0 = \sqrt{r^2 + z^2} \, ,$$

and we can rewrite Eq. 2.55 in polar coordinates as

$$\mathsf{F}\left\{\frac{1}{l_0^3}\right\} = \int_0^{2\pi} \int_0^\infty \frac{e^{-ikr\cos(\theta-\phi)}}{\left(r^2 + z^2\right)^{3/2}} \, r \, dr \, d\theta$$

$$= \int_0^\infty \frac{1}{\left(r^2 + z^2\right)^{3/2}} \left[\int_0^{2\pi} e^{-ikr\cos(\theta-\phi)} \, d\theta\right] r \, dr \, .$$

We now use Eq. 2.53 to replace the term in square brackets, giving

$$\mathsf{F}\left\{\frac{1}{l_0^3}\right\} = 2\pi \int_0^\infty \frac{J_0(kr)}{(r^2 + z^2)^{3/2}} \, r \, dr \ .$$

From Eq. 2.51, we can see that this is a Hankel transform of the function $(r^2 + z^2)^{-3/2}$, which has solution

$$\mathsf{H}\left\{\frac{1}{(r^2 + z^2)^{3/2}}\right\} = \frac{e^{-kz}}{z}$$

for $k > 0$ and $z > 0$ (Gradshteyn and Ryzhik 1980). Thus, the 2D Fourier transform of the cube of the reciprocal distance is

$$\mathsf{F}\left\{\frac{1}{l_0^3}\right\} = 2\pi \frac{e^{-kz}}{z} \ . \tag{2.56}$$

So in general, if a function $g(x, y)$ can be written in polar coordinates as $g(r)$, then its 2D Fourier transform is given by

$$\mathsf{F}\{g(x, y)\} = 2\pi \, \mathsf{H}\{g(r)\} \ .$$

Suppose now that we want to find the Fourier transform of the cube of the reciprocal distance between two points, P at (x, y, z) and Q at (x', y', z'), rather than just between P and the origin, where the distance between P and Q is

$$l = \sqrt{(x - x')^2 + (y - y')^2 + (z - z')^2} \ .$$

Instead of re-deriving the Fourier transform, we can just make use of the 2D shift theorem, Eq. 2.48, such that

$$\mathsf{F}\left\{\frac{1}{[(x - x')^2 + (y - y')^2 + (z - z')^2]^{3/2}}\right\} = e^{-i(k_x x' + k_y y')}$$

$$\times \ \mathsf{F}\left\{\frac{1}{[x^2 + y^2 + (z - z')^2]^{3/2}}\right\}$$

where the 2D Fourier transform is only concerned with the translation in the xy-plane, and not in the z-direction. Thus, from Eq. 2.56, we get

$$\mathsf{F}\left\{\frac{1}{l^3}\right\} = 2\pi \, e^{-i(k_x x' + k_y y')} \frac{e^{-k(z - z')}}{z - z'} \ , \tag{2.57}$$

where we must have $z > z'$ and $k > 0$, as noted above.

2.6.6   The 2D Discrete Fourier Transform

Consider a 2D space-domain sequence $g_{\xi\eta}$, where ξ is the index along the x-axis and η the index along the y-axis, with $\xi = 0, 1, \ldots, N_x - 1$, and $\eta = 0, 1, \ldots, N_y - 1$, where N_x is the number of observations in the x-direction, and N_y is the number of observations in the y-direction. Its 2D discrete Fourier transform is given by

$$G_{mn} = \sum_{\xi=0}^{N_x-1} \sum_{\eta=0}^{N_y-1} g_{\xi\eta}\, e^{-i(m\xi/N_x + n\eta/N_y)}\ ,$$

while the inverse 2D DFT is given by

$$g_{\xi\eta} = \frac{1}{N_x N_y} \sum_{m=0}^{N_x-1} \sum_{n=0}^{N_y-1} G_{mn}\, e^{i(m\xi/N_x + n\eta/N_y)}\ ,$$

where the 2D wavenumber indices have the values $m = 0, 1, \cdots, N_x - 1$, and $n = 0, 1, \ldots, N_y - 1$.

If the data grid spacing is Δx in the x-direction and Δy in the y-direction, then the discrete wavenumbers in the x- and y-directions are given by

$$k_{x,m} = m\,\Delta k_x\ , \qquad k_{y,n} = n\,\Delta k_y\ , \tag{2.58}$$

respectively, where

$$\Delta k_x = \frac{2\pi}{N_x \Delta x}\ , \qquad \Delta k_y = \frac{2\pi}{N_y \Delta y}$$

are the wavenumber spacings in each direction (cf. Eq. 2.29). Equations 2.58 also give the fundamental wavenumbers in the x- and y-directions (i.e. for $m = n = 1$) (see Sects. 2.2.1 and 2.4.2). The two Nyquist wavenumbers are given by

$$k_{N,x} = \frac{2\pi}{2\Delta x}\ , \qquad k_{N,y} = \frac{2\pi}{2\Delta y} \tag{2.59}$$

(cf. Eq. 2.45).

2.7   Summary

The continuous Fourier transform operates on continuous, aperiodic, infinitely long signals and produces a continuous, aperiodic, infinitely long spectrum. However, digital computers require digital data to process, so real-world continuous signals need to be truncated and sampled to give finite-length sequences. The sampling action forces both the sequence and its spectrum to become periodic. Unfortunately though, sampling can introduce spurious frequencies into the sampled sequence if the sampling is too infrequent, a phenomenon called aliasing. Truncation of the signal also causes problems, notably the loss of frequency resolution, spectral leakage and the appearance of spurious oscillations known as the Gibbs phenomenon. Fortunately, several methods exist to reduce these unwanted effects, such as zero padding, detrending, mirroring and windowing or tapering.

2.8 Further Reading

The topic of Fourier transforms has been written about extensively, and there is a multitude of texts of varying levels of difficulty for the interested reader. The material presented here is merely aimed at providing those unfamiliar with the subject with a primer in those aspects relevant to our topic, namely taking the fast Fourier transform of a 2D grid of data with the intent of calculating its spectrum.

Possibly the best introduction to Fourier transforms is the book by Bracewell (1986), though its treatment of the discrete versions is somewhat lacking. For that topic, one should turn first to Brigham (1988), but then to Percival and Walden (1993) which could be called the 'Bible' of spectral analysis. Then there are numerous books on mathematical physics which contain a chapter or more on Fourier transforms and the mathematics supporting it, examples being James (2011) and Stone and Goldbart (2009).

Appendix

The Dirichlet Kernel

Here a proof of Eq. 2.37 is given. Consider the geometric series in a variable, r:

$$S_N = \sum_{k=-n}^{n} r^k .$$

Expansion of the summation gives

$$S_n = r^{-n} + r^{-n+1} + \cdots + r^{-1} + r^0 + r^1 + \cdots + r^{n-1} + r^n ,$$

which, when multiplied by r gives

$$r S_n = r^{-n+1} + r^{-n+2} + \cdots + r^0 + r^1 + r^2 + \cdots + r^n + r^{n+1} .$$

If we subtract these two expressions, we get

$$S_n - r S_n = r^{-n} - r^{n+1} ,$$

which, after some rearrangement becomes

$$S_n = \frac{r^{-(n+\frac{1}{2})} - r^{n+\frac{1}{2}}}{r^{-1/2} - r^{1/2}} .$$

Now let $r = e^{ix}$ for any non-integer x. This gives

$$\sum_{k=-n}^{n} e^{ikx} = \frac{e^{-i(n+\frac{1}{2})x} - e^{i(n+\frac{1}{2})x}}{e^{-ix/2} - e^{ix/2}}$$

$$= \frac{-2i \sin\left[\left(n+\frac{1}{2}\right)x\right]}{-2i \sin\left(\frac{x}{2}\right)}$$

$$= \frac{\sin\left[\left(n+\frac{1}{2}\right)x\right]}{\sin\left(\frac{x}{2}\right)}$$

$$= D_n(x) .$$

References

Bracewell RN (1986) The Fourier transform and its applications. McGraw-Hill, New York

Brigham EO (1988) The fast Fourier transform and its applications. Prentice Hall, NJ

Gradshteyn IS, Ryzhik IM (1980) Table of integrals, series and products. Academic Press, New York

Harris FJ (1978) On the use of windows for harmonic analysis with the discrete Fourier transform. Proc IEEE 66:51–83

James JF (2011) A student's guide to Fourier transforms, 3rd edn. Cambridge University Press, Cambridge

Percival DP, Walden AT (1993) Spectral analysis for physical applications. Cambridge University Press, Cambridge

Singleton RC (1968) An algorithm for computing the mixed radix fast Fourier transform. IEEE Trans Audio Electroacoust AU-17(2):93–102

Stone M, Goldbart P (2009) Mathematics for physics. Cambridge University Press

Chapter 3
Multitaper Spectral Estimation

3.1 Introduction

In Sect. 2.5.1, we saw how truncation of a sequence distorts its spectrum. If $G_{1/\Delta t}$ is the spectrum of a sampled signal (the DTFT), and G_n the spectrum of the truncated, sampled signal (the DFT), then Eq. 2.38 told us that the DFT is a smeared version of the DTFT, reproduced here

$$G_n = \Delta t \, (G_{1/\Delta t} * D_m)(f_n) \, .$$

Thus, the smoothing that appears in the DFT is mathematically represented by convolution of the DTFT with a Dirichlet kernel. In general, any windowing function w_τ applied to a data sequence g_τ will distort its spectrum, as described in Sect. 2.5.6.4 and shown in Eq. 2.42, reproduced here:

$$g_\tau w_\tau \longleftrightarrow (G_{1/\Delta t} * W)(f_n) \, ,$$

where w_τ is a time-domain window function, and $W(f)$ is its spectrum. Recall that if w_τ is the rectangle function, then $W(f)$ will be the Dirichlet kernel.

This suggests that the *power spectrum* (Sect. 2.2.3) of a sequence should also be distorted by truncation or windowing, because it is computed from the real and imaginary parts of the Fourier transform. Hence, in this chapter we will continue the discussion of Sect. 2.5 and investigate in more depth techniques that can be used to get a more reliable power spectral estimate than simply squaring and summing the real and imaginary parts of a Fourier transform—as in Eqs. 2.4 or 2.8—and calling the result a 'power spectrum'.

© Springer Nature Switzerland AG 2022
J. Kirby, *Spectral Methods for the Estimation of the Effective Elastic Thickness of the Lithosphere*, Advances in Geophysical and Environmental Mechanics and Mathematics, https://doi.org/10.1007/978-3-031-10861-7_3

3.2   The Periodogram

As the first step, we should investigate just what is obtained when this simplistic approach is adopted. Let us define an *autocovariance sequence*, s_κ, of a sequence g_τ, where

$$s_\kappa \equiv \text{cov}\{g_\tau, g_{\tau+\kappa}\} = \text{E}\{(g_\tau - \mu)(g_{\tau+\kappa} - \mu)\}, \tag{3.1}$$

and where $\text{cov}\{X, Y\}$ is the covariance between any two sequences X and Y; the symbol $\text{E}\{X\}$ represents the *expected value* of a sequence X and is identical to its arithmetic mean; μ is the mean value of the sequence (i.e. $\mu = \text{E}\{g\}$); and κ is the *lag*, with $\kappa = 0, \pm 1, \ldots$. That is, s_κ is the covariance between g_τ and itself over a range of lags κ; it thus measures the degree of self-similarity within a sequence at large lags (think 'long wavelengths') and small lags (think 'short wavelengths').

We then note that the (DTFT) spectrum of the autocovariance is given by

$$S(f) = \sum_{\kappa=-\infty}^{\infty} s_\kappa \, e^{-2\pi i f \kappa \Delta t} \tag{3.2}$$

for $|f| \leq f_N$, the Nyquist frequency, while the inverse DTFT is

$$s_\kappa = \Delta t \int_{-f_N}^{f_N} S(f) \, e^{2\pi i f \kappa \Delta t} \, df \tag{3.3}$$

for $\kappa = 0, \pm 1, \ldots$ (Percival and Walden 1993). We call $S(f)$ the *true power spectrum* of g_τ. The power spectrum and the autocovariance sequence form a Fourier transform pair, thus

$$s_\kappa \longleftrightarrow S(f).$$

In practice, however, the sequence will be truncated to N observations. Experience should now tell us that the autocovariance of a truncated sequence might not be faithful to that of the infinite sequence from which it was extracted. In that case we write $\hat{s}_{p,\kappa}$ as an estimator of s_κ, and $\hat{S}_p(f)$ as an estimator of $S(f)$ (the reason for the subscript 'p' will become clear soon), and Eqs. 3.2 and 3.3 become, respectively,

$$\hat{S}_p(f) = \sum_{\kappa=-(N-1)}^{N-1} \hat{s}_{p,\kappa} \, e^{-2\pi i f \kappa \Delta t} \tag{3.4}$$

for $|f| \leq f_N$, and

$$\hat{s}_{p,\kappa} = \Delta t \int_{-f_N}^{f_N} \hat{S}_p(f) \, e^{2\pi i f \kappa \Delta t} \, df$$

for $\kappa = 0, \pm 1, \ldots, \pm(N-1)$. Furthermore, from Eq. 3.1, the autocovariance sequence of a finite sequence, g_τ, is also given by

$$\hat{s}_{p,\kappa} = \frac{1}{N} \sum_{\tau=0}^{N-1-|\kappa|} g_\tau \, g_{\tau+|\kappa|} \ , \tag{3.5}$$

if the sequence has a zero mean ($\mu = 0$). If we substitute Eq. 3.5 into Eq. 3.4, we obtain

$$\hat{S}_p(f) = \frac{1}{N} \sum_{\kappa=-(N-1)}^{N-1} \sum_{\tau=0}^{N-1-|\kappa|} g_\tau \, g_{\tau+|\kappa|} \, e^{-2\pi i f \kappa \Delta t} \ ,$$

which can be shown to be identical to

$$\hat{S}_p(f) = \frac{1}{N} \left| \sum_{\tau=0}^{N-1} g_\tau \, e^{-2\pi i f \tau \Delta t} \right|^2$$

(Percival and Walden 1993). The function $\hat{S}_p(f)$ is called the *periodogram* (hence the subscript p), and is simply the normalised sum of the squares of the real and imaginary components of the DFT of a sequence g_τ.

We should now ask the question, how good is the periodogram at estimating the true power spectrum? To answer that, first take the expected value of the truncated autocovariance sequence, Eq. 3.5, giving

$$\mathsf{E}\{\hat{s}_{p,\kappa}\} = \frac{1}{N} \sum_{\tau=0}^{N-1-|\kappa|} \mathsf{E}\{g_\tau \, g_{\tau+|\kappa|}\} \ .$$

If the signal has a zero mean value, then from Eq. 3.1 we have $\mathsf{E}\{g_\tau \, g_{\tau+|\kappa|}\} = s_\kappa$, the true autocovariance sequence. Making this substitution gives

$$\mathsf{E}\{\hat{s}_{p,\kappa}\} = \frac{1}{N} \sum_{\tau=0}^{N-1-|\kappa|} s_\kappa \ .$$

Now, since s_κ is independent of τ, we can write

$$\mathsf{E}\{\hat{s}_{p,\kappa}\} = \frac{1}{N} \left(\sum_{\tau=0}^{N-1-|\kappa|} 1 \right) s_\kappa$$

$$= \frac{1}{N} \left(N - |\kappa| \right) s_\kappa \ ,$$

or

$$\mathsf{E}\{\hat{s}_{p,\kappa}\} = \left(1 - \frac{|\kappa|}{N} \right) s_\kappa \ . \tag{3.6}$$

Thus, $\hat{s}_{p,\kappa}$ is a biased estimator of s_κ.[1] We then take the expected value of Eq. 3.4, giving

$$\mathsf{E}\{\hat{S}_p(f)\} = \sum_{\kappa=-(N-1)}^{N-1} \mathsf{E}\{\hat{s}_{p,\kappa}\} \, e^{-2\pi i f \kappa \Delta t} \, ,$$

and substitute Eq. 3.6, to get

$$\mathsf{E}\{\hat{S}_p(f)\} = \sum_{\kappa=-(N-1)}^{N-1} \left(1 - \frac{|\kappa|}{N}\right) s_\kappa \, e^{-2\pi i f \kappa \Delta t} \, .$$

If we now substitute s_κ from Eq. 3.3, we get

$$\mathsf{E}\{\hat{S}_p(f)\} = \sum_{\kappa=-(N-1)}^{N-1} \left(1 - \frac{|\kappa|}{N}\right) \left(\Delta t \int_{-f_N}^{f_N} S(f') \, e^{2\pi i f' \kappa \Delta t} \, df'\right) e^{-2\pi i f \kappa \Delta t}$$

$$= \Delta t \int_{-f_N}^{f_N} \sum_{\kappa=-(N-1)}^{N-1} \left(1 - \frac{|\kappa|}{N}\right) e^{2\pi i (f'-f)\kappa \Delta t} \, S(f') \, df' \, .$$

We then use the identity

$$\mathcal{F}_N(x) = \sum_{k=-(N-1)}^{N-1} \left(1 - \frac{|k|}{N}\right) e^{ikx}$$

(Percival and Walden 1993), where $\mathcal{F}_N(x)$ for some non-integer x is *Fejér's kernel*. This gives

$$\mathsf{E}\{\hat{S}_p(f)\} = \Delta t \int_{-f_N}^{f_N} \mathcal{F}_N(2\pi(f - f')\Delta t) \, S(f') \, df' \, , \tag{3.7}$$

where we have used the fact that $\mathcal{F}_N(x - y) = \mathcal{F}_N(y - x)$ for any (non-integer) x and y, because Fejér's kernel is an even function (Fig. 3.1). Now, Eq. 3.7 is a convolution integral, showing that the expected value of the periodogram is the true power spectrum convolved with Fejér's kernel, or

$$\mathsf{E}\{\hat{S}_p(f)\} = \Delta t \, (S * \mathcal{F}_N)(f) \, .$$

Fejér's kernel is the square of the Dirichlet kernel (Sect. 2.5.1)—as one might expect since we are now dealing with power rather than amplitude spectra—related to it by the identity

---

[1] A more detailed proof can be found in Percival and Walden (1993).

Fig. 3.1 Two Fejér kernels, $\mathcal{F}_N(f)$, for $N = 10$ (red line), and $N = 5$ (blue line), from Eq. 3.8, corresponding to the Dirichlet kernels in Fig. 2.15a

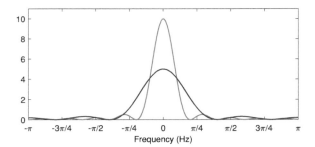

$$\mathcal{F}_N(x) \ = \ \frac{1}{N}\, D^2_{(N-1)/2}(x) \ = \ \frac{1}{N}\, \frac{\sin^2\left(\frac{Nx}{2}\right)}{\sin^2\left(\frac{x}{2}\right)} \ . \tag{3.8}$$

For very large N, Fejér's kernel acts like a Dirac delta function, giving, in theory,

$$\lim_{N\to\infty} \mathsf{E}\{\hat{S}_p(f)\} \ = \ S(f) \ .$$

That is, for long enough sequences, the periodogram gives an unbiased estimate of the true spectrum. However, N has to be exceptionally large for this to be true (Percival and Walden 1993). As N decreases to more realistic values (Fig. 3.1), the central lobe of Fejér's kernel broadens, causing smoothing and loss of resolution of the power spectrum, and its sidelobes increase in amplitude and width causing spectral leakage of power into adjacent (or distant) harmonics, as discussed in Sects. 2.5.2 and 2.5.3. Hence, the structure of the periodogram depends strongly on the sample size.

As discussed in Sect. 2.5.6.4, a method to reduce the bias caused by spectral leakage is the application of a window or taper to the sequence before taking its Fourier transform. Many of these windows are designed to decrease the amplitude of the sidelobes to levels below those of the Fejér kernel, though this comes at the cost of broadening the central lobe and thus smoothing the spectral estimate. Figure 3.2 compares the spectra of the Fejér kernel and a Hann window (Harris 1978). It is clear that the sidelobes of the Hann window are much smaller than those of the Fejér kernel, being up to 60 dB lower for the parameters used here (see caption). However, the central lobe of the Hann window is much broader than that of the Fejér kernel, with a 0.13 Hz half-width for the Hann compared to 0.063 Hz for the Fejér kernel.

When the window function is not a rectangle function, the estimate of the power spectrum of the windowed data sequence, $\hat{S}(f)$, is called the *modified periodogram* or *direct spectral estimator*. And it should be clear by now that, for a general window or taper, w_τ, with Fourier transform $W(f)$ and power spectrum $\mathcal{W}(f)$, the expected value of the modified periodogram is given by

$$\mathsf{E}\{\hat{S}(f)\} \ = \ \Delta t \int_{-f_N}^{f_N} \mathcal{W}(f - f')\, S(f')\, df' \ = \ \Delta t\, (S * \mathcal{W})(f) \ , \tag{3.9}$$

Fig. 3.2 Power spectra of a
Fejér kernel (red), and a
Hann taper (blue). For all
spectra, $N = 16$, $\Delta t = 1$ s
and $f_N = 0.5$ Hz. Note, 10
dB (decibels) represents a
factor of 10 difference in
power, 20 dB a factor of 100,
etc.

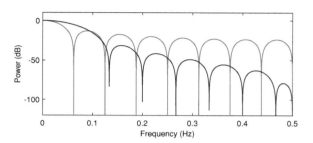

that is, the convolution of the true power spectrum with that of the window (cf.
Eq. 3.7). For example, if w_τ is the rectangle function, then $W(f)$ is the Dirichlet
kernel, $\mathcal{W}(f)$ is the Fejér kernel and $\mathsf{E}\{\hat{S}(f)\}$ is the periodogram.

Windows are designed for many purposes, but should generally yield spectral
estimates, $\hat{S}(f)$, with the properties listed below in mind (Percival and Walden 1993;
Simons et al. 2000). If many, say M, estimates have been made of the power spectrum
of a process with true spectrum $S(f)$, then the spectral estimates should be

1. Unbiased; that is, the mean of the estimates should approach the true spectrum;
   or

$$\mathsf{E}\{\hat{S}(f)\} = S(f) . \tag{3.10}$$

2. Well-behaved; the estimates should all be very similar, with zero variance; or

$$\lim_{M \to \infty} \mathsf{var}\{\hat{S}(f)\} = 0 . \tag{3.11}$$

3. Free of leakage; there should be no correlation between the power at different
   harmonics; or

$$\mathsf{cov}\{\hat{S}(f'), \hat{S}(f)\} = 0, \quad f' \neq f . \tag{3.12}$$

One set of windows that was designed to address these properties is discussed in
Sect. 3.3.

3.3 Slepian Tapers

The set of windows (or tapers) that form the basis of the 'multitaper method' of
spectral estimation are called *discrete prolate spheroidal sequences* (DPSSs), or
Slepian tapers, named after their inventor (Slepian 1978). Briefly—for it will be
discussed in detail in Sect. 3.4—the multitaper method forms the power spectrum of
a sequence by averaging a number of independent spectral estimates, as in Eq. 3.10,
each estimate being formed by tapering the sequence with one of the DPSSs. Due

to the particular construction of each DPSS, Eqs. 3.11 and 3.12 are also satisfied to varying degrees.

The DPSSs are the solutions to the so-called *spectral concentration problem* of signal processing, which seeks to find a window or set of windows that give the smallest amount of spectral leakage when computing an estimate of the power spectrum of a sequence. Since leakage depends on the size of the sidelobes of the window's power spectrum, the concentration problem stipulates that most of the window's energy should be concentrated in the central lobe of its power spectrum, leaving very little energy in the sidelobes. This can be expressed in two ways, below.

1. *Time- or index-limited*: For all sequences that are time-limited to the interval $[0, N - 1]$, find that particular sequence v_τ, with spectrum $V(f)$, that has the highest concentration of its energy in the frequency range $|f| \leq W$, where $W < f_N$, the Nyquist frequency. The fraction of energy lying in the range $[-W, W]$ is β^2, the frequency concentration measure, where

$$\beta^2(W) \equiv \int_{-W}^{W} |V(f)|^2 \, df \bigg/ \int_{-f_N}^{f_N} |V(f)|^2 \, df . \tag{3.13}$$

The numerator gives the energy lying in the frequency range $[-W, W]$, while the denominator gives the total energy in the sequence. We therefore want to maximise β^2.

2. *Band-limited*: For all sequences that are band-limited to $[-W, W]$, find that particular sequence v_τ that has the highest concentration of its energy in the (time) index range $[0, N - 1]$. The fraction of a signal's energy lying in an index range $[0, N - 1]$ is α^2, the time concentration measure, where

$$\alpha^2(N) \equiv \sum_{\tau=0}^{N-1} |v_\tau|^2 \bigg/ \sum_{\tau=-\infty}^{\infty} |v_\tau|^2 . \tag{3.14}$$

The numerator gives the energy lying within the sampled and truncated sequence, while the denominator gives the total energy in the sequence. We therefore want to maximise α^2.

Since the uncertainty relationship (Sect. 2.2.6) stipulates that a band-limited signal cannot also be time-limited, both α^2 and β^2 can together never be exactly 1 or exactly 0, though we will try to make them as close to 1 as possible.

As can be seen, the concentration problem is usually phrased in terms of the discrete-time Fourier transform (with discrete time data and continuous frequency data), though it can also be written in terms of the continuous and discrete Fourier transforms (Slepian 1983; Percival and Walden 1993). Although it might not seem so at first glance, both time-limited and band-limited approaches in all these domains give, essentially, the same answer (Simons and Dahlen 2006). Here, we will derive the sequences that satisfy both concentration problems, the DPSSs. We begin with the time-limited case, although the band-limited case appears most often in the literature and was the method first presented by Slepian (1978).

3.3.1 Time-Limited Concentration Problem

We begin with the numerator of Eq. 3.13, the energy of the band-limited spectrum, and write it as

$$\beta_{\text{num}}^2 = \int_{-W}^{W} V(f) \, V^*(f) \, df \, ,$$

where the $*$ denotes the complex conjugate. We then substitute Eq. 2.24, the DTFT, for $V(f)$, though here we stipulate that the sequence is index-limited (i.e. time-limited) to the range $[0, N-1]$ rather than $\pm\infty$, and adjust the summation limits of the DTFT accordingly. This gives

$$\beta_{\text{num}}^2 = \int_{-W}^{W} \left(\sum_{\tau'=0}^{N-1} v_{\tau'} \, e^{-2\pi i f \tau' \Delta t} \right) \left(\sum_{\tau=0}^{N-1} v_{\tau}^* \, e^{2\pi i f \tau \Delta t} \right) df$$

$$= \sum_{\tau'=0}^{N-1} \sum_{\tau=0}^{N-1} v_{\tau}^* \, v_{\tau'} \int_{-W}^{W} e^{2\pi i f (\tau - \tau') \Delta t} \, df \, .$$

We now use the identity Eq. 2.39 to write the numerator as

$$\beta_{\text{num}}^2 = 2W \sum_{\tau'=0}^{N-1} \sum_{\tau=0}^{N-1} v_{\tau}^* \, v_{\tau'} \, \text{sinc}[2\pi W (\tau - \tau') \Delta t] \, .$$

We next deal with the denominator of Eq. 3.13, the total energy in the spectrum. Here, we can use Parseval's theorem for the DTFT (Eq. 2.26) and write

$$\beta_{\text{den}}^2 = \int_{-f_N}^{f_N} |V(f)|^2 \, df = \Delta t^{-1} \sum_{\tau=0}^{N-1} |v_{\tau}|^2 \, ,$$

where, again, we require that the sequence is index-limited to the range $[0, N-1]$ rather than $\pm\infty$, and adjust the summation limits accordingly. We can now combine numerator and denominator to obtain

$$\beta^2 = \frac{2W \Delta t \; \sum_{\tau'=0}^{N-1} \sum_{\tau=0}^{N-1} v_{\tau}^* \, v_{\tau'} \, \text{sinc}[2\pi W (\tau - \tau') \Delta t]}{\sum_{\tau=0}^{N-1} |v_{\tau}|^2} \, . \tag{3.15}$$

Recall we need to find those functions, v_{τ}, that make β^2 as large as possible. In order to do this, we can use the calculus of variations. Essentially, this involves differentiating β^2 in Eq. 3.15 with respect to v and setting the result to equal zero, as one would do to find the turning point of any function. Doing so yields the equation

$$2W\Delta t \sum_{\tau'=0}^{N-1} v_{\tau'} \, \mathrm{sinc}[2\pi W(\tau - \tau')\Delta t] \; = \; \beta^2 \, v_\tau$$

for $\tau = 0, 1, \ldots, N-1$, or just

$$\sum_{\tau'=0}^{N-1} \frac{\sin[2\pi W(\tau - \tau')\Delta t]}{\pi(\tau - \tau')} \, v_{\tau'} \; = \; \beta^2 \, v_\tau \, . \tag{3.16}$$

Solutions to Eq. 3.16 give v_τ, the discrete prolate spheroidal sequences, the tapers that we will multiply the data by when we desire a power spectrum. Before describing the DPSSs, though, we will next show that the time-limited and band-limited problems are equivalent, and then discuss the DPSSs in Sect. 3.3.3.

3.3.2   Band-Limited Concentration Problem

Although it yields exactly the same result as the time-limited concentration problem, we will pursue a derivation of the solutions to the band-limited concentration problem as the literature is somewhat lacking in clarity. We begin with the numerator of Eq. 3.14, the energy of the truncated sequence, and write it as

$$\alpha_{\text{num}}^2 \; = \; \sum_{\tau=0}^{N-1} v_\tau \, v_\tau^* \, ,$$

where the $*$ denotes the complex conjugate. We then substitute Eq. 2.25, the inverse DTFT, for v_τ, though here we stipulate that the energy is concentrated between $\pm W$ rather than $\pm f_N$, and adjust the integration limits of the inverse DTFT accordingly. This gives

$$\alpha_{\text{num}}^2 = \sum_{\tau=0}^{N-1} \left(\Delta t \int_{-W}^{W} V(f) \, e^{2\pi i f \tau \Delta t} \, df \right) \left(\Delta t \int_{-W}^{W} V^*(f') \, e^{-2\pi i f' \tau \Delta t} \, df' \right)$$

$$= (\Delta t)^2 \int_{-W}^{W} \int_{-W}^{W} \left(\sum_{\tau=0}^{N-1} e^{2\pi i (f - f')\tau \Delta t} \right) V(f) \, V^*(f') \, df \, df' \, .$$

We now use the identity

$$\sum_{k=0}^{m} e^{ikx} = e^{imx/2} \frac{\sin\left(\frac{m+1}{2} x\right)}{\sin\left(\frac{x}{2}\right)} \equiv e^{imx/2} D_{m/2}(x) \qquad (3.17)$$

for any non-integer x (Percival and Walden 1993), where $D_m(x)$ is Dirichlet's kernel. Therefore

$$\sum_{\tau=0}^{N-1} e^{2\pi i(f-f')\tau \Delta t} = e^{i\pi(N-1)(f-f')\Delta t} D_{(N-1)/2}(2\pi(f - f')\Delta t) .$$

We can now write the numerator as

$$\alpha_{\text{num}}^2 = (\Delta t)^2 \int_{-W}^{W} \int_{-W}^{W} \left\{ e^{i\pi(N-1)(f-f')\Delta t} D_{(N-1)/2}(2\pi(f - f')\Delta t) \right.$$
$$\left. \times V(f) V^*(f') \right\} df \, df' .$$

We next deal with the denominator of Eq. 3.14, the energy in the infinitely long sequence. Here, we can use Parseval's theorem for the DTFT (Eq. 2.26) and write

$$\alpha_{\text{den}}^2 = \sum_{\tau=-\infty}^{\infty} |v_\tau|^2 = \Delta t \int_{-W}^{W} |V(f)|^2 df$$

where, again, we require that the energy in the spectrum be band-limited to $\pm W$, and adjust the integration limits accordingly.

We now introduce new functions, $U(f)$, called *discrete prolate spheroidal wave functions* (DPSWFs), which are related to the DTFT of v_τ ($V(f)$) by

$$U(f) = e^{i\pi(N-1)f\Delta t} V(f) . \qquad (3.18)$$

The DPSWFs are important functions to the theory of Slepian tapers, but as our study—elastic thickness estimation—has no direct need for them we will not pursue a full discussion. So, using Eq. 3.18 we can write the numerator and denominator as

$$\alpha_{\text{num}}^2 = (\Delta t)^2 \int_{-W}^{W} \int_{-W}^{W} D_{(N-1)/2}(2\pi(f - f')\Delta t) U(f) U^*(f') df \, df' ,$$

and

$$\alpha_{\text{den}}^2 = \Delta t \int_{-W}^{W} |U(f)|^2 df ,$$

giving

$$\alpha^2 = \frac{\Delta t \int_{-W}^{W} \int_{-W}^{W} D_{(N-1)/2}(2\pi(f - f')\Delta t) U(f) U^*(f') df \, df'}{\int_{-W}^{W} |U(f)|^2 df} . \qquad (3.19)$$

Equation 3.19 is the analogue of Eq. 3.15. Recall we need to find those functions, $U(f)$, that make α^2 as large as possible. In order to do this, we can use the calculus of variations, and differentiate α^2 in Eq. 3.19 with respect to U and set the result to equal zero. Doing so yields the equation

$$\Delta t \int_{-W}^{W} D_{(N-1)/2}(2\pi(f'-f)\Delta t)\, U(f)\, df \;=\; \alpha^2\, U(f')\,, \qquad (3.20)$$

or

$$\Delta t \int_{-W}^{W} \frac{\sin\left(N\pi(f'-f)\Delta t\right)}{\sin\left(\pi(f'-f)\Delta t\right)}\, U(f)\, df \;=\; \alpha^2\, U(f')\,.$$

This integral equation is of a type called a Fredholm equation of the second kind, and its solutions give the DPSWFs. However, here we are more concerned with the time-domain tapers, the DPSSs, so we now show how we can derive these from Eq. 3.20.

We now substitute Eq. 3.18 into Eq. 3.20, to get

$$\Delta t \int_{-W}^{W} D_{(N-1)/2}(2\pi(f'-f)\Delta t)\, e^{i\pi(N-1)f\Delta t}\, V(f)\, df \;=\; \alpha^2\, e^{i\pi(N-1)f'\Delta t}\, V(f')$$

or

$$\Delta t \int_{-W}^{W} \left\{ D_{(N-1)/2}(2\pi(f'-f)\Delta t)\, e^{i\pi(N-1)(f-f')\Delta t} \right\} V(f)\, df \;=\; \alpha^2\, V(f')\,.$$

Using Eq. 3.17 for the terms in the braces, { }, we can then write

$$\Delta t \int_{-W}^{W} \left\{ \sum_{\tau=0}^{N-1} e^{2\pi i(f-f')\tau\Delta t} \right\} V(f)\, df \;=\; \alpha^2\, V(f')\,,$$

or

$$\Delta t \sum_{\tau=0}^{N-1} \int_{-W}^{W} e^{2\pi i(f-f')\tau\Delta t}\, V(f)\, df \;=\; \alpha^2\, V(f')\,.$$

We now replace V by its inverse DTFT, Eq. 2.25 (over the truncation limits since it is concentrated there), giving

$$\Delta t \sum_{\tau=0}^{N-1} \int_{-W}^{W} e^{2\pi i(f-f')\tau\Delta t} \sum_{\tau'=0}^{N-1} v_{\tau'}\, e^{-2\pi i f\tau'\Delta t}\, df \;=\; \alpha^2 \sum_{\tau''=0}^{N-1} v_{\tau''}\, e^{-2\pi i f'\tau''\Delta t}\,,$$

which can be rewritten as

$$\Delta t \sum_{\tau=0}^{N-1} e^{-2\pi i f' \tau \Delta t} \sum_{\tau'=0}^{N-1} v_{\tau'} \int_{-W}^{W} e^{2\pi i f(\tau-\tau')\Delta t} \, df \; = \; \alpha^2 \sum_{\tau''=0}^{N-1} v_{\tau''} \, e^{-2\pi i f' \tau'' \Delta t} \; .$$

Then, using the identity Eq. 2.39 for the integral, we can write

$$\Delta t \sum_{\tau=0}^{N-1} e^{-2\pi i f' \tau \Delta t} \sum_{\tau'=0}^{N-1} v_{\tau'} \, 2W \, \mathrm{sinc}[2\pi W(\tau-\tau')\Delta t] \; = \; \alpha^2 \sum_{\tau''=0}^{N-1} v_{\tau''} \, e^{-2\pi i f' \tau'' \Delta t} \; .$$

Now, since the τ and τ'' are just dummy summation indices, we can replace the τ'' on the right-hand side with τ, and arrive at

$$2W \Delta t \sum_{\tau'=0}^{N-1} v_{\tau'} \, \mathrm{sinc}[2\pi W(\tau-\tau')\Delta t] \; = \; \alpha^2 \, v_\tau \, ,$$

for $\tau = 0, 1, \ldots, N-1$, or

$$\sum_{\tau'=0}^{N-1} \frac{\sin[2\pi W(\tau-\tau')\Delta t]}{\pi(\tau-\tau')} \, v_{\tau'} \; = \; \alpha^2 \, v_\tau \, . \tag{3.21}$$

Comparing Eqs. 3.16 and 3.21, we see that

$$\alpha^2 \; = \; \beta^2 \; = \; \lambda \, , \tag{3.22}$$

where we have introduced a new symbol, λ, to represent both of the concentration ratios.

The equality between α^2 and β^2 leads to an interesting conclusion: even though the uncertainty relation states that a function cannot be both time-limited and band-limited, the sequences that have most of their energy concentrated in the time interval given by the index range $[0, N-1]$ also have most of their energy concentrated in the bandwidth given by $\pm W$.

3.3.3   Discrete Prolate Spheroidal Sequences

Although there is only one solution to Eqs. 3.16 and 3.21, the DPSS that is the most concentrated in both time and frequency domains, we can generate further solutions that are only slightly less concentrated, each solution corresponding to a different but smaller value of λ. We can see this better if we write Eqs. 3.16 and 3.21 as a matrix eigenvalue equation,[2]

$$\mathbf{A}\mathbf{v}_k = \lambda_k \mathbf{v}_k , \tag{3.23}$$

where the index k is the *order*, and $k = 1, 2, \ldots, N$. When expanded, Eq. 3.23 is written as

$$A_{\tau,\tau'} \, v_{k,\tau'} = \lambda_k \, v_{k,\tau}$$

for $\tau, \tau' = 0, 1, \ldots, N - 1$. So there are N solutions to Eq. 3.23, corresponding to each value of k, although not all are useful, as we shall soon see. The λ_k are a set of N *eigenvalues*, being the concentration ratios (Eq. 3.22); the $\mathbf{v}_k$ are N *eigenvectors* (the DPSSs) corresponding to the N eigenvalues, each of length N; and the $\mathbf{A}$ is an $N \times N$ Toeplitz matrix[3] whose (τ, τ')th element is given by

$$A_{\tau,\tau'} = \frac{\sin[2\pi W (\tau - \tau')\Delta t]}{\pi(\tau - \tau')\Delta t} .$$

Equation 3.23 may be solved for the eigenvectors $v_{k,\tau}$ using any suitable matrix inversion algorithm on a computer. As mentioned, the solutions are called *discrete prolate spheroidal sequences* (DPSSs) and form the tapers that multiply the data sequence prior to Fourier transformation and thence estimation of the power spectrum. The discrete prolate spheroidal wave functions, $U_k(f)$, (Sect. 3.3.2) are often referred to as the *eigenfunctions*, and are related to the Fourier transforms of the DPSSs ($V_k(f)$) by Eq. 3.18.

We introduced the order k in Eq. 3.23, and stated that k denotes any one of the N possible solutions to that equation. Of course, as noted above, there is really only *one* solution to Eq. 3.23 because the concentration problem seeks to find *the most* concentrated sequence in both time and frequency domains, and not the second most, etc. However, it is very useful to be able to find the second most, third most, ..., concentrated solutions as these can be used in the multitaper method to great effect. This is because they are orthogonal to one another.[4] For any two DPSSs, $\mathbf{v}_j$ and $\mathbf{v}_k$, we have

$$\mathbf{v}_j^{\mathsf{T}} \mathbf{v}_k = \sum_{\tau=0}^{N-1} v_{j,\tau} \, v_{k,\tau} = \delta_{jk}$$

where δ_{jk} is the *Kronecker delta*,[5] and superscript T indicates the transpose vector. Because the DPSSs are orthogonal, power spectra computed using different DPSSs

---

[2] For those readers unsure of, or rusty on, eigenvectors and eigenvalues, see the Appendix to this chapter.

[3] A Toeplitz matrix is a matrix in which the values of each descending diagonal from left to right are constant.

[4] The concept of orthogonality is expanded upon in the Appendix to this chapter.

[5] The Kronecker delta, δ_{jk}, is very similar to the Dirac delta function; it has the value 1 if $j = k$, and 0 when $j \neq k$.

provide (almost) independent and uncorrelated estimates of the true power spectrum (see Sect. 3.4).

Thus, the most concentrated DPSS in both time and frequency domains is $\mathbf{v}_1$ ($= v_{1,\tau}$) with a concentration ratio of λ_1. We can then find a second index-limited DPSS, $\mathbf{v}_2$, with a concentration ratio of λ_2 that is orthogonal to $\mathbf{v}_1$ and has the highest concentration of energy in the frequency interval $\pm W$. And we can then find a third index-limited DPSS, $\mathbf{v}_3$, with a concentration ratio of λ_3 that is orthogonal to $\mathbf{v}_1$ and $\mathbf{v}_2$ and has the highest concentration of energy in the frequency interval $\pm W$, and so on, up to $\mathbf{v}_N$. The left-hand panels of Fig. 3.3 show some of the DPSSs in the time domain.

3.3.4 Resolution

Since the matrix $\mathbf{A}$, above, is a function of N, W and Δt, the eigenvectors and eigenvalues are functions of these parameters as well, and are characterised by the *time-bandwidth product*, NW, being

$$\text{NW} = NW\Delta t , \tag{3.24}$$

where $T = N\Delta t$ is the duration of the sequence, and W its bandwidth, being the width of the central lobe in the frequency domain.[6] Note the difference here between the product of N and W (NW) and the quantity NW (non-italicised font). This terminology is a product of information theorists' typical selection of $\Delta t = 1$, rendering NW $= NW$ by Eq. 3.24.

In practice, the DPSSs are obtained by solving Eq. 3.23 for a chosen integer value of NW ≥ 1. This value of NW then determines the bandwidth of the DPSSs—and thus the frequency-domain resolution of the eventual spectrum—calculated from Eq. 3.24 as

$$W = \frac{\text{NW}}{N\Delta t} . \tag{3.25}$$

Thus, the bandwidth is an integer multiple of the fundamental frequency, $1/(N\Delta t)$. The higher the time-bandwidth product NW, the higher the bandwidth, and the poorer the resolution; the lower the NW, the better the resolution.

This inverse relationship between NW and resolution can be seen in the right-hand panels of Fig. 3.3, which show plots of the power spectra of the DPSSs.[7] Recall from Sects. 2.5.2 and 3.2 that if a window function has a broad central lobe in its power spectrum, then the spectrum of the windowed signal/sequence is smoothed and there

---

[6] Strictly speaking, W is the half-bandwidth, as the total bandwidth extends from $-W$ to $+W$. Nevertheless, W is typically referred to as the 'bandwidth'. Note also that the eigenfunctions, U_k, can actually have multiple 'central lobes', as shown in the right-hand panels of Fig. 3.3.

[7] Note that, from Eq. 3.18, we have $|U_k(f)|^2 = |V_k(f)|^2$, so that Fig. 3.3 is also showing the spectra of the eigenfunctions.

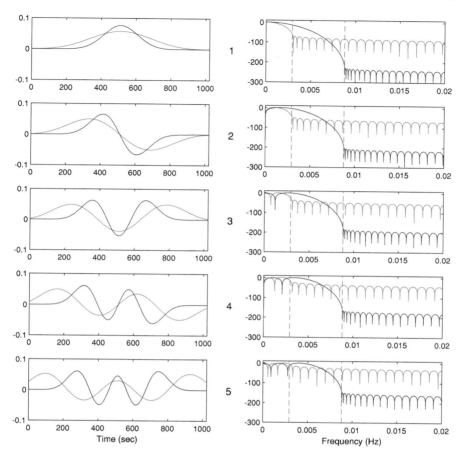

Fig. 3.3 Discrete prolate spheroidal sequences ('Slepian tapers'), $v_{k,\tau}$, in the time domain (left-hand panels), and their power spectra, $|V_k(f)|^2$, in decibels (right-hand panels). DPSSs for two values of the time-bandwidth product, NW, are shown: NW = 3 (red curves) and 9 (blue curves). All five useable tapers of the NW = 3 solution are plotted, while only the first five (of 17) of the NW = 9 solution are shown (taper orders, k, are shown in the centre). Data sequence parameters are $N = 1024$, $\Delta t = 1$ s, $f_N = 0.5$ Hz, giving the bandwidth W for NW = 3 as 0.0029 Hz, and 0.0088 Hz for NW = 9, from Eq. 3.25. These values of W are plotted as vertical dashed lines in the right-hand panels

is a loss of resolution. Thus, Fig. 3.3 shows that the higher-NW tapers (blue curves) have a broader central lobe than the lower-NW tapers (red curves) and will thus have a poorer frequency-domain resolution. Note that the order of the taper (the value of k in Eq. 3.23) does not affect the bandwidth significantly, except perhaps for the low-NW, high-order tapers (red curves with larger k in Fig. 3.3).

Also of note in the right-hand panels of Fig. 3.3 is the change in sidelobe amplitude for different values of NW and k. Tapers with lower values of NW have higher-amplitude sidelobes in their spectra, and hence more pronounced spectral leakage (Sects. 2.5.3 and 3.2). Furthermore, higher-order tapers have higher sidelobe levels. When these observations are combined, we see that the smallest sidelobe level occurs in low-order tapers with a large NW.

3.3.5 Eigenvalues

We have seen that, for a given NW, there are N solutions to Eq. 3.23: the eigenvectors $\mathbf{v}_k$, for order $k = 1, 2, \ldots, N$. These are, of course, the DPSSs or 'Slepian tapers'. Each solution has an associated eigenvalue, λ_k, which decreases in value with increasing k, thus

$$1 > \lambda_1 > \lambda_2 > \cdots > \lambda_N > 0 \, .$$

This decrease represents the decrease in the concentration ratio of each successive DPSS, and a corresponding increase in spectral leakage. Indeed, Bronez (1992) gives a measure of the leakage of a kth-order taper, for a given NW, as

$$L_k \; = \; 1 - \lambda_k \, , \tag{3.26}$$

reflecting the fact that any of the energy in the DPSWFs that is not within the bandwidth $\pm W$ constitutes leakage (because $1 - \lambda = 1 - \beta^2$ in Eq. 3.13).

Recall, from Eqs. 3.13, 3.14 and 3.22, that the eigenvalues should have values as close as possible to one, indicating strong concentration of the DPSSs' energy within the time and frequency limits. However, only the first $2NW\Delta t$ eigenvalues have values very close to one, after which they decrease rapidly in value to zero. This number, $2NW\Delta t$, is sometimes referred to as the *Shannon number*, and from Eq. 3.24 is equal to 2NW. In practice though, only the first $K = 2NW - 1$ tapers are used in conventional multitaper spectral estimation, whose eigenvalues are shown in Fig. 3.4.

Fig. 3.4 Eigenvalues (λ_k) for each order, k, of Slepian taper, for four NW values as indicated in the figure. All eigenvalues are shown, but the last usable one for a given NW (of order $k = 2NW - 1$) is circled

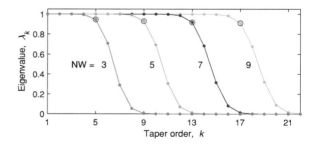

3.4   **Multitaper Spectral Estimation**

The multitaper method was developed by Thomson (1982) with the primary aim of estimating power spectra that were free from the bias caused by spectral leakage that is especially prevalent in periodograms. Of course, it was well-known that tapering or windowing the data to give a modified periodogram (Sect. 3.2) achieved this aim, though with the consequence of (1) losing, or severely suppressing, information at either end of the sequence, and (2) generating power spectra with a large variance. This latter issue was often corrected by frequency-domain convolution of the modified periodogram with a second window chosen for its smoothing properties.

Another approach to reducing variance is to use *moving window* methods, whereby the data sequence is segmented into many shorter subsequences, their periodograms or modified periodograms computed separately, and then all spectra averaged (Fig. 3.5). It is the averaging procedure that reduces the variance. Examples of such an approach are Bartlett's method, in which the moving windows do not overlap and are not tapered (thus averaging distinct periodograms), and Welch's method, in which the moving windows do overlap and may be tapered (for example with a Hann window). Even though it is not considered a particularly robust spectral estimator, an advantage of the Bartlett method is that the periodograms of each subsequence may be viewed as independent estimates of the true spectrum because they do not contain duplicated data, unlike the overlapping windows of the Welch method. Their average is therefore statistically more robust.

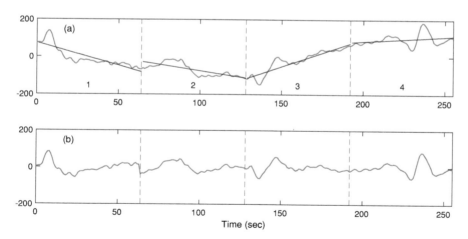

Fig. 3.5 Illustrating Bartlett's method of spectral estimation. **a** The time series is partitioned into four non-overlapping subsequences, 1–4. Note how, if each subsequence is detrended prior to Fourier transformation by subtraction of the black line, the low frequencies in the entire sequence are suppressed, and the sequence in **a** will appear to the spectral estimator as actually looking like the one in **b**

However, Thomson (1982) suggested that windowing the entire data sequence with a suite of Slepian tapers, and then averaging the separate spectral estimates (called *eigenspectra*), addressed all three issues. (1) The tapers' time- and band-limited concentration properties satisfied the spectral leakage problem, yielding bias-reduced spectra. (2) The higher-order DPSSs maintain large relative values close to the data edges (Fig. 3.6), preserving the information there. (3) Even though the tapers are applied to the entire sequence, independence of the separate eigenspectra is ensured because the DPSSs are orthogonal, and the more tapers that are used, the lower the variance of the averaged spectrum.

The multitaper method generally uses Slepian tapers, the DPSSs, though other sets of orthogonal window functions may be employed, such as the Hermite functions used by Simons et al. (2003). When DPSSs are used, one must first choose a value of the time-bandwidth product NW (Eq. 3.24), which implicitly sets the bandwidth of the tapers. One then computes the K tapers, $v_{k,\tau}$, where $1 \leq K \leq 2\text{NW} - 1$ and $k = 1, 2, \ldots, K$. Then, for a 1D sequence of N observations, g_τ, the sequence is multiplied by a taper of order k and the product Fourier transformed, as in

$$\check{G}_k(f) = \mathsf{F}\{g_\tau \, v_{k,\tau}\} \ .$$

The eigenspectrum at that taper is then given by

$$\hat{S}_k(f) = \left|\check{G}_k(f)\right|^2 \ .$$

In terms of Eq. 3.9, the expected value of the kth eigenspectrum is the convolution of the true spectrum, $S(f)$, with the kth eigenfunction, or

$$\mathsf{E}\{\hat{S}_k(f)\} = \Delta t \int_{-f_N}^{f_N} |U_k(f - f')|^2 \, S(f') \, df' = \Delta t \, (S * |U_k|^2)(f) \ .$$

Note that, although the eigenvectors and eigenfunctions are orthogonal, the eigen-spectra are not completely uncorrelated, only being asymptotically so (for very long sequences).

When all K eigenspectra have been computed, the average spectrum may be calculated from

$$\hat{S}(f) = \frac{1}{K} \sum_{k=1}^{K} \hat{S}_k(f) \ ,$$

or a weighted-average spectrum from

$$\hat{S}(f) = \frac{\sum_{k=1}^{K} \lambda_k \hat{S}_k(f)}{\sum_{k=1}^{K} \lambda_k} \ ,$$

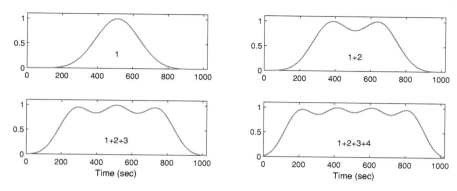

Fig. 3.6 Cumulative time-domain taper energy for the first four orders of DPSSs with NW = 3. The plots show the quantity $\sum_{k=1}^{K} v_{k,\tau}^2$ for the orders indicated on the panel (corresponding to $K = 1, 2, 3, 4$). As the number of tapers increases, the information towards the edges of the data sequence is weighted higher, ensuring that much more of the available information is used in the resulting average spectrum

in which the eigenspectra are weighted by the eigenvalues. The advantages of using multiple, orthogonal tapers over just a single taper are numerous. The inclusion of higher-order eigenspectra ensures data towards the edges of the sequence are given as much weight in the averaged spectrum as the data in the centre (Fig. 3.6), unlike with single tapers such as the Hann. Also, the data that are down-weighted or discarded by one taper are recovered by the other tapers (Fig. 3.3), so that the total energy in the tapers becomes more uniform (the curves in Fig. 3.6 become flatter and wider as K increases).

We have already discussed how, for a given NW, low-order tapers have better leakage-reduction properties than their higher-order relatives, shown in Fig. 3.3 by the decrease in sidelobe level with decreasing order. When the individual eigenspectra are averaged, the total leakage in the estimated spectrum is given as, from Eq. 3.26,

$$L = 1 - \frac{1}{K} \sum_{k=1}^{K} \lambda_k , \qquad (3.27)$$

(Bronez 1992) where K is the total number of tapers used for a given NW. Equation 3.27 is plotted in Fig. 3.7, which shows that, overall, leakage actually decreases with increasing NW. This might seem odd, considering that DPSSs with a large NW value possess more high-order tapers (with higher-amplitude sidelobes) than do low-NW tapers. However, according to this definition of spectral leakage, it is the concentration ratio, λ, that is used as the measure, rather than the amplitude of the sidelobes. And as Fig. 3.3 shows, the concentration of energy within the central lobe(s) ($\pm W$) improves with increasing NW.

Another advantage of using many tapers is that the variance of the average spectrum is reduced compared to that of the individual eigenspectra. When only a single

Fig. 3.7 DPSS spectral
leakage percentage as a
function of NW, from
Eq. 3.27 with $K = 2NW - 1$

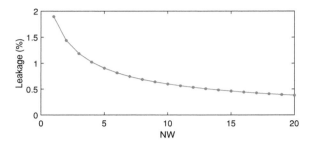

taper (or any single window function) is used, the sample size of the eigenspectrum
is effectively decreased because data at the edges are down-weighted; the variance
is thus increased, giving power spectra that appear 'noisy'. In contrast, using several
tapers has the effect of reducing the variance because more (independent) observations are included. The variance is further reduced because the sum of the tapers
gives a high weight to most of the data (Fig. 3.6). In fact, the variance of the average
spectrum is inversely proportional to the number of tapers used (K), or

$$\text{var}\{\hat{S}(f)\} = \frac{1}{K} \tag{3.28}$$

(Bronez 1992).

Thus, in multitaper spectral estimation there is a trade-off between resolution and
variance: using more tapers reduces the variance, but degrades the resolution because,
in order to get more tapers, one must choose a larger NW (and thus W) value. This
trade-off is seen in Fig. 3.8, which shows various power spectra of a synthetic time
series sequence with a random, fractal spectrum; such spectra plot as straight lines
with a negative gradient on log-power log-frequency axes (see figure caption). As
can be seen, there is more variation with frequency in the NW = 1 spectrum (which
uses but a single taper) than in the higher-NW spectra, which are smoother; the NW
= 1 spectrum is thus providing more resolution, perhaps, than the others, though its
similarity to the periodogram suggests that some of the 'detail' could possibly arise
from spectral leakage. The NW = 1 variance is certainly much larger than those of
the other multitapered spectra, whose variability diminishes with increasing NW.

One final observation, pertinent to elastic thickness estimation, is the tendency of
multitaper spectra to misrepresent the long wavelengths of data with red spectra.[8]
Certainly, multitaper spectra have a better resolution at the shorter wavelengths than
at long, as demonstrated by Kirby and Swain (2013) who compared spectra from
the multitaper method and azimuthally averaged Morlet ('fan') wavelets. This phenomenon is seen in Fig. 3.8 as a 'flattening' of the spectra at long wavelengths (low
frequencies), which becomes more pronounced as NW increases.

---

[8] A signal with a 'red' power spectrum has high power in its long-wavelength harmonics, and low
power in its short-wavelength harmonics. Fractally distributed data are examples of this. Taking
this further, 'blue' spectra possess low-power long-wavelength harmonics and high-power short-
wavelength harmonics; 'white noise' has equal power at all wavelengths.

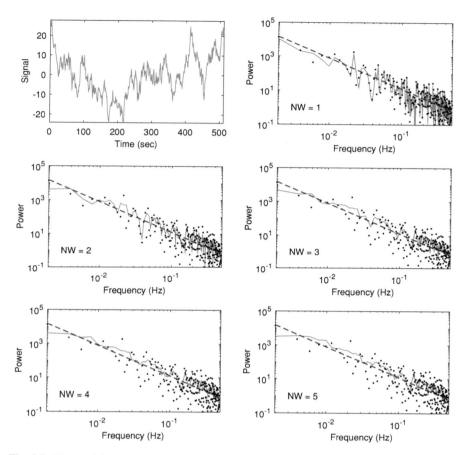

Fig. 3.8 The top-left panel shows a random, fractal time series generated using the 1D spectral synthesis method of Saupe (1988), with fractal dimension $D_F = 1.6$. The other panels show multitaper power spectra (red lines) of the time series at five different values of NW; in each case $2NW - 1$ tapers were used. The blue dashed line indicates the theoretical power spectrum of the fractal signal, $100 f^{-\beta}$, where f is frequency, and the spectral exponent, $\beta = 5 - 2D_F$. Also shown on each spectral plot is the periodogram (black dots)

3.5 Moving Windows

In Sect. 3.4 we saw how, in the Bartlett and Welch methods, power spectra are estimated in windows that are moved or translated across the entire sequence (Fig. 3.5). In those two methods, the spectra of the subsequences ('sub-spectra') are then averaged to form a single power spectrum for the entire sequence. If, however, the sub-spectra are not averaged but are rather plotted on a time-frequency plot at the time corresponding to the centre of each window, then an image of the time variation of the entire sequence's spectrum can be built. Such an image is called a *spectrogram*, and the procedure is known as the *windowed Fourier transform* (WFT), or *short-time Fourier transform* (STFT).

It should be clear by now, though, that the periodogram used in Bartlett's method, or even the Hann windows often applied in the Welch method, do not generally produce the best possible sub-spectra. This detriment is compounded by the fact that no spectral averaging can be performed within a single (moving) window as there is only one estimate of the sub-spectrum per window. However, if the data in each window are tapered with a suite of orthogonal tapers such as the DPSSs, and the resulting eigenspectra within each window averaged, then a more robust spectrogram results. Three examples of multitapered spectrograms are shown in Fig. 3.9, each one having a different moving window size. The signal under analysis is a single-frequency cosine where the frequency changes every 500 s. Immediately apparent is that the largest window has the best frequency resolution (Fig. 3.9a), with the spectral peaks being much more localised in frequency than those obtained with the smallest window, which are comparatively smeared (Fig. 3.9c). Large windows give a better frequency resolution because they contain more observations. Imagine a sequence of N points sampled at Δt seconds; from Eq. 2.29 its fundamental frequency (and frequency sampling interval) is $\Delta f_{\text{seq}} = 1/(N\Delta t)$. If a spectrogram of the sequence is desired, then a window of size $M < N$ must be selected, giving the fundamental frequency within the window as $\Delta f_{\text{win}} = 1/(M\Delta t)$, which is greater than the fundamental frequency of the whole sequence, or $\Delta f_{\text{win}} > \Delta f_{\text{seq}}$. Since Δf governs both the lowest frequency that can be resolved, and the resolution of the frequency domain, it should now be clear that larger windows give better spectra (larger bandwidth, finer resolution).

However, the largest window also has the worst time-localisation; even though the frequency changes are abrupt—at the 500 s, 1000 s and 1500 s points—the spectral peaks extend into adjacent segments much further than they do in the small-window spectrogram (compare Fig. 3.9a, c). Once again, the uncertainty relationship (Sect. 2.2.6) reveals itself: one cannot have high resolution in both time and frequency domains simultaneously.

An additional feature of the spectrogram is that its frequency resolution is frequency-independent. That is, the resolution of features in the spectrum, whether they be at low frequencies or high frequencies, is constant; this is seen in Fig. 3.9 by the constant 'thickness' of the high-power peaks for a given window size. This is the case for all windowed Fourier transform methods, and not just the Slepian multitaper method whose resolution is determined by the bandwidth W via the NW parameter (Sect. 3.3.4).

This frequency-independent resolution was sometimes seen as a disadvantage when interpreting spectrograms, and was one of the factors that caused the wavelet transform to be developed. Such topics will be explored further in Sect. 4.8.2.

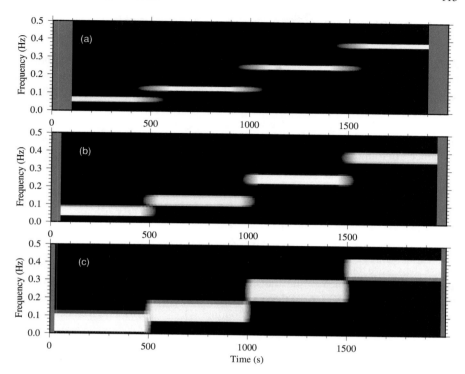

Fig. 3.9 Spectrogram (moving window multitaper power spectrum) of a cosine function with frequency 0.0625 Hz from 0–499 s, 0.125 Hz from 500–999 s, 0.25 Hz from 1000–1499 s and 0.375 Hz from 1500–1999 s. All spectra were computed using Slepian tapers with NW = 3. The sizes of the moving windows are **a** 200 s; **b** 100 s; **c** 50 s. The spacing between adjacent windows is 1 s, the same as the sampling interval of the data sequence. The Nyquist frequency is the same for all windows, being $1/(2\Delta t) = 0.5$ s. The grey scale is power in decibels, with light shades being high power, and dark shades low power. [Cf. Fig. 4.6.]

3.6 The 2D Multitaper Method

Following on from the extension from the 1D to the 2D Fourier transform, outlined in Sect. 2.6, we now extend the discussion on Slepian tapers and the multitaper method to the 2D case so that we can apply them to 2D gravity and topography data. We hence utilise the position vector $\mathbf{x} = (x, y)$ to coordinate the data in the space domain, and coordinate it in the Fourier domain with the wavenumber vector $\mathbf{k} = (k_x, k_y)$, where k_x is the wavenumber in the x-direction, and k_y is the wavenumber in the y-direction. As noted in Sect. 2.6.1, wavenumber is an angular frequency with units of radians per kilometre (rad/km).

3.6.1 2D Slepian Tapers

A rigorous derivation of 2D DPSSs would solve the concentration problem(s) in two dimensions, as has been done by Slepian (1964) and Simons (2010). However, as noted by Hanssen (1997) and Simons et al. (2000) for example, 2D tapers are readily formed by taking the outer product of two 1D tapers. For a 2D sequence of $N \times N$ observations, $g_{\xi\eta}$, where ξ is the index along the x-axis and η the index along the y-axis, and ξ, $\eta = 0, 1, \ldots, N-1$, the 2D DPSSs ($\mathbf{w}$) which will multiply the sequence are found from

$$\mathbf{w}_{jk} \;=\; \mathbf{v}_j \, \mathbf{v}_k^{\mathsf{T}} \;=\; \begin{pmatrix} v_{j,0} \\ v_{j,1} \\ \vdots \\ v_{j,N-1} \end{pmatrix} \begin{pmatrix} v_{k,0} & v_{k,1} & \cdots & v_{k,N-1} \end{pmatrix} \;,$$

where the $\mathbf{v}$'s are the 1D tapers (Eq. 3.23), j, $k = 1, 2, \ldots, N$ are the 1D taper orders and superscript T indicates the transpose vector. The product, $\mathbf{w}_{jk}$, is an $N \times N$ matrix, and there are $N \times N$ of them, examples being shown in Fig. 3.10. It can also be shown that the 2D tapers form an orthogonal set,

$$\mathbf{w}_{jk}^{\mathsf{T}} \, \mathbf{w}_{mn} \;=\; \delta_{jk} \, \delta_{mn} \;,$$

and that the 2D eigenvalues are simply the products of the corresponding 1D eigenvalues, or

$$\lambda_{jk} \;=\; \lambda_j \, \lambda_k$$

Fig. 3.10 2D Slepian tapers of NW $= 4$ in the space domain. Only the first three tapers in each direction are shown, indicated by their order in the x-direction $\times$ order in the y-direction. The horizontal axes are x- and y-coordinates

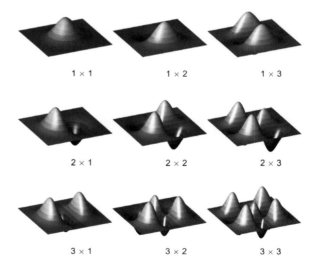

Fig. 3.11 The power spectra of the 2D Slepian tapers in Fig. 3.10. The contours are marked every 20 dB from -10 dB to -90 dB. The dashed lines show the bandwidth, $\pm W$, where $W = 0.0245$ rad/km. The abscissa and ordinate axes are x- and y-wavenumber (k_x and k_y), respectively

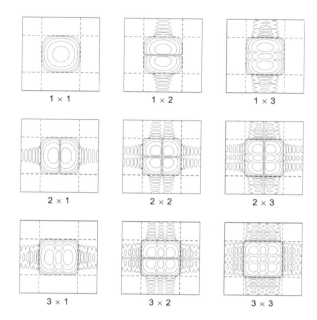

(Hanssen 1997). As for the 1D case, only the first $2NW - 1$ tapers in each direction are usable. For example, if $NW = 4$, then a maximum of seven tapers in each direction can be used, giving a maximum of 49 usable tapers in total.

The spectral response of the 2D DPSSs is shown in Fig. 3.11. As for the 1D case, the lower-order tapers are the most concentrated within the bandwidth $\pm W$, but the concentration depends upon the direction. Thus, a low-order $\times$ high-order 2D taper is concentrated along the k_x-axis, but not as much along the k_y-axis.

3.6.2 2D Average Spectrum

The average 2D multitaper spectrum is calculated similar to the 1D case (Sect. 3.4). The 2D data sequence, $g_{\xi\eta}$, is multiplied by each of the chosen 2D tapers and the 2D Fourier transform taken, giving

$$\check{G}_{jk}(\mathbf{k}) = \mathsf{F}\{g_{\xi\eta}\, w_{jk\xi\eta}\}\,,\tag{3.29}$$

which then forms the (j, k)th eigenspectrum thus

$$\hat{S}_{jk}(\mathbf{k}) = \left|\check{G}_{jk}(\mathbf{k})\right|^2.\tag{3.30}$$

Fig. 3.12 Various power
spectra of a random, fractal
surface of fractal dimension
2.4, generated using the 2D
spectral synthesis method of
Saupe (1988). Shown are the
periodogram, the averaged
multitaper spectrum and nine
eigenspectra at the taper
orders indicated. The Slepian
tapers had NW $= 3$ (giving a
maximum of 25 usable
tapers), but only three tapers
in each direction were used
to form the average.
Contours are every 10 dB,
and the axes are x- and
y-wavenumber over a range
$\pm|\mathbf{k}|_N/10$, where $|\mathbf{k}|_N$ is the
Nyquist wavenumber

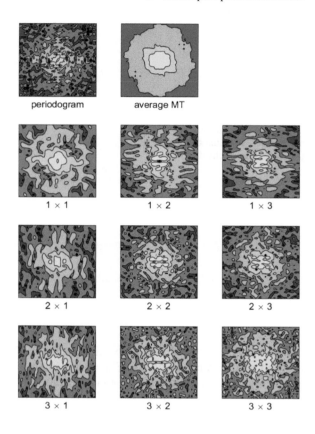

periodogram   average MT

1×1 1×2 1×3

2×1 2×2 2×3

3×1 3×2 3×3

When all K^2 eigenspectra (K in each direction) have been computed, the average
spectrum may be calculated from

$$\hat{S}(\mathbf{k}) \;=\; \frac{1}{K^2} \sum_{j=1}^{K} \sum_{k=1}^{K} \hat{S}_{jk}(\mathbf{k}) \;. \tag{3.31}$$

Figure 3.12 shows the averaged spectrum of a random, fractal surface, using three
Slepian tapers of NW $= 3$ in each direction (so $K^2 = 9$). The figure also shows the
nine eigenspectra that formed the average, and the periodogram for comparison.

3.6.3 Radially Averaged Power Spectrum

Very often, the spectrum of 2D data is displayed as a 1D graph, rather than a 2D contour plot such as those in Fig. 3.12. Doing this can aid a visual or even analytic interpretation of the spectrum, such as determining whether it is fractal, for example. One approach to generating such a 1D plot is to compute the *radially averaged power spectrum* (RAPS).

Here, we partition the (k_x, k_y)-plane into a number, M_a, of concentric annuli of equal width[9] $\Delta|\mathbf{k}|$, as shown in Fig. 3.13a. The annulus m has lower bound $(m - 1)\Delta|\mathbf{k}|$ and upper bound $m\Delta|\mathbf{k}|$ for integer m, where $1 \leq m \leq M_a$, and $M_a\Delta|\mathbf{k}| = |\mathbf{k}|_N$, where

$$|\mathbf{k}|_N = \frac{2\pi}{2\Delta x}$$

is the Nyquist wavenumber of the data sequence, assuming its grid spacing is Δx in both x- and y-directions (Eq. 2.59). Then, for every (k_x, k_y) pair that falls within a given annulus, we find the arithmetic mean of their wavenumber moduli $|\mathbf{k}| = (k_x^2 + k_y^2)^{1/2}$, and the arithmetic mean of all corresponding power estimates, $\hat{S}(k_x, k_y)$. Formally, the mean wavenumber assigned to an annulus m is given by

$$\overline{|\mathbf{k}|}_m = \frac{1}{N_m}\sum_{l=1}^{N_m} |\mathbf{k}|_l \equiv |\mathbf{k}| , \qquad (3.32)$$

and the RAPS is given by

$$\bar{S}(|\mathbf{k}|) = \frac{1}{N_m}\sum_{l=1}^{N_m} \hat{S}(\mathbf{k}_l) ,$$

where, for the lth (k_x, k_y) pair within the annulus, $\mathbf{k}_l = (k_x, k_y)_l$ is its wavenumber, $|\mathbf{k}|_l$ is its wavenumber modulus, $\hat{S}(\mathbf{k}_l)$ is its taper-averaged power estimate (Eq. 3.31) and N_m is the number of pairs in annulus m. The RAPS is shown as the red circles in Fig. 3.13b, which also shows the 2D power estimates, $\hat{S}(k_x, k_y)$, but plotted as a function of their wavenumber modulus, $|\mathbf{k}| = (k_x^2 + k_y^2)^{1/2}$ (rather than as a function of 2D wavenumber, which is shown in Fig. 3.13a).

Thus, it can be seen that the RAPS is actually an average around an annulus (i.e. an azimuthal average from 0° to 360°) rather than an average along a radius as its name might suggest. Hence, it suits data that are generally isotropic, with their 2D power spectra being similar in all directions. If the data are strongly anisotropic then the RAPS may provide misleading interpretations; it pays to plot the 2D spectrum as well as the RAPS.

---

[9] Other schemes are possible, such as annuli of equal area, or exponentially decreasing width.

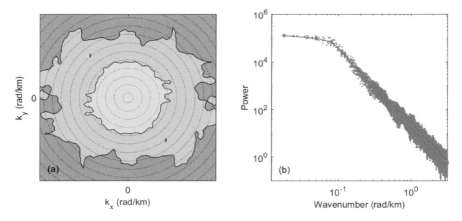

Fig. 3.13 **a** The averaged 2D multitaper power spectrum of a random, fractal surface, superimposed with the annuli used to compute the radially averaged 1D power spectrum (concentric circles). Other details as for Fig. 3.12. **b** The radially averaged power spectrum (red circles connected by line), and the actual 2D power estimates plotted as a function of their wavenumber modulus ($|\mathbf{k}|$) (grey dots)

As a final note, azimuthally averaging the 2D power spectrum over 360° gives sensible results because the power spectrum will have positive values at all wavenumbers. However, one must be careful if one wishes to average the real and imaginary parts of a Fourier transform separately. Many space- or time-domain data sets are real-valued and asymmetric, and such data have Hermitian Fourier transforms, with an even real part, and an odd imaginary part (Bracewell 1986). As noted in Sect. 2.2.3, when the data are 1D the negative frequencies contain redundant information; in the 2D case, the information in the lower two quadrants is redundant, as shown in Fig. 3.14a, b. The lower two quadrants of both real and imaginary parts are reflected and flipped versions of the upper two quadrants, though the imaginary part has its lower-quadrant values multiplied by -1. Thus, performing a 360° azimuthal average on the imaginary part of the Fourier transform will result in complete cancellation of its values, yielding a 'radially averaged' plot that is uniformly zero at all wavenumbers, as shown by the circles and solid line in Fig. 3.14d. The situation can be remedied by performing a 180° azimuthal average: only data in the upper two quadrants are used and no cancellation by the negative mirror-image occurs; a plot such as the black line in Fig. 3.14d will result. Note that the data redundancy in the real part does not present a problem, and 180° and 360° azimuthal averages yield the same result (Fig. 3.14c). Of course one could square both real and imaginary parts before performing an azimuthal average on each—and the cancellation of the imaginary part would not occur—but sometimes this is not fit for purpose, as will be seen in Sect. 4.12.1.

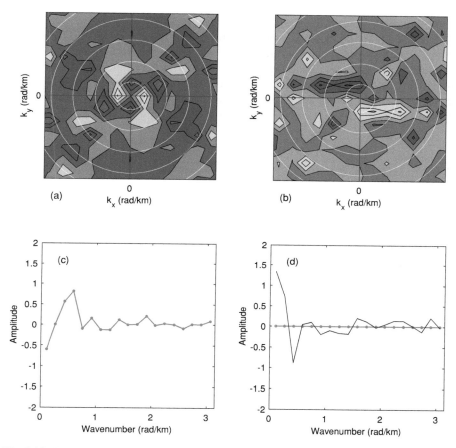

Fig. 3.14 The Fourier transform of some 2D real-valued data. The transform is a Hermitian function, with an even real part (**a**), and an odd imaginary part (**b**), both superimposed with the annuli used to calculate the RAPS as in Fig. 3.13; pink is positive, blue is negative. **c** The radially averaged real part from **a**. **d** The radially averaged imaginary part from **b** plotted as red circles connected by a red line. The black line shows the radial average over half-annuli, i.e. over the upper two quadrants only

3.7 Summary

The periodogram, which many people use as a power spectral estimate, provides a very biased estimate of the true power spectrum in most cases, containing a significant amount of spectral leakage. Application of a single-window function to the data sequence before Fourier transformation can reduce the leakage, but at the cost of increasing the variance of the resulting modified periodogram. The multitaper method was developed to both reduce leakage and variance, especially if the applied windows were a suite of discrete prolate spheroidal sequences, or 'Slepian tapers'.

The properties of the DPSSs depend mainly on their time-bandwidth product (NW) and the order of the taper. For a given taper order, low values of NW provide relatively good frequency-domain resolution but have higher spectral leakage; increasing NW degrades resolution but reduces the leakage. For a given NW value, low-order tapers have better (lower) leakage than do high-order tapers, which means that high-order tapers have slightly degraded resolution properties. Above all, the more tapers that are used, the smaller the variance of the power spectrum.

3.8 Further Reading

The multitaper method is widely used (although perhaps not as widely as it should be—periodograms persist), and most published studies that use it provide a short, referenced explanation of its workings. A few good examples in the geophysical context are Simons et al. (2000), Simons (2010) and Simons and Wang (2011).

The definitive paper on the discrete prolate spheroidal sequences is Slepian (1978), while that on the multitaper method is Thomson (1982). These are both quite mathematical, so the book by Percival and Walden (1993), which dilutes the maths with explanatory text, is an essential companion. Percival and Walden (1993) also covers the non-discrete (continuous) solutions to the concentration problem, as does Slepian (1983), and while these solutions are not as relevant to our purposes, they do provide a useful context of the concentration problem.

Appendix

Orthogonality

A useful coordinate system is one whose *basis vectors* are orthogonal, that is, at 90° to one another.[10] Consider the 2D Cartesian plane with coordinates (x, y) and basis (or 'unit') vectors $\hat{\mathbf{i}}$ and $\hat{\mathbf{j}}$, where

$$\hat{\mathbf{i}} \equiv \begin{pmatrix} 1 \\ 0 \end{pmatrix}, \qquad \hat{\mathbf{j}} \equiv \begin{pmatrix} 0 \\ 1 \end{pmatrix}.$$

Now, it is well known that when two vectors are orthogonal then they have a zero inner (or 'dot') product. For the two unit vectors of the Cartesian plane, we see that this is so, with

---

[10] That is not to say that non-orthogonal coordinate systems are never useful; while having limited applicability, they are sometimes very useful when solving certain problems.

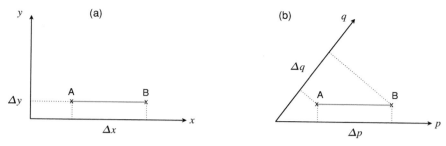

Fig. 3.15 a A 2D Cartesian coordinate system (x, y). In moving a distance Δx parallel to the x-axis, the change in the y-coordinate, Δy, is zero. **b** A skewed coordinate system (p, q), where a change in the p-coordinate also results in a change in the q-coordinate, even when the movement is parallel to the p-axis

$$\hat{\mathbf{i}} \cdot \hat{\mathbf{j}} \equiv \hat{\mathbf{i}}^T \hat{\mathbf{j}} = \begin{pmatrix} 1 & 0 \end{pmatrix} \begin{pmatrix} 0 \\ 1 \end{pmatrix} = (1 \times 0) + (0 \times 1) = 0 \,,$$

where superscript T indicates the transpose vector. The concept is shown schematically in Fig. 3.15a. In moving a distance Δx from point A to point B, parallel to the x-axis, one does not change one's y-coordinate, and $\Delta y = 0$. This is so because the basis vectors of the (x, y)-coordinate system are orthogonal (actually *orthonormal* because they are orthogonal and have unit length), and an x-movement is independent of a y-movement. If, however, the basis vectors are not orthogonal then the situation in Fig. 3.15b arises, where the (p, q)-coordinate system is skewed. Here, when one moves a distance Δp from point A to point B, parallel to the p-axis, one also moves in the q-direction, translating a distance Δq along the q-axis. Thus, a p-movement is not independent of a q-movement, and the basis vectors of such a system are non-orthogonal.

For a general N-dimensional space with N basis vectors $\mathbf{e}_j$ $(j = 1, \ldots, N)$, we would desire the basis vectors to form an orthonormal basis, such that their inner product obeys

$$\mathbf{e}_j^T \mathbf{e}_k = \delta_{jk} \,,$$

where δ_{jk} is the Kronecker delta, having the value 1 if $j = k$, and 0 when $j \neq k$. Thus, the vector **a** would be written as

$$\mathbf{a} = \sum_{j=1}^{N} a_j \mathbf{e}_j \,,$$

and the inner ('dot') product of two vectors **a** and **b** is written as

$$\mathbf{a} \cdot \mathbf{b} = \sum_{j=1}^{N} a_j b_j \,.$$

Functions can also form an orthonormal basis, and such functions are very efficient when combined to represent other functions. Just as for vectors, one can define an inner product of two orthonormal functions, written as

$$\langle \phi_m(t), \phi_n(t) \rangle \equiv \int_T \phi_m(t)\,\phi_n^*(t)\,dt = \delta_{mn}\,,$$

for the set of complex functions $\phi_m(t)$ which are orthonormal over the region T, where superscript * indicates the complex conjugate. For example, the complex exponentials of the Fourier transform form an orthonormal basis, one being able to show that

$$\langle e^{2\pi imt/T}, e^{2\pi int/T} \rangle = T\,\delta_{mn}\,.$$

Other examples are the associated Legendre functions of spherical harmonics, and of course the discrete prolate spheroidal sequences and wave functions. Thus, orthonormal basis functions act like orthonormal basis vectors, and they can be summed to give a representation of another function, $g(t)$, such as in

$$g(t) = \sum_{m=1}^{N} a_m \phi_m(t)\,,$$

for scalar coefficients a. Importantly, since the basis functions are orthogonal, each term $a_m \phi_m(t)$ is independent of the others, and the situation is therefore akin to Fig. 3.15a, replacing the coordinate axes x and y with ϕ_m and ϕ_n, respectively. But if the basis functions are non-orthogonal, then the situation is like Fig. 3.15b; the terms $a_m \phi_m(t)$ are not independent and there will be some component of $a_n \phi_n(t)$ present in $a_m \phi_m(t)$, for $m \neq n$.

Eigenvalues and Eigenvectors

If a linear transformation, represented by a matrix $\mathbf{A}$, acts on a vector ($\mathbf{v}$) such that the resultant vector ($\mathbf{Av}$) is simply a scaled version of the original vector ($\lambda\mathbf{v}$, for scalar λ), then the vector is said to be an *eigenvector* and the scalar an *eigenvalue* of $\mathbf{A}$, thus

$$\mathbf{Av} = \lambda\mathbf{v}\,. \tag{3.33}$$

In general, for an $n \times n$ matrix there will be n eigenvalues and n eigenvectors. Because λ is a scalar, the vector $\mathbf{Av}$ must be geometrically parallel to the vector $\mathbf{v}$; this is unusual because, in general, when a vector is multiplied by a matrix it changes direction (cf. rotation matrices).

The eigenvalues of A are calculated in the following way. Rewrite Eq. 3.33 as

$$(\mathbf{A} - \lambda\mathbf{I})\,\mathbf{v} = \mathbf{0}\;, \tag{3.34}$$

where $\mathbf{I}$ is the identity matrix and $\mathbf{0}$ is the zero vector. Equation 3.34 has a non-zero solution $\mathbf{v}$ when the determinant of its coefficient matrix is zero:

$$\det(\mathbf{A} - \lambda\mathbf{I}) = 0\;. \tag{3.35}$$

Equation 3.35 generates an nth-order polynomial with n solutions, λ_n, the eigenvalues of $\mathbf{A}$. The n eigenvectors are then found by substituting each of the n eigenvalues into Eq. 3.34 and solving for $\mathbf{v}$.

As an example, find the eigenvalues and eigenvectors of the 2×2 matrix,

$$\mathbf{A} = \begin{pmatrix} 2 & 1 \\ 1 & 2 \end{pmatrix}\;.$$

First, find the two eigenvalues using Eq. 3.35. The determinant of $\mathbf{A} - \lambda\mathbf{I}$ is

$$\begin{vmatrix} 2-\lambda & 1 \\ 1 & 2-\lambda \end{vmatrix} = (2-\lambda)^2 - 1 = \lambda^2 - 4\lambda + 3 = (\lambda - 1)(\lambda - 3) = 0\;.$$

The solutions of this quadratic equation are

$$\lambda_1 = 1\;, \qquad \lambda_2 = 3\;.$$

So the two eigenvalues of $\mathbf{A}$ are 1 and 3.

The eigenvectors are then found by substituting the eigenvalues into Eq. 3.34. For $\lambda_1 = 1$, we have

$$\left[\begin{pmatrix} 2 & 1 \\ 1 & 2 \end{pmatrix} - \begin{pmatrix} 1 & 0 \\ 0 & 1 \end{pmatrix} \right] \mathbf{v}_1 = \begin{pmatrix} 0 \\ 0 \end{pmatrix}\;,$$

or

$$\begin{pmatrix} 1 & 1 \\ 1 & 1 \end{pmatrix} \begin{pmatrix} v_{1,1} \\ v_{1,2} \end{pmatrix} = \begin{pmatrix} 0 \\ 0 \end{pmatrix}\;,$$

giving

$$v_{1,1} + v_{1,2} = 0 \qquad \text{and} \qquad v_{1,1} + v_{1,2} = 0\;,$$

or

$$v_{1,1} = -v_{1,2}\;.$$

Therefore, the first eigenvector is

$$\mathbf{v}_1 \;=\; s_1 \begin{pmatrix} 1 \\ -1 \end{pmatrix} ,$$

where s_1 is any non-zero scalar.

For $\lambda_2 = 3$, we have

$$\left[\begin{pmatrix} 2 & 1 \\ 1 & 2 \end{pmatrix} - 3 \begin{pmatrix} 1 & 0 \\ 0 & 1 \end{pmatrix} \right] \mathbf{v}_2 \;=\; \begin{pmatrix} 0 \\ 0 \end{pmatrix} ,$$

or

$$\begin{pmatrix} -1 & 1 \\ 1 & -1 \end{pmatrix} \begin{pmatrix} v_{2,1} \\ v_{2,2} \end{pmatrix} \;=\; \begin{pmatrix} 0 \\ 0 \end{pmatrix} ,$$

giving

$$-v_{2,1} + v_{2,2} = 0 \qquad \text{and} \qquad v_{2,1} - v_{2,2} = 0 ,$$

or

$$v_{2,1} \;=\; v_{2,2} .$$

Therefore, the second eigenvector is

$$\mathbf{v}_2 \;=\; s_2 \begin{pmatrix} 1 \\ 1 \end{pmatrix} ,$$

where s_2 is any non-zero scalar.

References

Bracewell RN (1986) The Fourier transform and its applications. McGraw-Hill, New York

Bronez TP (1992) On the performance advantage of multitaper spectral analysis. IEEE Trans Signal Process 40:2941–2946

Hanssen A (1997) Multidimensional multitaper spectral estimation. Signal Process 58:327–332

Harris FJ (1978) On the use of windows for harmonic analysis with the discrete Fourier transform. Proc IEEE 66:51–83

Kirby JF, Swain CJ (2013) Power spectral estimates using two-dimensional Morlet-fan wavelets with emphasis on the long wavelengths: jackknife errors, bandwidth resolution and orthogonality properties. Geophys J Int 194:78–99

Percival DP, Walden AT (1993) Spectral analysis for physical applications. Cambridge University Press, Cambridge

Saupe D (1988) Algorithms for random fractals. In: Peitgen H-O, Saupe D (eds) The science of fractal images. Springer, New York, pp 71–136

Simons FJ (2010) Slepian functions and their use in signal estimation and spectral analysis. In: Freeden W, Nashed MZ, Sonar T (eds) Handbook of geomathematics. Springer, Berlin, pp 891–923

Simons FJ, Dahlen FA (2006) Spherical Slepian functions and the polar gap in geodesy. Geophys J Int 166:1039–1061

Simons FJ, Wang DV (2011) Spatiospectral concentration in the Cartesian plane. Int J Geomath 2:1–36

Simons FJ, Zuber MT, Korenaga J (2000) Isostatic response of the Australian lithosphere: estimation of effective elastic thickness and anisotropy using multitaper spectral analysis. J Geophys Res 105(B8):19,163–19,184

Simons FJ, van der Hilst RD, Zuber MT (2003) Spatiospectral localization of isostatic coherence anisotropy in Australia and its relation to seismic anisotropy: implications for lithospheric deformation. J Geophys Res 108(B5):2250. https://doi.org/10.1029/2001JB000704

Slepian D (1964) Prolate spheroidal wave functions, Fourier analysis, and uncertainty–IV: extensions to many dimensions; generalized prolate spheroidal functions. Bell Syst Tech J 43:3009–3057

Slepian D (1978) Prolate spheroidal wave functions, Fourier analysis, and uncertainty–V: the discrete case. Bell Syst Tech J 57:1371–1430

Slepian D (1983) Some comments on Fourier analysis, uncertainty and modeling. SIAM Rev 25:379–393

Thomson DJ (1982) Spectrum estimation and harmonic-analysis. Proc IEEE 70:1055–1096

Chapter 4
The Continuous Wavelet Transform

4.1 Introduction

Estimates of the effective elastic thickness (T_e) from gravity and topography data are made in the frequency domain by comparing spectral quantities called the *admittance* and *coherence* (Chap. 5) against those predicted by some model of plate flexure. In the early days of spectral T_e estimation—the 1970s and early 1980s—data would be extracted and gridded over a 1D profile or 2D rectangular region of interest, and the Fourier transform employed to determine a single admittance or coherence spectrum. This would then give a single T_e estimate, because the Fourier transform can only provide a single spectrum for a given data sequence. If maps of the spatial variation of T_e are required, then clearly sub-sets of the data must be extracted, or 'windowed', and admittance and coherence spectra computed in each window.

However, windowing data has its problems, besides those discussed in Sect. 2.5. In order to properly map spatial variations in T_e, the window needs to be small; otherwise, there is a risk of over-smoothing any variations actually present. But if the window is too small, there is a risk of it not capturing the long wavelengths of the topography and gravity field associated with flexure.

This is where the wavelet transform proves useful. The wavelet transform enables a power spectrum to be calculated at each and every location of a specified signal, i.e. power as a function of both time and frequency (or space and wavenumber), without the need for windowing. This advantage over the Fourier transform is achieved through the use of basis functions that are localised in both time and frequency—*wavelets*—rather than the infinitely repeating sines and cosines of the Fourier transform which have no localisation in time.

Some readers may be aware that the wavelet transform comes in two versions: the *continuous wavelet transform* (CWT) that we will use in this book and a *discrete wavelet transform*. But unlike the continuous and discrete Fourier transforms, where the words 'continuous' and 'discrete' refer to the data (whether they are represented by a continuous signal or a discrete sequence), when used with the wavelet

J. Kirby, *Spectral Methods for the Estimation of the Effective Elastic Thickness of the Lithosphere*, Advances in Geophysical and Environmental Mechanics and Mathematics, https://doi.org/10.1007/978-3-031-10861-7_4

transform, these words refer to the basis functions, the wavelets. The distinction between discrete and continuous wavelets is that the former are *orthogonal*, while continuous wavelets are *non-orthogonal*. Orthogonality is discussed in the Appendix to Chap. 3, but, briefly, orthogonal functions use precisely the minimum amount of information needed to completely model a signal or sequence with no information loss. In contrast, non-orthogonal functions can also model a signal but they do so with an arbitrarily large amount of information, generating redundancy. This redundancy is not necessarily a bad thing as it can increase the detail in a model, but it does result in models with large file sizes.

So, to reflect the fact that the CWT is frequently applied to discrete data sequences but can also be applied to continuous signals, in this chapter, we will follow the bulk of the wavelet literature and use the notation $g(t)$, or $g(\mathbf{x})$, to represent the data being analysed, whether it be in the form of a continuous signal or a discrete sequence.

4.2 The 1D Continuous Wavelet Transform

As we have seen, the multitaper method is well-suited to spectral estimation because the Slepian tapers are specifically designed to minimise spectral leakage. However, in order to reveal the variations in spectral power of *non-stationary* signals, where the harmonic content of the signal may vary over time or space, one must adopt a moving window technique. And as noted in Sect. 3.5, the choice of window size can affect results.

The wavelet transform was developed—by Grossmann and Morlet (1984)—largely to provide time-varying power spectra without the need for windowing. Such a feat is achieved by convolving a signal, say $g(t)$, with a *wavelet*, $\psi(t)$ (or more accurately, with the complex conjugate of a scaled wavelet), to give *wavelet coefficients*, $\widetilde{G}(s, t)$, where

$$\widetilde{G}(s, t) = (g * \psi_{s,t'}^*)(t) , \tag{4.1}$$

or written in full form as

$$\widetilde{G}(s, t) = \frac{1}{\sqrt{s}} \int_{-\infty}^{\infty} g(t') \, \psi^*\left(\frac{t - t'}{s}\right) dt' ,$$

where t' is the translation parameter of the convolution, and s is the *scale* of the wavelet (Sect. 4.4). The wavelet coefficients contain the harmonic information of the signal as a function of time; that is, at a specific time t_0, the wavelet spectrum of the signal is $\widetilde{G}(s, t_0)$, with the frequency dependence of the wavelet spectrum being provided by the wavelet's scale.

The scaling of the wavelets is crucial to the wavelet transform. When convolved with a signal, large-scale wavelets reveal low-frequency harmonics, while small-scale wavelets reveal high-frequency harmonics. This is seen in Fig. 4.1, where a signal (in the left-hand column) is convolved with four scaled wavelets (the middle

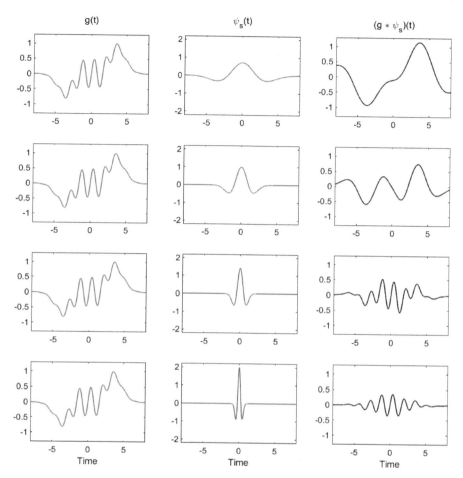

Fig. 4.1 Schematic diagram of the wavelet transform of a function, $g(t)$ (first column), as time-domain convolution with a Derivative of Gaussian (DoG) wavelet, $\psi_s(t)$ (second column), at four scales, with the largest scale in the top row, and the smallest in the bottom row. The third column shows the wavelet coefficients, $\widetilde{G}(s, t) = (g * \psi_s^*)(t)$, at the corresponding scale. See also Fig. 2.6

column), giving the wavelet coefficients at the four scales (right-hand column). The large-scale wavelet (top row) yields just the low-frequency, long-period harmonics of the signal. As the wavelet scale decreases, the harmonic content of the wavelet coefficients becomes higher frequency, until the smallest-scale wavelet (bottom row) yields just the highest-frequency, shortest-period harmonics. Hence, it can be seen that the wavelet transform is a form of a band-pass filter.

The wavelet transform is able to express spectra as a function of time because, first, its basis functions, the wavelets, are localised in time; that is, their energy is concentrated over a finite duration and they only 'sample' a finite sub-set of the signal. Second, the translation of the wavelet over the signal, via the convolution integral, enables a region-by-region analysis of the signal. Then, the variation of scale at each time observation allows the signal's frequency characteristics to be revealed. Hence, the wavelet coefficients, $\widetilde{G}(s,t)$, provide the spectrum at each time location. In contrast, the basis functions of the Fourier transform are the sines and cosines of the complex exponential, Eq. 2.6, which are of infinite extent and certainly not time-localised. Hence, in the Fourier transform, there is no information about the time location of particular harmonics, and the Fourier spectrum, $G(f)$, is a function of frequency only.

As noted, the wavelet transform is a convolution of a signal with the complex conjugate of a scaled wavelet. Unfortunately, convolution is a computationally demanding and time-consuming procedure. A more efficient way to compute the convolution is to make use of the convolution theorem (Sect. 2.2.8) and its statement that performing convolution in the time domain is the same as taking the inverse Fourier transform of a product of Fourier transforms. Thus, Eq. 4.1 can be written as

$$\widetilde{G}(s,t) \;=\; \mathsf{F}^{-1}\big\{G(f)\,\Psi_s^*(f)\big\}\;, \tag{4.2}$$

where $G(f)$ is the Fourier transform of the signal, and $\Psi_s(f)$ is the Fourier transform of a scaled wavelet.

4.3 Continuous Wavelets

Not every function can be used as a wavelet, even if it is time-localised. Here we will investigate what constitutes a wavelet, specifically a continuous wavelet.

4.3.1 Properties of Continuous Wavelets

Wavelets are often defined in both time and frequency domains, though some wavelets only have a time-domain formula, and others can only be written in the frequency domain. In the notation used in this book, $\psi(t)$ is the wavelet function in the time domain, while $\Psi(f)$ is the wavelet's frequency-domain representation. These two hence form a Fourier transform pair:

$$\psi(t) \;\longleftrightarrow\; \Psi(f)\;.$$

Functions must fulfil three mathematical properties in order to be classified as 'wavelets' and used in a wavelet transform:

1. Wavelets must satisfy the *admissibility condition*:

$$\int_{-\infty}^{\infty} \frac{|\Psi(f)|^2}{|f|} \, df \ < \ \infty \,. \tag{4.3}$$

Equation 4.3 is only satisfied when

$$\Psi(0) = 0 \,.$$

If a function's Fourier transform is zero at the zero frequency ('DC value', Sect. 2.4.3), then the function has a zero mean value, or

$$\int_{-\infty}^{\infty} \psi(t) \, dt \ = \ 0 \,.$$

Thus, wavelets must have a zero mean value.
2. Wavelets must have finite energy:

$$\int_{-\infty}^{\infty} |\psi(t)|^2 dt \ = \ \int_{-\infty}^{\infty} |\Psi(f)|^2 df \ < \ \infty \,,$$

where the equality arises from Parseval's theorem (Sect. 2.2.5). In other words, they are *localised* in both time and frequency domains.
3. In addition, wavelets should fulfil a third condition if one needs to study the behaviour of the Mth derivative of a signal:

$$\int_{-\infty}^{\infty} t^m \, \psi(t) \, dt \ = \ 0 \,.$$

That is, they should have m vanishing moments, where $0 \le m \le M$.

4.3.2 The Derivative of Gaussian Wavelet

The family of wavelets obtained by differentiating a Gaussian function is called *derivative of Gaussian* (DoG) wavelets. In the time domain, they are given by

$$\psi(t) \ = \ (-1)^{m/2} \frac{d^m}{dt^m} \left(e^{-t^2/2} \right) \,,$$

where m is the 'order' of the wavelet, a positive integer. DoG wavelets are real-valued if the order m is even, and purely imaginary if m is odd, though the latter have limited applicability. One of the most commonly used is the second-order wavelet, often called the 'Mexican hat' wavelet, and shown in Fig. 4.2. In the time domain, its equation is

$$\psi(t) \ = \ (1 - t^2) \, e^{-t^2/2} \,.$$

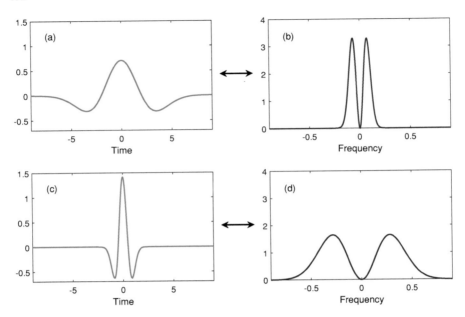

Fig. 4.2 Second-order ($m = 2$) 1D derivative of Gaussian (DoG) wavelets in the **a** and **c** time domain, and **b** and **d** frequency domain. **a** and **b** show large-scale wavelets, while **c** and **d** show small-scale wavelets. The double-headed arrows indicate Fourier transform pairs

In the frequency domain, the DoG wavelets are given by

$$\Psi(f) = (2\pi f)^m e^{-2\pi^2 f^2} .$$

The order of this wavelet governs its time- and frequency-domain behaviour. Increasing the order improves the frequency-domain resolution (the wavelets become narrower and more localised in this domain), but degrades its resolution in the time domain where the wavelets become broader and have more oscillations. Decreasing m has the opposite effect.

For our purposes, (even-order) DoG wavelets have limited applicability because they are real-valued, and in order to generate the phase information needed to compute a coherency, complex-valued wavelet coefficients are essential. Nevertheless, even though they will not be used further in this book, DoG wavelets are mentioned to highlight the differences between real and complex wavelets.

4.3.3 The 1D Morlet Wavelet

The Morlet wavelet is a Gaussian-modulated complex exponential and thus is a complex-valued wavelet. In the time domain, its form is

$$\psi(t) = \left(e^{2\pi i f_0 t} - e^{-2\pi^2 f_0^2}\right) e^{-t^2/2}, \tag{4.4}$$

with Fourier transform

$$\Psi(f) = e^{-2\pi^2(f-f_0)^2} - e^{-2\pi^2(f^2+f_0^2)}, \tag{4.5}$$

which may also be written as

$$\Psi(f) = e^{-2\pi^2(f^2+f_0^2)} \left(e^{4\pi^2 f_0 f} - 1\right). \tag{4.6}$$

The second, negative terms in Eqs. 4.4 and 4.5 are correction terms to ensure that the wavelet satisfies the admissibility condition of having zero mean value (Eq. 4.3). The Morlet wavelet in both domains is shown in Fig. 4.3.

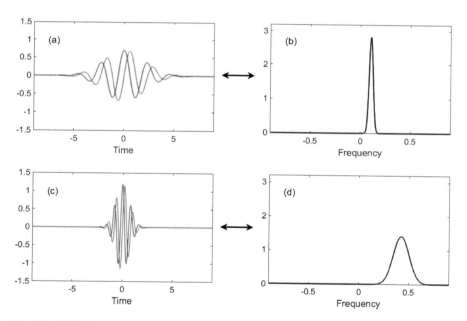

Fig. 4.3 1D Morlet wavelets of $f_0 = 0.849$ in the **a** and **c** time domain (with real part red, and imaginary part green lines), and **b** and **d** frequency domain. **a** and **b** show large-scale wavelets, while **c** and **d** show small-scale wavelets. The double-headed arrows indicate Fourier transform pairs

The parameter that controls the Morlet wavelet's time- and frequency-domain resolution (akin to the order, m, of the DoG wavelet) is the *central frequency*, f_0, which can adopt any positive real number value, $0 < f_0 < \infty$. A commonly used value for the central frequency is $f_0 \approx 0.849$, the significance of which will be explained below. When the central frequency has values $f_0 \geq 0.849$, the correction term in Eq. 4.4 has a value $e^{-2\pi^2 f_0^2} \leq 6 \times 10^{-7}$. This is considered small enough to be negligible, and such Morlet wavelets are often represented by their *simple* forms, being

$$\psi(t) = e^{2\pi i f_0 t} e^{-t^2/2} \qquad (4.7)$$

in the time domain, and

$$\Psi(f) = e^{-2\pi^2 (f - f_0)^2} \qquad (4.8)$$

in the frequency domain. To distinguish between the full form of the Morlet wavelet, given by Eqs. 4.4 and 4.5, and the simple forms, the former will often be referred to as the *complete* Morlet wavelet.

One can use the simple wavelet to understand the central frequency, whose values are related to the ratio between the amplitude of the first sidelobe of the real part of the Morlet wavelet and the amplitude of its central peak (Fig. 4.4). From the real part of Eq. 4.7, this sidelobe occurs at a time such that $2\pi f_0 t = 2\pi$, or $t = 1/f_0$. At this time, the Gaussian envelope has an amplitude a fraction q ($0 < q < 1$) of its maximum amplitude (which is 1), or, again from Eq. 4.7,

$$e^{-1/(2f_0^2)} = q ,$$

giving

$$f_0 = \sqrt{\frac{-1}{2 \ln q}} . \qquad (4.9)$$

Commonly chosen values of f_0 and their relationship with q are shown in Table 4.1. For instance, when the first sidelobe of the real part of the wavelet is a sixteenth of the amplitude of the central peak, then f_0 must have a value of 0.425.

Figure 4.4 illustrates the effect that the central frequency value has upon time- and frequency-domain resolution. The Morlet wavelet in Fig. 4.4a, b has a central frequency of $f_0 = 1.201$ and a scale of 4 s. It is broad in the time domain and narrow in the frequency domain; its time resolution is hence poor, but its frequency resolution good. In contrast Fig. 4.4c, d shows a Morlet wavelet with $f_0 = 0.425$, but in order for this wavelet to resolve the same frequency as the $f_0 = 1.201$ wavelet (i.e. have a frequency-domain peak at the same frequency), it needs to have a smaller scale (1.41 s). And at this scale, the $f_0 = 0.425$ wavelet is narrower in the time domain and broader in the frequency domain than the $f_0 = 1.201$ wavelet; its time resolution is better, but its frequency resolution worse.

So, if one wants to prioritise time resolution (at the expense of frequency resolution), enabling a more accurate time location of features in the spectrum, then one

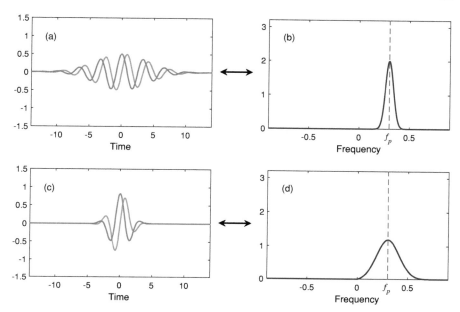

Fig. 4.4 As Fig. 4.3, but now **a** and **b** show 1D Morlet wavelets with a central frequency of $f_0 = 1.201$ ($q = 1/\sqrt{2}$) at a scale of 4 s. **c** and **d** show wavelets with a central frequency of $f_0 = 0.425$ ($q = 1/16$) at a scale of 1.41 ($\sqrt{2}$) s. Both wavelets have the same peak frequency, f_p

Table 4.1 Values of the central frequency, f_0, in relation to q, the ratio between the amplitude of the first sidelobe of the real part of the Morlet wavelet, and the amplitude of its central peak. Also shown are the corresponding values of the *central wavenumber*, $k_0 = 2\pi f_0$, which will be used when considering the 2D spatially dependent Morlet wavelet (Sect. 4.11)

q	f_0	k_0
$2^{-1/4}$	1.699	10.673
$\frac{1}{\sqrt{2}}$	1.201	7.547
$\frac{1}{2}$	0.849	5.336
$\frac{1}{4}$	0.601	3.773
$\frac{1}{8}$	0.490	3.081
$\frac{1}{16}$	0.425	2.668

chooses smaller f_0 wavelets. But if one is more interested in nailing the frequency of spectral features, then larger values of f_0 are in order. In the wavelet literature, the choice $f_0 = 0.849$ is often made as giving a reasonable resolution in both time and frequency domains, with the bonus that the simple, rather than complete, Morlet wavelet can be used.

The attractiveness of the Morlet wavelet lies in its ability to almost replicate the Fourier-derived power spectrum, especially for wavelets with a larger central frequency value (Kirby 2005). This feature arises because the wavelets are, in essence, sine and cosine waves like the Fourier basis functions. Furthermore, they are complex wavelets, generating complex-valued wavelet coefficients from which quantities like the wavelet coherency and phase can be derived. Their complex form in the time domain implies that their Fourier transforms are asymmetric (Fig. 4.3b, d), unlike real-valued wavelets, such as the even-order DoG wavelets, which are even functions in both time and frequency domains (Fig. 4.2). The asymmetry of the Morlet wavelets makes them *directional*, a property that becomes highly useful when they are generalised to two dimensions (Sect. 4.11).

4.4   Wavelet Scales

The utility of the wavelet transform lies in the scaling of the wavelets. As shown in Figs. 4.1, 4.2 and 4.3, the scale determines the width ('dilation') of the wavelet in the time domain, and hence determines the resolution of the wavelet transform. Large-scale wavelets are broad and span most of the signal when convolved with it; they hence reveal long wavelengths in the data. Small-scale wavelets are comparatively narrow and pick out the signal's short wavelengths.

Wavelets are defined algebraically by a *mother wavelet*, $\psi(t)$, such as Eq. 4.7, the 1D Morlet mother wavelet. The wavelets used in the wavelet transform (Eq. 4.1) are scaled and translated versions of these, called *daughter wavelets*, $\psi_{s,t'}(t)$. The relationship between mother and daughter wavelets in the time domain is given by, for 1D wavelets,

$$\psi_{s,t'}(t) \; = \; \frac{1}{\sqrt{s}} \; \psi\left(\frac{t-t'}{s}\right) \, , \tag{4.10}$$

where s is the wavelet scale, and t' is the translation parameter in the convolution. For example, from Eqs. 4.4 and 4.10, the 1D daughter Morlet wavelet is

$$\psi_{s,t'}(t) \; = \; s^{-1/2} \left(e^{2\pi i f_0 (t-t')/s} - e^{-2\pi^2 f_0^2}\right) e^{-(t-t')^2/2s^2} \, . \tag{4.11}$$

In the frequency domain, the mother-daughter relationship is given by, again for 1D wavelets,

$$\Psi_{s,t'}(f) \; = \; \sqrt{s} \; \Psi(sf) \, e^{-2\pi i f t'} \, ,$$

where the exponential term is the Fourier-domain representation of the translation (the 'shift theorem', Sect. 2.2.4) and does not affect the amplitude of the wavelet, only its phase. Therefore, daughter wavelets in the frequency domain are usually written without it as

$$\Psi_s(f) \; = \; \sqrt{s} \; \Psi(sf) \, . \tag{4.12}$$

For example, the 1D daughter Morlet wavelet is

$$\Psi_s(f) = \sqrt{s}\left(e^{-2\pi^2(sf-f_0)^2} - e^{-2\pi^2(s^2f^2+f_0^2)}\right), \tag{4.13}$$

from Eqs. 4.5 and 4.12.

The scaling of the wavelets in the frequency domain has the opposite effect on their shape to the scaling in the time domain, as can be seen in Figs. 4.2b, d and 4.3b, d. Large-scale wavelets are narrow functions in the frequency domain, while small-scale wavelets are relatively broad. This affects the resolution properties of wavelets. At large scales (low frequencies), wavelets tend to have a good frequency precision (because they are very localised, Fig. 4.3b), but a poor temporal precision (because they are very broad, Fig. 4.3a). In contrast, at small scales (high frequencies), they tend to have a poor frequency precision (because they are very broad, Fig. 4.3d), but a good temporal precision (because they are very localised, Fig. 4.3c). Mathematical support of these statements is given in Sect. 4.7.

The range of scales should be chosen to suit the signal being analysed, governed by the signal's bandwidth and Nyquist rate (Sect. 2.3.3). If, after sampling, the minimum and maximum wavelengths present in the sequence are λ_{min} and λ_{max}, respectively, then the corresponding minimum and maximum scales are given by, for the Morlet wavelet,

$$s_{min} = f_0\,\lambda_{min}, \qquad s_{max} = f_0\,\lambda_{max}, \tag{4.14}$$

where f_0 is the Morlet wavelet's central frequency. The relationships expressed in Eq. 4.14 will be derived in Sect. 4.6.

The wavelet scales are traditionally calculated according to an *octave decomposition*, that is, successive scales double in value. In practice, the octave may be subdivided into δs *voices*, being the number of scales per octave, or the number of scales it takes to double the frequency. Typically, $4 \le \delta s \le 8$ is chosen (e.g. Torrence and Compo 1998). In this fashion, the jth scale is given by

$$s_j = s_{min}\,2^{j/\delta s}, \tag{4.15}$$

where $j = 0, 1, \ldots, N_s - 1$, and N_s is the total number of scales, given by

$$N_s = 1 + \left\lfloor \delta s\,\log_2\left(\frac{s_{max}}{s_{min}}\right)\right\rfloor,$$

where $\lfloor A\rfloor$ is the 'floor' operator defined in Sect. 2.4.3.

4.5   Normalisation

Wavelets may be *normalised* to preserve either the energy or amplitude of the signal across the wavelet transform. The choice of normalisation depends upon the signal's bandwidth, with energy normalisation being more appropriate for broadband (especially fractal) signals, while amplitude normalisation is more suited to narrowband signals (Ridsdill-Smith 2000). Here we choose energy normalisation, as our signals—gravity and topography—are fractal.[1]

4.5.1   Time-Domain Normalisation

Energy normalisation requires that the wavelet has unit energy at each scale, or

$$\int_{-\infty}^{\infty} |\psi_{s,t'}(t)|^2 \, dt = 1 . \tag{4.16}$$

Starting with the 1D daughter Morlet wavelet, Eq. 4.11, if we introduce a constant term, c_1, to ensure Eq. 4.16 is satisfied, we get

$$\psi_{s,0}(t) = c_1 \, s^{-1/2} \left(e^{2\pi i f_0 t/s} - e^{-2\pi^2 f_0^2} \right) e^{-t^2/2s^2} .$$

Without loss of generality, we have set $t' = 0$ because the integral is made over an infinite region and the particular location of the Morlet wavelet will not matter. The aim here is to determine c_1. We can then show that

$$|\psi_{s,0}(t)|^2 = c_1^2 \, s^{-1} \left[1 + e^{-4\pi^2 f_0^2} - 2 \, e^{-2\pi^2 f_0^2} \cos(2\pi f_0 t/s) \right] e^{-t^2/s^2} ,$$

and can now write Eq. 4.16 as

$$c_1^{-2} = s^{-1} \left(1 + e^{-4\pi^2 f_0^2} \right) \int_{-\infty}^{\infty} e^{-t^2/s^2} \, dt$$
$$- 2s^{-1} e^{-2\pi^2 f_0^2} \int_{-\infty}^{\infty} \cos(2\pi f_0 t/s) \, e^{-t^2/s^2} \, dt .$$

The integrals are solved using the identities Eqs. 4.60a and 4.60b, giving, after some algebra,

$$c_1^{-2} = \sqrt{\pi} \left(1 - 2 \, e^{-3\pi^2 f_0^2} + e^{-4\pi^2 f_0^2} \right) ,$$

---

[1] A fractal signal has a power spectrum that decays with frequency as $f^{-\beta}$, where β is the spectral exponent—see Sect. 10.2 for further information.

and

$$c_1 = \pi^{-1/4} \left(1 - 2\,e^{-3\pi^2 f_0^2} + e^{-4\pi^2 f_0^2} \right)^{-1/2},$$

which is independent of scale. The terms with the exponentials are small. When $f_0 = 0.3$, we have $c_1 = 1.0603\,\pi^{-1/4}$, but when $f_0 = 0.849$, the two exponential terms sum to $< 10^{-8}$, so it is common to write

$$c_1 \approx \pi^{-1/4}. \tag{4.17}$$

4.5.2 Frequency-Domain Normalisation

A similar procedure is followed to ensure that the frequency-domain wavelets also have unit energy, or

$$\int_{-\infty}^{\infty} |\Psi_s(f)|^2\, df = 1. \tag{4.18}$$

For the 1D Morlet daughter wavelet, Eq. 4.13, we can show that

$$|\Psi_s(f)|^2 = c_2^2\, s\, e^{-4\pi^2 f_0^2} \left(e^{-4\pi^2(s^2 f^2 - 2 s f_0 f)} - 2\,e^{-4\pi^2(s^2 f^2 - s f_0 f)} + e^{-4\pi^2 s^2 f^2} \right),$$

and can now write Eq. 4.18 as

$$c_2^{-2} = s\,e^{-4\pi^2 f_0^2} \left\{ \int_{-\infty}^{\infty} e^{-4\pi^2 s^2 f^2 + 8\pi^2 s f_0 f}\, df \right.$$
$$- 2 \int_{-\infty}^{\infty} e^{-4\pi^2 s^2 f^2 + 4\pi^2 s f_0 f}\, df$$
$$\left. + \int_{-\infty}^{\infty} e^{-4\pi^2 s^2 f^2}\, df \right\}.$$

The integrals are solved using the identity Eq. 4.60a, giving, after some rearrangement,

$$c_2^{-2} = \frac{1}{2\sqrt{\pi}} \left(1 - 2\,e^{-3\pi^2 f_0^2} + e^{-4\pi^2 f_0^2} \right),$$

or

$$c_2 = \pi^{1/4} \sqrt{2} \left(1 - 2\,e^{-3\pi^2 f_0^2} + e^{-4\pi^2 f_0^2} \right)^{-1/2}.$$

As with time-domain normalisation, if the exponential terms are neglected, we find

$$c_2 \approx \pi^{1/4} \sqrt{2}. \tag{4.19}$$

4.5.3 *Practical Normalisation*

The normalisation constants above were computed assuming wavelets of infinite extent. In practice, however, the finite size of the data sequence means that the wavelet's energy is not constant over scales, especially for small-scale wavelets. Thus, when applied to discretely sampled sequences, Eq. 4.16 becomes

$$\sum_t \left| \psi_{s,t'}(t) \right|^2 = 1$$

and the normalised time-domain wavelet is given by

$$\overline{\psi_{s,t'}(t)} = \psi_{s,t'}(t) \sqrt{\frac{1}{\sum_t |\psi_{s,t'}(t)|^2}} .$$

If one performs the wavelet transform in the frequency domain, Eq. 4.18 becomes

$$\sum_f \left| \Psi_s(f) \right|^2 = N$$

where N is the number of data records in the sequence being analysed. The normalised frequency-domain wavelet is given by

$$\overline{\Psi_s(f)} = \Psi_s(f) \sqrt{\frac{N}{\sum_f |\Psi_s(f)|^2}} .$$

4.6 Equivalent Fourier Frequency

Figures 4.2b, d and 4.3b, d show that wavelets are essentially band-pass filters, suppressing harmonics that do not fall within the range of the curve of the frequency-domain wavelet. Therefore, it is useful to be able to relate the wavelet scale to harmonic frequency so that one can identify the harmonics—as would be calculated by a Fourier-based method—that are filtered out.

There are several methods available to perform the mapping from scale to an *equivalent Fourier frequency*, f_e. The simplest approach is to take the frequency at which the daughter wavelet has a maximum value as being representative of the passband. This is computed from the location in the frequency domain where the gradient of the Fourier transform of the scaled wavelet is zero (ignoring the trivial case of a possible zero gradient at the zero frequency). This is called the *peak frequency* method.

Another widely used method is to find the wavelet power spectrum of a sine wave of known frequency; the scale where the power spectrum is a maximum corresponds

Fig. 4.5 The power spectrum of a 1D Morlet wavelet, i.e. $\Psi_s^2(f)$. The peak frequency is f_p, shown at the vertical dashed line. The thin, solid, vertical lines show the extents $f_p \pm \sigma_\Psi$

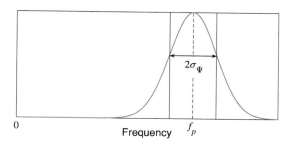

to the sine wave's frequency and gives the equivalent Fourier frequency for the wavelet. This is called the *maximum-power frequency* method. However, as this method is more appropriate when the signal has a narrow bandwidth, and gravity field and topography are invariably broadband signals, it will not be used in this work.

The peak frequency of a wavelet is the frequency of its turning point in the frequency domain, calculated from

$$\frac{\partial \Psi_s(f)}{\partial f}\bigg|_{f_p} = 0 \,,$$

where f_p is the peak frequency, as shown in Fig. 4.5. For example, take the normalised daughter Morlet wavelet, Eq. 4.13:

$$\Psi_s(f) = \pi^{1/4}\sqrt{2s}\left(e^{-2\pi^2(sf-f_0)^2} - e^{-2\pi^2(s^2f^2+f_0^2)}\right)$$

(including the normalisation constant, Eq. 4.19). Its derivative can be shown to be

$$\frac{\partial \Psi_s(f)}{\partial f} = 4\pi^2\pi^{1/4}s\sqrt{2s}\left[sf\left(e^{-4\pi^2sf_0f} - 1\right) + f_0\right]e^{-2\pi^2(sf-f_0)^2}$$

which is zero—at frequency f_p—when

$$sf_p\left(e^{-4\pi^2sf_0f_p} - 1\right) + f_0 = 0$$

or

$$f_p\left(1 - e^{-4\pi^2sf_0f_p}\right) = \frac{f_0}{s} \,. \tag{4.20}$$

Equation 4.20 must be solved for f_p by a root-finding algorithm such as the Newton-Raphson method.[2] Iterative solutions such as Newton-Raphson need an initial

---

[2] If one has an equation $g(x) = 0$ and wishes to find the values of x that satisfy this equation, one can use an iterative approach. First, make an estimate, x_1. A better estimate of the solution is

estimate, which can be provided by the peak frequency of the simple Morlet wavelet, whose normalised daughter wavelet is, from Eqs. 4.8, 4.12 and 4.19,

$$\Psi_s(f) = \pi^{1/4}\sqrt{2s}\, e^{-2\pi^2(sf-f_0)^2} \, .$$

Differentiating this gives

$$\frac{\partial \Psi_s(f)}{\partial f} = 4\pi^2 \pi^{1/4} s \sqrt{2s}\, (f_0 - sf)\, e^{-2\pi^2(sf-f_0)^2} \, ,$$

which is zero at the peak frequency, or

$$f_0 - sf_p = 0 \, .$$

Thus, the peak frequency f_p of the simple Morlet wavelet is

$$f_p = \frac{f_0}{s} \, , \tag{4.21}$$

which can safely be used as the peak frequency of the complete Morlet wavelet even when f_0 is relatively small, as shown below.

When the Newton-Raphson method is employed to solve Eq. 4.20 for f_p, if Eq. 4.21 is taken as the first estimate in the iteration, the next estimate can be shown to be

$$f_p = \frac{f_0}{s}\left[1 + \left(e^{4\pi^2 f_0^2} + 4\pi^2 f_0^2 - 1\right)^{-1}\right] \, .$$

The fractional error made in using Eq. 4.21 as an approximation to Eq. 4.20 is thus given by the term

$$\left(e^{4\pi^2 f_0^2} + 4\pi^2 f_0^2 - 1\right)^{-1} \, ,$$

which is independent of scale, and, for most values of f_0, negligible. For example, when $f_0 = 0.418$, the error is 0.1% but decreases sharply as f_0 increases.

To recap, here we use the peak frequency of the Morlet wavelet as its equivalent Fourier frequency, or

$$f_e = f_p \, . \tag{4.22}$$

Furthermore, the wavelet scale can also be related to an *equivalent Fourier wavelength*, λ_e, found from

$$\lambda_e = \frac{1}{f_e} = \frac{s}{f_0} \tag{4.23}$$

for the Morlet wavelet, where we have used Eq. 4.21.

---

given by $x_2 = x_1 - g(x_1)/g'(x_1)$, where $g'(x)$ is the derivative of $g(x)$. The procedure can then be iterated to find x_3, and so on. This is the Newton-Raphson method.

4.7 Wavelet Resolution

It is useful to be able to find expressions for the resolution capabilities of wavelets in both time and frequency domains, so that we can be confident (or not) about elastic thickness estimates (which depend on wavenumber resolution), or the spatial variations of elastic thickness (which depend on spatial resolution). Resolution can be measured in several ways (Percival and Walden 1993; Kirby and Swain 2008, 2011), but here we will derive expressions for the variance of the Morlet wavelet, which then gives the standard deviation, a measure of precision.

We saw in Sect. 4.4 how large-scale wavelets have a good frequency localisation but a poor temporal localisation, while small-scale wavelets have a poor frequency localisation but a good temporal localisation. These properties are in line with the 'uncertainty relation' of signal processing (Sect. 2.2.6). For wavelets, this relation is mathematically represented by

$$\sigma_\psi \, \sigma_\Psi \; \geq \; \frac{1}{4\pi} \, , \tag{4.24}$$

where σ_ψ is the standard deviation of the wavelet in the time domain, and σ_Ψ is the standard deviation in the frequency domain (cf. Eq. 2.11). Thus there exists an inverse proportionality between temporal and frequency precision with all wavelets. The standard deviations in Eq. 4.24 are calculated from the variances of the wavelet equations. However, since ψ is a complex function, say $g(u)$ for some dummy variable u, its variance is calculated from its squared modulus rather than the function itself, using

$$\sigma_g^2 \; = \; \int_{-\infty}^{\infty} (u - u_0)^2 \, |g(u)|^2 \, du \tag{4.25}$$

(Percival and Walden 1993; Antoine et al. 2004). The standard deviation is the positive square root of the variance and is used to quantify the precision. The parameter u_0 is the centre of gravity of the function, or mean location on the u-axis. Without loss of generality, we can choose $u_0 = 0$ in Eq. 4.25, which will be used to calculate the precision of the simple Morlet wavelet in both time and frequency domains.

4.7.1 Time-Domain Resolution

Here we will derive an expression for the standard deviation of the 1D simple Morlet wavelet in the time domain. Forming a daughter wavelet from Eqs. 4.7 and 4.10, we can set $t' = 0$ because the integral is made over an infinite region and the particular location of the Morlet wavelet will not affect the result. Including the normalisation constant, Eq. 4.17, we can show that

$$|\psi_{s,0}(t)|^2 \; = \; \pi^{-1/2} \, s^{-1} \, e^{-t^2/s^2}$$

at scale s, giving the variance integral, Eq. 4.25, as

$$\sigma_\psi^2 = \pi^{-1/2} s^{-1} \int_{-\infty}^{\infty} t^2 e^{-t^2/s^2} \, dt .$$

The integral is solved using the identity Eq. 4.60c, and we find

$$\sigma_\psi^2 = \frac{s^2}{2} .$$

Therefore, the precision (standard deviation) of the simple Morlet wavelet in the time domain is given by

$$\sigma_\psi = \frac{s}{\sqrt{2}} , \qquad (4.26)$$

which is scale-dependent, as expected.

It is also useful to express Eq. 4.26 in terms of the (equivalent Fourier) frequency (f_e). If we substitute Eqs. 4.21 and 4.22 into Eq. 4.26, we find that the standard deviation of the 1D Morlet wavelet is

$$\sigma_\psi = \frac{f_0}{f_e \sqrt{2}} . \qquad (4.27)$$

Equation 4.27 tells us that Morlet wavelets with small central frequencies (f_0) give better time-domain precision (small standard deviation) than do large-f_0 wavelets. It also tells us that spectral features with high (equivalent Fourier) frequencies are better located in the time domain than are low-frequency features, facts already established qualitatively in Sect. 4.4.

4.7.2 Frequency-Domain Resolution

In the frequency domain, the square of the normalised simple 1D daughter Morlet wavelet is, from Eqs. 4.8, 4.12 and 4.19,

$$\Psi_s^2(f) = 2s \sqrt{\pi} e^{-4\pi^2 s^2 f^2} ,$$

where we have set the central frequency $f_0 = 0$ because we want the variance of the Gaussian about the peak frequency (see Fig. 4.5). The variance integral, Eq. 4.25, is thus

$$\sigma_\psi^2 = 2s \sqrt{\pi} \int_{-\infty}^{\infty} f^2 e^{-4\pi^2 s^2 f^2} \, df .$$

Using the identity Eq. 4.60c to solve the integral, we arrive at

$$\sigma_\psi^2 = \frac{1}{8\pi^2 s^2} \, .$$

Therefore, the precision of the simple Morlet wavelet in the frequency domain is given by

$$\sigma_\psi = \frac{1}{2\pi s \sqrt{2}} \, , \tag{4.28}$$

which is scale-dependent, as expected.

Again, it is useful to express Eq. 4.28 in terms of the (equivalent Fourier) frequency (f_e). If we substitute Eqs. 4.21 and 4.22 into Eq. 4.28, we find that the standard deviation of the 1D Morlet wavelet is

$$\sigma_\psi = \frac{1}{2\pi\sqrt{2}} \frac{f_e}{f_0} \, . \tag{4.29}$$

Equation 4.29 tells us that Morlet wavelets with large central frequencies (f_0) give better frequency-domain precision (small standard deviation) than do small-f_0 wavelets. It also tells us that spectral features with low (equivalent Fourier) frequencies are better resolved than are high-frequency features, facts already established qualitatively in Sect. 4.4.

For interest, we can now test the uncertainty relationship, Eq. 4.24. From Eqs. 4.26 and 4.28, we obtain

$$\sigma_\psi \sigma_\psi = \frac{s}{\sqrt{2}} \frac{1}{2\pi s \sqrt{2}} = \frac{1}{4\pi} \, ,$$

which obeys Eq. 4.24.

4.8   Wavelet Power Spectra

The wavelet transform gives time-varying power spectra without the need for windowing, unlike Fourier-based methods (Sect. 3.5). Such spectra are called *local*, but the wavelet transform can also provide *global* spectra, which are time-averaged local spectra and comparable to the classic Fourier power spectrum.

4.8.1   *Local Scalograms*

The wavelet coefficients $\widetilde{G}(s, t)$, as computed by Eqs. 4.1 or 4.2, are most often visually displayed as a wavelet power spectrum, or *scalogram*, especially if they are complex since the real and imaginary parts are combined into a single real-valued quantity, the power. The scalogram of a signal $g(t)$ is computed from

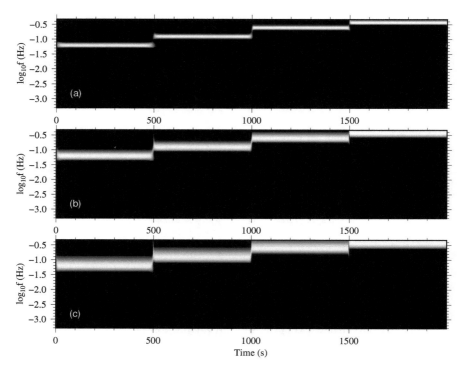

Fig. 4.6 Morlet wavelet scalograms as functions of equivalent Fourier frequency, at three values of the central frequency, f_0: **a** 1.201, **b** 0.601 and **c** 0.425. The signal is a cosine with a frequency of 0.0625 Hz from 0–499 s, 0.125 Hz from 500–999 s, 0.25 Hz from 1000–1499 s and 0.375 Hz from 1500–1999 s. The grey scale is power in decibels, with light shades being high power, and dark shades low power. Note that the wavelets were amplitude, rather than energy, normalised since the signal is narrowband (see Sect. 4.5). [Cf. Fig. 3.9.]

$$S(s, t) = \left| \widetilde{G}(s, t) \right|^2 . \tag{4.30}$$

Scale may be converted to equivalent Fourier frequency using Eq. 4.21 for the Morlet wavelet. Examples of the scalograms of two time series are shown in Figs. 4.6 and 4.7. The signal in Fig. 4.6 is a cosine function with a frequency that changes four times over its duration, getting higher at each step (see figure caption). Immediately apparent is that the spectral peaks are frequency-localised much better with the larger-f_0 wavelet (Fig. 4.6a) than with the smaller-f_0 wavelet (Fig. 4.6c), as expected (Sect. 4.7). Contrary to expectations, though, the larger-f_0 wavelet also appears to do a reasonable job of time-localising the frequency changes in the signal, though it must be appreciated that the cosine's harmonics do not have long periods compared to the length of the signal (the signal's duration is 2000 s, while the longest-period cosine is in the first segment with a period of 16 s). We would expect low-frequency harmonics to have relatively poor time-localisation (Eq. 4.27), but it must be said

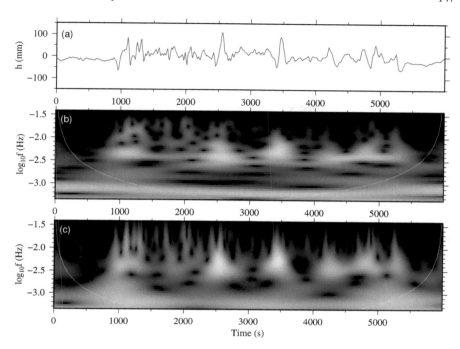

Fig. 4.7 A time series of height variations (**a**), and its Morlet wavelet scalograms as functions of equivalent Fourier frequency, at two values of the central frequency, f_0: **b** 0.849 and **c** 0.425. The grey scale is $\log_{10}$ of power, with light shades being high power, and dark shades low power. The light blue curves in **b** and **c** mark the limit of the cones of influence, calculated from Eq. 4.33

that 0.0625 Hz is not really that low compared to the fundamental frequency of the signal (0.0005 Hz), so its time resolution is not poor.

Figure 4.7 shows the time-localisation properties rather better. The high-power features in the lower-f_0 scalogram (Fig. 4.7c) are narrower and more well-defined in time than for the higher-f_0 scalogram in Fig. 4.7b, and in both images, the high-frequency harmonics are more time-localised than are the low-frequency harmonics. The reverse is true in the frequency domain. The scalogram with the higher f_0 value, Fig. 4.7b, has a better frequency resolution than that with the lower f_0 value, Fig. 4.7c: the high-power features in Fig. 4.7b are more concentrated and well-defined in the frequency direction than they are in Fig. 4.7c where they are smeared, particularly at the high frequencies.

4.8.2 Heisenberg Boxes

The time- and frequency-domain properties of the wavelet scalogram and windowed Fourier transform (WFT) spectrogram (Sect. 3.5) can be explained and compared schematically using *Heisenberg boxes*. Figure 4.8 shows the time-frequency plane tiling of Heisenberg boxes for the wavelet transform, and two versions of the WFT, one where the frequency resolution is prioritised, and the other where time resolution is prioritised. Heisenberg boxes are a manifestation of the reciprocal relationship between temporal and frequency precision expressed in the uncertainty relation, Eqs. 2.11 and 4.24, reproduced here as

$$\sigma_t \, \sigma_f \; \geq \; \frac{1}{4\pi} \, ,$$

where σ_t is the standard deviation in the time coordinate, and σ_f is the standard deviation in the frequency coordinate. Thus, for a given Heisenberg box, the longer the side length, the less precise the function in that domain. Note that the area of the boxes must obey the uncertainty relation: boxes cannot be shrunk on both sides to become vanishingly small. If one side shrinks (increasing precision in that domain), the other side must expand (decreasing precision in the other domain).

In Sects. 4.4 and 4.7, we established that wavelets always have a good frequency resolution at low frequencies (large-scale wavelets), but a poor one at high frequencies (small-scale wavelets); accordingly, they always have a poor time resolution at low frequencies (large-scale wavelets), but a good one at high frequencies (small-scale wavelets). Thus, their Heisenberg boxes always appear as shown in Fig. 4.8a, where the low-frequency (large-scale) boxes are extended in the time axis, and the high-frequency (small-scale) boxes are extended in the frequency axis. Note that the frequency-axis length of the boxes in Fig. 4.8a doubles with increasing frequency, reflecting the octave decomposition choice of scale in Eq. 4.15. It should also be appreciated that scalogram plots like Figs. 4.6 and 4.7 typically plot frequency logarithmically which causes the high frequencies to become compressed relative to the low frequencies. This means that, in Fig. 4.6, the spectral peaks (the white bars) all appear to have the same 'thickness', or the same frequency precision. If the frequency axis were linear rather than logarithmic, then the white bars would get thicker with increasing frequency, reflecting the reality.

In contrast, the resolution of the WFT is determined by the chosen—and then fixed—window size, which means that the shape of the Heisenberg boxes does not change over the time-frequency plane. In Fig. 4.8b, the WFT window size is chosen such that the WFT has a better resolution/precision in the frequency coordinate than in the time coordinate: the boxes are extended in the time direction for all frequencies. In Fig. 4.8c, the WFT window size is chosen such that the WFT has a better resolution/precision in the time coordinate than in the frequency coordinate: the boxes are extended in the frequency direction at all frequencies. These phenomena are seen clearly in Fig. 3.9. For a given window size, the dimensions of the spectral peaks (the white bars) are independent of frequency. The large window (Fig. 3.9a) has a good frequency precision but a poor time precision at all frequencies, and vice versa for the small window (Fig. 3.9c), as already discussed in Sect. 3.5.

So the difference between the wavelet transform and WFT is that the WFT Heisenberg boxes always preserve their shape, while the wavelet boxes adjust to frequency.

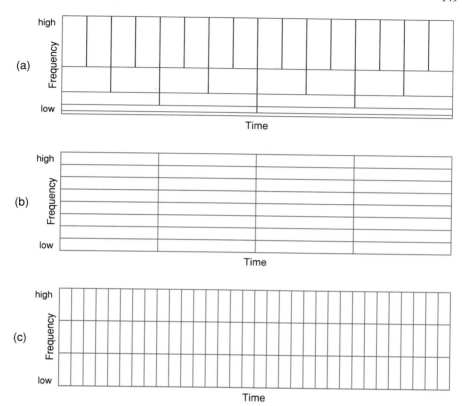

Fig. 4.8 Tiling of the time-frequency plane for **a** the wavelet transform, **b** the windowed Fourier transform with preferential frequency resolution and **c** the windowed Fourier transform with preferential time resolution. Each rectangle is a Heisenberg box, with dimensions $2\sigma_t$ along the time axis, and $2\sigma_f$ along the frequency axis

In other words, the frequency resolution of the WFT is independent of frequency, while the frequency resolution of the wavelet transform is better at low frequencies than at high. Alternatively, the time resolution of the WFT is also independent of frequency, while the time resolution of the wavelet transform is better at high frequencies than at low. This property of the wavelet transform has been viewed as being more 'natural', if one would expect short-wavelength features to be better located than long-wavelength ones, thinking about it in terms of a space, rather than time, coordinate.

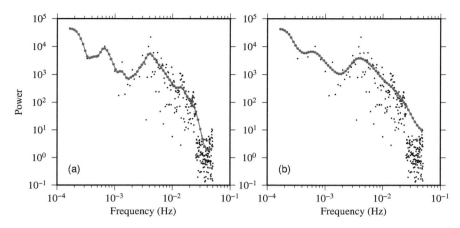

Fig. 4.9 Morlet wavelet global scalograms (red circles and lines) of the time series in Fig. 4.7a, at two values of the central frequency, $f_0 = $ **a** 0.849 and **b** 0.425. Also plotted as black dots are the periodogram estimates

4.8.3   Global Scalograms

The wavelet transform was devised in order to be able to estimate the power spectrum as a function of time and frequency—rather than just frequency as with the Fourier power spectrum—so it might seem rather pointless to compute a time-averaged scalogram, or *global scalogram*. Nevertheless, many studies have done so (Torrence and Compo 1998; Kirby and Swain 2013), and such a quantity is computed from

$$S(s) \; = \; \frac{1}{N_t} \sum_t S(s, t) \, ,$$

where N_t is the number of records in the time series and $S(s, t)$ is the scalogram from Eq. 4.30. Scale may be converted to equivalent Fourier frequency using Eq. 4.21. The global scalograms of the data in Fig. 4.7a are shown in Fig. 4.9. Again, the better frequency resolution of the larger-f_0 wavelet is evident, with the spectrum of the smaller-f_0 wavelet being much smoothed and lacking detail.

4.9   Cone of Influence (CoI)

The *cone of influence* (CoI) defines a region of the time-frequency space where the wavelet coefficients are potentially contaminated by edge effects due to discontinuities at the data boundaries. However, if the edges exhibit only minor-amplitude discontinuities—for example, naturally or due to a procedure such as mirroring (Sect. 2.5.6)—then the wavelet coefficients might not be altered.

The half-width of the CoI (t_c) is given by the time difference from the peak of the time-domain daughter wavelet to the point at which the wavelet (or its envelope) decays to a fraction q_{coi} of its peak amplitude ($0 < q_{coi} < 1$). Thus,

$$\psi_{s,0}(t_c) = q_{coi} \, \psi_{s,0}(0) \tag{4.31}$$

where the peak is taken to occur at time $t = 0$, and without loss of generality, we can consider a daughter wavelet with no translation ($t' = 0$). Some studies take $q_{coi} = e^{-1}$ and call it the *e-folding time*, but other values are accepted, e.g. $q_{coi} = 0.5$ (Kirby and Swain 2013).

The CoI of the 1D simple Morlet wavelet is readily calculated. Take the normalised simple daughter wavelet, obtained from Eqs. 4.7, 4.12 and 4.17, at $t' = 0$:

$$\psi_{s,0}(t) = \pi^{-1/4} \, s^{-1/2} \, e^{2\pi i f_0 t/s} \, e^{-t^2/2s^2} \; .$$

At its peak, we have

$$\psi_{s,0}(0) = \pi^{-1/4} \, s^{-1/2} \; .$$

Therefore, from Eq. 4.31, we find that

$$\pi^{-1/4} \, s^{-1/2} \, e^{2\pi i f_0 t_c/s} \, e^{-t_c^2/2s^2} = q_{coi} \, \pi^{-1/4} \, s^{-1/2} \; .$$

So, just focussing on the Gaussian envelope, we see that

$$e^{-t_c^2/2s^2} = q_{coi} \; ,$$

which gives the CoI as

$$t_c = s \sqrt{-2 \ln q_{coi}}$$

at a scale s. Thus, the CoI is proportional to the wavelet scale. If we choose $q_{coi} = e^{-1}$, then we find the e-folding time of the Morlet wavelet is

$$t_c = s \sqrt{2} \; . \tag{4.32}$$

If we need to represent the CoI as a function of equivalent Fourier frequency (f_e), then we substitute Eq. 4.21, giving

$$t_c = \frac{f_0 \sqrt{2}}{f_e} \; . \tag{4.33}$$

Examples of the CoI are shown in Fig. 4.7, though because the two ends of the time series have approximately equal values, the edge effects are minimal and contamination of the wavelet coefficients across the scalograms is negligible.

4.10 The 2D Continuous Wavelet Transform

As noted in Sect. 2.6, much of the data used in modelling the spatial variations of effective elastic thickness are 2D and of course functions of space rather than time. Thus, as outlined in Sects. 2.6.1 and 2.6.6, we will proceed by swapping 1D frequency (f) with 2D wavenumber ($\mathbf{k}$), remembering that $k = 2\pi f$ in the 1D case. We will also coordinate the data with the position vector $\mathbf{x} = (x, y)$ in the space domain, and the wavenumber vector $\mathbf{k} = (k_x, k_y)$ in the spatial frequency domain, where k_x is the wavenumber in the x-direction, and k_y is the wavenumber in the y-direction.

Much of the theory of the 2D CWT and 2D wavelets is common with the 1D cases, so the reader jumping straight to this section is encouraged to jump back to the beginning of this chapter.

The 2D continuous wavelet transform is defined similarly to the 1D CWT, but now, being 2D functions, the wavelets can also be *rotated*, as well as scaled and translated. Thus, the 2D CWT is the convolution of the signal, $g(\mathbf{x})$, with a set of scaled and rotated wavelets, giving wavelet coefficients

$$\widetilde{G}(s, \mathbf{x}, \alpha) \;=\; (g * \psi^*_{s, \mathbf{x}', \alpha})(\mathbf{x}) \;, \tag{4.34}$$

(cf. Eq. 4.1) or written in full as

$$\widetilde{G}(s, \mathbf{x}, \alpha) \;=\; s^{-1} \int_{\mathbb{R}^2} g(\mathbf{x}') \, \psi^* \left(\Omega(\alpha) \, \frac{\mathbf{x} - \mathbf{x}'}{s} \right) d^2\mathbf{x}' \;,$$

where $\mathbb{R}^2$ is the real number plane (see Sect. 2.6.4). Note that the scale, s, remains '1D'. The rotation property of 2D wavelets now enables features at different azimuths within the signal to be revealed. The rotation matrix, $\Omega(\alpha)$, for positive-anticlockwise rotations of the mother wavelet, belongs to the SO(2) Lie group, with

$$\Omega(\alpha) \;=\; \begin{pmatrix} \cos\alpha & \sin\alpha \\ -\sin\alpha & \cos\alpha \end{pmatrix}$$

(Farge 1992), where α is the rotation angle determining the resolving azimuth of the wavelet and is the positive-anticlockwise angle made with the positive x-axis. In two dimensions, the daughter wavelets, $\psi_{s, \mathbf{x}', \alpha}(\mathbf{x})$, are dilated, translated and rotated versions of the mother wavelet, $\psi(\mathbf{x})$, or

$$\psi_{s, \mathbf{x}', \alpha}(\mathbf{x}) \;=\; s^{-1} \psi \left(\Omega(\alpha) \, \frac{\mathbf{x} - \mathbf{x}'}{s} \right) \tag{4.35}$$

(cf. Eq. 4.10). However, not all wavelets are directional, such as the 2D version of the DoG wavelet (Sect. 4.3.2) which is isotropic and thus unaffected by rotation. As will be explained soon, we will not be concerned with isotropic wavelets in this book.

As with the 1D case—indeed perhaps more so owing to the larger volume of data—a more efficient way to compute the convolution in Eq. 4.34 is to use the Fourier transform, via

$$\widetilde{G}(s, \mathbf{x}, \alpha) = \mathsf{F}^{-1}\{G(\mathbf{k}) \, \Psi^*_{s,\alpha}(\mathbf{k})\} \tag{4.36}$$

(cf. Eq. 4.2), where $\Psi_{s,\alpha}(\mathbf{k})$ is the Fourier transform of a daughter wavelet. The wavenumber-domain relationship between mother and daughter wavelets is given by, for 2D wavelets,

$$\Psi_{s,\alpha}(\mathbf{k}) = s \, \Psi(s \, \Omega(\alpha) \, \mathbf{k}) \tag{4.37}$$

(cf. Eq. 4.12).

4.11 The 2D Morlet Wavelet

4.11.1 Governing Equations

The Morlet wavelet in two dimensions has very similar properties to its 1D version, but it has a directional dependence, given by its azimuth, α, the positive-anticlockwise angle made with the positive x-axis. In the space domain, its formula is

$$\psi_\alpha(\mathbf{x}) = \left(e^{i\mathbf{k}_0 \cdot \mathbf{x}} - e^{-|\mathbf{k}_0|^2/2}\right) e^{-|\mathbf{x}|^2/2} \tag{4.38}$$

(cf. Eq. 4.4), shown in Fig. 4.10. The *central wavenumber* vector is $\mathbf{k}_0 = (k_{x0}, k_{y0})$, where

$$k_{x0} = k_0 \cos \alpha \,, \qquad k_{y0} = k_0 \sin \alpha \,, \qquad k_0 \equiv |\mathbf{k}_0| = (k_{x0}^2 + k_{y0}^2)^{1/2} \,.$$

Values of the central wavenumber are calculated as for the central frequency of the 1D Morlet wavelet (Sect. 4.3.3), because that derivation depends on the Gaussian envelope, and the 2D envelope is just the rotation of the 1D envelope. Hence, taking into account the fact that we are now using wavenumber rather than frequency, so that $k_0 = 2\pi f_0$, Eq. 4.9 becomes

$$|\mathbf{k}_0| = \pi\sqrt{\frac{-2}{\ln q}} \,.$$

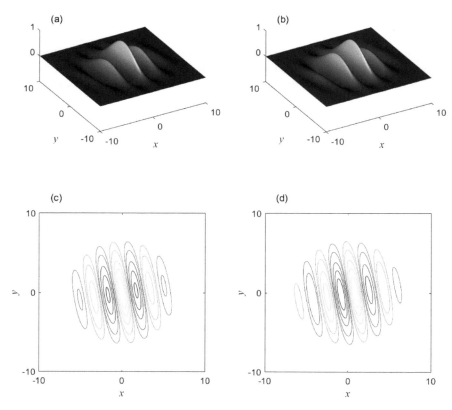

Fig. 4.10 A 2D Morlet wavelet of azimuth $\alpha = 10°$ in the space domain in **a** and **b** perspective, and **c** and **d** plan views. **a** and **c** show the real part of the wavelet, and **b** and **d** show its imaginary part

Commonly used values of $k_0 \equiv |\mathbf{k}_0|$ are given in Table 4.1. The similarity of the 1D and 2D envelopes also means the cone of influence of the 1D and 2D wavelets is identical, given by Eq. 4.32.

In the wavenumber domain, the 2D Morlet wavelet is given by

$$\Psi_\alpha(\mathbf{k}) = e^{-|\mathbf{k}-\mathbf{k}_0|^2/2} - e^{-(|\mathbf{k}|^2+|\mathbf{k}_0|^2)/2} \tag{4.39}$$

(cf. Eq. 4.5), shown in Fig. 4.11. The effect of azimuth upon the space- and wavenumber-domain wavelets is shown in Fig. 4.12. It is often useful to write Eq. 4.39 in polar coordinates, (k, θ). Letting $k_x = k \cos\theta$, $k_y = k \sin\theta$ and $k \equiv |\mathbf{k}| = (k_x^2 + k_y^2)^{1/2}$, Eq. 4.39 can be written as

$$\Psi_\alpha(k, \theta) = e^{-(k^2+k_0^2)/2} \left(e^{k_0 k \cos(\theta-\alpha)} - 1 \right). \tag{4.40}$$

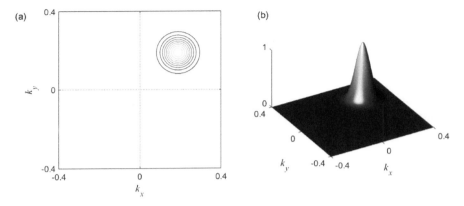

Fig. 4.11 A 2D Morlet wavelet of azimuth $\alpha = 45°$ in the wavenumber domain, in **a** plan and **b** perspective views. The axes are x and y wavenumber (rad km^{-1})

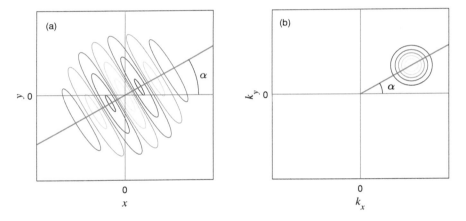

Fig. 4.12 A 2D Morlet wavelet of azimuth α in the **a** space (real part only plotted) and **b** wavenumber domains. The red lines show the azimuth of the wavelet

Note the similarity between Eqs. 4.40 and 4.6: if one substitutes $k = 2\pi f$ and $k_0 = 2\pi f_0$ into Eq. 4.40 and chooses $\theta = \alpha$, then one obtains Eq. 4.6. That is, the line from the origin along the azimuth $\theta = \alpha$ through the 2D wavelet (Fig. 4.12b) gives the 1D wavelet.

Indeed, it is this similarity that means that many of the properties of the 1D Morlet wavelet can be extended to the 2D wavelet. So not only are the equations for the central wavenumber and cone of influence similar, as noted above, but so is the equation for the peak wavenumber, k_p, which we use as the equivalent Fourier wavenumber, k_e, or

$$k_e = k_p = \frac{|\mathbf{k}_0|}{s} \tag{4.41}$$

(cf. Eqs. 4.21 and 4.22), while the equivalent Fourier wavelength, λ_e, for the Morlet wavelet is found from

$$\lambda_e = \frac{2\pi}{k_e} = \frac{2\pi s}{|\mathbf{k}_0|} \tag{4.42}$$

(cf. Eq. 4.23).

The simple 2D Morlet wavelet is also constructed in the same way as the 1D version, without the correction terms in Eqs. 4.38 and 4.39. In the space domain, the simple wavelet is given by

$$\psi_\alpha(\mathbf{x}) = e^{i\mathbf{k}_0 \cdot \mathbf{x}} \, e^{-|\mathbf{x}|^2/2} \,, \tag{4.43}$$

while in the wavenumber domain, its form is

$$\Psi_\alpha(\mathbf{k}) = e^{-|\mathbf{k} - \mathbf{k}_0|^2/2} \,. \tag{4.44}$$

Other equations describing properties of the 2D Morlet wavelet have slight differences to the 1D versions, and these will be re-derived, following.

4.11.2   2D Normalisation

Normalisation formulae of 1D wavelets are based upon the wavelet's energy being unity in the two domains, and in the 1D case, this is expressed by a calculation of the area under the wavelet curve (Sect. 4.5). In two dimensions, the area under a curve becomes a volume under a surface, leading to slight differences between 1D and 2D normalisation formulae.

Space-Domain Normalisation

The energy normalisation requirement of unit energy at each scale implies that, in the space domain

$$\int_{\mathbb{R}^2} |\psi_{s,\mathbf{x}',\alpha}(\mathbf{x})|^2 \, d^2\mathbf{x} = 1 \,. \tag{4.45}$$

As in the 1D case, we introduce a constant term, c_1, to the 2D Morlet wavelet, Eq. 4.38, and set $\mathbf{x}' = (0, 0)$ without loss of generality because the integral is made over an infinite region and the particular location of the Morlet wavelet will not affect the result. Applying Eq. 4.35 gives the squared daughter wavelet as

$$|\psi_{s,0,\alpha}(\mathbf{x})|^2 = c_1^2 \, s^{-2} \left[1 + e^{-|\mathbf{k}_0|^2} - 2\, e^{-|\mathbf{k}_0|^2/2} \cos(\mathbf{k}_0 \cdot \mathbf{x}/s) \right] e^{-|\mathbf{x}|^2/s^2} \,.$$

We can now write Eq. 4.45 as

$$c_1^{-2} = s^{-2}\left(1 + e^{-|\mathbf{k}_0|^2}\right) \int_{\mathbb{R}^2} e^{-|\mathbf{x}|^2/s^2} \, d^2\mathbf{x}$$

$$- 2s^{-2}e^{-|\mathbf{k}_0|^2/2} \int_{\mathbb{R}^2} \cos(\mathbf{k}_0 \cdot \mathbf{x}/s) \, e^{-|\mathbf{x}|^2/s^2} \, d^2\mathbf{x}$$

$$\equiv s^{-2}\left(1 + e^{-|\mathbf{k}_0|^2}\right) I_1 - 2s^{-2}e^{-|\mathbf{k}_0|^2/2} I_2 . \tag{4.46}$$

The first integral, I_1, is solved by transforming to polar coordinates (r, θ). Letting $|\mathbf{x}| \equiv r = (x^2 + y^2)^{1/2}$, $x = r \cos \theta$ and $y = r \sin \theta$, we get

$$I_1 = \int_0^{2\pi} \int_0^{\infty} r \, e^{-r^2/s^2} \, dr \, d\theta$$

(Stewart 1999). The r-integral on the right-hand side is solved using the identity Eq. 4.60d, giving

$$I_1 = \frac{2\pi s^2}{2} = \pi s^2 .$$

To solve the second integral, I_2, we can choose, without loss of generality, $\alpha = 0°$ because the integral is made over an infinite region, and the orientation of the Morlet wavelet will not affect the result. This means that

$$\mathbf{k}_0 \cdot \mathbf{x} = |\mathbf{k}_0| x \cos \alpha + |\mathbf{k}_0| y \sin \alpha = |\mathbf{k}_0| x .$$

Hence

$$I_2 = \int_{\mathbb{R}^2} \cos(|\mathbf{k}_0| x/s) \, e^{-|\mathbf{x}|^2/s^2} \, d^2\mathbf{x}$$

$$= \left(\int_{-\infty}^{\infty} \cos(|\mathbf{k}_0| x/s) \, e^{-x^2/s^2} \, dx\right) \left(\int_{-\infty}^{\infty} e^{-y^2/s^2} \, dy\right)$$

using Fubini's theorem (Stewart 1999). Then using the identities Eqs. 4.60a and 4.60b, we find

$$I_2 = \pi s^2 e^{-|\mathbf{k}_0|^2/4} .$$

Hence, Eq. 4.46 now becomes

$$c_1^{-2} = \pi \left(1 - 2e^{-3|\mathbf{k}_0|^2/4} + e^{-|\mathbf{k}_0|^2}\right) ,$$

giving the time-domain normalisation constant as

$$c_1 = \pi^{-1/2} \left(1 - 2e^{-3|\mathbf{k}_0|^2/4} + e^{-|\mathbf{k}_0|^2}\right)^{-1/2} .$$

As in the 1D case, it is common to neglect the exponential terms and write

$$c_1 \approx \pi^{-1/2} .$$

(4.47)

Wavenumber-Domain Normalisation

The volume under the wavelet surface should have unit energy, or

$$\int_{\mathbb{R}^2} |\Psi_{s,\alpha}(\mathbf{k})|^2 \, d^2\mathbf{k} = 1 .$$

(4.48)

Expanding Eq. 4.39 and forming its daughter wavelet using Eq. 4.37 gives, when squared,

$$|\Psi_{s,\alpha}(\mathbf{k})|^2 = c_2^2 \, s^2 \, e^{-|\mathbf{k}_0|^2} \left(e^{-s^2 k_x^2 + 2s k_x k_{x0}} \, e^{-s^2 k_y^2 + 2s k_y k_{y0}} \right.$$
$$\left. - 2 \, e^{-s^2 k_x^2 + s k_x k_{x0}} \, e^{-s^2 k_y^2 + s k_y k_{y0}} + e^{-s^2 k_x^2} \, e^{-s^2 k_y^2} \right).$$

We can now write Eq. 4.48 as

$$c_2^{-2} = s^2 \, e^{-|\mathbf{k}_0|^2} \left\{ \int_{-\infty}^{\infty} \int_{-\infty}^{\infty} e^{-s^2 k_x^2 + 2s k_x k_{x0}} \, e^{-s^2 k_y^2 + 2s k_y k_{y0}} \, dk_x \, dk_y \right.$$
$$- 2 \int_{-\infty}^{\infty} \int_{-\infty}^{\infty} e^{-s^2 k_x^2 + s k_x k_{x0}} \, e^{-s^2 k_y^2 + s k_y k_{y0}} \, dk_x \, dk_y$$
$$\left. + \int_{-\infty}^{\infty} \int_{-\infty}^{\infty} e^{-s^2 k_x^2} \, e^{-s^2 k_y^2} \, dk_x \, dk_y \right\}$$
$$\equiv s^2 \, e^{-|\mathbf{k}_0|^2} \, (I_1 - 2 I_2 + I_3) .$$

(4.49)

In the first integral, I_1, the k_x and k_y integrals can be separated using Fubini's theorem (Stewart 1999), giving

$$I_1 = \left(\int_{-\infty}^{\infty} e^{-s^2 k_x^2 + 2s k_x k_{x0}} \, dk_x \right) \left(\int_{-\infty}^{\infty} e^{-s^2 k_y^2 + 2s k_y k_{y0}} \, dk_y \right)$$

each of which can be solved using the identity Eq. 4.60a, giving

$$I_1 = \frac{\pi}{s^2} \, e^{|\mathbf{k}_0|^2} .$$

The second and third integrals, I_2 and I_3, are solved in similar fashions, giving

$$I_2 = \frac{\pi}{s^2} \, e^{|\mathbf{k}_0|^2/4} ,$$

and

$$I_3 = \frac{\pi}{s^2} .$$

Substituting I_1, I_2 and I_3 into Eq. 4.49 gives

$$c_2^{-2} = \pi \left(1 - 2e^{-3|\mathbf{k}_0|^2/4} + e^{-|\mathbf{k}_0|^2} \right)$$

and thus

$$c_2 = \pi^{-1/2} \left(1 - 2e^{-3|\mathbf{k}_0|^2/4} + e^{-|\mathbf{k}_0|^2} \right)^{-1/2} .$$

Again, we can neglect the exponential terms and write

$$c_2 \approx \pi^{-1/2} . \tag{4.50}$$

Practical Normalisation

As discussed in Sect. 4.5, the finite size of the data sequence requires a practical implementation of normalisation on a computer. In the space domain, this takes the form

$$\overline{\psi_{s,\mathbf{x}',\alpha}(\mathbf{x})} = \psi_{s,\mathbf{x}',\alpha}(\mathbf{x}) \sqrt{\frac{1}{\sum_{\mathbf{x}} |\psi_{s,\mathbf{x}',\alpha}(\mathbf{x})|^2}} ,$$

while in the wavenumber domain, the normalisation is given by

$$\overline{\Psi_{s,\alpha}(\mathbf{k})} = \Psi_{s,\alpha}(\mathbf{k}) \sqrt{\frac{N_x N_y}{\sum_{\mathbf{k}} |\Psi_{s,\alpha}(\mathbf{k})|^2}} ,$$

where N_x and N_y are the number of records in each dimension of the 2D data sequence.

4.11.3 2D Resolution

For a 2D complex-valued function, $g(u, v)$ with a mean location in the (u, v)-plane of $(0, 0)$, its variance is given by

$$\sigma_g^2 = \int_{-\infty}^{\infty} \int_{-\infty}^{\infty} (u^2 + v^2) |g(u, v)|^2 \, du \, dv \tag{4.51}$$

(cf. Eq. 4.25).

Space-Domain Resolution

Here we will derive an expression for the standard deviation of the 2D simple daughter Morlet wavelet in the space domain. We begin with Eq. 4.43, form the daughter wavelet using Eq. 4.35 and set $\mathbf{x}' = (0, 0)$ as in Sect. 4.11.2. Including the normalisation constant, Eq. 4.47, we can show that

$$|\psi_{s,0,\alpha}(\mathbf{x})|^2 = \pi^{-1} s^{-2} e^{-|\mathbf{x}|^2/s^2}$$

at scale s, giving the variance integral, Eq. 4.51, as

$$\sigma_\psi^2 = \pi^{-1} s^{-2} \int_{-\infty}^{\infty} \int_{-\infty}^{\infty} (x^2 + y^2) e^{-(x^2+y^2)/s^2} \, dx \, dy \, .$$

The integral may be solved by transforming to polar coordinates. Letting $|\mathbf{x}| \equiv r = (x^2 + y^2)^{1/2}$, $x = r \cos \theta$ and $y = r \sin \theta$, it becomes

$$\sigma_\psi^2 = \pi^{-1} s^{-2} \int_0^{2\pi} \int_0^{\infty} r^3 e^{-r^2/s^2} \, dr \, d\theta$$

(Stewart 1999). The r-integral is solved using the identity Eq. 4.60d, giving

$$\sigma_\psi^2 = s^2.$$

Therefore, the precision (standard deviation) of the simple 2D Morlet wavelet in the space domain is given by

$$\sigma_\psi = s$$

(cf. Eq. 4.26).

Wavenumber-Domain Resolution

In the wavenumber domain, the square of the normalised simple Morlet wavelet, Eq. 4.44, is, setting $k_{x0} = k_{y0} = 0$ as we did in Sect. 4.7.2,

$$|\Psi_{s,\alpha}(\mathbf{k})|^2 = \pi^{-1} s^2 e^{-s^2|\mathbf{k}|^2} \, ,$$

using Eq. 4.37 to form its daughter wavelet and including the normalisation constant, Eq. 4.50. The variance integral, Eq. 4.51, is thus

$$\sigma_\psi^2 = \pi^{-1} s^2 \int_{-\infty}^{\infty} \int_{-\infty}^{\infty} (k_x^2 + k_y^2) e^{-s^2(k_x^2+k_y^2)} \, dk_x \, dk_y \, .$$

The integral may be solved by transforming to polar coordinates. Letting $|\mathbf{k}| \equiv k = (k_x^2 + k_y^2)^{1/2}$, $k_x = k \cos\theta$ and $k_y = k \sin\theta$, it becomes

$$\sigma_\psi^2 = \pi^{-1} s^2 \int_0^{2\pi} \int_0^\infty k^3 e^{-s^2 k^2} \, dk \, d\theta$$

(Stewart 1999). The k-integral is solved using the identity Eq. 4.60d, giving

$$\sigma_\psi^2 = s^{-2} .$$

Therefore, the precision (standard deviation) of the simple 2D Morlet wavelet in the wavenumber domain is given by

$$\sigma_\psi = s^{-1} \tag{4.52}$$

(cf. Eq. 4.28).

4.12 The Fan Wavelet Method

The fan wavelet is a superposition of 2D Morlet wavelets of differing azimuth. However, this wavelet per se is not used to compute the admittance or coherency, but rather we construct a fan-like superposition of 2D Morlet wavelet auto- and cross-scalograms. Nevertheless, this section will begin with a discussion of the fan wavelet and its particular geometry.

4.12.1 The Fan Wavelet

As will be seen in Chap. 5, the 2D coherency requires spectra with complex harmonic coefficients for its computation. And unless we specifically require directional coherency, then the harmonic coefficients should be isotropic. These two requirements—complex yet isotropic coefficients—are satisfied by Fourier coefficients (whether multitapered or otherwise), but not by wavelet coefficients, at least in a straightforward manner. On the one hand, the 2D Mexican hat wavelet, for example, is an isotropic wavelet, but it is real-valued, yielding real harmonic (wavelet) coefficients. On the other hand, the 2D Morlet wavelet is complex—yielding complex harmonic coefficients—yet it is anisotropic.

The fan wavelet was developed by Kirby and Swain (2004) and Kirby (2005) to give wavelet coefficients that are both complex and isotropic, so that a directionless coherence may be formed. The key lies in the recognition that isotropy can be achieved by performing the analyses over an azimuthal range of 180° rather than

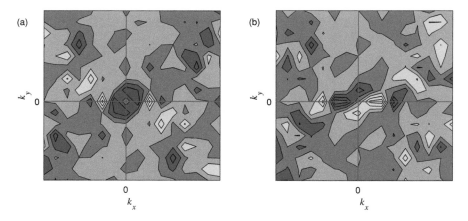

Fig. 4.13 The Fourier transform of some 2D real-valued data. The transform is a Hermitian function, with an even real part (**a**) and an odd imaginary part (**b**). The colour scale range is ± 1 units, with pinks positive and blues negative

360° as might be expected. If one is only concerned with orientations and not with directions—as we are here—then north and south, or north-east and south-west, are identical orientations. This reduces the number of azimuths to analyse by one-half.

Consider the plots in Fig. 4.13, which show the real and imaginary parts of the Fourier transform of some 2D real-valued data. As noted in Sects. 2.2.3 and 3.6.3, asymmetric real-valued functions have Hermitian Fourier transforms (Bracewell 1986), with an even real part and an odd imaginary part. This means that all the essential information about the Fourier transform is contained in just two of the four quadrants in Fig. 4.13; the information contained in the other two quadrants is redundant. Hence, the lower two quadrants of the real part (Fig. 4.13a) are merely reflected and flipped versions of the upper two quadrants. The same applies for the imaginary part, except the lower two quadrants have their values further multiplied by -1. Here we choose the upper two quadrants (defined by $k_y \geq 0$) on which to perform the analysis, but any other two adjacent quadrants would suffice (e.g. the two defined by $k_x \geq 0$).

Thus, complex, isotropic wavelet coefficients can be obtained by averaging the wavelet transforms from a complex, directional wavelet (such as the 2D Morlet) over 180°. Kirby (2005) also showed that if the range of azimuths extends beyond 180° up to 360°, then the resulting wavelet coefficients lose power in the imaginary component and eventually become real-valued.

Figure 4.14 shows two geometries of fan-like superpositions of Morlet wavelets, in the wavenumber domain. The geometry in Fig. 4.14a is called the 'minimum curvature' superposition because the sum of the wavelets produces a surface whose crest is flat. The geometry in Fig. 4.14b is calculated under the assumption that the azimuthally adjacent Morlet wavelets are almost orthogonal (in the sense described

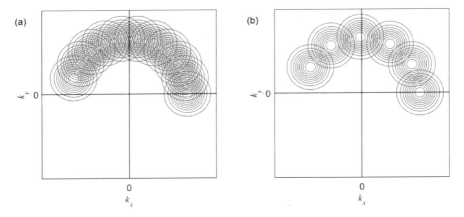

Fig. 4.14 Showing the superposition of Morlet wavelets in the fan geometry, in the wavenumber domain. The Morlet wavelets have central wavenumber $|\mathbf{k}_0| = 5.336$. **a** Minimum curvature superposition ($q_\alpha = 0.75$ in Eq. 4.53) of 11 Morlet wavelets with $\delta\alpha \approx 16.3°$. **b** Orthogonal superposition ($q_\alpha = e^{-1}$ in Eq. 4.53) of 6 Morlet wavelets with $\delta\alpha \approx 30.5°$

in Sect. 4.1), although of course Morlet wavelets can never be truly orthogonal as they are continuous, and not discrete, wavelets (see also Sect. 5.5.1).

An expression for the angular increment ($\delta\alpha$) between adjacent Morlet wavelets in the fan can be derived from the polar-coordinate equation for the complete Morlet wavelet in the wavenumber domain, Eq. 4.40, in the following manner. First, without loss of generality, we can choose a daughter wavelet with an azimuth $\alpha = 0°$, with equation

$$\Psi_{0°}(k, \theta) \;=\; s\, e^{-(s^2 k^2 + k_0^2)/2} \left(e^{s k_0 k \cos\theta} \;-\; 1 \right).$$

We then find an expression for the value of the wavelet along the locus of the peak value, which will occur at wavenumber $k = k_p = k_0/s$ (Eq. 4.41) for any angle θ. This is given by

$$\Psi_{0°}(k_p, \theta) \;=\; s\, e^{k_0^2(\cos\theta - 1)} \;-\; s\, e^{-k_0^2}.$$

At some angle $\theta = \theta_{q_\alpha}$ along the locus, the wavelet has a value of q_α ($0 < q_\alpha < 1$) times its maximum value (which is s). Thus, we have

$$e^{k_0^2(\cos\theta_{q_\alpha} - 1)} \;-\; e^{-k_0^2} \;=\; q_\alpha,$$

which we can readily solve for θ_{q_α}. But since the angular separation between the peaks of adjacent Morlet wavelets overlapping at θ_{q_α} is $\delta\alpha = 2\theta_{q_\alpha}$, we find

$$\delta\alpha \;=\; 2\cos^{-1}\left[1 + |\mathbf{k}_0|^{-2} \ln\left(q_\alpha + e^{-|\mathbf{k}_0|^2} \right) \right] \tag{4.53}$$

(in radians), where we replace k_0 with $|\mathbf{k}_0|$. Note that Kirby and Swain (2013) provided the solution for the simple Morlet wavelet, which does not have the $e^{-|\mathbf{k}_0|^2}$ term. Kirby and Swain (2013) also showed that $q_\alpha = e^{-1}$ gives the 'orthogonal' superposition, while $q_\alpha = 0.75$ gives the minimum curvature superposition. Note that Eq. 4.53 gives equivalent results to Eq. (18) of Kirby (2005).

The number of Morlet wavelets constituting the fan (N_α) is then calculated from

$$N_\alpha = \text{nint}\left(\frac{\Delta\alpha}{\delta\alpha}\right) , \tag{4.54}$$

where nint indicates the operation of taking the nearest integer, and $\Delta\alpha$ is the total azimuthal extent of the fan, usually taken as $\Delta\alpha = 180°$ for isotropy. Note, though, that the Morlet wavelets in Fig. 4.14 do not appear to completely span $180°$. This is designed so because of the Hermitian nature of the Fourier transform; if there were a wavelet with azimuth $180°$, it would 'cancel out' the imaginary components of the wavelet with azimuth $0°$ (Fig. 4.13b).

Due to limitations on the accuracy of Eq. 4.41—which is derived from the simple, rather than complete, Morlet wavelet—Kirby and Swain (2011) suggested restricting values of the central wavenumber to $|\mathbf{k}_0| \geq 2.668$.

4.12.2 Fan Wavelet Transform

The fan wavelet itself is not used in T_e estimation. This is because the admittance and coherency formulae (Chap. 5) require averages of several estimates of the auto- and cross-spectra of two signals, and the fan wavelet created from the sum of the Morlet wavelet superpositions in Fig. 4.14 is but a single wavelet and cannot be used to form averages. This can be shown by considering a 'fan wavelet transform'. Working in the Fourier, rather than space, domain, the isotropic fan wavelet is given by the superposition

$$\Psi_s^F(\mathbf{k}) = \frac{1}{N_\alpha} \sum_{j=1}^{N_\alpha} \Psi_{s,\alpha_j}^M(\mathbf{k}) , \tag{4.55}$$

where $\Psi_{s,\alpha_j}^M(\mathbf{k})$ is the 2D Morlet wavelet at azimuth α_j given by Eq. 4.40, and N_α is chosen such that the superposition is isotropic. Fan wavelet coefficients of a signal $g(\mathbf{x})$ would then be given by Eq. 4.36 as

$$\widetilde{G}^F(s, \mathbf{x}) = \mathsf{F}^{-1}\{G(\mathbf{k})\,\Psi_s^{F*}(\mathbf{k})\} . \tag{4.56}$$

If we substitute Eq. 4.55 into Eq. 4.56 and interchange the ordering of Fourier transformation and summation (because the Fourier transform is a linear operator), we find

$$\widetilde{G}^F(s, \mathbf{x}) = \frac{1}{N_\alpha} \sum_{j=1}^{N_\alpha} \mathsf{F}^{-1}\left\{ G(\mathbf{k})\, \Psi_{s,\alpha_j}^{M\,*}(\mathbf{k}) \right\} ,$$

or

$$\widetilde{G}^F(s, \mathbf{x}) = \frac{1}{N_\alpha} \sum_{j=1}^{N_\alpha} \widetilde{G}^M(s, \mathbf{x}, \alpha_j) , \qquad (4.57)$$

where the $\widetilde{G}^M(s, \mathbf{x}, \alpha_j)$ are the Morlet wavelet coefficients at azimuth α_j. So the fan wavelet transform can be viewed as either the CWT of a signal with a fan wavelet (Eq. 4.56), or the superposition of a series of Morlet wavelet transforms in a fan geometry (Eq. 4.57).

The Morlet wavelet coefficients in Eq. 4.57 can be determined from either Eq. 4.34 using space-domain convolution, or Eq. 4.36 via the Fourier transform. Note, however, that Kirby and Swain (2013) found that when the CWT is computed using the Fourier transform, very large-scale wavelets are not represented faithfully upon the wavenumber grid, imparting errors due to misrepresented azimuths. Kirby and Swain (2013) recommend using space-domain convolution at such scales, even though this is computationally more expensive.

However, while the fan wavelet transform is useful for many applications, it is not for our purposes, and we must go further and look at how to best form a power spectrum.

4.12.3 Fan Wavelet Power Spectra

For most wavelets, a 2D scalogram is computed from the modulus-squared of the wavelet coefficients,

$$S(s, \mathbf{x}, \alpha) = \left| \widetilde{G}(s, \mathbf{x}, \alpha) \right|^2 ,$$

where the α term is present if the wavelet is anisotropic, and not present if the wavelet is isotropic. Thus, one could form a fan wavelet scalogram by taking the modulus-squared of the fan wavelet coefficients computed using the approach in Eq. 4.56, to get

$$S(s, \mathbf{x}) = \left| \widetilde{G}^F(s, \mathbf{x}) \right|^2 = \left| \mathsf{F}^{-1}\{ G(\mathbf{k})\, \Psi_s^{F\,*}(\mathbf{k}) \} \right|^2 .$$

However, such an approach does not lend itself to forming multiple estimates of the auto- and cross-spectra—and thence the averages required to compute the coherency and admittance—because there is only one estimate of the power spectrum.

Instead, the construction of the fan wavelet scalogram (and thus the auto- and cross-spectra that will be discussed in Sect. 5.4.2) is based on the approach in Eq. 4.57, but instead of averaging Morlet wavelet coefficients, we average a series of Morlet wavelet scalograms in a fan geometry, as in

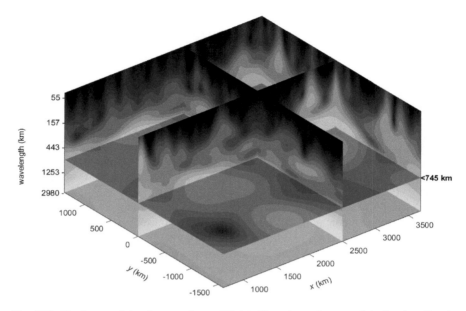

Fig. 4.15 The fan wavelet scalogram of some 2D data (free-air gravity anomaly), showing slices in the xs-, ys- and xy-planes. The wavelet scale has been converted to equivalent Fourier wavelength using Eq. 4.42. The colour scale is $\log_{10}$ of power, with light shades being high power and dark shades low power

$$S(s, \mathbf{x}) \;=\; \frac{1}{N_\alpha} \sum_{j=1}^{N_\alpha} \left| \widetilde{G}^M(s, \mathbf{x}, \alpha_j) \right|^2 . \tag{4.58}$$

Thus, the fan wavelet scalogram is obtained from the Morlet wavelet coefficients, rather than from the fan wavelet coefficients. An example of a fan wavelet scalogram is shown in Fig. 4.15.

If desired, a fan wavelet global scalogram of the signal is then determined by averaging the fan wavelet scalogram over the whole space domain, at each scale, or

$$S(s) \;=\; \frac{1}{N_{\mathbf{x}}} \sum_{\mathbf{x}} S(s, \mathbf{x}) , \tag{4.59}$$

where $N_{\mathbf{x}} = N_x N_y$, and N_x and N_y are the number of records in each dimension of the 2D data sequence.

4.13 Summary

The wavelet transform provides a power spectrum, or 'scalogram', as a function of both time and frequency, without the need to use moving windows. Many types and families of wavelets exist, but for the purposes of estimating elastic thickness, the Morlet wavelet is ideal. This continuous wavelet is a Gaussian-modulated complex exponential function and, of all wavelets, most closely replicates the Fourier-derived power spectrum, an essential property when needing to compare observed spectra with theoretical functions. The wavelet transform can be evaluated either through time-domain convolution of a signal with a wavelet or by multiplication of their Fourier transforms in the frequency domain, the latter method being faster to implement.

The 2D Morlet wavelet forms the basis of the fan wavelet transform. This was developed in order to obtain wavelet coefficients that were both isotropic and complex; usually isotropic wavelets are real-valued, while complex wavelets are anisotropic. The method also allows for the averaging required when computing the admittance and coherence between two signals.

4.14 Further Reading

The discussion on wavelets presented here is very much aimed at the reader who knows of their existence and purpose but lacks the finer details. It is intended to get such a reader up-to-speed with the fundamental mathematics supporting wavelet theory, without overdoing it and covering things that aren't pertinent, or getting too bogged down in the maths. The discussion is also tailored to the Morlet wavelet almost exclusively, as this is the only wavelet used in the fan wavelet T_e estimation method.

Therefore, for the wavelet novice wishing a broader treatment of wavelet transforms, the book by Addison (2017) and the journal article by Torrence and Compo (1998) cannot be bettered, in my opinion. If one then wishes to explore the mathematics more (but not too) robustly, then the book by Antoine et al. (2004) and the article by Farge (1992) are essential reading.

Finally, if the reader really wants to get his or her hands dirty with wavelets, then there exists a host of literature on the topic. These are mostly intended for mathematicians and scientists or engineers with a very strong mathematical background, but those I have found useful, to varying degrees, are the books by Chui (1992) and Percival and Walden (2000).

Appendix

Standard Definite Integrals

The following solutions of standard definite integrals are used in this chapter.

$$\int_{-\infty}^{\infty} e^{-a^2 x^2 + bx}\, dx = \frac{\sqrt{\pi}}{a}\, e^{b^2/4a^2} \tag{4.60a}$$

$$\int_{-\infty}^{\infty} \cos(b\, x)\, e^{-a^2 x^2}\, dx = \frac{\sqrt{\pi}}{a}\, e^{-b^2/4a^2} \tag{4.60b}$$

$$\int_{-\infty}^{\infty} x^2\, e^{-a^2 x^2}\, dx = \frac{\sqrt{\pi}}{2a^3} \tag{4.60c}$$

$$\int_{0}^{\infty} x^{2n+1}\, e^{-a^2 x^2}\, dx = \frac{n!}{2a^{2n+2}} \tag{4.60d}$$

for dummy variable x, integer $n = 0, 1, 2, \ldots$, and any real numbers $a > 0$ and b (Abramowitz and Stegun 1972).

References

Abramowitz M, Stegun IA (1972) Handbook of mathematical functions. Dover, New York

Addison PS (2017) The illustrated wavelet transform handbook, 2nd edn. CRC Press, Boca Raton

Antoine J-P, Murenzi R, Vandergheynst P, Ali ST (2004) Two-dimensional wavelets and their relatives. Cambridge University Press, Cambridge

Bracewell RN (1986) The Fourier transform and its applications. McGraw-Hill, New York

Chui CK (1992) An introduction to wavelets. Academic Press, San Diego

Farge M (1992) Wavelet transforms and their applications to turbulence. Annu Rev Fluid Mech 24:395–457

Grossmann A, Morlet J (1984) Decomposition of Hardy functions into square integrable wavelets of constant shape. SIAM J Math Anal 15:723–736

Kirby JF (2005) Which wavelet best reproduces the Fourier power spectrum? Comput Geosci 31:846–864

Kirby JF, Swain CJ (2004) Global and local isostatic coherence from the wavelet transform. Geophys Res Lett 31:L24608. https://doi.org/10.1029/2004GL021569

Kirby JF, Swain CJ (2008) An accuracy assessment of the fan wavelet coherence method for elastic thickness estimation. Geochem Geophys Geosyst 9:Q03022. https://doi.org/10.1029/2007GC001773, (Correction, Geochem Geophys Geosyst 9:Q05021. https://doi.org/10.1029/2008GC002071, 2008)

Kirby JF, Swain CJ (2011) Improving the spatial resolution of effective elastic thickness estimation with the fan wavelet transform. Comput Geosci 37:1345–1354

Kirby JF, Swain CJ (2013) Power spectral estimates using two-dimensional Morlet-fan wavelets with emphasis on the long wavelengths: jackknife errors, bandwidth resolution and orthogonality properties. Geophys J Int 194:78–99

Percival DP, Walden AT (1993) Spectral analysis for physical applications. Cambridge University Press, Cambridge

Percival DP, Walden AT (2000) Wavelet methods for time series analysis. Cambridge University Press, Cambridge

Ridsdill-Smith TA (2000) The application of the wavelet transform to the processing of aeromagnetic data. PhD thesis, University of Western Australia

Stewart J (1999) Calculus, 4th edn. Brooks/Cole

Torrence C, Compo GP (1998) A practical guide to wavelet analysis. Bull Am Meteorol Soc 79:61–78

Chapter 5
Admittance, Coherency and Coherence

5.1 Introduction

The admittance and coherency are frequency-domain measures of the strength of
the relationship between two signals, with the coherence being derived from the
coherency. In the context of effective elastic thickness (T_e) estimation, the two signals of particular interest are the gravity anomaly and the topography,[1] because, as
will be explained in Sect. 5.2, topography has a strong influence on gravity, and the
relationship between the two is highly dependent upon T_e. Indeed, T_e values can be
estimated by comparing the admittance or coherency observed from actual data with
predictions made by theoretical models of plate flexure, with T_e adjusted until there
is a good fit between observation and prediction.

Therefore it is crucial to be able to calculate the admittance and coherency accurately, because mistakes made in this procedure will result in incorrect estimates of
T_e being made. For instance, as we saw in Chap. 3, the periodogram gives a biased
estimate of the true spectrum, and better estimates can be made by using tapers,
especially Slepian tapers. Hence, in this chapter the basics of the admittance and
coherency will be introduced, before proceeding to discussion of their practical estimation using multitapers and wavelets.

5.2 The Admittance

5.2.1 The Earth's Response to Loading

The density contrast between rock and air ($\sim$2000–3000 kg m^{-3}), or rock and water
($\sim$1000–2000 kg m^{-3}), is larger than any other density contrast between adjacent
layers within the Earth's crust or uppermost mantle. As gravitation is dependent
upon density and its spatial distribution, it is appealing to express the gravity field

---

[1] Throughout this book, 'topography' will refer to the surface mapped out by the interface between
solid rock and air/water/ice.

© Springer Nature Switzerland AG 2022
J. Kirby, *Spectral Methods for the Estimation of the Effective Elastic Thickness
of the Lithosphere*, Advances in Geophysical and Environmental Mechanics
and Mathematics, https://doi.org/10.1007/978-3-031-10861-7_5

of the Earth as a function of the topography. To do this we use the theory of *linear time-invariant (LTI) systems* (Bracewell 1986; Percival and Walden 1993; Bendat and Piersol 2000). Such a system can be described by (at least) three functions: an input, h (in our case the topography); an output, g (the gravity field[2]); and an *impulse response*, q, which characterises the properties of interest of the system. In our case—effective elastic thickness estimation—these properties would include T_e itself, the densities of the various crust and mantle layers and the depths to their interfaces, the values of the elastic constants governing plate flexure, and any other physical properties that may influence the topography such as mantle temperature variations (generating convection currents and thence topographic uplift or subsidence), or asthenospheric viscosity (which influences glacial isostatic adjustment rates). In the context of the relationship between gravity and topography, the impulse response q is known as the *isostatic response function*.

Under LTI system theory, the output function is expressed as the convolution of the input function with the impulse response, or

$$g(\mathbf{x}) = q(\mathbf{x}) * h(\mathbf{x})$$

(cf. Eq. 2.23), working in two spatial dimensions. The function q is called the impulse response because if h were a Dirac delta function (an impulse), then g would look like q, since convolution of a function with a delta function yields the (time-shifted) function (Sect. 2.3.1). However, the actual gravity field of the Earth also contains signals generated by sources unconnected with the topography. For example, due to a combination of erosion and uplift/subsidence there may exist lateral density changes—due to compositional variations of rocks within the crust or mantle—that are isostatically compensated yet do not affect the topography; or deep convection currents within the lower mantle and core that give rise to density variations but not measurable changes in the topography (Kirby 2019). Such signals may be designated *noise* in that they are not represented in or quantified by the model embodied in the isostatic response function, and they are independent of the topography. Thus, including a noise term, n, gives us an equation for the full LTI system as

$$g(\mathbf{x}) = q(\mathbf{x}) * h(\mathbf{x}) + n(\mathbf{x}), \tag{5.1}$$

where $n(\mathbf{x})$ is the *gravitational noise*, or sometimes *geological noise* (Munk and Cartwright 1966; Dorman and Lewis 1970; Lewis and Dorman 1970).

If we wish to estimate those properties of the Earth upon which it depends, we now need a way to find q. Evidently we would need to measure both the topography and gravity field in order to solve Eq. 5.1 for $q(\mathbf{x})$, but given the wealth of data now available for both these quantities this is not an obstacle. We would also need to know, or at least estimate, the gravitational noise, a much more difficult proposition. Fortunately, one can avoid having to do this by transforming to the wavenumber

---

[2] In practice the gravity anomaly is used, and this may be of free-air or Bouguer type—see Chap. 8.

domain and making use of the convolution theorem (Sect. 2.2.8). Taking the Fourier transform of Eq. 5.1 gives

$$G(\mathbf{k}) = Q(\mathbf{k}) H(\mathbf{k}) + N(\mathbf{k}). \tag{5.2}$$

The Fourier transform of the impulse response function $q(\mathbf{x})$ is called the *admittance*, $Q(\mathbf{k})$. In LTI system theory Q is an example of a *transfer function* or *frequency filter* (Sect. 2.3.5). In other words, the gravity field is treated as a filtered version of the topography, plus some noise. Now, in order to remove the noise term from the solution, we make use of its definition, above: it is defined as being that part of the gravity field that is uncorrelated with the topography. If we multiply Eq. 5.2 by the complex conjugate of the topography, H^*, and then conduct some averaging scheme, we get

$$\langle G(\mathbf{k}) H^*(\mathbf{k}) \rangle = \langle Q(\mathbf{k}) H(\mathbf{k}) H^*(\mathbf{k}) \rangle + \langle N(\mathbf{k}) H^*(\mathbf{k}) \rangle,$$

where the angle brackets denote the averaging process. If the gravitational noise is indeed uncorrelated with the topography, then, on average, we should expect $\langle N(\mathbf{k}) H^*(\mathbf{k}) \rangle = 0$, giving

$$\langle G(\mathbf{k}) H^*(\mathbf{k}) \rangle = \langle Q(\mathbf{k}) H(\mathbf{k}) H^*(\mathbf{k}) \rangle. \tag{5.3}$$

The desired properties of the Earth, and therefore the particular nature of the averaging scheme, now become important. Conventionally, the Earth is assumed to have an isotropic response to loading which would imply an isotropic impulse response and therefore an isotropic admittance; hence the averaging scheme would likely be isotropic. However, several recent studies have investigated anisotropy in the flexural rigidity, meaning the averaging scheme would need to preserve any directional variations in the response function. Therefore we will keep things general for now, and assume that the averaging scheme is such that it is possible to write $\langle QHH^* \rangle = Q\langle HH^* \rangle$ in Eq. 5.3, giving

$$Q(\mathbf{k}) = \frac{\langle G(\mathbf{k}) H^*(\mathbf{k}) \rangle}{\langle H(\mathbf{k}) H^*(\mathbf{k}) \rangle}. \tag{5.4}$$

Depending on the chosen averaging scheme, the admittance may be either isotropic, $Q(|\mathbf{k}|)$, or anisotropic, $Q(\mathbf{k})$. Commonly used averaging schemes are discussed in detail in Sects. 5.4.1 and 5.4.2.

The quantity $\langle HH^* \rangle$ in Eq. 5.4 is of course the power spectrum of the topography, also called its *auto-spectrum*. The quantity $\langle GH^* \rangle$ is the *cross-spectrum* between gravity and topography and measures the degree of similarity between two signals. It is equal to the Fourier transform of the *cross-correlation* between two space-domain signals, or

$$(h \star g)(\mathbf{x}) \quad \longleftrightarrow \quad G(\mathbf{k}) H^*(\mathbf{k}),$$

where the $\star$ indicates cross-correlation (Bracewell 1986; Percival and Walden 1993). Cross-correlation is similar to convolution. For two real-valued 1D functions p and q, their cross-correlation is given by

$$(p \star q)(t) \equiv \int_{-\infty}^{\infty} p(\tau) \, q(t + \tau) \, d\tau$$

for $|t| < \infty$, where τ is a dummy variable of integration. Note the similarity with the equation for convolution, Eq. 2.13 in Sect. 2.2.8. However, unlike convolution, cross-correlation is not commutative ($p \star q \neq q \star p$).

5.2.2 The Complex Admittance

The isostatic response function, $q(\mathbf{x})$ in Eq. 5.1, must be real-valued because both gravity and topography are real-valued. However, even though q must be real, the admittance can be a complex variable. This is so because, while $\langle H H^* \rangle$ is real-valued, $\langle G H^* \rangle$ is complex. We can show this by writing the gravity and topography Fourier transforms as $G(\mathbf{k}) = G_R(\mathbf{k}) + i G_I(\mathbf{k})$ and $H(\mathbf{k}) = H_R(\mathbf{k}) + i H_I(\mathbf{k})$ (where subscripts R and I indicate the real and imaginary parts), and substituting them into Eq. 5.4, giving

$$\begin{aligned} Q(\mathbf{k}) &= \frac{\langle (G_R + i G_I)(H_R - i H_I) \rangle}{\langle (H_R + i H_I)(H_R - i H_I) \rangle} \\ &= \frac{\langle (G_R H_R + G_I H_I) + i(G_I H_R - G_R H_I) \rangle}{\langle H_R^2 + H_I^2 \rangle}, \end{aligned} \tag{5.5}$$

dropping the dependence on $\mathbf{k}$ from the right-hand side. Thus, the admittance is complex. Furthermore, we can also show that, if $g(\mathbf{x})$ and $h(\mathbf{x})$ are real-valued signals then the admittance is Hermitian, with an even real part and an odd imaginary part (Sects. 2.2.3 and 3.6.3). A proof is simple. If g and h are real-valued then their Fourier transforms, $G(\mathbf{k})$ and $H(\mathbf{k})$ respectively, are Hermitian functions (Bracewell 1986). Then, in Eq. 5.5, we replace real terms with the symbol E for an even function, and imaginary terms with the symbol O for an odd function, giving

$$Q(\mathbf{k}) = \frac{\langle (EE + OO) + i(OE - EO) \rangle}{\langle EE + OO \rangle}.$$

We can now use the well-known properties of the arithmetic of even and odd functions[3] to obtain

---

[3] $EE = E, OO = E, EO = O, E \pm E = E, O \pm O = O, E/E = E,$ and $O/E = O.$

$$Q(\mathbf{k}) = \frac{\langle E \rangle + i\langle O \rangle}{\langle E \rangle} = E + iO.$$

Thus the admittance is Hermitian, and a Hermitian $Q(\mathbf{k})$ gives a real-valued $q(\mathbf{x})$ upon inverse Fourier transformation (Bracewell 1986). Note that if the averaging scheme is angular in nature—such as annular or azimuthal—the averaging of the imaginary term, $\langle O \rangle$, must be performed over 180° rather than 360° so that cancellation of the imaginary component does not occur (see Sects. 3.6.3 and 4.12.1, especially Figs. 3.14 and 4.13, for more details). Figure 5.1a and b show the real and imaginary parts, respectively, of the 2D admittance between two synthetic surfaces. The surfaces were generated using the method described in the Appendix to this chapter. Figure 5.1c, d show the corresponding 1D profiles after radial averaging over 180° (Sect. 5.4.1).

Having established that the admittance is a Hermitian function, we will now take some time to explore the significance of its imaginary part, and pursue this in Sects. 5.2.3 and 5.3.3. Remember that the admittance is a transfer function from an input signal (topography, h) to an output signal (gravity, g). If it is complex it will, in general, produce an output with a different magnitude and phase to those of the input, even if the system is noise-free. Consider such a noise-free system, i.e., Eq. 5.2 without the $N(\mathbf{k})$ term, given by

$$G'(\mathbf{k}) = Q(\mathbf{k}) H(\mathbf{k}). \tag{5.6}$$

That is, G' is that part of the gravity field that can be predicted from the topography using the admittance. If we write both the admittance and topography in their polar forms, as $Q(\mathbf{k}) = |Q(\mathbf{k})|e^{i\varphi_Q(\mathbf{k})}$ and $H(\mathbf{k}) = |H(\mathbf{k})|e^{i\varphi_H(\mathbf{k})}$, respectively, we see that

$$G'(\mathbf{k}) = |Q(\mathbf{k})| |H(\mathbf{k})| \, e^{i[\varphi_Q(\mathbf{k}) + \varphi_H(\mathbf{k})]}.$$

Thus, the phase of G' is $\varphi_{G'} = \varphi_Q + \varphi_H$, which even in a noise-free system is different to the phase of the topography (by an amount φ_Q, the phase of the admittance). Figure 5.2 shows such a scenario. Dealing with one dimension for simplicity, Fig. 5.2a shows a random, fractal time series, $h(t)$, generated using the spectral synthesis method of Saupe (1988), while Fig. 5.2b shows two transfer functions, $Q_1(f)$ and $Q_2(f)$. Both transfer functions have identical even, real parts, but different imaginary parts; the imaginary part of Q_1 is an odd function, while that of Q_2 is zero. Hence, $Q_1(f)$ is complex and Hermitian, while $Q_2(f)$ is real-valued only. Figure 5.2c shows $g_1(t)$, the signal generated by applying the transfer function Q_1 to the time series, and while they look similar, Fig. 5.2e shows a significant difference between $h(t)$ and $g_1(t)$, especially at the higher frequencies where Q_1 has values close to zero. This difference is highlighted in Fig. 5.2g, a plot of the phase difference between $h(t)$ and $g_1(t)$, where it can be seen that $g_1(t)$ possesses harmonics that are not in phase with those in $h(t)$.

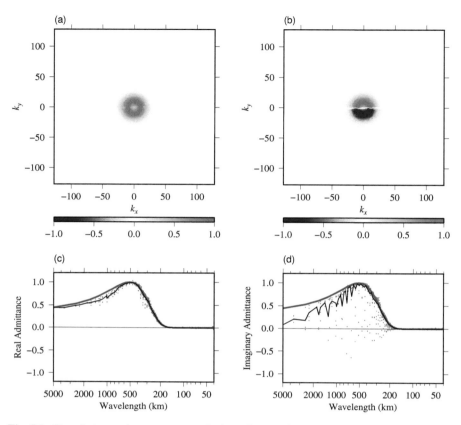

Fig. 5.1 The admittance between two synthetic surfaces, estimated using the multitaper method (NW = 3, $K = 3$). The **a** real and **b** imaginary parts of the 2D admittance $Q(\mathbf{k})$; axes are x and y wavenumber in units of 1.23×10^{-3} rad/km. **c** and **d** show the radially-averaged (1D) versions of the 2D estimates in (**a**) and (**b**), respectively, plotted as black lines; the grey dots show the actual 2D estimates in the upper two quadrants ($k_y \geq 0$) plotted as a function of wavelength ($\lambda = 2\pi/|\mathbf{k}|$); the red lines show the admittance used to generate the synthetic models, Eq. 5.42

Turning to the case of a real-valued transfer function, $Q_2(f)$, Fig. 5.2d shows $g_2(t)$, the signal generated by applying Q_2 to the time series. At first glance, g_2 looks much more different from h than does g_1, as g_2 is very smooth; the difference plot in Fig. 5.2f seems to verify this. However, when one studies the phase difference between $h(t)$ and $g_2(t)$ in Fig. 5.2h, it can be seen that these two signals have identical phase. Therefore the difference exhibited in Fig. 5.2f is purely an amplitude difference caused by the exponentially-decreasing transfer function; there is no imaginary part to Q_2 to cause a further phase difference. Returning to two dimensions, if we let Q in Eq. 5.6 be real-valued, we have $|Q| = Q_R$ and $\varphi_Q = 0$, giving

$$G'(\mathbf{k}) = Q_R(\mathbf{k}) |H(\mathbf{k})| e^{i\varphi_H(\mathbf{k})}. \tag{5.7}$$

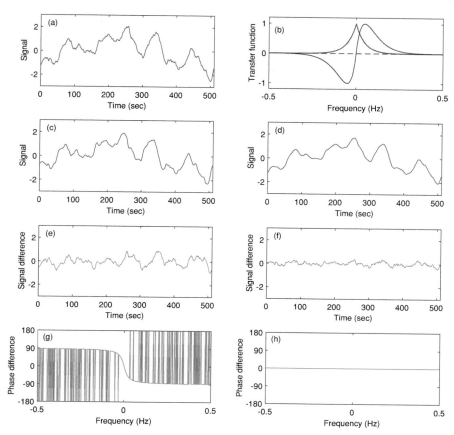

Fig. 5.2 **a** A random, fractal time series, $h(t)$. **b** Two transfer functions, $Q_1(f)$ and $Q_2(f)$: both have real parts given by the green line ($\mathsf{Re}\,Q_1(f) = \mathsf{Re}\,Q_2(f) = e^{-a|f|}$, where f is frequency, and $a = 20$ s); the imaginary part of Q_1 is given by the solid blue line ($\mathsf{Im}\,Q_1(f) = f\,e^{-a|f|}$); the imaginary part of Q_2 is given by the dashed blue line ($\mathsf{Im}\,Q_2(f) = 0$, $\forall f$). **c** The signal $g_1(t)$, where $g_1(t) = \mathsf{F}^{-1}\{Q_1(f)\,H(f)\}$. **d** The signal $g_2(t)$, where $g_2(t) = \mathsf{F}^{-1}\{Q_2(f)\,H(f)\}$. **e** The difference $h(t) - g_1(t)$. **f** The difference $h(t) - g_2(t)$. **g** The phase difference, $\varphi_H(f) - \varphi_{G_1}(f)$. **h** The phase difference, $\varphi_H(f) - \varphi_{G_2}(f)$. The Nyquist frequency is 0.5 Hz

This equation shows that, if the admittance is not complex but is real-valued, then the output signal, G', will have an identical phase to the input signal, H, as shown graphically for 1D data in Fig. 5.2h.

This might seem like a trivial point, but it shows that there may be two reasons why the actual gravity field and topography might have *random phase*, which is generally what occurs in nature. That is, the random phase difference could arise from either (1) noise (the $n(\mathbf{x})$ term in Eq. 5.1), or (2) no noise but the Earth possessing a complex, Hermitian admittance. Since the classical flexural models (Part III) predict a real-valued admittance ($Q_{pr,R}$), one could conceivably perform the operation

$F^{-1}\{Q_{pr,R}H\}$ to obtain a predicted gravity field, G' as in Eq. 5.7, which would of course be in phase with the topography. One could then subtract this from the actual gravity field,[4] point to the difference, and say 'look, there is noise'. However, if one were able to model the Earth's actual (complex) admittance function, Q_{pr}, then the operation $F^{-1}\{Q_{pr}H\}$ would recover the actual gravity field, and there would be no need to invoke noise. This is not to say that gravitational noise does not exist in nature—it may well do and this was a topic of some controversy recently (Sect. 9.6)— but the reasoning presented here shows that one doesn't necessarily need noise to explain any phase difference between the actual gravity field and topography that may be present.

5.2.3 Admittance Phase

The realisation that the admittance is complex encourages us to calculate its phase, defined by

$$\varphi_Q(\mathbf{k}) = \left| \tan^{-1}\left(\frac{Q_I(\mathbf{k})}{Q_R(\mathbf{k})} \right) \right|, \tag{5.8}$$

from Eq. 2.7. As calculated here,[5] the phase will take values over the range $\pm 180°$, but as we will see below an admittance phase of, say, $67°$ contains the same information— for our purposes—as an admittance phase of $-67°$, which is why we take the absolute value in Eq. 5.8.

We can show this equivalence of positive and negative admittance phases by writing an expression for the admittance phase in terms of the gravity and topography phases. If we assume a noise-free system, then Eq. 5.2 can be written

$$Q(\mathbf{k}) = \frac{G(\mathbf{k})}{H(\mathbf{k})}. \tag{5.9}$$

Writing the gravity and topography Fourier transforms as $G(\mathbf{k}) = |G(\mathbf{k})|e^{i\varphi_G(\mathbf{k})}$ and $H(\mathbf{k}) = |H(\mathbf{k})|e^{i\varphi_H(\mathbf{k})}$, respectively, Eq. 5.9 becomes

$$Q(\mathbf{k}) = \frac{|G(\mathbf{k})|\, e^{i\varphi_G(\mathbf{k})}}{|H(\mathbf{k})|\, e^{i\varphi_H(\mathbf{k})}} = \frac{|G(\mathbf{k})|}{|H(\mathbf{k})|}\, e^{i\,[\varphi_G(\mathbf{k}) - \varphi_H(\mathbf{k})]}.$$

We now write the phase difference between the gravity and topography signals as

$$\delta(\mathbf{k}) = \varphi_G(\mathbf{k}) - \varphi_H(\mathbf{k}) \tag{5.10}$$

---

[4] This is one way of forming the isostatic anomaly.
[5] Using the `atan2` function in a computer programming language, rather than `atan`.

giving

$$Q(\mathbf{k}) = \frac{|G(\mathbf{k})|}{|H(\mathbf{k})|} \, e^{i\delta(\mathbf{k})} = \frac{|G(\mathbf{k})|}{|H(\mathbf{k})|} [\cos\delta(\mathbf{k}) + i\,\sin\delta(\mathbf{k})].$$

Substituting this into Eq. 5.8 (without taking the absolute value) we obtain

$$\tan\varphi_Q(\mathbf{k}) = \frac{\sin\delta(\mathbf{k})}{\cos\delta(\mathbf{k})} = \tan\delta(\mathbf{k})$$

or

$$\varphi_Q(\mathbf{k}) = \delta(\mathbf{k}).$$

This shows that the phase on the admittance is equal to the phase difference between the gravity and topography signals. And if we are not concerned whether the gravity phase lags the topography phase, or vice versa, then the sign of the phase difference is not important, justifying the taking of the absolute value in Eq. 5.8.

Even if noise is present in the system, the admittance phase is still a good measure of the phase difference between the gravity and topography signals. If we substitute the polar forms of the gravity and topography spectra into Eq. 5.4 instead of Eq. 5.9, we find

$$Q(\mathbf{k}) = \frac{\langle |G(\mathbf{k})||H(\mathbf{k})| \, e^{i\delta(\mathbf{k})}\rangle}{\langle |H(\mathbf{k})|^2\rangle},$$

which gives, when substituted into Eq. 5.8,

$$\tan\varphi_Q(\mathbf{k}) = \frac{\langle |G(\mathbf{k})||H(\mathbf{k})| \sin\delta(\mathbf{k})\rangle}{\langle |G(\mathbf{k})||H(\mathbf{k})| \cos\delta(\mathbf{k})\rangle}. \tag{5.11}$$

Figure 5.3 shows the admittance phase between the same two synthetic surfaces used to generate Fig. 5.1. Now, both the real and imaginary parts of the admittance function used to generate $G(\mathbf{k})$ from $H(\mathbf{k})$ are the same Gaussian functions (see the Appendix to this chapter and Fig. 5.1c, d). Therefore the ratio $Q_I(\mathbf{k})/Q_R(\mathbf{k})$ in Eq. 5.8 ought to be 1 at all wavelengths, giving a constant phase of 45°. This is seen in Fig. 5.3, but only at wavelengths longer than approximately 200 km; at shorter wavelengths the phase is random. A glance at Fig. 5.1c, d shows that, at wavelengths less than approximately 200 km, the real part of the admittance tends to zero, meaning that the ratio $Q_I(\mathbf{k})/Q_R(\mathbf{k})$ becomes unstable and the phase becomes very sensitive to changes in both Q_R and Q_I. This is especially pronounced when the signals being analysed have random phase, as these synthetic models do.

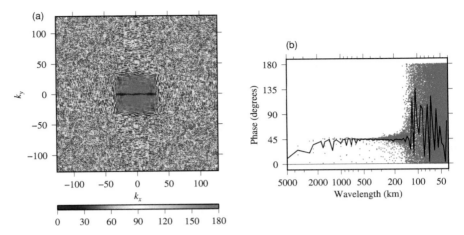

Fig. 5.3 **a** The 2D admittance phase, $\varphi_Q(\mathbf{k})$, from the admittance data in Fig. 5.1a, b, and Eq. 5.8; axes are x and y wavenumber in units of 1.23×10^{-3} rad/km. **b** The radially-averaged admittance phase (black line); the grey dots show the actual 2D estimates in the upper two quadrants ($k_y \geq 0$) from Fig. 5.3a, plotted as a function of wavelength ($\lambda = 2\pi/|\mathbf{k}|$)

5.3 The Coherency and Coherence

5.3.1 The Coherency

The other measure commonly inverted against theoretical flexure models is the coherence. However, some recent studies (Kirby 2014) have based their work on a more fundamental quantity from which the coherence is derived, the *coherency*. Unlike the admittance, which is a transfer function estimated through statistical methods, the coherency is itself a statistical measure, being essentially the Pearson product-moment correlation coefficient calculated in the frequency/wavenumber domain, using the formula

$$\Gamma(\mathbf{k}) = \frac{\langle G(\mathbf{k}) H^*(\mathbf{k}) \rangle}{\langle G(\mathbf{k}) G^*(\mathbf{k}) \rangle^{1/2} \langle H(\mathbf{k}) H^*(\mathbf{k}) \rangle^{1/2}}, \tag{5.12}$$

where the angle brackets denote the averaging schemes discussed in Sects. 5.4.1 and 5.4.2. Note the similarity with the admittance, Eq. 5.4. Both have the term $\langle GH^* \rangle$ in their numerator, and both have real-valued denominators, meaning that the coherency is also complex. However, while the admittance is unbounded, the real and imaginary components of the coherency take values between ± 1, or

$$-1 \leq \mathsf{Re}[\Gamma(\mathbf{k})] \leq 1, \qquad -1 \leq \mathsf{Im}[\Gamma(\mathbf{k})] \leq 1,$$

with the condition that

$$0 \leq |\Gamma(\mathbf{k})| \leq 1.$$

Values of $+1$ indicate perfect positive correlation between the two signals, values of -1 indicate perfect negative correlation, while a zero value indicates no correlation at all.

Figure 5.4a and b show the real and imaginary parts, respectively, of the 2D coherency between two synthetic surfaces. The surfaces were generated using the method described in the Appendix to this chapter. Figure 5.4c, d show the corresponding 1D profiles after radial averaging over $180°$ (Sect. 5.4.1). The radially-averaged, 1D real coherency agrees with the theoretical coherence used to generate the synthetic models: the harmonics are coherent (coherency of 1) at wavelengths above the transition wavelength (200 km), and incoherent (coherency of 0) below it. Both real and imaginary 2D coherency show much scatter at wavelengths below the transition wavelength. This, of course, is expected: the signals are incoherent at these wavelengths and the random phases of the two fractal surfaces are preserved.

Because the cross-spectrum is normalised by the (square roots of the) gravity and topography power spectra in Eq. 5.12, the coherency between two signals is more a measure of their phase relationship than of the relationship between their amplitudes (visible in Fig. 5.4 in the large amount of short-wavelength scatter). This can be seen by substituting $G(\mathbf{k}) = |G(\mathbf{k})|e^{i\varphi_G(\mathbf{k})}$ and $H(\mathbf{k}) = |H(\mathbf{k})|e^{i\varphi_H(\mathbf{k})}$ into Eq. 5.12, giving

$$\Gamma(\mathbf{k}) = \frac{\langle |G(\mathbf{k})||H(\mathbf{k})| e^{i\delta(\mathbf{k})} \rangle}{\langle |G(\mathbf{k})|^2 \rangle^{1/2} \langle |H(\mathbf{k})|^2 \rangle^{1/2}},$$

where $\delta(\mathbf{k})$ is the phase difference between the gravity and topography signals, given by Eq. 5.10. If we abandon the rules of statistics for a moment and ignore the averaging, then we can cancel some terms and obtain the relationship

$$\Gamma(\mathbf{k}) \sim e^{i\delta(\mathbf{k})},$$

showing that the coherency is primarily a measure of phase differences (Kirby and Swain 2009). It is also straightforward to see that the phase on the coherency is identical to the phase on the admittance, or

$$\varphi_\Gamma(\mathbf{k}) = \tan^{-1}\left(\frac{\Gamma_I(\mathbf{k})}{\Gamma_R(\mathbf{k})}\right) = \varphi_Q(\mathbf{k}). \tag{5.13}$$

This is so because of the equivalence $\Gamma_I/\Gamma_R = Q_I/Q_R$. Thus $\varphi_\Gamma(\mathbf{k})$ is also described by Eq. 5.11. As with the admittance phase, it is customary to take the absolute value of Eq. 5.13.

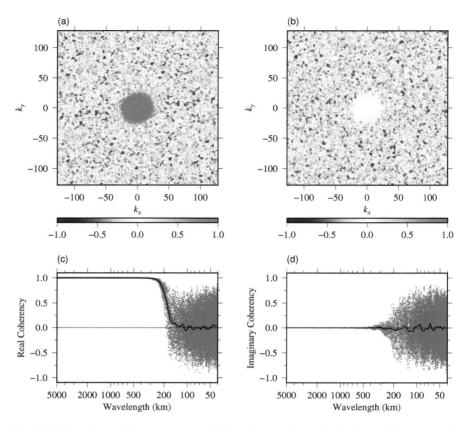

Fig. 5.4 The coherency between two synthetic surfaces, estimated using the multitaper method (NW = 3, K = 3). The **a** real and **b** imaginary parts of the 2D coherency $\Gamma(\mathbf{k})$; axes are x and y wavenumber in units of 1.23×10^{-3} rad/km. **c** and **d** show the radially-averaged (1D) versions of the 2D estimates in (**a**) and (**b**), plotted as black lines; the grey dots show the actual 2D estimates in the upper two quadrants ($k_y \geq 0$) plotted as a function of wavelength ($\lambda = 2\pi/|\mathbf{k}|$); the red line in (**c**) shows the square root of the coherence used to generate the synthetic models (Eq. 5.43)

5.3.2   The Coherence

The coherence is used more commonly than the coherency, certainly in flexural studies. It is the modulus-squared of the coherency, that is

$$\gamma^2(\mathbf{k}) \equiv |\Gamma(\mathbf{k})|^2,$$

or

$$\gamma^2(\mathbf{k}) = \frac{\left|\left\langle G(\mathbf{k})\, H^*(\mathbf{k})\right\rangle\right|^2}{\left\langle G(\mathbf{k})\, G^*(\mathbf{k})\right\rangle \left\langle H(\mathbf{k})\, H^*(\mathbf{k})\right\rangle}, \qquad (5.14)$$

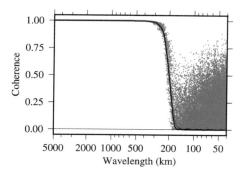

Fig. 5.5 The coherence corresponding to the coherency in Fig. 5.4. The black line is the radially-averaged (1D) coherence; the grey dots show the actual 2D estimates in the upper two quadrants ($k_y \geq 0$) plotted as a function of wavelength ($\lambda = 2\pi/|\mathbf{k}|$); the red line shows the coherence used to generate the synthetic models (Eq. 5.43)

and is hence real-valued. Its values are limited to

$$0 \leq \gamma^2(\mathbf{k}) \leq 1.$$

Thus, it is another way of expressing the degree of correlation between gravity and topography at different wavelengths. But the coherence can also be thought of as a measure of the fraction of energy in the gravity field that can be predicted from the topography using the admittance. Now, that part of the gravity field that can be predicted from the topography using the admittance is just $G'(\mathbf{k}) = Q(\mathbf{k})H(\mathbf{k})$, from Eq. 5.6. Thus, if the coherence is 0.75 at a particular wavenumber, then 75% of the energy in $G(\mathbf{k})$ at that wavenumber arises from $G'(\mathbf{k})$. Or, as Munk and Cartwright (1966) put it, the coherent energy fraction at wavenumber $\mathbf{k}$ is $\gamma^2(\mathbf{k})\langle|G(\mathbf{k})|^2\rangle$, while the incoherent energy fraction is $[1 - \gamma^2(\mathbf{k})]\langle|G(\mathbf{k})|^2\rangle$, if we assume that $\langle|G(\mathbf{k})|^2\rangle$ is a measure of the energy in the observed gravity field.

Figure 5.5 shows the coherence corresponding to the coherency in Fig. 5.4. It is important to note that not only does the coherence not preserve information about the real and imaginary components of the coherency, but it also removes information about whether the correlation is positive or negative. Despite these shortcomings, use of the coherence over the coherency persists.

Unlike the admittance, the coherence (and coherency) can be positively biased by noise. We can see this by separating the gravity into a term that is correlated with the topography ($G' = QH$, from Eq. 5.6) and a term that is not (N), which is just Eq. 5.2 ($G = QH + N$). If we substitute this into Eq. 5.14 we get

$$\gamma^2(\mathbf{k}) = \frac{\left|\langle QHH^* + NH^*\rangle\right|^2}{\langle Q^2HH^* + QHN^* + QNH^* + NN^*\rangle \langle HH^*\rangle},$$

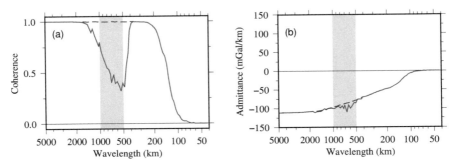

Fig. 5.6 The effect of noise on the **a** coherence, and **b** admittance (real part). The dashed blue lines show the multitaper (NW = 3, K = 3) coherence and admittance between two random, fractal surfaces, while the solid red lines show those quantities when one surface had noise added to it. The noise field was itself a random, fractal surface, bandpass filtered between 500 and 1000 km wavelength (shown as the grey shaded region), with a standard deviation 33% that of the field to which it was added

dropping the dependence on **k** from the right-hand side. As we did in Sect. 5.2.1, we can set $\langle N H^* \rangle = \langle H N^* \rangle = 0$ because the noise and topography are statistically uncorrelated, giving

$$\gamma^2(\mathbf{k}) = \frac{\left|\langle Q H H^* \rangle\right|^2}{\langle Q^2 H H^* + N N^* \rangle \langle H H^* \rangle},$$

or, using $G' = QH$ again,

$$\gamma^2(\mathbf{k}) = \frac{\left|\langle G' H^* \rangle\right|^2}{\langle G' G'^* + N N^* \rangle \langle H H^* \rangle}.$$

Because it is always positive, the $\langle N N^* \rangle$ term in the denominator will reduce the coherence from what it would have been otherwise, potentially leading to erroneous results. Figure 5.6 shows the effect of such noise upon the coherence and admittance. As expected, the coherence suffers significantly from the presence of the noise, while the admittance is only slightly affected and even then its overall shape is preserved. Importantly, though, both quantities are only affected within the wavenumber range corresponding to the bandwidth of the noise spectrum (shown by the grey shading).

5.3.3   The Complex Coherency

We saw in Fig. 5.6a how noise reduces the coherence within the wavelength band in which it exists. Sometimes, though, the coherence decreases 'naturally', that is, as predicted by a model. The question then becomes, how would one distinguish a noise-based reduction in coherence from a natural decrease? We can simulate this by generating two synthetic surfaces with a known coherence between them using the method described in the Appendix to this chapter. The known coherence is an inverted Gaussian function, Eq. 5.44, which exhibits a decrease in coherence around 200 km wavelength (the dashed blue line in Fig. 5.7a). That is, the model has a naturally-lowered coherence, which is picked up quite well by the analysis method. When noise is added to one of the surfaces, this also lowers the observed coherence (Fig. 5.7b). So the coherence alone cannot tell us which wavenumber band corresponds to the noise, and which to the natural decrease. Evidently, some other measure is required to distinguish between natural and noise-induced decoherence.

Kirby and Swain (2009) proposed an investigation of the imaginary part of the coherency to resolve this. Motivated by flexural modelling—which predicts a real-valued coherency—they reasoned that only the real part of the observed coherence should be inverted against predicted models to recover T_e, rather than the 'whole' coherence which merges the real and imaginary parts into a single quantity (Eq. 5.14).

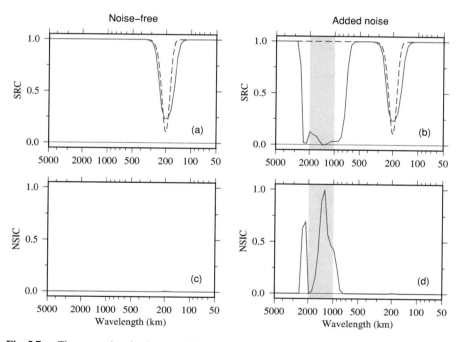

Fig. 5.7 **a** The squared real coherency (SRC) between two synthetic surfaces, estimated using the fan wavelet method with $|\mathbf{k}_0| = 7.547$ (solid red line), and the coherence function used to generate the synthetic surfaces (dashed blue line, Eq. 5.44). **b** The SRC between the two surfaces when one surface had noise added to it (the grey shaded region shows the bandwidth of the noise field). **c** The normalised squared imaginary coherency (NSIC) corresponding to the noise-free model in (**a**). **d** The NSIC corresponding to the added-noise model in (**b**)

Then, if there is any non-zero signal present in the observed imaginary component of the coherency, it must not be part of the flexural process (if the model is accurate) and should indicate the presence of noise.

The observation that flexural modelling predicts only a real-valued coherency (and therefore admittance, due to the common $\langle GH^* \rangle$ term) is a statement that the two surfaces (gravity and topography) are in phase at all wavelengths (Sect. 5.2.2). Therefore any part of one signal that is not in phase with the other signal would be classified as noise, according to the model. This reasoning is borne out by considering two surfaces with a variable degree of correlation between them. Take two random, uncorrelated surfaces, $u(\mathbf{x})$ and $v(\mathbf{x})$, and create a third surface, $v'(\mathbf{x})$, from them by

$$v'(\mathbf{x}) \; = \; R\, u(\mathbf{x}) \; + \; v(\mathbf{x})\sqrt{1 - R^2}, \tag{5.15}$$

where R is the correlation coefficient, with $-1 \le R \le 1$ (Macario et al. 1995). Taking the Fourier transform and multiplying both sides by $U^*(\mathbf{k})$ to form the cross-spectrum, we get

$$V'(\mathbf{k})\, U^*(\mathbf{k}) \; = \; R\,|U(\mathbf{k})|^2 \; + \; V(\mathbf{k})\, U^*(\mathbf{k})\sqrt{1 - R^2}. \tag{5.16}$$

When $R = 1$, Eq. 5.15 tells us that $v'(\mathbf{x}) = u(\mathbf{x})$, and therefore that $v'(\mathbf{x})$ and $u(\mathbf{x})$ are 100% correlated. Equation 5.16 thus becomes

$$V'(\mathbf{k})\, U^*(\mathbf{k}) \; = \; |U(\mathbf{k})|^2,$$

which is just the power spectrum of $u(\mathbf{x})$ and is thus real-valued at all wavenumbers. That is, the imaginary component of the cross-correlation between two correlated surfaces is always zero.

Next, to show that the cross-spectrum between two uncorrelated surfaces can never have a zero imaginary part (in general), we use the proof of reductio ad absurdum, or proof by contradiction. When $R = 0$, Eq. 5.15 shows that $v'(\mathbf{x}) = v(\mathbf{x})$, and therefore that $v'(\mathbf{x})$ and $u(\mathbf{x})$ are 0% (randomly) correlated (because v and u are). Equation 5.16 becomes

$$V'(\mathbf{k})\, U^*(\mathbf{k}) \; = \; V(\mathbf{k})\, U^*(\mathbf{k}).$$

The imaginary part of the cross-spectrum is thus

$$\begin{aligned}
\mathsf{Im}[V'(\mathbf{k})\, U^*(\mathbf{k})] &= \mathsf{Im}[V(\mathbf{k})\, U^*(\mathbf{k})] \\
&= V_I(\mathbf{k})\, U_R(\mathbf{k}) \; - \; V_R(\mathbf{k})\, U_I(\mathbf{k}).
\end{aligned}$$

This tells us that the only way that $\mathsf{Im}[V'U^*]$ can be zero, in general, is when $V_I\, U_R = V_R\, U_I$, or

$$\frac{V_I(\mathbf{k})}{V_R(\mathbf{k})} \; = \; \frac{U_I(\mathbf{k})}{U_R(\mathbf{k})}.$$

Taking the inverse tangent of both sides (cf. Eq. 2.7) shows that the two surfaces need to have identical phases, or

$$\varphi_V(\mathbf{k}) = \varphi_U(\mathbf{k}).$$

Therefore, the imaginary component of the cross-correlation between two uncorrelated surfaces can only be zero when the surfaces are correlated—a contradiction. Thus $\mathsf{Im}[V'U^*]$ can never be zero, except perhaps at isolated wavenumbers or spatial locations where the random nature of the phases of the surfaces means that there is a non-zero probability of some degree of correlation occurring there.

In conclusion, a zero imaginary coherency (or admittance) is indicative of correlation between two signals; a non-zero imaginary coherency (or admittance) reveals the presence of uncorrelated harmonics, which, in the context of flexural modelling, can indicate the existence of noise.[6]

To avoid unifying the real and imaginary parts when computing the coherence via Eq. 5.14, we can define the *squared real coherency* (SRC) as

$$\Gamma_R^2(\mathbf{k}) = [\mathsf{Re}\,\Gamma(\mathbf{k})]^2, \tag{5.17}$$

and the *squared imaginary coherency* (SIC) as

$$\Gamma_I^2(\mathbf{k}) = [\mathsf{Im}\,\Gamma(\mathbf{k})]^2,$$

with

$$|\Gamma(\mathbf{k})|^2 = \Gamma_R^2(\mathbf{k}) + \Gamma_I^2(\mathbf{k}) = \gamma^2(\mathbf{k}). \tag{5.18}$$

However, Kirby and Swain (2009) noticed that when the overall coherence is low the SIC becomes difficult to interpret, so they normalised the SRC and SIC by the total coherence to form quantities that represent the relative power in the real and imaginary parts of the coherency. These are the *normalised squared real coherency* (NSRC), given by

$$\overline{\Gamma_R^2}(\mathbf{k}) = \frac{\Gamma_R^2(\mathbf{k})}{|\Gamma(\mathbf{k})|^2},$$

and the *normalised squared imaginary coherency* (NSIC), given by

$$\overline{\Gamma_I^2}(\mathbf{k}) = \frac{\Gamma_I^2(\mathbf{k})}{|\Gamma(\mathbf{k})|^2}. \tag{5.19}$$

It is then easy to see that

$$\overline{\Gamma_R^2}(\mathbf{k}) + \overline{\Gamma_I^2}(\mathbf{k}) = 1.$$

---

[6] It indicates noise, rather than a Hermitian admittance/coherency with a non-zero imaginary part, because classical flexural models predict a real-valued admittance/coherency (Sects. 9.3.1 and 9.3.3, though see also Sect. 9.7).

Furthermore, using the trigonometric identities $\cos^2\theta = 1/(1 + \tan^2\theta)$ and $\sin^2\theta = \tan^2\theta/(1 + \tan^2\theta)$, and Eq. 5.13 for the coherency phase, $\varphi_\Gamma(\mathbf{k})$, it is straightforward to show that

$$\overline{\Gamma_R^2}(\mathbf{k}) \;=\; \cos^2\varphi_\Gamma(\mathbf{k}), \qquad\qquad \overline{\Gamma_I^2}(\mathbf{k}) \;=\; \sin^2\varphi_\Gamma(\mathbf{k}).$$

Thus, the NSRC and NSIC are alternative expressions of the coherency phase.

Returning to Fig. 5.7, we can see that the NSIC distinguishes between the naturally-reduced and noise-reduced SRCs. Even though the SRC is low at 200 km wavelength in the noise-free model (Fig. 5.7a) its corresponding NSIC shows no signal (Fig. 5.7c). When noise is added, however, the NSIC shows a distinctive signal at the wavelengths of the noise bandwidth, but no signal at the 200 km wavelength of the natural decoherence (Fig. 5.7d).

We will return to the topic of noise—in a flexural context—in Sects. 9.6 and 11.9.

5.4 Practical Estimation of the Admittance and Coherency

Equations 5.4 and 5.12 are instructive, but not that useful for the practical estimation of the observed admittance and coherency as they stand—unless one is satisfied with a periodogram-style, direct spectral estimate of these quantities (not recommended; see Sect. 3.2). In this section, therefore, we will discuss the estimation of the observed admittance and coherency using multitapers and wavelets. It is therefore essential to have read Chaps. 3 and 4 before proceeding further.

5.4.1 Using Multitapers

5.4.1.1 2D Multitaper Admittance and Coherency

In Sect. 3.6.2 we went through the procedure of how to form 2D eigenspectra and an averaged power spectrum using multiple, Slepian tapers. We can apply the same method when constructing cross-spectra and of course the auto-spectra (another name for the power spectra).

First, one selects a value of the time-bandwidth product, NW, to achieve a desired resolution and variance, and then chooses the number of tapers, K, to be used, where $1 \le K \le 2\text{NW} - 1$. Then take the space-domain gravity and topography sequences, $g_{\xi\eta}$ and $h_{\xi\eta}$ ($\xi, \eta = 0, 1, \ldots, N-1$), respectively, multiply them by each taper, $\mathbf{w}_{jk} = w_{jk\xi\eta}$ ($j, k = 1, 2, \ldots, K$), and take the Fourier transform, or

$$\check{G}_{jk}(\mathbf{k}) \;=\; \mathsf{F}\big\{g_{\xi\eta}\, w_{jk\xi\eta}\big\}, \tag{5.20a}$$

$$\check{H}_{jk}(\mathbf{k}) \;=\; \mathsf{F}\big\{h_{\xi\eta}\, w_{jk\xi\eta}\big\}, \tag{5.20b}$$

as in Eq. 3.29. The (j, k)th cross- and auto-eigenspectra are then, from Eq. 3.30,

$$\hat{S}_{jk}^{(gh)}(\mathbf{k}) = \check{G}_{jk}(\mathbf{k})\,\check{H}_{jk}^{*}(\mathbf{k}), \tag{5.21a}$$

$$\hat{S}_{jk}^{(gg)}(\mathbf{k}) = \check{G}_{jk}(\mathbf{k})\,\check{G}_{jk}^{*}(\mathbf{k}), \tag{5.21b}$$

$$\hat{S}_{jk}^{(hh)}(\mathbf{k}) = \check{H}_{jk}(\mathbf{k})\,\check{H}_{jk}^{*}(\mathbf{k}). \tag{5.21c}$$

The average cross- and auto-spectra are then calculated by averaging the K^2 cross- and auto-eigenspectra over the tapers, or

$$S_{gh}(\mathbf{k}) = \frac{1}{K^2}\sum_{j=1}^{K}\sum_{k=1}^{K}\hat{S}_{jk}^{(gh)}(\mathbf{k}), \tag{5.22a}$$

$$S_{gg}(\mathbf{k}) = \frac{1}{K^2}\sum_{j=1}^{K}\sum_{k=1}^{K}\hat{S}_{jk}^{(gg)}(\mathbf{k}), \tag{5.22b}$$

$$S_{hh}(\mathbf{k}) = \frac{1}{K^2}\sum_{j=1}^{K}\sum_{k=1}^{K}\hat{S}_{jk}^{(hh)}(\mathbf{k}), \tag{5.22c}$$

following Eq. 3.31. Finally, the 2D admittance and coherency are determined from the ratios

$$Q(\mathbf{k}) = \frac{S_{gh}(\mathbf{k})}{S_{hh}(\mathbf{k})}$$

and

$$\Gamma(\mathbf{k}) = \frac{S_{gh}(\mathbf{k})}{\sqrt{S_{gg}(\mathbf{k})\,S_{hh}(\mathbf{k})}},$$

respectively. The coherence is merely $\gamma^2(\mathbf{k}) = |\Gamma(\mathbf{k})|^2$.

The multitaper method with $NW = 3$ and $K = 3$ was used to generate the 2D admittance in Fig. 5.1a, b, the 2D admittance phase in Fig. 5.3a, and the 2D coherency in Fig. 5.4a, b.

5.4.1.2 Radially-Averaged Multitaper Admittance and Coherency

If 1D profiles are desired, then one must radially average[7] the 2D cross- and auto-spectral estimates and then take their ratio, rather than radially averaging the 2D admittance and coherency themselves, because the averaging schemes implied in Eqs. 5.4 and 5.12 act on the numerators and denominators separately. For reasons that

---

[7] As noted in Sect. 3.6.3 this type of averaging is actually performed azimuthally, around annuli in the wavenumber domain, despite its name.

will become clear in Sect. 5.5.3.2, we vary the procedure in Sect. 3.6.3 and perform the radial averaging before the taper averaging (the order in which averaging is performed having no effect upon the final estimate itself). Hence, having computed the cross- and auto-eigenspectra (Eqs. 5.21), we assign each of them to an annulus in the wavenumber domain and average over all estimates that fall in that annulus, giving the radially-averaged cross- and auto-eigenspectra as

$$\bar{S}_{jk}^{(gh)}(|\mathbf{k}|) = \frac{1}{N_m} \sum_{l=1}^{N_m} \hat{S}_{jk}^{(gh)}(\mathbf{k}_l), \tag{5.23a}$$

$$\bar{S}_{jk}^{(gg)}(|\mathbf{k}|) = \frac{1}{N_m} \sum_{l=1}^{N_m} \hat{S}_{jk}^{(gg)}(\mathbf{k}_l), \tag{5.23b}$$

$$\bar{S}_{jk}^{(hh)}(|\mathbf{k}|) = \frac{1}{N_m} \sum_{l=1}^{N_m} \hat{S}_{jk}^{(hh)}(\mathbf{k}_l), \tag{5.23c}$$

where N_m is the number of (k_x, k_y) pairs in annulus m, $\mathbf{k}_l = (k_x, k_y)_l$ is the wavenumber of the lth pair within the annulus, and $|\mathbf{k}|$ is the mean wavenumber modulus of the annulus (Eq. 3.32). We then average the radially-averaged cross- and auto-eigenspectra over their tapers, as

$$\bar{S}_{gh}(|\mathbf{k}|) = \frac{1}{K^2} \sum_{j=1}^{K} \sum_{k=1}^{K} \bar{S}_{jk}^{(gh)}(|\mathbf{k}|), \tag{5.24a}$$

$$\bar{S}_{gg}(|\mathbf{k}|) = \frac{1}{K^2} \sum_{j=1}^{K} \sum_{k=1}^{K} \bar{S}_{jk}^{(gg)}(|\mathbf{k}|), \tag{5.24b}$$

$$\bar{S}_{hh}(|\mathbf{k}|) = \frac{1}{K^2} \sum_{j=1}^{K} \sum_{k=1}^{K} \bar{S}_{jk}^{(hh)}(|\mathbf{k}|). \tag{5.24c}$$

Then, the radially-averaged 1D admittance is

$$Q(|\mathbf{k}|) = \frac{\bar{S}_{gh}(|\mathbf{k}|)}{\bar{S}_{hh}(|\mathbf{k}|)},$$

while the coherency is

$$\Gamma(|\mathbf{k}|) = \frac{\bar{S}_{gh}(|\mathbf{k}|)}{\sqrt{\bar{S}_{gg}(|\mathbf{k}|)\, \bar{S}_{hh}(|\mathbf{k}|)}}. \tag{5.25}$$

The coherence is merely $\gamma^2(|\mathbf{k}|) = |\Gamma(|\mathbf{k}|)|^2$. Note that, as discussed in Sect. 3.6.3, the annular averaging in Eqs. 5.23 must be performed over the upper two wavenumber-domain quadrants only, rather than over all four quadrants, to avoid cancellation of the imaginary components of the cross-eigenspectra, which are Hermitian. In that case, N_m is the number of (k_x, k_y) pairs in the half-annulus m.

Examples of the 1D radially-averaged admittance are shown in Fig. 5.1c, d, the admittance phase in Fig. 5.3b, the coherency in Fig. 5.4c and d, and the coherence in Fig. 5.5.

5.4.1.3 Multitapered Moving Windows

The moving window technique, introduced in Sect. 3.5, may be used to obtain spatial variations in admittance and coherency. First, a value of NW must be chosen, together with K, the number of tapers in each direction. Next the window size must be selected, as well as the spacing between adjacent windows (which is usually chosen to be the grid spacing of the data). Then, within each window, the radially-averaged admittance and/or coherency is estimated using the approach of Sect. 5.4.1.2, and assigned to the coordinates of the centre of the window. Thus one obtains $Q(|\mathbf{k}|, \mathbf{x})$ and $\Gamma(|\mathbf{k}|, \mathbf{x})$.

An example of the real part of the moving-window multitaper coherency is shown in Fig. 5.8. In this model, the coherency is spatially dependent, with the transition from -1 coherency to zero and positive values occurring at smaller wavenumbers in the centre of the model, and at larger wavenumbers towards the edges. As explained in Sect. 3.5, Fig. 5.8 shows how the window size affects the bandwidth and wavenumber resolution of the estimates, and their spatial resolution and extent. The wavenumber-domain properties are also determined by the values of NW and K. For example, larger window sizes allow for more wavenumber estimates and a better resolution in that domain compared to smaller windows, but—if the amount of data in each window is to be preserved—sacrifice more data at the edges and have a poorer resolution in the space domain than smaller windows.

5.4.2 Using the Wavelet Transform

The strength of the wavelet transform lies in its ability to map spatial variations in admittance and coherency. Of course this may be done using moving, multitapered windows (e.g. Fig. 5.8), but the fact remains that the wavelet transform is better suited to this, having a scale-dependent resolution (Fig. 4.8) and not requiring the construction of moving windows. It also only involves the arbitrary selection of one constant ($|\mathbf{k}_0|$ for the Morlet wavelet), compared to three for the moving-window multitaper method (NW, K, and window size).

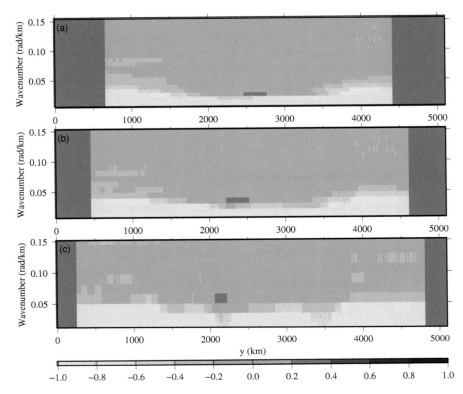

Fig. 5.8 Slices in the ys-plane of the real part of the coherency of some synthetic data, using the multitaper method with NW = 3 and $K = 3$, and moving windows of sizes **a** 1400×1400 km, **b** 1000×1000 km, and **c** 600×600 km. See also Fig. 5.9

5.4.2.1 Local Wavelet Admittance and Coherency

Here we will apply the procedure used to form a fan wavelet power spectrum that was introduced in Sect. 4.12.3, and generalise it to the computation of the cross-spectrum. First, a value of the central wavenumber of the Morlet wavelet, $|\mathbf{k}_0|$, must be selected, according to the desired space- versus wavenumber-domain resolution (Sect. 4.3.3). One then evaluates the wavelet coefficients using one of two methods. Both methods begin with the space-domain gravity and topography data, $g(\mathbf{x})$ and $h(\mathbf{x})$, respectively.[8] If computational speed is required then one takes the Fourier transform of the gravity and topography data, giving $G(\mathbf{k})$ and $H(\mathbf{k})$, respectively, and then determines their Morlet wavelet coefficients at an azimuth α from Eq. 4.36, or

---

[8] Recall (Sect. 4.1) that with the wavelet transform we use the notation $g(\mathbf{x})$ (a continuous data signal), rather than $g_{\xi\eta}$ (a discrete data sequence), even though the data are discretely sampled.

$$\widetilde{G}(s, \mathbf{x}, \alpha) = \mathsf{F}^{-1}\{G(\mathbf{k})\, \Psi^*_{s,\alpha}(\mathbf{k})\}, \tag{5.26a}$$

$$\widetilde{H}(s, \mathbf{x}, \alpha) = \mathsf{F}^{-1}\{H(\mathbf{k})\, \Psi^*_{s,\alpha}(\mathbf{k})\}, \tag{5.26b}$$

where $\Psi_{s,\alpha}(\mathbf{k})$ is the Fourier transform of the 2D daughter Morlet wavelet given by Eqs. 4.37 and 4.40. Alternatively—and if the very long wavelengths are of particular interest (Sect. 4.12.2)—one could use space-domain convolution, Eq. 4.34, to get the wavelet coefficients:

$$\widetilde{G}(s, \mathbf{x}, \alpha) = (g * \psi^*_{s,\mathbf{x}',\alpha})(\mathbf{x}), \tag{5.27a}$$

$$\widetilde{H}(s, \mathbf{x}, \alpha) = (h * \psi^*_{s,\mathbf{x}',\alpha})(\mathbf{x}), \tag{5.27b}$$

where $\psi_{s,\mathbf{x}',\alpha}(\mathbf{x})$ is the space-domain 2D daughter Morlet wavelet given by Eqs. 4.35 and 4.38. One can then form the Morlet wavelet cross- and auto-scalograms at azimuth α through

$$S_{gh}(s, \mathbf{x}, \alpha) = \widetilde{G}(s, \mathbf{x}, \alpha)\, \widetilde{H}^*(s, \mathbf{x}, \alpha), \tag{5.28a}$$

$$S_{gg}(s, \mathbf{x}, \alpha) = \widetilde{G}(s, \mathbf{x}, \alpha)\, \widetilde{G}^*(s, \mathbf{x}, \alpha), \tag{5.28b}$$

$$S_{hh}(s, \mathbf{x}, \alpha) = \widetilde{H}(s, \mathbf{x}, \alpha)\, \widetilde{H}^*(s, \mathbf{x}, \alpha). \tag{5.28c}$$

The fan wavelet cross- and auto-scalograms are then found by averaging the Morlet wavelet cross- and auto-scalograms over azimuth in a fan geometry (Sect. 4.12.1), using Eq. 4.58, or

$$S_{gh}(s, \mathbf{x}) = \frac{1}{N_\alpha} \sum_{j=1}^{N_\alpha} S_{gh}(s, \mathbf{x}, \alpha_j), \tag{5.29a}$$

$$S_{gg}(s, \mathbf{x}) = \frac{1}{N_\alpha} \sum_{j=1}^{N_\alpha} S_{gg}(s, \mathbf{x}, \alpha_j), \tag{5.29b}$$

$$S_{hh}(s, \mathbf{x}) = \frac{1}{N_\alpha} \sum_{j=1}^{N_\alpha} S_{hh}(s, \mathbf{x}, \alpha_j). \tag{5.29c}$$

Finally, the 3D[9] wavelet admittance and coherency are determined by

$$Q(s, \mathbf{x}) = \frac{S_{gh}(s, \mathbf{x})}{S_{hh}(s, \mathbf{x})}$$

---

[9] '3D' because the wavelet coefficients are functions of two space dimensions and one scale dimension.

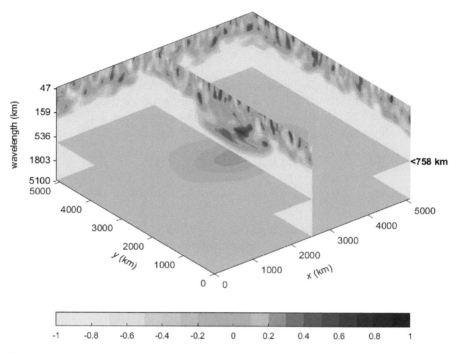

Fig. 5.9 The real part of the 3D wavelet coherency ($|\mathbf{k}_0| = 5.336$) of some synthetic data, showing slices in the ys- and xy- planes. The wavelet scale has been converted to equivalent Fourier wavelength using Eq. 4.42. See also Fig. 5.8

and

$$\Gamma(s, \mathbf{x}) = \frac{S_{gh}(s, \mathbf{x})}{\sqrt{S_{gg}(s, \mathbf{x}) \, S_{hh}(s, \mathbf{x})}}, \qquad (5.30)$$

respectively. The coherence is merely $\gamma^2(s, \mathbf{x}) = |\Gamma(s, \mathbf{x})|^2$.

An example of the real part of the 3D wavelet coherency is shown in Fig. 5.9. In this model, the coherency is spatially dependent, with the transition from -1 coherency to zero and positive values occurring at longer wavelengths in the centre of the model, and at shorter wavelengths towards the edges.

Finally, when comparing the wavelet admittance and coherency with theoretical predictions of these quantities derived in the Fourier domain, one must convert the wavelet scale (s) to an equivalent Fourier wavenumber (k_e) or wavelength (λ_e) (Sect. 4.6). For the 2D Morlet wavelets in the fan this is done using Eqs. 4.41 and 4.42, reproduced here:

$$k_e = \frac{|\mathbf{k}_0|}{s}, \qquad \lambda_e = \frac{2\pi s}{|\mathbf{k}_0|}.$$

5.4.2.2 Global Wavelet Admittance and Coherency

To obtain the (1D) global admittance and coherency, one must spatially average the 3D fan wavelet cross- and auto-scalograms and then take their ratio, rather than averaging the 3D admittance and coherency themselves, because the averaging schemes implied in Eqs. 5.4 and 5.12 act on the numerators and denominators separately. Using Eq. 4.59, we spatially average the fan wavelet cross- and auto-scalograms in Eqs. 5.29 over all space-domain locations, giving the global cross- and auto-scalograms as

$$S_{gh}(s) = \frac{1}{N_{\mathbf{x}}} \sum_{\mathbf{x}} S_{gh}(s, \mathbf{x}), \tag{5.31a}$$

$$S_{gg}(s) = \frac{1}{N_{\mathbf{x}}} \sum_{\mathbf{x}} S_{gg}(s, \mathbf{x}), \tag{5.31b}$$

$$S_{hh}(s) = \frac{1}{N_{\mathbf{x}}} \sum_{\mathbf{x}} S_{hh}(s, \mathbf{x}), \tag{5.31c}$$

where $N_{\mathbf{x}} = N_x N_y$ and $\sum_{\mathbf{x}} = \sum_x \sum_y$. The global admittance and coherency are then obtained from

$$Q(s) = \frac{S_{gh}(s)}{S_{hh}(s)},$$

and

$$\Gamma(s) = \frac{S_{gh}(s)}{\sqrt{S_{gg}(s)\,S_{hh}(s)}}, \tag{5.32}$$

respectively. The global coherence is merely $\gamma^2(s) = |\Gamma(s)|^2$.

Figure 5.10 shows the global coherency (solid line) corresponding to the model in Fig. 5.9. However, when the spectra are highly spatially variable, such as in this example, the global scalograms are perhaps not so useful. Of more interest perhaps are 1D profiles extracted directly from 3D quantities, also shown (as magenta and

Fig. 5.10 1D wavelet coherency profiles extracted from Fig. 5.9, at a point near its centre (magenta line) and at a point towards its edges (green line). Also shown is the global coherency (black line)

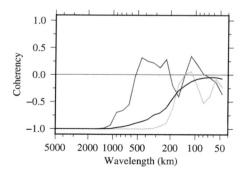

green lines) in Fig. 5.10. That is, for a specified spatial location $\mathbf{x}_0 = (x_0, y_0)$, a 1D profile of admittance or coherency, $Q(s, \mathbf{x}_0)$ and $\Gamma(s, \mathbf{x}_0)$, can be extracted at that location.

5.4.2.3 Partial Spatial Averaging

The spatial averaging need not encompass the whole study area, however. Rather than averaging the fan wavelet cross- and auto-scalograms over every spatial location at a given scale, these scalograms can be averaged within a limited spatial area (a *block*) and mapped to the centre of that block. For example, the data and their wavelet coefficients may be given on a 20×20 km grid in the space domain, but the spatial averaging performed over a block of 10×10 grid cells, giving a new spatially-averaged grid cell size of 200×200 km. If $\mathbf{x}$ represents coordinates in the original space-domain grid, and $\bar{\mathbf{x}}$ represents coordinates of the blocks in the spatially-averaged grid, then the spatially-averaged cross- and auto-scalograms are obtained by spatially averaging the fan wavelet cross- and auto-scalograms in Eqs. 5.29, thus:

$$S_{gh}(s, \bar{\mathbf{x}}) = \frac{1}{N_{\bar{\mathbf{x}}}} \sum_{\bar{\mathbf{x}}} S_{gh}(s, \mathbf{x}), \tag{5.33a}$$

$$S_{gg}(s, \bar{\mathbf{x}}) = \frac{1}{N_{\bar{\mathbf{x}}}} \sum_{\bar{\mathbf{x}}} S_{gg}(s, \mathbf{x}), \tag{5.33b}$$

$$S_{hh}(s, \bar{\mathbf{x}}) = \frac{1}{N_{\bar{\mathbf{x}}}} \sum_{\bar{\mathbf{x}}} S_{hh}(s, \mathbf{x}), \tag{5.33c}$$

where $N_{\bar{\mathbf{x}}} = N_{\bar{x}} N_{\bar{y}}$ is the number of original-grid cells in a single spatially-averaged block, and the notation $\sum_{\bar{\mathbf{x}}}$ means sum over the elements within a block. The spatially-averaged admittance and coherency are then defined by

$$Q(s, \bar{\mathbf{x}}) = \frac{S_{gh}(s, \bar{\mathbf{x}})}{S_{hh}(s, \bar{\mathbf{x}})}$$

and

$$\Gamma(s, \bar{\mathbf{x}}) = \frac{S_{gh}(s, \bar{\mathbf{x}})}{\sqrt{S_{gg}(s, \bar{\mathbf{x}}) \, S_{hh}(s, \bar{\mathbf{x}})}},$$

respectively. The spatially-averaged coherence is merely $\gamma^2(s, \bar{\mathbf{x}}) = |\Gamma(s, \bar{\mathbf{x}})|^2$.

Figure 5.11a shows a slice in the xy-plane through the 3D wavelet coherency of North America on the original 20×20 km grid, while Fig. 5.11b shows a slice at the same equivalent Fourier wavelength through the 200×200 km spatially-averaged coherency. If spatial resolution is of relatively low importance, spatial averaging is useful for reducing file sizes and for decreasing 'noise' due to high spatial variability in the spectra.

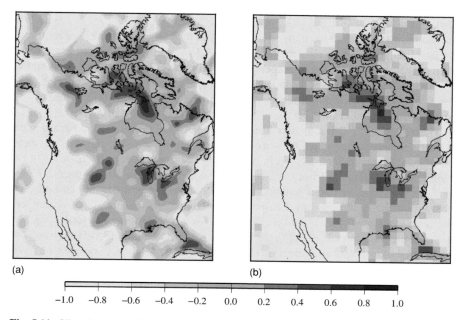

Fig. 5.11 Slices in the xy-plane through the 3D real coherency between North American gravity and topography data, at 222 km equivalent Fourier wavelength with $|\mathbf{k}_0| = 5.336$. **a** No spatial averaging; **b** 200 km spatial averaging

5.5 Errors on the Admittance and Coherence

All scientific studies should provide errors on measurements and quantities derived from them, or at least some discussion of the error budget of the system. The errors provide a measure of the reliability of the quantities and thence of the conclusions of the study. In this section we will look at two methods to estimate errors on the admittance and coherency/coherence.

5.5.1 Independent Estimates

One of the fundamental requirements of statistical error analysis is that the observations for which errors are required are statistically independent of one another. In the context of admittance and coherency estimation—where averaging is required—this property of independence must be held by the various estimates being averaged.

Consider the 2D multitaper method. First, the elements of the 2D spectrum, $G(k_x, k_y)$, are independent because the basis functions of the Fourier transform—the sines and cosines in the complex exponential—are orthogonal functions (see the Appendix to Chap. 3); second, the Slepian tapers are also orthogonal. Thus, each element of the tapered spectrum, $\check{G}_{jk}(k_x, k_y)$, is independent of $\check{G}_{j'k'}(k'_x, k'_y)$. So when the 2D multitaper admittance and coherency and their error estimates are required, the averaging takes place over the tapers (Eqs. 5.22), and the number of independent estimates is the total number of tapers, K^2. If radial averaging takes place, then the averaging is performed over the tapers and the (k_x, k_y) pairs in the annulus (Eqs. 5.23 and 5.24), so the number of independent estimates is the product of the number of tapers and the number of (k_x, k_y) pairs in the (half-)annulus m, or $K^2 N_m$.

The situation is different with the fan wavelet method, however, because the Morlet wavelets comprising the fan are continuous wavelets, and continuous wavelets are non-orthogonal (Sect. 4.1). This means that the wavelet coefficients cannot be completely independent from one another. Consider two normalised space-domain Morlet wavelets at different scales (s and a), locations ($\mathbf{t}$ and $\mathbf{b}$), and azimuths (α and β): $\psi_{s,\mathbf{t},\alpha}(\mathbf{x})$ and $\psi_{a,\mathbf{b},\beta}(\mathbf{x})$. Let the wavelets have the same central wavenumber, $|\mathbf{k}_0|$; one wavelet will have the central wavenumber vector $\mathbf{k}_0 = |\mathbf{k}_0|(\cos\alpha, \sin\alpha)$, and the other will have $\mathbf{l}_0 = |\mathbf{k}_0|(\cos\beta, \sin\beta)$. Kirby and Swain (2013) showed that the 'degree of orthogonality' between the two Morlet wavelets is a complex variable given by

$$O_{st\alpha} = \frac{2sa}{s^2 + a^2} \, e^{-[|\mathbf{b}-\mathbf{t}|^2 + |s\mathbf{l}_0 - a\mathbf{k}_0|^2]/[2(s^2+a^2)]} \, e^{i(s\mathbf{k}_0 + a\mathbf{l}_0)\cdot(\mathbf{b}-\mathbf{t})/(s^2+a^2)}.$$

It can be shown that $0 < |O_{st\alpha}| \leq 1$, meaning that the Morlet wavelet can never be truly orthogonal (which would require $|O_{st\alpha}| = 0$). This means that the wavelet coefficients $\widetilde{G}(s, \mathbf{t}, \alpha)$ are, in general, not independent from $\widetilde{G}(a, \mathbf{b}, \beta)$. Nevertheless, one can find values of s and a, $\mathbf{t}$ and $\mathbf{b}$, and α and β, that give low values to $O_{st\alpha}$ (<0.1, say), making the Morlet wavelets 'quasi-orthogonal'.

Kirby and Swain (2013) found that azimuthal quasi-orthogonality can be achieved by requiring Morlet wavelets to be azimuthally separated by an angle $\delta\alpha$, where $\delta\alpha$ is determined by setting $q_\alpha = e^{-1}$ in Eq. 4.53. So when using the fan wavelet method for local admittance/coherence computation, the averaging of the Morlet wavelet cross- and auto-scalograms [$S_{gh}(s, \mathbf{x}, \alpha)$, etc.] takes place over azimuth (Eqs. 5.29), and the number of (quasi-)independent estimates is just the number of Morlet wavelet azimuths comprising the fan (N_α, found from Eq. 4.54), providing the above condition regarding $\delta\alpha$ is met.

When the global admittance or coherency/coherence and their errors are computed, the averaging takes place over both azimuth and space (Eqs. 5.29 and 5.31). However, while the Morlet wavelet cross- and auto-scalograms [$S_{gh}(s, \mathbf{x}, \alpha)$, etc.] will be quasi-independent in an azimuth sense, not all of them will be independent in a spatial sense, which means that spatial averaging in the sense of Eq. 5.31 utilises non-independent estimates and should not be used to estimate errors. Instead, Kirby and Swain (2013) found that if the spatial separation between spatially-adjacent wavelets is twice the e-folding distance ($2s\sqrt{2}$ at a scale s, from Eq. 4.32) then the cross- and auto-scalograms will be quasi-independent in a spatial sense ($|O_{st\alpha}| < 0.1$). But rather than choosing specific scalogram estimates a distance $2s\sqrt{2}$ apart, one can partially-spatial average the cross- and auto-scalograms into spatial blocks of size

$2s\sqrt{2}$ square, the averaging being based on Eqs. 5.33. Then, the number of independent cross- and auto-scalogram estimates in a spatial sense will just be the number of blocks spanning the area, $M_{\bar{x}}(s)$, which is scale-dependent owing to the scale-dependence of block size. So in a space-azimuth sense, the number of independent cross- and auto-scalogram estimates is given by the product $N_\alpha M_{\bar{x}}(s)$. Figure 4.8a provides a schematic indication of the concept of block sizes, albeit in one dimension.

Note that the requirement of independent estimates applies only for error estimation; the admittance and coherency estimates themselves can be computed using non-independent cross- and auto-scalograms, even though this implies redundancy.

5.5.2 Errors from Analytic Formulae

Munk and Cartwright (1966) provide a formula for the standard deviation of the admittance as

$$\sigma_Q(\kappa) = |Q(\kappa)| \sqrt{\frac{\gamma^{-2}(\kappa) - 1}{2N_\kappa}},$$

while Bendat and Piersol (2000) give the formula for the standard deviation of the coherence as

$$\sigma_{\gamma^2}(\kappa) = \left[1 - \gamma^2(\kappa)\right] \sqrt{\frac{2\gamma^2(\kappa)}{N_\kappa}}. \tag{5.34}$$

In these equations, κ is used to represent a wavenumber, whether it is $\mathbf{k}$ the 2D wavenumber vector or $|\mathbf{k}|$ the radially-averaged wavenumber used in the multitaper method, or k_e the equivalent Fourier wavenumber used in the fan wavelet method.

The symbol N_κ is the number of independent estimates of the admittance/coherence at the wavenumber κ, discussed in Sect. 5.5.1. When 2D multitaper error estimates are required, $N_\kappa = K^2$, the number of tapers. For radially-averaged multitaper errors, $N_\kappa = K^2 N_m$, the product of the number of tapers and the number of (k_x, k_y) pairs in the (half-)annulus m. If local wavelet admittance/coherence errors are required, $N_\kappa = N_\alpha$, the number of Morlet wavelet azimuths comprising the fan. For the global wavelet method we have $N_\kappa = N_\alpha M_{\bar{x}}(s)$, the product of the number of Morlet wavelet azimuths and the number of spatial blocks at scale s (see Sect. 5.5.1).

Note that one can use Eq. 5.34 to compute the errors on the squared real coherency (SRC, Eq. 5.17), by simply replacing the $\gamma^2(\kappa)$ by $\Gamma_R^2(\kappa)$. The same applies to the squared imaginary coherency.

5.5.3 Jackknife Error Estimates

Another method widely used to compute error estimates is through *jackknifing* (Thomson and Chave 1991; Thomson 2007). The general principle of jackknifing can be found in the Appendix to this chapter. Kirby and Swain (2013) show how the jackknife method may be applied to multitaper and wavelet power spectra to obtain errors on those quantities, but Thomson and Chave (1991) and Thomson (2007) present a discussion on its application to the admittance, coherency and coherence, summarised here, beginning with the 2D multitaper method (Sect. 5.4.1) as an example.

5.5.3.1 2D Multitaper Jackknife Errors

Since the Slepian tapers are orthogonal, there are K^2 independent estimates of the cross- and auto-spectra, one for each of the $K \times K$ tapers: these are the (j, k)th cross- and auto-eigenspectra from Eqs. 5.21, which we now write as a sequence $\hat{S}_i^{(gh)}(\mathbf{k})$ ($i = 1, \ldots, K^2$, where $i = j + (k - 1)K$, and $j, k = 1, 2, \ldots, K$), and similarly for the two auto-spectra. The delete-one mean is then formed by summing over all tapers except one, or

$$\bar{S}_{gh,\backslash l}(\mathbf{k}) = \frac{1}{K^2 - 1} \sum_{i=1(i \neq l)}^{K^2} \hat{S}_i^{(gh)}(\mathbf{k}), \tag{5.35}$$

($l = 1, \ldots, K^2$) by analogy with Eq. 5.22a. The same procedure can be applied to the two auto-spectra, to get $\bar{S}_{gg,\backslash l}(\mathbf{k})$ and $\bar{S}_{hh,\backslash l}(\mathbf{k})$. Note that when there is only one taper ($K^2 = 1$), the jackknife method cannot be used. In a slight departure from the general outline of jackknife errors in the Appendix, we then form a 'delete-one mean admittance' through

$$\bar{Q}_{\backslash l}(\mathbf{k}) = \frac{\bar{S}_{gh,\backslash l}(\mathbf{k})}{\bar{S}_{hh,\backslash l}(\mathbf{k})}, \tag{5.36}$$

and a mean of these delete-one means by

$$\bar{Q}_{\backslash \bullet}(\mathbf{k}) = \frac{1}{K^2} \sum_{l=1}^{K^2} \bar{Q}_{\backslash l}(\mathbf{k}).$$

The jackknife variance of the admittance is then given by

$$\mathrm{var}\{\bar{Q}\}(\mathbf{k}) = \frac{K^2 - 1}{K^2} \sum_{l=1}^{K^2} \left[\bar{Q}_{\backslash l}(\mathbf{k}) - \bar{Q}_{\backslash \bullet}(\mathbf{k}) \right]^2,$$

and the jackknife error on the admittance by its square root, or

$$\sigma_Q(\mathbf{k}) = \sqrt{\text{var}\{\bar{Q}\}(\mathbf{k})}.$$

As one might expect, the same procedure can be applied to obtain jackknife errors on the coherency, coherence, SRC and SIC. Instead of Eq. 5.36, one computes

$$\overline{\Gamma}_{\backslash l}(\mathbf{k}) = \frac{\bar{S}_{gh,\backslash l}(\mathbf{k})}{\sqrt{\bar{S}_{gg,\backslash l}(\mathbf{k})\,\bar{S}_{hh,\backslash l}(\mathbf{k})}},$$

for the coherency, and

$$\overline{\gamma^2}_{\backslash l}(\mathbf{k}) = \frac{|\bar{S}_{gh,\backslash l}(\mathbf{k})|^2}{\bar{S}_{gg,\backslash l}(\mathbf{k})\,\bar{S}_{hh,\backslash l}(\mathbf{k})},$$

for the coherence, and then proceeds as for the admittance, above.

5.5.3.2 Radially-Averaged Multitaper Jackknife Errors

When radially-averaged multitaper error estimates are required, the independent, deleted observations are the radially-averaged cross- and auto-eigenspectra from Eqs. 5.23, which we, again, write as a sequence: $\bar{S}_i^{(gh)}(|\mathbf{k}|)$ ($i = 1, \ldots, K^2$), and similarly for the two auto-spectra. The equivalent of Eq. 5.35 in this case is obtained from Eq. 5.24a, or

$$\bar{S}_{gh,\backslash l}(|\mathbf{k}|) = \frac{1}{K^2 - 1} \sum_{i=1(i\neq l)}^{K^2} \bar{S}_i^{(gh)}(|\mathbf{k}|)$$

($l = 1, \ldots, K^2$). One then proceeds with Eq. 5.36 to arrive at $\sigma_Q(|\mathbf{k}|)$. Note that here we treat the radially-averaged cross- and auto-eigenspectra as the independent estimates, rather than the individual estimates within the annulus at each taper. Hence the number of independent estimates is K^2 rather than the $K^2 N_m$ used in the analytic formula.

5.5.3.3 Local Wavelet Jackknife Errors

Jackknife error estimates on the local wavelet admittance and coherency are obtained in a similar fashion. Here, though, the deleted observations are the Morlet wavelet cross- and auto-scalograms from Eqs. 5.28, $S_{gh}(s, \mathbf{x}, \alpha_i)$ ($i = 1, \ldots, N_\alpha$), and similarly for the two auto-spectra, provided N_α is calculated as described in Sect. 5.5.1. The equivalent of Eq. 5.35 here is obtained from Eq. 5.29a, or

$$S_{gh,\backslash l}(s, \mathbf{x}) = \frac{1}{N_\alpha - 1} \sum_{i=1(i \neq l)}^{N_\alpha} S_{gh}(s, \mathbf{x}, \alpha_i),$$

($l = 1, \ldots, N_\alpha$) finding $S_{gg,\backslash l}(s, \mathbf{x})$ and $S_{hh,\backslash l}(s, \mathbf{x})$ in a similar fashion. A delete-one mean admittance is then found from

$$\bar{Q}_{\backslash l}(s, \mathbf{x}) = \frac{S_{gh,\backslash l}(s, \mathbf{x})}{S_{hh,\backslash l}(s, \mathbf{x})}. \tag{5.37}$$

One then proceeds as in Sect. 5.5.3.1, forming a mean of these delete-one mean admittances by

$$\bar{Q}_{\backslash \bullet}(s, \mathbf{x}) = \frac{1}{N_\alpha} \sum_{l=1}^{N_\alpha} \bar{Q}_{\backslash l}(s, \mathbf{x}),$$

the jackknife variance of the admittance by

$$\mathrm{var}\{\bar{Q}\}(s, \mathbf{x}) = \frac{N_\alpha - 1}{N_\alpha} \sum_{l=1}^{N_\alpha} \left[\bar{Q}_{\backslash l}(s, \mathbf{x}) - \bar{Q}_{\backslash \bullet}(s, \mathbf{x}) \right]^2,$$

and the jackknife error on the admittance by

$$\sigma_Q(s, \mathbf{x}) = \sqrt{\mathrm{var}\{\bar{Q}\}(s, \mathbf{x})}.$$

The same procedure can be applied to obtain jackknife errors on the coherency, coherence, SRC and SIC. Instead of Eq. 5.37, one computes

$$\overline{\Gamma}_{\backslash l}(s, \mathbf{x}) = \frac{\bar{S}_{gh,\backslash l}(s, \mathbf{x})}{\sqrt{\bar{S}_{gg,\backslash l}(s, \mathbf{x}) \, \bar{S}_{hh,\backslash l}(s, \mathbf{x})}},$$

for the coherency, and

$$\overline{\gamma^2}_{\backslash l}(s, \mathbf{x}) = \frac{|\bar{S}_{gh,\backslash l}(s, \mathbf{x})|^2}{\bar{S}_{gg,\backslash l}(s, \mathbf{x}) \, \bar{S}_{hh,\backslash l}(s, \mathbf{x})},$$

for the coherence, and then proceeds as for the admittance, above.

A comparison of analytic-formula errors against jackknifed error estimates is shown in Fig. 5.12. Here, 1D profiles of the SRC and its errors were extracted from local wavelet estimates at a location $\mathbf{x}_0$ in North America, being $\Gamma_R^2(s, \mathbf{x}_0)$ and $\sigma_{\Gamma_R^2}(s, \mathbf{x}_0)$, respectively. The figure shows that the jackknife errors are compatible with the errors derived from the analytic formula, Eq. 5.34. Kirby and Swain (2014) provide a further comparison of analytic and jackknife admittance errors using both multitapers and wavelets.

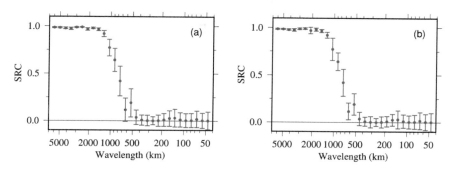

Fig. 5.12 The wavelet squared real coherency between the Bouguer gravity anomaly and topography at a location in central North America, showing **a** errors from the analytic formula, Eq. 5.34, and **b** jackknife error estimates

5.5.3.4 Global Wavelet Jackknife Errors

The procedure to estimate global wavelet jackknife error estimates devised by Kirby and Swain (2013) is slightly different due to the fact that the wavelet coefficients are not mutually independent at all spatial locations, as described in Sect. 5.5.1. Hence, the deleted observations are not the fan wavelet cross- and auto-scalograms [$S_{gh}(s, \mathbf{x})$, etc.], but rather partially-spatially-averaged Morlet wavelet cross- and auto-scalograms defined by

$$S_{gh}(s, \bar{\mathbf{x}}, \alpha) = \frac{1}{N_{\bar{\mathbf{x}}}(s)} \sum_{\bar{\mathbf{x}}} S_{gh}(s, \mathbf{x}, \alpha),$$

$$S_{gg}(s, \bar{\mathbf{x}}, \alpha) = \frac{1}{N_{\bar{\mathbf{x}}}(s)} \sum_{\bar{\mathbf{x}}} S_{gg}(s, \mathbf{x}, \alpha),$$

$$S_{hh}(s, \bar{\mathbf{x}}, \alpha) = \frac{1}{N_{\bar{\mathbf{x}}}(s)} \sum_{\bar{\mathbf{x}}} S_{hh}(s, \mathbf{x}, \alpha),$$

where the $S_{gh}(s, \mathbf{x}, \alpha)$, etc., are given by Eqs. 5.28. The partial spatial averaging (Sect. 5.4.2.3) is performed over spatial blocks of size $2s\sqrt{2}$ square, giving $N_{\bar{\mathbf{x}}}(s)$ original-grid cells per spatial block and $M_{\bar{\mathbf{x}}}(s)$ blocks at scale s (Sect. 5.5.1). Therefore, in a space-azimuth sense there are a total of $N_\chi(s) = N_\alpha M_{\bar{\mathbf{x}}}(s)$ independent cross- and auto-scalogram estimates, $S_{gh}(s, \bar{\mathbf{x}}, \alpha)$, etc., with N_α calculated as described in Sect. 5.5.1. $N_\chi(s)$ is scale-dependent.

To form a delete-one estimate, let $S_{gh,i}(s, \bar{\mathbf{x}}, \alpha)$ be the ith cross-scalogram at scale s, averaged spatial block $\bar{\mathbf{x}}_m$ ($m = 1, \ldots, M_{\bar{\mathbf{x}}}(s)$), and azimuth α_n ($n = 1, \ldots, N_\alpha$), where i is a unique identifier of m and n; i.e. $S_{gh,i}(s, \bar{\mathbf{x}}, \alpha) \equiv S_{gh}(s, \bar{\mathbf{x}}_m, \alpha_n)$, where $i = 1, \ldots, N_\chi(s)$. The delete-one equivalent of Eq. 5.35 is hence

$$S_{gh,\backslash l}(s) = \frac{1}{N_\chi(s) - 1} \sum_{i=1(i \neq l)}^{N_\chi(s)} S_{gh,i}(s, \bar{\mathbf{x}}, \alpha),$$

($l = 1, \ldots, N_\chi(s)$) and similarly for the two auto-spectra, the summation occurring over space and azimuth together. This gives a 'delete-one mean admittance' as

$$\bar{Q}_{\backslash l}(s) = \frac{S_{gh,\backslash l}(s)}{S_{hh,\backslash l}(s)}.$$

The usual procedure is then followed. The mean of these delete-one mean admittances is calculated by

$$\bar{Q}_{\backslash \bullet}(s) = \frac{1}{N_\chi(s)} \sum_{l=1}^{N_\chi(s)} \bar{Q}_{\backslash l}(s),$$

with the jackknife variance of the admittance being

$$\text{var}\{\bar{Q}\}(s) = \frac{N_\chi(s) - 1}{N_\chi(s)} \sum_{l=1}^{N_\chi(s)} \left[\bar{Q}_{\backslash l}(s) - \bar{Q}_{\backslash \bullet}(s) \right]^2.$$

Finally, the jackknife error on the admittance is given by

$$\sigma_Q(s) = \sqrt{\text{var}\{\bar{Q}\}(s)}.$$

One follows a similar procedure to obtain jackknife errors on the global wavelet coherency, coherence, squared real coherency, and squared imaginary coherency.

5.6 Wavenumber/Wavelength Uncertainty

As discussed in Chaps. 2, 3 and 4, the frequency/wavenumber resolution of the spectral analysis methods is not perfect. This imparts an uncertainty to the wavenumber location of each spectral estimate, in addition to the uncertainty/error in the value of the estimate discussed in Sect. 5.5.

5.6.1 Slepian Tapers

In Sect. 3.3.4 we saw how Eq. 3.25 gave us the bandwidth, W, of the discrete prolate spheroidal sequences (the Slepian tapers) and therefore the frequency-domain resolution of the power spectrum calculated from them. If we write Eq. 3.25 in terms of the wavenumber (angular spatial frequency) it becomes

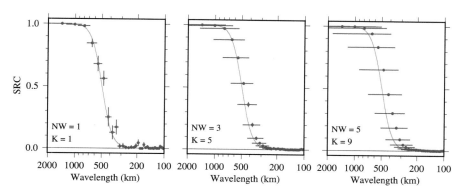

Fig. 5.13 The multitaper squared real coherency between some synthetic data for various values of time-bandwidth product, NW, and number of tapers, K. The wavelength errors are from Eq. 5.39, while SRC errors are jackknife error estimates. The green curve in each panel shows the reference coherence of the synthetic model

$$W = 2\pi \frac{\text{NW}}{N \Delta x},$$ (5.38)

measured in rad/km, where NW is the time-bandwidth product, N is the number of data points in the signal/sequence, and Δx is the space-domain sampling interval. Hence, the resolution is wavenumber-independent, is proportional to NW, and inversely proportional to the data length ($L = N \Delta x$). So low values of NW provide better wavenumber-domain resolution than do higher values.

Figure 5.13 shows the squared real coherency (SRC) of some synthetic data, where the horizontal bars on the SRC estimates show the uncertainty in their wavelength. If the uncertainty in the wavenumber is $k \pm W$, then we can write the upper and lower limits of uncertainty in wavelength, λ, as

$$\lambda_+ = \frac{2\pi}{k - W}, \qquad \lambda_- = \frac{2\pi}{k + W},$$ (5.39)

respectively, where W is given by Eq. 5.38. So for wavelength $\lambda = 2\pi/k$, its limits of uncertainty are $\lambda_- \leq \lambda \leq \lambda_+$. As expected, the wavenumber-domain resolution of the SRC estimates is dependent upon NW, and while it is independent of wavenumber, the logarithmic scale of wavelength in Fig. 5.13 makes the resolution appear wavelength-dependent.

Also of note in Fig. 5.13 is the decrease in SRC error as NW increases. This is actually a consequence of an increase in the number of tapers, K, with the maximum allowable number of tapers being $K \leq 2\text{NW} - 1$ (Sect. 3.3.5). As shown by Eq. 3.28, the variance of the average spectrum is inversely proportional to the number of tapers used, or

$$\text{var}\{\hat{S}(k)\} = \frac{1}{K}.$$

Thus, once again, we see the trade-off between resolution and variance: using more tapers reduces the variance but degrades the resolution by virtue of requiring a larger NW value (Sect. 3.4).

5.6.2 Fan Wavelet

In Sect. 4.11.3 we developed an expression for the variance of a 2D Morlet wavelet in the wavenumber domain, and used its square root, the standard deviation, as a measure of its resolution, being

$$\sigma_\psi = s^{-1}$$

(Eq. 4.52). If we substitute for scale from Eq. 4.41, we obtain an expression for the standard deviation in terms of the equivalent Fourier wavenumber, k_e, as

$$\sigma_\psi = \frac{k_e}{|\mathbf{k}_0|}, \tag{5.40}$$

where $|\mathbf{k}_0|$ is the central wavenumber of the Morlet wavelet. So, unlike the Slepian tapers whose resolution is wavenumber-independent, the wavenumber uncertainty of wavelets is proportional to wavenumber, the uncertainty increasing (or resolution decreasing) at higher wavenumbers. It is also inversely proportional to central wavenumber, with high-$|\mathbf{k}_0|$ wavelets possessing better resolution in the wavenumber domain.

In terms of wavelength, if the uncertainty in the equivalent Fourier wavenumber is $k_e \pm \sigma_\psi$, then we can write the upper and lower limits of uncertainty in equivalent Fourier wavelength, λ_e, as

$$\lambda_{e+} = \frac{2\pi}{k_e - \sigma_\psi}, \qquad \lambda_{e-} = \frac{2\pi}{k_e + \sigma_\psi}, \tag{5.41}$$

respectively, where σ_ψ is given by Eq. 5.40. So, for equivalent Fourier wavelength $\lambda_e = 2\pi/k_e$, its limits of uncertainty are $\lambda_{e-} \le \lambda_e \le \lambda_{e+}$. The upper and lower limits of equivalent Fourier wavelength uncertainty for wavelet SRC estimates are plotted in Fig. 5.14. Although the dependence of wavelength upon uncertainty is not visually apparent (due to the logarithmic axes), the dependence of central wavenumber is, with low-$|\mathbf{k}_0|$ wavelets possessing poorer wavenumber-domain resolution.

5.7 Summary

The admittance and coherency are the fundamental observables of spectral elastic thickness estimation methods, describing the frequency-domain relationship between

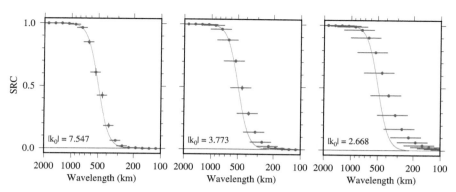

Fig. 5.14 As Fig. 5.13 but for the fan wavelet method with the indicated $|\mathbf{k}_0|$ values. The wavelength errors are from Eq. 5.41, while SRC errors are jackknife error estimates

gravity and topography data. They are complex quantities, and while their real parts provide the information necessary to estimate elastic thickness, their imaginary parts are useful in the identification of noise. However, their estimation is not trivial, and this chapter described methods to best undertake this task. The multitaper method may be used to obtain 2D plots of these quantities, or perhaps more usefully 1D radially-averaged profiles. This method can also be used to map spatial variations in admittance and coherence through the use of moving windows, but the fan wavelet method is more suited to this as it is precisely what the wavelet transform was developed for. Errors on the admittance and coherence can readily be estimated using both methods, through analytic formulae or application of the jackknife method.

5.8 Further Reading

Texts on LTI system theory and transfer functions (the admittance) abound, particularly in the electrical engineering literature. Books devoted to the coherence and coherency are less common, but those that discuss coherence will generally also discuss admittance, and Bendat and Piersol (2000) is such a book. Percival and Walden (1993), of course, is an excellent read and has a whole chapter on LTI systems, but does not cover coherence.

As one might expect, there are a multitude of articles in the journal literature, but in our context, three that stand out are Munk and Cartwright (1966) and Dorman and Lewis (1970) for their early use of admittance, and Simons et al. (2000) who go into detail on multitaper estimation of the coherence. But for further information on the admittance, coherency and coherence not covered here, especially their historical development and usage within flexural studies, the review article by Kirby (2014) is worth a read, while the articles by Simons and Olhede (2013) and Audet (2014) also contain review sections.

Appendix

Synthetic Modelling and Testing

Any newly developed method should be tested to see how well it will perform when used 'in anger'. This is often achieved by generating synthetic models with known parameters, and then applying the new method to the models to see if it can retrieve these parameters. In this chapter we want to see whether the admittance and coherency estimated using multitapers or wavelets are meeting expectations or are wildly off. To this end we generate some synthetic models with known ('predicted') admittance and coherency. Then we compute the admittance and coherency of the synthetic models using multitapers and/or wavelets, and compare those with the theoretical predictions.

The synthetic models we use here are random, fractal surfaces, generated using the spectral synthesis method of Saupe (1988). For more information see Sect. 10.2.

Synthetic Admittance

Two grids with a known admittance between them are computed in the following manner. First generate a random, fractal surface, $h(\mathbf{x})$, using the method of Saupe (1988). Then take its Fourier transform, and multiply that by the known admittance, $Q_0(\mathbf{k})$, to form a second spectrum, $G(\mathbf{k})$, where

$$G(\mathbf{k}) \;=\; Q_0(\mathbf{k})\,H(\mathbf{k}).$$

Here we use the isotropic function

$$Q_0(\mathbf{k}) \;=\; e^{-(|\mathbf{k}|-k_0)^2/10}\left[1 + i\,\mathrm{sgn}(k_y)\right] \tag{5.42}$$

as the known admittance, where $\mathrm{sgn}(x)$ is the sign function[10] whose presence ensures that the imaginary part of Q_0 is an odd function, and therefore that Q_0 is Hermitian. Equation 5.42 for $k_0 = 2\pi/\lambda_0$ with $\lambda_0 = 500$ km is shown in Fig. 5.1c, d.

Synthetic Coherence

Two grids with a known coherence between them are computed using the approach used by Lowry and Smith (1994), in the following manner. First generate two random, fractal surfaces, $h(\mathbf{x})$ and $p(\mathbf{x})$, using the method of Saupe (1988). Then take their Fourier transforms and from $P(\mathbf{k})$ create a new spectrum, $N(\mathbf{k})$, with the same phase

---

[10] The sign, or 'signum', function, $\mathrm{sgn}(x)$, has the value $+1$ when $x > 0$, 0 when $x = 0$, and -1 when $x < 0$.

as $P(\mathbf{k})$ but with amplitude

$$|N(\mathbf{k})| = \frac{|H(\mathbf{k})|}{|P(\mathbf{k})|} \sqrt{\frac{1 - \gamma_0^2(\mathbf{k})}{\gamma_0^2(\mathbf{k})}},$$

where the known coherence is $\gamma_0^2(\mathbf{k})$. The new surface is thus given by $N(\mathbf{k}) = |N(\mathbf{k})|e^{i\varphi_P(\mathbf{k})}$. Then take the inverse Fourier transform of $N(\mathbf{k})$ to get the surface $n(\mathbf{x})$, and form a third surface $g(\mathbf{x})$ by

$$g(\mathbf{x}) = h(\mathbf{x}) + n(\mathbf{x}).$$

To create the synthetic models used in Figs. 5.4 and 5.5, the known coherence is given by the *logistic function*

$$\gamma_0^2(\mathbf{k}) = \frac{1}{1 + e^{(|\mathbf{k}|-k_t)/2}}, \tag{5.43}$$

for $k_t = 2\pi/\lambda_t$ with $\lambda_t = 200$ km, shown in Fig. 5.5. The nature of the logistic function means that the surface $n(\mathbf{x})$ has almost zero amplitude at long wavelengths and high amplitude at short wavelengths. When this is added to $h(\mathbf{x})$ to get $g(\mathbf{x})$, at long wavelengths the amplitudes and phases of $g(\mathbf{x})$ are dominated by those of $h(\mathbf{x})$ so there is high coherence between $g(\mathbf{x})$ and $h(\mathbf{x})$. At short wavelengths the amplitudes and phases of $g(\mathbf{x})$ are dominated by those of $n(\mathbf{x})$ so there is low coherence between $g(\mathbf{x})$ and $h(\mathbf{x})$.

The synthetic models used in Fig. 5.7 were generated with

$$\gamma_0^2(\mathbf{k}) = 1 - 0.9\,e^{-(|\mathbf{k}|-k_0)^2/4}, \tag{5.44}$$

as the known coherence, for $k_0 = 2\pi/\lambda_0$ with $\lambda_0 = 200$ km.

Jackknife Error Estimates

The jackknife method (Efron and Gong 1983; Thomson 2007) allows for an estimation of the errors of a set of data using only the data themselves—no theoretical equations are involved. The only condition is that there must be repeated, independent observations of the same data.

By way of an example, consider a set of N independent observations of some time series $y_i(t)$ $(i = 1, \ldots, N)$. Begin by removing the first time-series observation $(y_1(t))$ from the set, then computing a mean value at each value of t from the remaining $N - 1$ observations. Repeat this, removing the second observation $(y_2(t))$, then the third $(y_3(t))$, etc., until there are N of these so-called 'delete-one' means. In general, for a deleted observation y_j, the delete-one mean is

$$\bar{y}_{\backslash j}(t) \;=\; \frac{1}{N-1} \sum_{i=1(i \neq j)}^{N} y_i(t),$$

where $j = 1, \ldots, N$, which can be shown to also be given by

$$\bar{y}_{\backslash j}(t) \;=\; \frac{1}{N-1} \left(\sum_{i=1}^{N} y_i(t) \;-\; y_j(t) \right).$$

Next, take the mean of all the delete-one means, or

$$\bar{y}_{\backslash \bullet}(t) \;=\; \frac{1}{N} \sum_{j=1}^{N} \bar{y}_{\backslash j}(t),$$

which is also equal to the mean of the observations themselves, or

$$\bar{y}_{\backslash \bullet}(t) \;=\; \bar{y}(t) \;=\; \frac{1}{N} \sum_{i=1}^{N} y_i(t).$$

Finally, the jackknife variance of the delete-one means is obtained by subtracting the mean of all the delete-one means from each delete-one mean, then squaring and summing, using the formula

$$\mathrm{var}\{\bar{y}\}(t) \;=\; \frac{N-1}{N} \sum_{j=1}^{N} \left[\bar{y}_{\backslash j}(t) - \bar{y}_{\backslash \bullet}(t) \right]^2,$$

where the $N-1$ term reflects the fact that the delete-one means are not independent (Thomson and Chave 1991; Thomson 2007). The jackknife error estimate is provided by the corresponding standard deviation, through

$$\sigma_y(t) \;=\; \sqrt{\mathrm{var}\{\bar{y}\}(t)}.$$

References

Audet P (2014) Toward mapping the effective elastic thickness of planetary lithospheres from a spherical wavelet analysis of gravity and topography. Phys Earth Planet Inter 226:48–82

Bendat JS, Piersol AG (2000) Random data: analysis and measurement procedures, 3rd edn. Wiley, New York

Bracewell RN (1986) The Fourier transform and its applications. McGraw-Hill, New York

Dorman LM, Lewis BTR (1970) Experimental isostasy, 1: theory of the determination of the Earth's isostatic response to a concentrated load. J Geophys Res 75:3357–3365

Efron B, Gong G (1983) A leisurely look at the bootstrap, the jackknife, and cross-validation. Am Stat 37:36–48

Kirby JF (2014) Estimation of the effective elastic thickness of the lithosphere using inverse spectral methods: the state of the art. Tectonophys 631:87–116

Kirby JF (2019) On the pitfalls of Airy isostasy and the isostatic gravity anomaly in general. Geophys J Int 216:103–122

Kirby JF, Swain CJ (2009) A reassessment of spectral T_e estimation in continental interiors: the case of North America. J Geophys Res 114(B8), B08401. https://doi.org/10.1029/2009JB006356

Kirby JF, Swain CJ (2013) Power spectral estimates using two-dimensional Morlet-fan wavelets with emphasis on the long wavelengths: jackknife errors, bandwidth resolution and orthogonality properties. Geophys J Int 194:78–99

Kirby JF, Swain CJ (2014) The long wavelength admittance and effective elastic thickness of the Canadian shield. J Geophys Res Solid Earth 119:5187–5214

Lewis BTR, Dorman LM (1970) Experimental isostasy, 2: An isostatic model for the USA derived from gravity and topography data. J Geophys Res 75:3367–3386

Lowry AR, Smith RB (1994) Flexural rigidity of the Basin and Range–Colorado Plateau–Rocky Mountain transition from coherence analysis of gravity and topography. J Geophys Res 99(B10):20,123–20,140

Macario A, Malinverno A, Haxby WF (1995) On the robustness of elastic thickness estimates obtained using the coherence method. J Geophys Res 100(B8):15,163–15,172

Munk WH, Cartwright DE (1966) Tidal spectroscopy and prediction. Philos Trans R Soc Lond A 259:533–581

Percival DP, Walden AT (1993) Spectral analysis for physical applications. Cambridge University Press, Cambridge

Saupe D (1988) Algorithms for random fractals. In: Peitgen H-O, Saupe D (eds) The science of fractal images. Springer, New York, pp 71–136

Simons FJ, Olhede SC (2013) Maximum-likelihood estimation of lithospheric flexural rigidity, initial-loading fraction and load correlation, under isotropy. Geophys J Int 193:1300–1342

Simons FJ, Zuber MT, Korenaga J (2000) Isostatic response of the Australian lithosphere: estimation of effective elastic thickness and anisotropy using multitaper spectral analysis. J Geophys Res 105(B8):19,163–19,184

Thomson DJ (2007) Jackknifing multitaper spectrum estimates. IEEE Signal Process Mag 24:20–30

Thomson DJ, Chave AD (1991) Jackknifed error estimates for spectra, coherences, and transfer functions. In: Haykin S (ed) Advances in spectrum analysis and array processing, vol 1. Prentice Hall, Englewood Cliffs, pp 58–113

Chapter 6
Map Projections

6.1 Introduction

The spectral estimation topics discussed in Part II are given in planar coordinate systems: $\mathbf{x} = (x, y)$ in the space domain, and $\mathbf{k} = (k_x, k_y)$ in the wavenumber domain. However, gravity and topography data are collected on the Earth's surface—a curved surface—and are located in terms of their latitude (ϕ) and longitude (λ). While techniques exist to obtain spectra of data coordinated on the surface of a sphere, notably spherical harmonics, this book focusses on the plane.

Geodetically, the Earth is best modelled by an *oblate ellipsoid of revolution* rather than a sphere: its equatorial semi-major axis is approximately 21 km larger than its polar semi-minor axis, the equatorial bulge being caused by the planet's rotation. However, for certain problems of mathematics or physics, their solutions in ellipsoidal coordinate systems are significantly more difficult to obtain than the approximate, equivalent spherical solutions; hence the widespread use of spherical, rather than ellipsoidal, harmonics when modelling planetary geoids, for example. Roughly speaking, the error incurred when making a spherical approximation to an ellipsoidal planet is of the order of the planet's geodetic flattening, or ∼0.3% for the Earth. The error incurred when making a planar approximation of a spherical or ellipsoidal surface is the topic of this chapter.

The problem of accurately representing Earth-surface features on a flat map dates from the time of the ancient Greeks, who devised at least five projections, all of which are still in use today (for example, the orthographic, stereographic and gnomonic). The problem became more pressing during the age of global exploration, from the sixteenth through to the eighteenth centuries, when the famous Mercator and Lambert conic conformal were born. Since the nineteenth century dozens, even hundreds, of projections have been developed, sometimes for a single application only, never to be used again.

The problem is akin to peeling an orange, say, or better still, a tennis ball—if one can peel that—and trying to press the skin flat upon a table top. The orange

© Springer Nature Switzerland AG 2022
J. Kirby, *Spectral Methods for the Estimation of the Effective Elastic Thickness of the Lithosphere*, Advances in Geophysical and Environmental Mechanics and Mathematics, https://doi.org/10.1007/978-3-031-10861-7_6

peel rips; the rubber of the tennis ball stretches, or part of it may be compressed, and tearing may also occur. The same happens with a map projection, though now mathematically. Latitudes and longitudes are squeezed to nothingness or stretched out to infinity; countries and continents are distorted into gruesome shapes and even ripped asunder. If Lovecraft would write a geometric horror story, it would involve map projections in a big way.

The aim, then, is to minimise this distortion, though just as it is impossible to peel an orange and press its skin flat without distortion, it is impossible to devise a map projection that possesses no distortion at all. The map projection must therefore be designed with a particular purpose in mind so that, while some qualities and properties of Earth-surface features are sacrificed, others will be faithfully preserved on the map. In the rest of this chapter we will look at these properties and their distortions, always with a mind on map projections that will give the most accurate spectra using planar spectral estimation methods.

6.2   Types of Map Projection

The many different map projections are mostly variations on nine themes, there being three shapes of surface on which to project, and three properties to conserve during projection (only one of which may be chosen at a time).

6.2.1   Developable Surfaces

The simplest map projection is also the oldest, the *gnomonic* projection, devised in the sixth century BCE by the Greek philosopher and mathematician Thales. Here, the map is lain flat on the sphere (or ellipsoid) touching it at a tangent point, which is usually chosen to be the origin of the projected coordinates. In Fig. 6.1 the tangent point is chosen to be the North Pole but it should be located at the centre of the region one wants to map. Then, the map is made by projecting features on the surface of the sphere (e.g. point A in Fig. 6.1) along their radius vector to a corresponding point (B) on the map. If the plane of Fig. 6.1 is extended out of the page into three dimensions then both latitudes and longitudes can be projected in this fashion.

There are, naturally, limitations. First, the projection cannot map a whole hemisphere. If point A lies on the equator in Fig. 6.1, then its radius vector is parallel to the map, and point B will lie an infinite distance from the origin. Second, the distortion can be seen to increase with increasing distance from the origin. The three radius vectors to the right of the origin all map out equal distances (l) on the map, though the corresponding surface spherical distances of the sectors (s) decrease as θ increases. So, on a gnomonic map, at locations close to the origin, 1 cm on the map might represent 100 km on the Earth's surface (giving a map scale of 1:10,000,000 there),

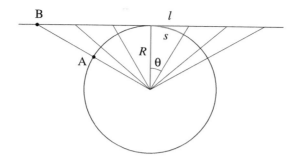

Fig. 6.1 The principle of an azimuthal map projection, in this case the *gnomonic*. Points on the sphere are mapped onto the map plane along radii (e.g. A to B). Using $s = R\theta$ and $l = R\tan\theta$, one can relate distances on the map plane to equivalent distances on the sphere, giving $l = s\tan\theta/\theta$

but further away 1 cm might only represent 50 km on the planet's surface (giving a map scale of 1:5,000,000 at that point).[1]

The projection in Fig. 6.1 is of a type called an *azimuthal* projection because it uses a flat map sheet on which to make the projection. There are many different azimuthal projections though, and one way of devising new ones is to vary the location of the projection point. When this point is at the centre of the sphere, as in Fig. 6.1, the gnomonic projection results. If the projection point occurs at the antipode to the tangent point (which would be the South Pole in Fig. 6.1) then we obtain the *stereographic* projection. Extend the projection point to infinity in the other direction, 'above' the tangent point, and the *orthographic* projection is formed, which is how the Earth looks from deep space, given a powerful telescope.

The map sheet, however, does not need to be flat when making the projection and can be rolled into different shapes to surround the sphere, increasing the area that can be mapped without significant distortion. Besides a flat sheet, the other two most common geometries are a cylinder and a cone, giving *cylindrical* and *conic* map projections. These sheets are called *developable surfaces* and each has advantages and disadvantages depending on the purpose of the map projection, discussed as we proceed through the chapter. As a general—but not exclusive—rule, cylindrical projections tend to have a rectangular *graticule* (the collective term for parallels of latitude and meridians of longitude), such as the Mercator projection shown in Fig. 6.2a. Conic projections resemble an expanded cone with straight-line meridians and circular parallels, for instance the Albers projection shown in Fig. 6.2b. And azimuthal projections tend to be plotted as circles, like the azimuthal equidistant projection shown in Fig. 6.2c, though their graticule geometry depends upon the location of the projection point.

The developable surface can penetrate the surface of the sphere. Projections such as that shown in Figs. 6.1 and 6.3a are called *tangent projections* because they touch

---

[1] As will be seen in Sect. 6.2.2, this means that the gnomonic projection is not an *equidistant* projection. Its forte, however, is that any straight line drawn between two points on the map plots the shortest distance (the *geodesic*) between the points.

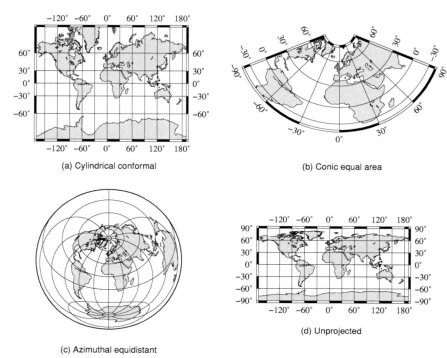

(a) Cylindrical conformal

(b) Conic equal area

(c) Azimuthal equidistant

(d) Unprojected

Fig. 6.2 Four map projections: **a** cylindrical conformal (Mercator); **b** conic equal area (Albers); **c** azimuthal equidistant; **d** 'unprojected' (plate carrée)

the sphere at one point only, or in the case of cylindrical or conic projections along a single line (such as a preferred parallel or meridian). Projections where the developable surface penetrates the surface of the sphere are called *secant* projections (Fig. 6.3b). As will be discussed in Sect. 6.3, the points or lines where the developable surface intersects the sphere have no distortion, meaning that secant projections have a greater area where the distortion is low, compared to tangent projections.

The developable surface can be oriented in space so that the point or lines where it touches or intersects the sphere or ellipsoid—giving zero distortion—lie in, or close to, the region to be mapped. If the cylinder or cone is wrapped around the globe such that its axis of symmetry is parallel to the globe's north-south axis, as in Figs. 6.3 and 6.4a, then this style of projection is called *normal*. Alternatively, if the cylinder or cone is wrapped around the globe such that its axis of symmetry is perpendicular to the globe's north-south axis, and parallel to the equatorial plane, then one obtains a *transverse* projection (Fig. 6.4b). So, in a tangent cylindrical projection of the sphere, a normal projection has the equator as its line of zero distortion, while a transverse projection takes a meridian as the zero distortion line. It is also possible to choose the developable surface to have any orientation, giving *oblique* projections.

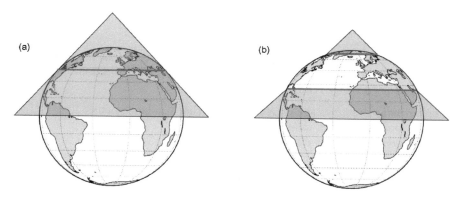

Fig. 6.3 A conic projection in **a** tangent, and **b** secant form. The tangent projection has one standard parallel (the red line) where the cone touches the sphere, and where the distortion is zero. The secant projection has two standard parallels where the cone penetrates the sphere

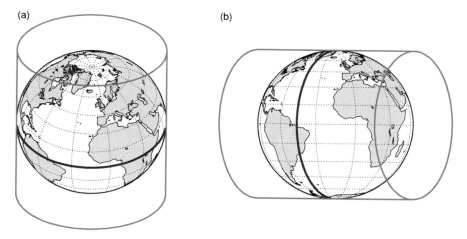

Fig. 6.4 A tangent cylindrical developable surface in **a** normal, and **b** transverse orientation. The blue lines are the lines of zero distortion, where the cylinder touches the sphere

Once the projection has been performed, and the cone or cylinder unrolled, the result is a flat sheet—the map. The curvilinear coordinates of latitude and longitude are now replaced by rectangular Cartesian coordinates referred to as *eastings* (x or E) and *northings* (y or N). In general, the relationship between the four coordinates is not simple, with $x = f_1(\phi, \lambda)$ and $y = f_2(\phi, \lambda)$, for some algebraic functions f_1 and f_2. The reverse transformations, or 'de-projections', are usually even more complicated, with $\phi = g_1(x, y, \phi, \lambda)$ and $\lambda = g_2(x, y, \phi, \lambda)$, in general, for some functions g_1 and g_2. They thus often require the use of iteration for their solution, or truncated infinite series approximations.

6.2.2   Projection Classes

There are three properties that can be preserved when making the projection from the sphere or ellipsoid to the map plane, which identify the projection's *class*. The resulting projections are referred to as:

1. *conformal*—preserves angles, and therefore shape and directions[2];
2. *equal area*—preserves areas;
3. *equidistant*—preserves distances.[3]

For example, if two roads on the Earth's surface cross each other at an angle of 75°, then their projections on a conformal map will also cross at 75°, no matter their overall orientation, nor whereabouts on the map they are located. Alternatively, if the ratio between the areas of two countries is 2:1 in reality, on an equal area map projection the ratio of the projected areas will also be 2:1. In general, though, only one of these properties—angle, area or distance—can be preserved, and the other two must be sacrificed. This leads to large distortions appearing on some maps. For instance the requirement of conformality in the Mercator projection (Fig. 6.2a)— and another enforced property whereby straight lines on a Mercator map represent paths on the Earth that maintain a constant azimuth along their length[4]—leads to grossly exaggerated areas as one approaches the poles, which cannot be mapped at all (Sect. 6.3).

Figure 6.2 shows some combinations of developable surface and class: cylindrical conformal (Mercator), conic equal area (Albers), azimuthal equidistant, and a so-called 'unprojected' projection. This latter example is actually a special case of the cylindrical equidistant projection, called the 'plate carrée', but since its eastings and northings are merely latitudes and longitudes scaled by a constant factor, it is looked down upon by the other, more sophisticated projections and deemed to be 'unprojected'. Nevertheless, it does preserve distance, albeit only along meridians. The poles, however, plot as straight lines.

6.3   Distortion

Distortion is a feature of all map projections. No projection exists that is conformal, equal area and equidistant at once; one of these properties must be selected at the expense of the other two, depending upon the purpose of the map. It therefore

---

[2] In fact many azimuthal projections—whether conformal, equal area or equidistant—will preserve directions, but only relative to one or two points.

[3] This applies only to distances in one selected direction, such as those measured along a meridian, for example. This has often led to the elementary mistake that equidistant maps are best for navigation; in fact this benefit belongs to conformal maps with their preservation of angles (bearings).

[4] This feature was designed by Gerardus Mercator to aid navigation using sixteenth century instruments. Such lines are called *rhumb lines* or *loxodromes*.

becomes useful, and necessary, to quantify the amount of distortion on one's map, so that errors, inaccuracies and limitations may be highlighted.

6.3.1 Scale Factors

The most useful measure of distortion is the *point scale factor*, typically given as two quantities; the scale factor along a meridian (h), and that along a parallel (k). Both numbers are dimensionless and take values in the range $0 < h, k < \infty$. In general, h and k are functions of latitude and longitude (or eastings and northings), and points or lines of zero distortion have $h = 1$ and/or $k = 1$.

The scale factors are defined as follows. Consider two points, P and Q, on the surface of the sphere or ellipsoid, which are separated by a geodesic distance δs, with Q at an azimuth α_{PQ} with respect to P (Fig. 6.5).[5] Under a map projection their projected locations are P′ and Q′, separated by a chord distance δl. The point scale factor, μ, at the geodetic coordinates of P and in the direction α_{PQ} is given by the ratio of the chord distance on the map between P′ and Q′ to the geodesic distance on the ellipsoid between P and Q, in the limit that Q approaches P, or

$$\mu(\phi_P, \lambda_P, \alpha_{PQ}) = \lim_{\delta s \to 0} \frac{\delta l}{\delta s} = \frac{dl}{ds}. \tag{6.1}$$

If P and Q lie on the same meridian ($\alpha_{PQ} = 0°$) one obtains the meridional scale factor, $h(\phi_P, \lambda_P)$; if P and Q lie on the same parallel ($\alpha_{PQ} = 90°$) one obtains the parallel scale factor, $k(\phi_P, \lambda_P)$. Thus, in general, the scale factor depends not only upon the location (ϕ, λ) of the point of interest, but also upon the direction (α) one is facing.

6.3.2 Cylindrical Projections

Specific formulae for h and k vary from projection to projection, as one would expect. As an example, we will consider the cases of some cylindrical projections, and use the spherical approximation as ellipsoidal formulae are slightly more complicated. We will also assume a normal cylinder (Fig. 6.4a), giving a rectangular graticule (e.g. Fig. 6.2a, d). The meridional scale factor is obtained from Eq. 6.1 by letting $dl = dy$

---

[5] Here we use the definitions of azimuth and bearing from geodetic surveying. Consider two points on the ellipsoid, P and Q, and their projected versions on the map, P′ and Q′. The *azimuth*, α, of point Q from point P is the positive-clockwise angle from geodetic north at P (the northerly meridian) to the geodesic connecting P to Q. The *bearing*, β, of point Q′ from point P′ is the positive-clockwise angle from map north at P′ to the *chord* connecting P′ to Q′. The chord is the straight-line distance on the map between P′ and Q′.

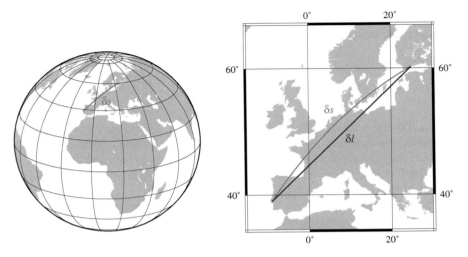

Fig. 6.5 The red line in both plots shows the great circle route (the geodesic) between Lisbon and Helsinki. In the Mercator projection on the right, this is taken to be the geodesic distance, δs, while the blue line is the chord distance, δl. Both are greatly exaggerated, and as per Eq. 6.1, one lets $\delta s \to 0$

(a change in northing only) and $ds = R\,d\phi$ (the arc-length subtended at the centre of a sphere of radius R by a small angle $d\phi$), giving

$$h = \frac{1}{R}\frac{dy}{d\phi}. \tag{6.2}$$

The parallel scale factor is also obtained from Eq. 6.1, by letting $dl = dx$ (a change in easting only) and $ds = R\cos\phi\,d\lambda$ (the arc-length subtended at the centre of a sphere of radius R by a small angle $d\lambda$, taking into account the convergence of the meridians as one approaches the poles), giving

$$k = \frac{1}{R\cos\phi}\frac{dx}{d\lambda}. \tag{6.3}$$

6.3.2.1   Cylindrical Equidistant

Having established the scale factor formulae for a cylindrical projection (Eqs. 6.2 and 6.3), we next choose its class. The simplest case is the equidistant projection, whose formulae are given in Snyder (1987), for example. Cylindrical equidistant eastings are found from

$$x = R\,(\lambda - \lambda_0)\,\cos\phi_s, \tag{6.4}$$

where λ_0 is the *central meridian* or longitude origin, and ϕ_s is the *standard parallel*, the latitude where the normal cylinder touches the sphere in tangent projections (in secant projections there will be two standard parallels, one at $+\phi_s$, the other at $-\phi_s$ degrees latitude). The cylindrical equidistant northings are given by

$$y = R\phi, \tag{6.5}$$

for ϕ in radians. The plate carrée 'projection' (Fig. 6.2d) is the special case where the standard parallel is the equator, $\phi_s = 0$.

Then, the meridional scale factor is found by applying Eq. 6.2 to Eq. 6.5, giving

$$h = 1.$$

The parallel scale factor is found by applying Eq. 6.3 to Eq. 6.4, giving

$$k = \frac{\cos \phi_s}{\cos \phi}. \tag{6.6}$$

This shows how equidistant projections only have that property (equal distances) in one direction, along a meridian in this case of the normal cylindrical equidistant.

6.3.2.2 Cylindrical Equal Area

We next turn to the cylindrical equal area projection, devised by Johann Heinrich Lambert—a titan of the world of map projections—in 1772 (Snyder 1987). In normal orientation, its eastings are given by

$$x = R(\lambda - \lambda_0) \cos \phi_s, \tag{6.7}$$

identical to the cylindrical equidistant eastings of Eq. 6.4, where λ_0 is the central meridian, and ϕ_s the standard parallel of the projection. Its northings are found from

$$y = \frac{R \sin \phi}{\cos \phi_s}. \tag{6.8}$$

Applying Eq. 6.2 to Eq. 6.8 yields the meridional scale factor as

$$h = \frac{\cos \phi}{\cos \phi_s},$$

while the parallel scale factor is found by applying Eq. 6.3 to Eq. 6.7, giving

$$k = \frac{\cos \phi_s}{\cos \phi}. \tag{6.9}$$

In contrast to the cylindrical equidistant projection, both scale factors of the cylindrical equal area projection vary with latitude. Furthermore, they are reciprocals, with

$$hk = 1. \tag{6.10}$$

Equation 6.10 is characteristic of all equal area projections, no matter the developable surface.

6.3.2.3   Cylindrical Conformal

As an example of the cylindrical conformal projection we will study the Mercator projection, devised in 1569 by Gerardus Mercator as a navigational aid (Snyder 1987). Its eastings are calculated using the formula

$$x = R(\lambda - \lambda_0), \tag{6.11}$$

where λ_0 is the central meridian, while its northings are found from

$$y = R \ln \left[\tan \left(\frac{\phi}{2} + \frac{\pi}{4} \right) \right]. \tag{6.12}$$

Note that the standard parallel in the Mercator projection is the equator. As before, we find the meridional scale factor by applying Eq. 6.2 to Eq. 6.12, giving

$$h = \sec \phi,$$

while the parallel scale factor is found by applying Eq. 6.3 to Eq. 6.11, giving

$$k = \sec \phi \tag{6.13}$$

Again, both scale factors vary with latitude, but this time they are equal, with

$$h = k. \tag{6.14}$$

Equation 6.14 is characteristic of all conformal projections, no matter the developable surface, just as Eq. 6.10 is characteristic of all equal area projections.

6.3.3   Tissot's Indicatrix

To recap, we have found that the meridional and parallel scale factors have the following relationships in normal cylindrical projections:

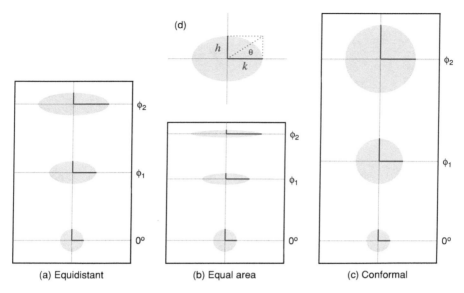

Fig. 6.6 Meridional (h) and parallel (k) scale factors (blue axes), and the corresponding Tissot's indicatrix (blue ellipses), at three latitudes (the equator and two higher latitudes) on an arbitrary meridian, for three normal cylindrical map projections; **a** equidistant, **b** equal area, and **c** conformal. All parallel scale factors vary as $k = \sec \phi$ (for a standard parallel on the equator), whereas the meridional scale factors have values $h = 1$ for the equidistant projection, $h = \cos \phi$ for the equal area projection, and $h = \sec \phi$ for the conformal projection. Panel **d** shows a schematic indicatrix referencing its parameters

1. equidistant: $h = 1$;
2. equal area: $hk = 1$;
3. conformal: $h = k$.

In other orientations besides normal, or with other developable surfaces, the two scale factors may be expressed in other directions besides along meridians and parallels. For example, with some azimuthal and conic projections, h' can represent the scale factor along a line radiating outwards from the origin, while k' gives the scale factor perpendicular to that, as in a polar coordinate system.

Returning to the normal cylindrical case, we see from Eqs. 6.6, 6.9 and 6.13 that $k = \sec \phi$ for all classes of projection (for a standard parallel on the equator: $\phi_s = 0°$). Consider first the equidistant projection, where we have $h = 1$ at any latitude. This is shown in Fig. 6.6a by the magnitude of the axis h having the same value at the equator and at all higher latitudes. Thus, as one proceeds up the meridian, the parallels of latitude in the graticule have an equal spacing on normal cylindrical equidistant projections, with both poles being plottable. However, the magnitude of the axis k increases as latitude increases, varying as $k = \sec \phi$, as just noted. On the equator itself one has $k = 1$, or zero distortion, as expected because the cylinder is tangent here. So, on a sphere of radius 6371 km, one traverses approximately 54' of longitude

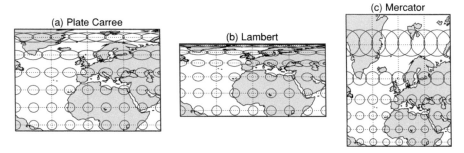

Fig. 6.7 Tissot's indicatrices for three map projections: **a** plate carrée (cylindrical equidistant); **b** Lambert (cylindrical equal area); **c** Mercator (cylindrical conformal)

in moving 100 km along the equator. However, at a latitude of 60°, we have $k = 2$ and a movement of 100 km along this parallel sees one traverse twice the longitude, or 108′. And at a latitude of 89°59′, the value of k is approximately 3438 and a journey of 100 km along this parallel takes one around the globe about eight and a half times. Finally, at the pole the distortion is infinite ($k = \infty$) and it is mapped as an extended straight line. In summary, the meridians of longitude in the graticule of a normal cylindrical equidistant projection are straight and parallel to one another. This is so because while meridians on a sphere converge to the poles as the cosine of latitude, $\cos \phi$, the parallel scale factor varies as $\sec \phi$ and the stretching parallels 'push' the meridians apart at the same rate as they converge: thus they maintain a constant map-distance between them.

If the two scale factors, h and k, are treated as the semi-minor and semi-major axes of an ellipse, then one obtains *Tissot's indicatrix* or the *ellipse of distortion*. This geometric figure is plotted on maps as an ellipse or circle, showing the amount of distortion in a certain direction at each part of the map. At the point(s) where the developable surface touches or penetrates the sphere or ellipsoid both scale factors are unity ($h = k = 1$) and the indicatrix is a circle of radius one, such as on the equator in Fig. 6.6a. The actual Tissot's indicatrices of the cylindrical equidistant projection with its standard parallel on the equator (the plate carrée) are shown in Fig. 6.7a. The ellipses replicate those in Fig. 6.6a, being stretched longitudinally (but not latitudinally) as the north pole is approached.

In the cylindrical equal area projection, the parallel scale factor is again $k = \sec \phi$, shown in Figs. 6.6b and 6.7b. However, the meridional scale factor has a reciprocal relationship, $h = \cos \phi$, decreasing in size as latitude increases. This means that, as the north (or south) pole is approached, the Tissot's indicatrices become stretched longitudinally, but compressed latitudinally, such that the product hk always remains one (Eq. 6.10). Since the area of an ellipse with semi-minor axis h and semi-major axis k is given by πhk, this means that area is preserved everywhere on the map, hence the name 'equal area projection'.

Finally, in the cylindrical conformal projection, $h = k$ according to Eq. 6.14, and both scale factors vary as $\sec \phi$. This dependency means, again, that the parallels on a conformal projection become extended by an amount $k = \sec \phi$ as one approaches the poles, explained above. However, while in the equidistant projection the parallels are separated by equal intervals (because $h = 1$), and in the equal area projection the parallels are separated by decreasing intervals (because $h = \cos \phi$), in the conformal projection the spacing between the parallels increases as one approaches the poles, with the meridians being stretched by an amount $h = \sec \phi$. This means, first, that the Tissot's indicatrices are always circles with an increasing radius and hence area, implying that neither distance nor area are preserved across the map (Figs. 6.6c and 6.7c). Second, since the angle θ (Fig. 6.6d) between the hypotenuse of the indicatrix and the h axis does not change (unlike the same angle in equidistant or equal area indicatrices), angle is preserved in conformal projections but not in the others. Third, as one approaches the poles, the meridians become so stretched that finally, at the poles, one has $h = \infty$: they are an infinite map-distance from the equator and cannot actually be plotted. This is why Mercator maps of the world—so frustratingly common—show Canada, Greenland, Russia and Antarctica as being so huge compared to countries closer to the equator (Fig. 6.2a). In reality the surface area of India is 1.5 times greater than that of Greenland, not about ten times smaller as it appears on a Mercator map of the world.

6.4 Which Projection?

How long is a piece of string? Which map projection to choose for a project depends strongly upon the purpose of the map, and the preference of the user. There are dozens of different map projections, and when one tweaks their parameters—such as location of the origin or the standard parallels—the choices run into the hundreds. Somewhat crude rules of thumb say that if one is mapping regions in the tropics then normal cylindrical projections work well, temperate latitudes call for conic projections, while polar regions demand the use of azimuthal projections. Or that conic projections should be used to map regions with large east-west extents, transverse cylindrical projections are best for areas that are predominantly north-south in their aspect, while oblique projections are particularly suited to narrow countries that are aligned neither north-south nor east-west (think New Zealand).

Then there is class to consider. For instance, conformal maps are considered to be better for navigation as they preserve angles, an important measurement even in today's world of satellite positioning and navigation. Equal area maps are frequently chosen by social scientists seeking to compare population density, or some other quantity given per square kilometre. And azimuthal equidistant maps are ideal when one is plotting airline routes from a carrier's hub, or comparing distances from an earthquake epicentre.

The size of the region also plays a role: secant projections should probably be used when mapping large regions to maximise the area of low distortion, while with smaller regions one could probably get away with a tangent projection. And then there are issues like 'is this map just going to be used to plot a nice diagram, or are

Fig. 6.8 A sinusoidal 'egg box' on the sphere, with a wavelength of 20° in latitude and longitude. Orthographic projection, with a 20° graticule

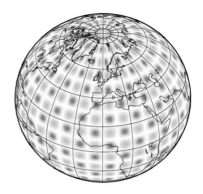

rigorous calculations going to be based on it?' If the former then projection from a sphere will be adequate; if the latter then an ellipsoid is probably the best model.

Here, however, we will focus the purpose of taking the Fourier or wavelet transform of 2D georeferenced data. That is, we desire a projection that will preserve—as faithfully as possible—the harmonic frequency content of data collected and coordinated on the surface of the Earth. This is so because the effective elastic thickness (T_e) estimation method of inverting admittance and coherency is sensitive to wavelengths of features in the gravity field and topography. Get the wavelengths wrong and the T_e estimate will likely be wrong.

First, one should choose an ellipsoid as the projection datum. For the purposes of T_e estimation it probably doesn't matter that much whether the spherical or ellipsoidal equations of a projection are used, as the errors incurred in choosing the former over the latter will be swamped by errors from other sources. Nevertheless, ellipsoidal equations are just as available as their spherical counterparts, and the fact remains that gravity and topography data are usually referenced to a geodetic datum (i.e. one that uses an ellipsoid as its reference surface).

Then one has to choose from the three developable surfaces and the three projection classes. Consider the data in Fig. 6.8, an 'egg box' geometry of sine waves with wavelengths of 20° along a parallel, and 20° along a meridian. Here the data are displayed in an orthographic projection, which is a perspective projection showing how the data would actually look from deep space. On a sphere of radius 6371 km, the wavelength along a meridian corresponds to approximately 2220 km wavelength everywhere on the meridian since parallels are equally spaced on a sphere. The wavelength along a parallel corresponds to 2220 km wavelength at the equator, but due to meridian convergence varying as cos ϕ, this wavelength decreases to 1110 km at 60° latitude, and to zero at the poles. We thus desire a map projection that reproduces the observed pattern as best as possible.

One might think an equidistant projection would be best at preserving wavelength, but recall that distance is only preserved in one direction in such projections: along a meridian, a parallel, or some other direction if an oblique developable surface is used (Sect. 6.3.2.1). This is clear in Fig. 6.9a, which shows part of the egg box data

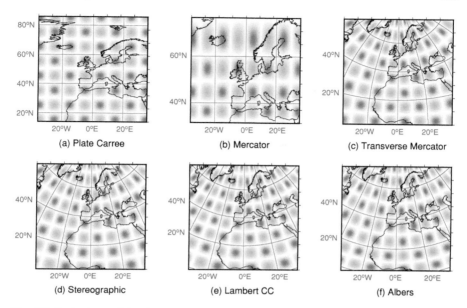

Fig. 6.9 Projections of the data in Fig. 6.8: **a** plate carrée (cylindrical equidistant); **b** Mercator (cylindrical conformal); **c** transverse Mercator (cylindrical conformal); **d** stereographic (azimuthal conformal); **e** Lambert conic conformal; **f** Albers (conic equal area). Each grid is 8000 km easting × 8000 km northing in size (tickmarks every 1000 km)

in plate carrée 'projection', where distance is preserved along a northing (meridian) but not an easting (parallel). Because the projection is simply a scaling of latitudes and longitudes (Eqs. 6.4 and 6.5)—essentially treating them as 2D plane Cartesian coordinates—the graticule is rectangular, the eastings follow the parallels, the northings follow the meridians, and the egg box bumps come out as square anomalies of equal dimensions in easting and northing, with a wavelength of approximately 2220 km in both dimensions all over the map. This is acceptable for the wavelength along a northing, but not along an easting which we would expect to decrease proportionally to $\cos \phi$, as noted above. Of course, one could change the orientation of the cylinder (to transverse) so that distance was preserved along an easting, but this would simply mean that distance along a northing would no longer be preserved.

The effect is worse with the Mercator projection, a normal cylindrical conformal projection where the graticule is again rectangular, the eastings follow the parallels, and the northings follow the meridians (Fig. 6.9b). As with the plate carrée, the easting wavelength remains at 2220 km at all latitudes instead of decreasing with increasing latitude, because both projections have $k = \sec \phi$. Furthermore, the northing wavelength of the Mercator-projected data increases with latitude (because $h = \sec \phi$), from 2220 km at the equator, to 4440 km at 60 °N, as seen in Fig. 6.9b. So it does seem that the normal cylindrical projections are not suited to mapping data at high latitudes, at least as far as good reproduction of wavelength is concerned.

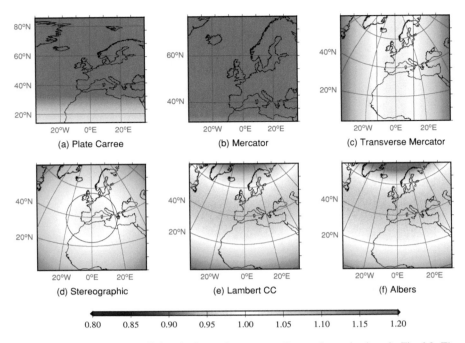

Fig. 6.10   Values of the parallel scale factor, k, corresponding to the projections in Fig. 6.9. The black lines show the contours of zero distortion, $h = k = 1$ (which lie on the equator in the plate carrée and Mercator projections). The equations for k for each projection were taken from Snyder (1987).

One can rotate the Mercator cylinder by $90°$, however, so that its line of tangency leaves the equator and instead follows a meridian. One can also make the projection secant, giving two lines of zero distortion (Fig. 6.10c). This is the transverse Mercator projection, shown in Fig. 6.9c. Being a conformal projection the Tissot's indicatrices of the transverse Mercator are circles, and will remain relatively small if the region is not too large in an east-west direction. Notably in Fig. 6.9c, the shape of the egg box anomalies in the north of the region is preserved, being narrow in 'width' and long in 'height' as the meridians converge. The same is true for the remaining projections in Fig. 6.9, which are azimuthal (the stereographic), or conic (the Lambert and Albers). As shown in Fig. 6.10c–f, these projections are secant because they penetrate the surface of the Earth; the cylindrical and conic projections have two lines of zero distortion and the azimuthal projection has a ring of zero distortion. Furthermore, their origins or standard parallels are chosen to be central to the area of interest, rather than far away on the equator as one must do with the plate carrée and Mercator projections. The value of the point scale factors of these latter two projections at $70°N$ is approximately 2.9, an unacceptable value.

The effect of the choice of map projection upon T_e estimates is shown in Fig. 6.11. All results were obtained from the coherence between the same Bouguer gravity

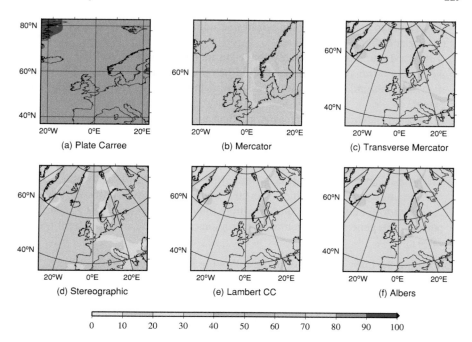

Fig. 6.11 Effective elastic thickness (in km) of Europe and the North Atlantic using data projected using: **a** plate carrée (cylindrical equidistant); **b** Mercator (cylindrical conformal); **c** transverse Mercator (cylindrical conformal); **d** stereographic (azimuthal conformal); **e** Lambert conic conformal; **f** Albers (conic equal area). Each grid is 5100 km easting × 5100 km northing in size (tickmarks every 1000 km)

anomaly and topography data sets, which differ slightly in their area depending upon the chosen projection. Those projections where the developable surface lies very close to the surface of the Earth and the lines of zero distortion cover a large area—the transverse Mercator, stereographic, Lambert conic conformal and Albers, shown in Fig. 6.10c–f—all have similar spatial T_e distributions. The two projections that use the normal cylinder—the plate carrée and Mercator—generally show higher T_e values, especially in the north of the region. As shown in Fig. 6.9, the apparent wavelengths of the egg box anomalies increase northwards, and as will be explained in Part III, a longer wavelength of transition from coherent to incoherent gravity and topography signifies a larger elastic thickness.

To conclude, the choice of developable surface is somewhat arbitrary but should be guided as follows.

1. Use a secant projection with its lines of zero distortion covering a large proportion of the study area, for example splitting it almost equally into regions of $k < 1$ and $k > 1$, as in Fig. 6.10c–f.
2. If the area is almost square then an azimuthal projection with its origin in the centre of the area is appropriate. Azimuthal projections are particularly useful as

they can map any region of the world (including the poles and equator) without risking coordinate singularities.

3. If the study area is considerably longer in its north-south aspect than it is east-west, then a transverse cylinder should be chosen. Such projections are able to map the poles and equator without distortion, but cannot be too wide in longitude extent.

4. If the study area is broader east-west than north-south use a conic projection, particularly in temperate latitudes. Conic projections are able to map the poles and equator, but not in the same map as the distortion at these locations will be too high.

5. The size of area to be mapped is limited by the choice of developable surface and the amount of acceptable distortion. Ideally, one would not want the point scale factor to deviate from unity by more than a few percent ($0.97 < k < 1.03$, perhaps). In general, continental-sized regions seem to be an upper limit (Audet 2014).

Concerning the class of projection, the following notes should be considered.

1. Conformal projections are probably most suited to T_e estimation, where wavelength in all directions should be preserved. Conformal projections preserve angle and their Tissot's indicatrices remain circles at all points of the map (their radii can be kept small by a suitable choice of origin or standard parallel location; see above). In the annular averaging used to form a radially-averaged power spectrum, or the azimuthal averaging of the fan wavelet method, isotropy of wavelength is crucial.

2. The Tissot's indicatrices of equidistant and equal area projections are not isotropic, becoming more and more elliptical with distance from the lines of zero distortion. This would lead to distortions in wavelength and direction over and above the distortions of a conformal projection (where the indicatrix might expand, but at least the expansion would be equal in all directions).

3. Where possible, choose the ellipsoidal versions of the projection formulae, rather than their spherical counterparts, using the GRS80 ellipsoid (Moritz 2000) or WGS84 (NGA 2014). These reference ellipsoids are very similar in size and shape: their semi-major axes are identical and their semi-minor axes differ by 0.105 mm.

The projections shown in Figs. 6.9c–e, 6.10c–e, and 6.11c–e, namely the transverse Mercator, stereographic, and Lambert conic conformal are all particularly useful. First, they are all conformal. Second, the transverse Mercator and stereographic are able to map any location on a planet, though the Lambert conic conformal should not be used to map the equator or poles. Third, their distortions are not excessive, given a reasonably-sized area. Indeed, Audet (2014) compared T_e results obtained from both spherical and planar fan wavelet methods applied to continent-sized regions, where he used the transverse Mercator projection to project the geodetically-coordinated data onto the plane. He found a <5% relative difference in T_e values, mostly concentrated at the data set edges.

6.5 Data Area Considerations

The algorithms used to compute the multitaper or wavelet admittance or coherency use the fast Fourier transform (FFT, Sect. 2.4.3), which requires that the data be given on a *complete and regular grid*. In two dimensions, the regularity condition means that such a grid must be rectangular (Cartesian) with constant spacing in both dimensions. The completeness condition means that there can be no data gaps: all grid cells must have a value.

Although gravity and topography data are generally collected at irregularly-distributed locations, most models are supplied on a grid (or in a form that can be readily converted to a grid). However, this grid is most often geodetically-coordinated, with a constant and equal spacing in latitude and longitude (i.e. $\Delta\phi = \Delta\lambda$), like the data shown in Fig. 6.8 for example. And as we have seen, we should not apply the 2D Fourier transform—which assumes planar coordinates—to data coordinated with curvilinear coordinates. Therefore the gravity and topography data that are provided on the geodetic grid must be interpolated onto a grid that is regular in Cartesian coordinates, for example using a map projection together with some interpolation method (giving grids such as those in Fig. 6.9).

It is advisable that the 2D Cartesian grid be approximately square ($N_x \sim N_y$), and have equal grid spacing in x and y directions ($\Delta x = \Delta y$). These two suggestions will ensure that the fundamental wavenumbers in the x- and y-directions are similar in value [$k_{x,1} = 2\pi/(N_x \Delta x)$ and $k_{y,1} = 2\pi/(N_y \Delta y)$, from Eq. 2.58], and that the two Nyquist wavenumbers are identical [$k_{N,x} = 2\pi/(2\Delta x)$ and $k_{N,y} = 2\pi/(2\Delta y)$, from Eq. 2.59].

6.5.1 Data Area Size

While the maximum size of the data area is governed by the amount of distortion one is prepared to accept from the map projection (Sect. 6.4), its minimum size depends upon the expected highest value of T_e. When a point load is applied to the surface of an elastic plate the resulting deflection has the form of a damped sinusoid (Sect. 7.7), whose wavelength (the *flexural wavelength*) and amplitude depend on T_e. Weak plates with low elastic thickness deflect much more than stronger plates, but the deflection is localised around the location of the load; with strong plates the deflection is distributed over greater distances. This means that, in order to capture the entire deflection signature, the data area needs to be at least as wide as the longest expected flexural wavelength. From the theory in Sect. 7.7, we can derive a 'rule of thumb' relating T_e to flexural wavelength (λ_F), writing Eq. 7.85 as

$$\lambda_F \approx 36.2\, T_e^{3/4}, \tag{6.15}$$

Fig. 6.12 The relationship between effective elastic thickness (T_e) and flexural wavelength (λ_F), from Eq. 6.15. One can use this information to estimate the minimum size of data area required to capture the longest anticipated flexural wavelength

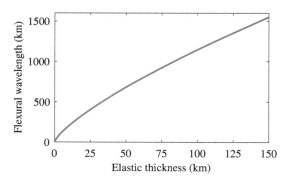

for subaerial (continental) loading environments, where T_e and λ_F must have units of kilometres. A graph of Eq. 6.15 is shown in Fig. 6.12. For example, if a continental region contains a province with an elastic thickness of 100 km, then the associated flexural wavelength is approximately 1145 km. The study area should therefore be at least 1145 km in each dimension (rounded up to 1200 km, say), and the high-T_e province should be located in its approximate centre, if possible.

6.5.2 Grid Spacing

At the other end of the spectrum, the shortest wavelength present in the Cartesian grid is its Nyquist wavelength, given by Eq. 2.46, or

$$\lambda_N = 2\Delta x, \qquad (6.16)$$

if we choose $\Delta x = \Delta y$, as noted above. Hence our choice of grid spacing governs the smallest resolvable flexural wavelength, and thence T_e.

Now, Eq. 6.15 tells us that the flexural wavelength for a region in a state of Airy isostasy (where $T_e = 0$ km) is zero, implying that we would need a grid with a Nyquist wavelength of zero in order to resolve it, which is obviously not achievable. So we must be realistic and choose a 'meaningful' grid spacing. As will be seen in Sect. 11.8, typical errors on T_e estimated from spectral methods can be as low as 1–2 km, but seldom smaller, so there is a good chance that one would estimate the elastic thickness of a region in a state of Airy isostasy as 1 km, suggesting a flexural wavelength of approximately 36 km by Eq. 6.15. Taking this minimum-present flexural wavelength as our Nyquist wavelength, we can then use Eq. 6.16 to determine an appropriate Cartesian grid spacing of 18 km (rounded up to 20 km, say). Thus, in order to resolve an elastic thickness of 1 km, one needs gravity and topography data given on a grid with a maximum grid spacing of 20 km; if the grid spacing is any bigger then one starts to miss larger flexural wavelengths and elastic thickness values.[6]

However, gravity and topography data are often provided on a grid with a much smaller grid spacing than the 20 km quoted above. Furthermore, they are given on geodetic grids: many current models are supplied on a 1′ grid in latitude and longitude (1′ being 1.853 km on the equator), and several topography models are given on even finer grids, such as 15″. But such small grid spacings are completely unnecessary for our purposes; if we used a grid spacing of 2 km, for example, then from Eqs. 6.16 and 6.15 we could resolve elastic thicknesses as low as 53 m in principle, an unrealistic level of accuracy given typical errors. Furthermore, when applying the fan wavelet or multitaper methods to data given over an area of 1000×1000 km, say, such small grid spacings would require an unreasonably huge amount of computer memory and processing time.

If we want to create a coarse Cartesian grid from a high-resolution geodetic grid, we must do it in such a way as to avoid aliasing (Sects. 2.3.3 and 2.6.2). As an example, consider the case where the geodetic grid has a 1′ cell size, and we desire a 20 km grid spacing on the Cartesian grid. First, we must find out (or assume) the bandwidth of the geodetic grid. If the data are given on a 1′ grid, then from Eq. 6.16 the Nyquist wavelength of this grid is 2′, and there will be no harmonics with wavelengths less than this value. Therefore we can treat the high-resolution geodetic grid as a continuous signal with minimum wavelength $\lambda_{min} = 2'$. Of course, it may be the case that the data are provided on a 1′ grid, but that the minimum wavelength is higher than 2′, say 10′. However, this is unlikely because it is reasonable to assume that the architects of the model will use the largest possible grid spacing to save computer memory. As explained in Sects. 2.3.3 and 2.6.2, a signal can be perfectly reconstructed from a sequence sampled from it if the sample spacing is less than or equal to half the minimum wavelength present in the signal, or

$$\Delta x \ \le \ \frac{\lambda_{min}}{2}, \tag{6.17}$$

which is called the *Nyquist criterion* (Eq. 2.43). So, if a model has a minimum wavelength of 10′, there is no point in providing it on a 1′ grid—a 5′ grid will suffice, using Eq. 6.17.

So, treating the high-resolution 1′ geodetic grid as a continuous signal with minimum wavelength of 2′, in order to avoid aliasing we must satisfy the Nyquist criterion, Eq. 6.17, and sample it at an interval of 1′ or less. But we actually want a coarser—not finer—grid, with a spacing of 20 km. On the equator, 20 km is approximately 10.8′, so merely selecting or interpolating values at every 10′ or so from a grid with a minimum wavelength of 2′ will introduce spurious long-wavelength harmonics into the sampled sequence through aliasing. Therefore, to obtain a 10′ sampled grid, Eq. 6.17 tells us that we need to increase the minimum wavelength present in the geodetic-grid

---

[6] In Sect. 11.2.1 we will perform a similar calculation using the coherence transition wavelength instead of the flexural wavelength. The former is roughly three-quarters the size of the latter.

Fig. 6.13 a A
high-resolution geodetic grid
of some data. **b** The same
grid low-pass filtered to
avoid aliasing, then
interpolated at the black
dots, which are located at the
nodes of a map projected
grid

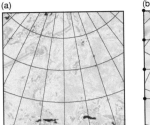
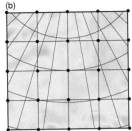

data to 20′ or higher. One can do this using a low-pass anti-aliasing filter (Sect. 2.3.5)
or by averaging the high-resolution grid values into 10′ cells. If the high-resolution
data are given in the form of spherical harmonic coefficients then one can expand
these to a lower harmonic degree and order than the model maximum. For instance,
a degree-1080 expansion will give data with a minimum wavelength of 20′, which
can be represented on a 10′ grid without aliasing (see Sect. 11.2.3).

An added complication arises from the need to perform a map projection, so that
the sampled grid is Cartesian rather than geodetic. This extra step does not pose
a huge challenge, though, as shown in Fig. 6.13. Some high-resolution data given
on a geodetic grid are shown in Fig. 6.13a, while Fig. 6.13b shows how a complete
and regular Cartesian grid might be obtained. First, the geodetic-grid data could be
low-pass filtered so that their minimum wavelength will conform to the Nyquist
criterion, Eq. 6.17, given the desired Cartesian-grid cell size. Software such as the
Generic Mapping Tools (GMT) (Wessel et al. 2019) is useful as it calculates spherical
distances accounting for meridian convergence. The filtered data can then be map-
projected and values interpolated at the nodes of a regular Cartesian grid (Fig. 6.13b).

6.6   Summary

When planar spectral analysis and estimation methods are applied to data collected
and coordinated on the surface of a planet, those data must be projected onto a plane
using a map projection. Unfortunately, the projection inevitably causes distortions
to appear in the projected data. Therefore map projections are designed to minimise
distortion, but can only do so to distances, areas or angles, it being mathematically
impossible to preserve all three in one projection. Projections that preserve only
distance are called equidistant projections, those that preserve only area are equal
area projections, while those that preserve only angle are called conformal. There
are also three types of developable surface on which to make the projection: a flat
surface, a cylinder and a cone, giving azimuthal, cylindrical and conic projections,
respectively. Furthermore, the surface can just graze, or can penetrate the planet's
surface, giving tangent or secant projections, respectively, the latter generally having
lower overall distortion. The amount of distortion present in a projection can be

quantified using the point scale factor, often visually portrayed in an ellipse called the Tissot's indicatrix.

For the purposes of spectral estimation, wavelength should be preserved in all directions. Only conformal projections can do this as they preserve angle irrespective of the location on the map or the compass direction of the angle. In contrast, equidistant projections will only preserve distance in one selected direction and sometimes only if the trajectory passes through a single location on the map. Equal area projections preserve areas but can turn a circular feature into an extended linear anomaly, for example. Azimuthal, cylindrical and conic developable surfaces are all acceptable for spectral estimation, but the stereographic (azimuthal conformal), transverse Mercator (transverse cylindrical conformal) and Lambert conic conformal are perhaps the most useful. Some combinations of class and surface are unadvisable though, such as the Mercator projection, a normal cylindrical conformal projection with very high distortion near the poles, or the plate carrée, a cylindrical equidistant 'projection' that merely treats geodetic coordinates as 2D Cartesian.

The study region should be approximately square and small enough to prevent excessive distortion. However, it must be large enough to capture the largest flexural wavelength present in the region, which is dictated by the largest expected elastic thickness. The data grid spacing needs to be small enough to resolve the smallest flexural wavelength, but does not need to be as small as the grid spacing in modern global gravity and topography models, some of which are provided on $15''$ (460 m at the equator) or even $1''$ (31 m) grids. A usable grid spacing of the map-projected data would lie in the range 5–20 km.

6.7 Further Reading

The book by Snyder (1987) is almost a 'Bible' of map projections, containing both projection theory and the formulae of over two dozen projections in both spherical and ellipsoidal form. Two other good books that delve into the mathematics of map projections are those by Richardus and Adler (1972) and Maling (1992). For a less mathematical introduction to map projections and coordinate datums, Iliffe and Lott (2008) is recommended. The history of map projections—annotated with equations—can be found in the excellent Snyder (1993).

References

Audet P (2014) Toward mapping the effective elastic thickness of planetary lithospheres from a spherical wavelet analysis of gravity and topography. Phys Earth Planet Inter 226:48–82

Iliffe N, Lott R (2008) Datums and map projections for remote sensing, GIS and surveying, 2nd edn. Whittles Publishing, Dunbeath

Maling DH (1992) Coordinate systems and map projections, 2nd edn. Pergamon Press, Oxford

Moritz H (2000) Geodetic Reference System 1980. J Geod 74:128–133

NGA (2014) Department of Defense World Geodetic System 1984: Its definition and rela-
tionships with local geodetic systems. National Geospatial-Intelligence Agency, Arnold,
Missouri. https://earth-info.nga.mil/GandG/publications/NGA_STND_0036_1_0_0_WGS84/
NGA.STND.0036_1.0.0_WGS84.pdf. Accessed 3 Jun 2020

Richardus P, Adler RK (1972) Map projections for geodesists, cartographers and geographers.
North-Holland, Amsterdam

Snyder JP (1987) Map projections—a working manual. US Geological Survey Professional Paper
1395, Washington DC

Snyder JP (1993) Flattening the Earth: two thousand years of map projections. University of Chicago
Press, Chicago

Wessel P, Luis JF, Uieda L, Scharroo R, Wobbe F, Smith WHF, Tian D (2019) The Generic Mapping
Tools version 6. Geochem Geophys Geosyst 20:5556–5564

Part III
Flexure

Part III of the book is all about T_e estimation using the coherence and admittance. Chapter 7 begins with elasticity theory and ends up with the partial differential equation for flexure of a thin, elastic plate with spatially variable flexural rigidity. From that, the biharmonic equation for uniform rigidity is obtained, which is used to describe scenarios of flexure under surface and internal loading of the lithospheric plate. Next, Chap. 8 concerns the other half of the data needed for T_e estimation: gravity. After an introduction to gravity potential theory, the predicted gravity anomalies corresponding to the flexural scenarios in Chap. 7 are provided. Chapter 8 concludes with expressions for the admittances of these scenarios.

Chapter 9 describes the T_e-estimation method of load deconvolution, sometimes called 'Forsyth's method', or 'the coherence method'. It describes in detail the steps taken to construct maps of the spatial variation of T_e using both multitaper and fan wavelet methods. It also addresses some more recent developments surrounding correlated signals and noise. Every analysis method needs testing for its accuracy, and Chap. 10 covers the generation of synthetic gravity and topography models—with a known T_e—for that purpose. Besides providing equations to create synthetic models with uniform elastic thickness, a finite difference solution of the partial differential equation with spatially variable flexural rigidity is provided. Finally, Chap. 11 presents some examples of T_e estimation using actual data; some contemporary data sources and models are also discussed. This chapter concludes with a head-to-head of multitaper and wavelet methods.

Chapter 7
Loading and Flexure of an Elastic Plate

7.1 Introduction

When subject to typical loads acting over long timescales, it has been shown that
the lithosphere will generally deform like a *thin, elastic plate*. 'Typical loads' are
those generated by most tectonic processes, volcanism, rifting, sedimentation, etc.
and 'long timescales' are of the order of a million years or greater. The evidence
supporting this can be as simple as the observation that the surface of the lithosphere
has structure that has persisted for up to hundreds of millions of years: if the plates
were made up of some material that did not behave elastically, then the observed
structure would have long since disappeared under its own weight, consumed—
essentially—by gravity.

In order to study plate deformation, we turn to the fields of mechanical and civil
engineering. One can think of myriad scenarios where sheets or slabs of metal,
concrete or some other material are routinely flexed and deformed, and it is essential
for the architects and engineers to know just how much stress and strain the structure
can accommodate before it fails or breaks. Such knowledge is embodied in partial
differential equations (PDEs) of plate flexure, whose exact form is governed not only
by the material of the plate, but also by its size and shape.

To that end, plates are categorised by their thickness relative to their lateral dimensions, because the equations describing the flexure of a plate whose thickness is much
less than its lateral extent—so that the plate is more of a sheet or membrane—are
quite different to those describing a thick plate such as a sidewalk paving stone.
In some of these thickness scenarios certain forces are negligible and can safely be
approximated or even ignored, while in others they must be considered and modelled
as faithfully as possible.

The material comprising the plate also governs the form of the PDEs. When subject
to deforming forces, materials are usually classed by their *rheology*, being the branch
of physics concerned with the deformation and flow of materials. Common rheologies
studied in engineering and geophysics are *elastic*, *plastic* and *viscoelastic*. Elastic
materials return to their original shape after the applied load is released, while plastic
materials will remain deformed. Viscoelastic materials do not undergo deformation
instantaneously—unlike elastic materials—but rather creep into their deformed state

© Springer Nature Switzerland AG 2022
J. Kirby, *Spectral Methods for the Estimation of the Effective Elastic Thickness
of the Lithosphere*, Advances in Geophysical and Environmental Mechanics
and Mathematics, https://doi.org/10.1007/978-3-031-10861-7_7

over time; yet they will return to their original shape upon removal of the applied force, albeit slowly. Each rheology thus leads to quite different PDEs governing its flexure.

In this book we concern ourselves only with thin plates possessing an elastic rheology, though a handful of flexural studies have considered thick plates and viscoelastic plates (Kirby 2014). We also note that the spectral method of elastic thickness estimation presented in this book works best when the region being studied has a 'continuous', unbroken lithosphere, and is reasonably far from the edges of the tectonic plate. While many studies have applied this method to 'broken' plates, containing for example a subduction zone or major fault, such scenarios cannot readily be examined using Fourier-based methods. This is because the solution of the plate flexure PDE is performed in the wavenumber domain and in this domain one cannot specify spatial locations, such as the location of a fault. Additionally, a break in the plate represents a discontinuity, and as noted in Sect. 2.5.4, discontinuities are not dealt with very well by the Fourier transform. Nevertheless, readers are encouraged to experiment and push (methodological) boundaries, while noting the caveats.

7.2   Thin, Elastic Plate Flexure

In this section, we first derive a set of PDEs describing the behaviour of a thin plate being deformed by an applied load. We then use the conditions of elasticity theory to derive the plate flexure PDE for a thin, elastic plate. The derivations can be made easier—and can give a PDE that is more readily solvable—by making certain assumptions and approximations. The main three of these are:

1. The plate is a *thin plate*: its thickness, h, is small compared to its lateral dimensions (L), with $0.02 < h/L < 0.1$.
2. The plate is an *elastic plate*, obeying the equations of linear elasticity.
3. Deflections, v, of the plate are small compared to its thickness, with $v/h < 0.1$.

In the engineering literature, thin plates are those whose thickness-to-width ratio is as described above; thinner plates ($h/L < 0.02$) are referred to as 'membranes' and have no flexural rigidity, supporting their load through in-plate tension; moderately thick plates ($0.1 < h/L < 0.2$) make fewer approximations in the derivation of their equations; while thick plates ($h/L > 0.2$) make no approximations but their PDEs are very difficult to solve, at least algebraically. The assumptions 1–3, above, generate further conditions and approximations that will be required in our derivations; we will refer back to them throughout this section. They are:

4. The *neutral surface* of the plate remains unstrained during bending.
5. The slopes of the deflected neutral surface remain small (a consequence of Assumption 3).
6. Plane cross-sections through the plate that are perpendicular to the neutral surface before flexure, remain plane and perpendicular to the deflected neutral surface during and after bending (a consequence of Assumptions 1 and 3).

7. Transverse stresses[1] are neglected, and only stresses that are parallel to the plane of the plate are considered.

Terms that the reader may be unfamiliar with at present will be explained in this section.

7.2.1 Thin Plates

When a plate is subject to an applied, downward force the plate bends at the point(s) of application, and stresses generated within the plate create reactive forces and moments.[2] If the plate remains motionless after the force has acted it is said to be in a state of *static equilibrium*. Thus, by balancing the applied and induced forces and moments, we can derive a partial differential equation governing the flexure of the plate. Note that we will adopt the convention that the positive z-direction is upwards (so that continental topography has a positive elevation), whereas most engineering texts use a positive-downwards convention. We also employ a positive-anticlockwise angle convention.

Figure 7.1 shows a small element of a rectangular plate subject to an applied, downward, distributed, transverse force, $p_z(x, y)$, upon its upper surface. The element has horizontal dimensions dx and dy, and thickness h. Under thin plate theory, the applied force generates three types of reactions: transverse shear forces, bending moments and twisting moments. Shear forces act to distort the material by shearing, where one face is pushed up and the opposite face pushed down. Bending moments can be imagined by the effort involved in trying to snap a 30 cm ruler by bending it using both hands, while the action one employs when wringing out a wet cloth corresponds to a twisting moment.

Our first PDE is derived by considering the shear forces induced in the element by the applied force p_z. Shear forces act on opposite faces in opposite directions, as shown in Fig. 7.1. So on the right-hand rear face there is a shear force Q_x. Moving a small distance dx to the opposite (left-hand near-side) face, we use a Taylor series expansion over the distance dx to evaluate the corresponding shear force there, or

$$Q_x + \frac{\partial Q_x}{\partial x} dx + \frac{1}{2} \frac{\partial^2 Q_x}{\partial x^2} dx^2 + \cdots .$$

Because dx is small (and dx^2 much smaller) we use only the first two (linear) terms of the expansion. And since the force might vary over the length, dy, of each face, we will consider forces per unit length. So the shear force per unit length on the right-hand rear face is $Q_x dy$, while that on the left-hand near-side face is $[Q_x + (\partial Q_x / \partial x) dx] dy$. Using similar reasoning we see that the other two shear

---

[1] A *transverse* stress is one in which the direction of the stress is perpendicular to the long axes of the plate. Some engineering texts refer to this—somewhat confusingly—as a 'lateral' stress.

[2] *Moments* are explained in the Appendix to this chapter.

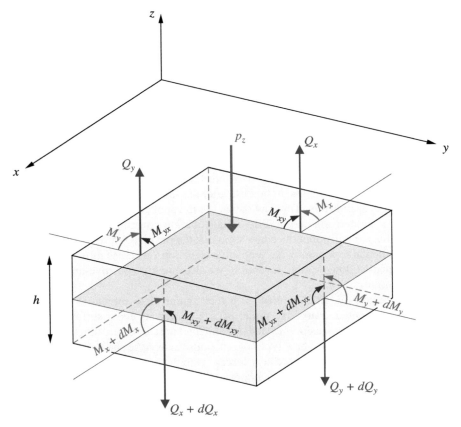

Fig. 7.1 An element of a thin plate subject to a transverse force, p_z. The neutral surface (in darker grey) bisects the element's thickness. The element has dimensions dx, dy, and thickness h. The resultant shear forces (magenta) are Q_x and Q_y, and we use the notation $dQ_x = (\partial Q_x/\partial x)\,dx$ and $dQ_y = (\partial Q_y/\partial y)\,dy$. The bending moments (green) are M_x and M_y, where $dM_x = (\partial M_x/\partial x)\,dx$ and $dM_y = (\partial M_y/\partial y)\,dy$. The twisting moments (blue) are M_{xy} and M_{yx}, with $dM_{xy} = (\partial M_{xy}/\partial x)\,dx$ and $dM_{yx} = (\partial M_{yx}/\partial y)\,dy$

forces per unit length are $Q_y dx$ and $[Q_y + (\partial Q_y/\partial y)\,dy]\,dx$. Finally, we must take into account the applied force itself, which over the element has a value (force per unit area) of $p_z\,dx\,dy$. We can now sum all the forces, assigning positive or negative signs according to the directions of their vectors in Fig. 7.1, giving

$$Q_x dy - \left(Q_x + \frac{\partial Q_x}{\partial x}\,dx \right) dy + Q_y dx - \left(Q_y + \frac{\partial Q_y}{\partial y}\,dy \right) dx - p_z\,dx\,dy = 0$$

or

$$\frac{\partial Q_x}{\partial x} + \frac{\partial Q_y}{\partial y} = -p_z \, . \tag{7.1}$$

The second differential equation is derived by considering all moments around the y-axis (looking in the direction of positive y), that is, those acting in the xz-plane. As shown in Fig. 7.1, there is a clockwise bending moment, M_x, and a clockwise twisting moment, M_{yx}, acting on the two far faces of the element. As above, these act over the whole of their corresponding faces, so we must consider rather the moments per unit length and multiply each by the length of their respective face, giving $M_x dy$ and $M_{yx} dx$. Moving small distances dx and dy to the opposite (near-side) faces, respectively, we use a Taylor series expansion to evaluate the corresponding moments, so that

$$M_x + \frac{\partial M_x}{\partial x} dx + \frac{1}{2} \frac{\partial^2 M_x}{\partial x^2} dx^2 + \cdots$$

gives the bending moment at the left-hand near-side face, while

$$M_{yx} + \frac{\partial M_{yx}}{\partial y} dy + \frac{1}{2} \frac{\partial^2 M_{yx}}{\partial y^2} dy^2 + \cdots$$

gives the twisting moment at the right-hand near-side face. As before, we use only the first two (linear) terms of the expansions. Both these moments are anticlockwise: a bending moment per unit length $[M_x + (\partial M_x/\partial x) dx] dy$, and a twisting moment per unit length $[M_{yx} + (\partial M_{yx}/\partial y) dy] dx$. Finally, we must consider the moments in the xz-plane generated by the shearing forces. At the right-hand rear face the force per unit length is $Q_x dy$, acting a distance $dx/2$ from the centre of the element, and is clockwise (with respect to the centre, looking in the positive y direction): its moment is thus $Q_x dy\, dx/2$. At the opposite (left-hand near-side) face, the force per unit length is $[Q_x + (\partial Q_x/\partial x) dx] dy$, again acting a distance $dx/2$ from the centre of the element, and is also clockwise (with respect to the centre): its moment is thus $[Q_x + (\partial Q_x/\partial x) dx] dy\, dx/2$. Summing these moments using a positive-anticlockwise convention gives

$$\left(M_x + \frac{\partial M_x}{\partial x} dx\right) dy - M_x\, dy + \left(M_{yx} + \frac{\partial M_{yx}}{\partial y} dy\right) dx$$

$$- M_{yx}\, dx - \left(Q_x + \frac{\partial Q_x}{\partial x} dx\right) dy\, \frac{dx}{2} - Q_x\, dy\, \frac{dx}{2} = 0.$$

If we neglect the term $(\partial Q_x/\partial x) dx^2 dy/2$, since dx^2 is very small, we obtain the differential equation

$$\frac{\partial M_x}{\partial x} + \frac{\partial M_{yx}}{\partial y} = Q_x. \tag{7.2}$$

The third differential equation is derived as for the second, above, but instead considering all moments around the x-axis, that is, those acting in the yz-plane. Using similar reasoning, we can derive

$$\left(M_y + \frac{\partial M_y}{\partial y}\,dy\right) dx \;-\; M_y\,dx \;+\; \left(M_{xy} + \frac{\partial M_{xy}}{\partial x}\,dx\right) dy$$

$$-\;M_{xy}\,dy \;-\; \left(Q_y + \frac{\partial Q_y}{\partial y}\,dy\right) dx\,\frac{dy}{2} \;-\; Q_y\,dx\,\frac{dy}{2} \;=\; 0\,,$$

or

$$\frac{\partial M_y}{\partial y} \;+\; \frac{\partial M_{xy}}{\partial x} \;=\; Q_y\,. \tag{7.3}$$

If we now substitute Eqs. 7.2 and 7.3 into Eq. 7.1, we obtain the partial differential equation describing the relationship between the two orthogonal bending moments M_x and M_y, and the twisting moment, M_{xy}, induced by a transverse force, p_z, applied to the surface of a thin plate, as

$$\frac{\partial^2 M_x}{\partial x^2} \;+\; 2\,\frac{\partial^2 M_{xy}}{\partial x\,\partial y} \;+\; \frac{\partial^2 M_y}{\partial y^2} \;=\; -p_z\,, \tag{7.4}$$

where we have used $M_{yx} = M_{xy}$. We will call this the *thin plate PDE*.

Equation 7.4 describes a thin plate but is independent of rheology, which only enters the formulation when we need to relate the moments to actual deflections. Here we proceed by choosing an elastic rheology to find the deflections, but first we must introduce the theory of linear elasticity.

7.2.2   Elasticity: Stress and Strain

One's first encounter with studies of elasticity is often with a stretched spring. If an elastic spring is extended by an amount x from its relaxed length by a force F, then *Hooke's law* states that, for relatively small extensions, the extension is proportional to the force, or

$$F = kx\,, \tag{7.5}$$

where k is the 'spring constant', a property of the material of the spring. Importantly, when the force is removed, the spring returns to its relaxed length exactly, obeying the definition of an elastic (as opposed to a plastic) substance. However, this simple, linear equation merely scratches the surface of the vast topic of elasticity. For example, for large extensions Eq. 7.5 gains nonlinear terms, and when the extension is very large, the material might not behave elastically any longer and may even break. Furthermore, when the elastic substance is a three-dimensional solid which can be stretched, squeezed and sheared in all its dimensions, the linear, one-dimensional form of Hooke's law in Eq. 7.5 is just one piece of the puzzle, as it does not describe the possible extensions or contractions that may occur in perpendicular directions within the material.

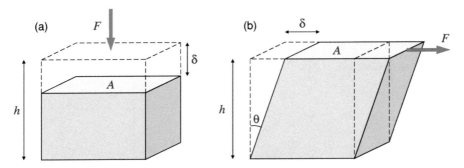

Fig. 7.2 a Normal stress and strain, and **b** shear stress and strain of an elastic solid. A force F (red arrow) is applied to the top of a cube (dashed line) of side length h; in **a** the force is perpendicular to the area A, in **b** the force is parallel to the area A and the base of the cube remains fixed. The grey shaded shapes show the new geometries of the cubes after deformation. In **a**, the normal stress is $\sigma = F/A$, and the normal strain is $\varepsilon = \delta/h$; in **b**, the shear stress is $\tau = F/A$, and the shear strain is $\gamma = \delta/h = \tan\theta$

A more general form of Hooke's law than Eq. 7.5 is

$$\text{stress} \ = \ \text{modulus} \ \times \ \text{strain} \ . \qquad (7.6)$$

In Eq. 7.6, the 'modulus' is a constant property of the solid, discussed in Sect. 7.2.3. The *stress* is defined as the force per unit area applied to a particular dimension of the solid, or

$$\text{stress} \ = \ \frac{\text{force}}{\text{area}} \ .$$

The *strain* is the ratio of the change in length to the original length of the solid in a particular dimension, or

$$\text{strain} \ = \ \frac{\text{change in length}}{\text{original length}} \ .$$

So stress is effectively a pressure and has units of pascals (newtons per square metre), while strains are dimensionless. Two types of stress and strain are illustrated in Fig. 7.2.

Equation 7.6 describes a state of *linear elasticity* in which stress is proportional to strain. Importantly for studies of thin plate flexure, if we consider only small deflections of a thin plate (Assumptions 1 and 3 in our list) we are able to use linear elastic theory (Assumption 2), ensuring that Hooke's law applies. As we shall see, this greatly facilitates both the derivation and solution of a differential equation of plate flexure; the alternatives—thick plate flexure or large-deflection flexure—are orders of magnitude harder.

Fig. 7.3 An element of an
elastic solid, showing the
normal stresses (σ, red) and
shear stresses (τ, blue)
acting on each visible face

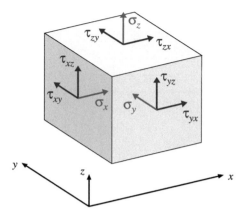

When Hooke's law is written in its most general form for a 3D solid, the 'stress'
and 'strain' in Eq. 7.6 are both second-order 3×3 tensors,[3] while the 'modulus' is
a fourth-order $3 \times 3 \times 3 \times 3$ tensor, the 'elasticity tensor'. However, when applied to
small deformations of a thin plate this degree of complexity is unnecessary: the stress
and strain tensors can be written as 6×1 vectors, while the elasticity tensor becomes
a 6×6 matrix, or

$$\begin{pmatrix} \sigma_x \\ \sigma_y \\ \sigma_z \\ \tau_{xy} \\ \tau_{xz} \\ \tau_{yz} \end{pmatrix} = \begin{pmatrix} \lambda + 2\mu & \lambda & \lambda & 0 & 0 & 0 \\ \lambda & \lambda + 2\mu & \lambda & 0 & 0 & 0 \\ \lambda & \lambda & \lambda + 2\mu & 0 & 0 & 0 \\ 0 & 0 & 0 & \mu & 0 & 0 \\ 0 & 0 & 0 & 0 & \mu & 0 \\ 0 & 0 & 0 & 0 & 0 & \mu \end{pmatrix} \begin{pmatrix} \varepsilon_x \\ \varepsilon_y \\ \varepsilon_z \\ \gamma_{xy} \\ \gamma_{xz} \\ \gamma_{yz} \end{pmatrix}, \qquad (7.7)$$

which we shall call the *generalised Hooke's law*. Here, the σ are normal stresses, the
ε are normal strains, the τ are shear stresses, the γ are shear strains, and λ and μ are
the *Lamé parameters* (the elastic moduli of Eq. 7.6).

Stresses and strains are classified as either *normal* or *shear*. A normal stress
is a force that acts in a direction that is perpendicular to one face of an element,
serving to compress or extend the element; a normal strain represents the amount of
compression or extension in that direction (Fig. 7.2a). Therefore, normal stresses and
strains are categorised by just one subscript, showing that the force or extension acts
in one direction only (so σ_x acts along the x-axis, that is, on an area in the yz-plane,
as shown in Fig. 7.3).

In contrast, a shear stress is a force that acts in a direction that is parallel to one face
of an element, trying to rip the element in two; a shear strain represents the amount of

---

[3] Tensors are mathematical constructions that have their own rules of algebra and calculus. A
second-order tensor resembles an $n \times m$ matrix, but a third-order tensor would be 'written' as an
$n \times m \times l$ 'matrix'. They are very useful and powerful, but we will not be concerned with tensors
here.

compression or extension in that direction (Fig. 7.2b). Therefore, shear stresses and strains are categorised by two subscripts, where the first subscript identifies the plane that is being sheared (so that the first subscript axis is perpendicular to that plane), while the second subscript denotes the direction in which the force acts. For example, the symbol τ_{zx} implies that the stress is shearing the plane that is perpendicular to the z-axis (the xy-plane) in the x direction, while τ_{zy} tells us that the stress is again shearing the plane that is perpendicular to the z-axis (the xy-plane), but now in the y direction (Fig. 7.3).

7.2.3 The Elastic Moduli

The Lamé parameters in Eq. 7.7 are measures of a material's resistance to deformation in different directions. Since stress is measured in units of pascals (Pa), and strain is dimensionless, the Lamé parameters also have units of pascals. The parameter λ is the *bulk modulus* and is a measure of a material's resistance to compression in all directions at once. For example, water and rubber both have a bulk modulus of approximately 2 GPa, while granite has a value of approximately 50 Pa and steel 160 GPa.

The Lamé parameter μ is often given the symbol G and is the *modulus of rigidity* or *shear modulus*, being the material's resistance to shearing forces. It is defined as the ratio of shear stress to shear strain in a material. Steel, for example, has a shear modulus of approximately 80 GPa, granite's value is 24 Gpa, while sandstone has a shear modulus of only 0.4 GPa. Water cannot be sheared as it is a liquid, and has a shear modulus of zero.

However, these parameters are more commonly represented in terms of two other elastic constants, *Young's modulus* and *Poisson's ratio*. Young's modulus (E) measures a material's resistance to deformation along a single axis ('uniaxial stress') and is computed from the Lamé parameters using

$$E = \frac{\mu(3\lambda + 2\mu)}{\lambda + \mu} . \tag{7.8}$$

Young's modulus is also measured in pascals, with some typical values being 200 GPa for steel, 50 GPa for granite, 20 GPa for sandstone and 7 GPa for rubber.

Poisson's ratio (ν), on the other hand, is dimensionless as it is the (negative of the) ratio of transverse strain to axial strain in a material. That is, when a material is stretched along one (axial) direction, it gets thinner in the other two perpendicular (transverse) directions, and Poisson's ratio measures the strain response to this stress in all directions. It is related to the Lamé parameters by

$$\nu = \frac{\lambda}{2(\lambda + \mu)} . \tag{7.9}$$

Values of Poisson's ratio are generally in a range 0–0.5 for most materials, with cork having a value of zero, and rubber 0.5. Granite has a Poisson's ratio in a range 0.2–0.3, steel's value is around 0.3, while sandstone has a Poisson's ratio of approximately 0.2.

In most studies of T_e estimation the values used for Young's modulus and Poisson's ratio of lithospheric rocks (whether continental or oceanic) are $E = 100$ GPa and $\nu = 0.25$, respectively. Note also that elasticity theory allows both these 'constants' to be anisotropic, having different values along different axes, though we do not consider anisotropy in this book.

7.2.4 Plane Stress

The topic of the elastic deformation of solids has many applications in geology and geophysics beyond elastic plate flexure, such as faulting and the propagation of seismic waves. To reduce complexity when solving problems in such fields it is commonplace to impose certain conditions or to make approximations regarding the nature of stress, strain and elasticity in different scenarios. Some common approximations made in the Earth sciences (and many other disciplines) are the states of uniaxial stress, uniaxial strain, pure shear, plane strain and plane stress. Of these, only plane stress concerns us.

One can begin by recognising that the generalised Hooke's law, Eq. 7.7, embodies two sets of equations, as it combines the equations of normal stress and strain and shear stress and strain into a single matrix equation. If we take the normal stress and strain components of Eq. 7.7, we obtain

$$
\begin{pmatrix} \sigma_x \\ \sigma_y \\ \sigma_z \end{pmatrix} = \begin{pmatrix} \lambda + 2\mu & \lambda & \lambda \\ \lambda & \lambda + 2\mu & \lambda \\ \lambda & \lambda & \lambda + 2\mu \end{pmatrix} \begin{pmatrix} \varepsilon_x \\ \varepsilon_y \\ \varepsilon_z \end{pmatrix} , \tag{7.10}
$$

which has no dependence on shear stress or strain. The inverse of Eq. 7.10 is given by

$$
\begin{pmatrix} \varepsilon_x \\ \varepsilon_y \\ \varepsilon_z \end{pmatrix} = \frac{1}{E} \begin{pmatrix} 1 & -\nu & -\nu \\ -\nu & 1 & -\nu \\ -\nu & -\nu & 1 \end{pmatrix} \begin{pmatrix} \sigma_x \\ \sigma_y \\ \sigma_z \end{pmatrix} , \tag{7.11}
$$

using Eqs. 7.8 and 7.9.

A material is said to be under *plane stress* if the normal stress is zero in a particular direction, but not in the two orthogonal directions. For small deflections of thin plates the direction of zero stress is the z direction, perpendicular to the plane of the plate, or

$$
\sigma_x \neq 0 , \qquad \sigma_y \neq 0 , \qquad \sigma_z = 0 .
$$

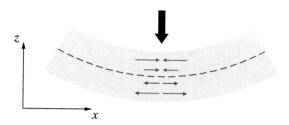

Fig. 7.4 Cross-section in the xz-plane through a plate deflected by a load (black arrow), showing the neutral surface as a dashed black line. The flexure induces stresses, σ_x, which are compressive (red arrows) above the neutral surface and extensional (blue arrows) below it. The curvature has been exaggerated

As shown in Fig. 7.4, when an external transverse load is applied to the surface of a thin plate, its top half undergoes compression ($\sigma_x > 0$) while its bottom half undergoes extension ($\sigma_x < 0$); the neutral surface must therefore be unstrained ($\sigma_x = 0$; Assumption 4). So while there exist horizontal stresses within the plate, there are no vertical stresses (despite the transverse external load) because the plate material is not being deformed in this direction and maintains its thickness (Assumption 7). Furthermore, because the plate is thin, we can assume that this state of $\sigma_z = 0$ extends throughout its thickness.

Thus, substitution of $\sigma_z = 0$ into Eq. 7.11 gives the strains

$$\varepsilon_x = \frac{\sigma_x - \nu\sigma_y}{E}, \quad \varepsilon_y = \frac{\sigma_y - \nu\sigma_x}{E},$$

which can be rearranged to give the stresses

$$\sigma_x = \frac{E}{1 - \nu^2}\left(\varepsilon_x + \nu\varepsilon_y\right), \quad \sigma_y = \frac{E}{1 - \nu^2}\left(\varepsilon_y + \nu\varepsilon_x\right). \tag{7.12}$$

7.2.5 Bending Moments

We must now find a way to relate the deflections of the plate (v) to the bending moments M_x and M_y so that we can develop Eq. 7.4 further. We do this by considering the stresses and strains within a small element of a flexed plate. Figure 7.5a shows the overall geometry of a plate undergoing flexure, identifying a small fibre a distance z below the neutral surface, i.e. in the extensional part of the plate (see Fig. 7.4). Note that curvature, angles and lengths have been exaggerated considerably in Fig. 7.5; in reality the fibre is infinitesimally thin and short, with its length $dx \ll L$ the length of the plate (so that $d\theta \ll \theta_x$), and its thickness $dz \ll h$ the thickness of the plate.

Figure 7.5b shows an enlargement of a small element, ABCD, of the unflexed plate. The solid line AB is the small fibre of length dx, a distance z below the neutral

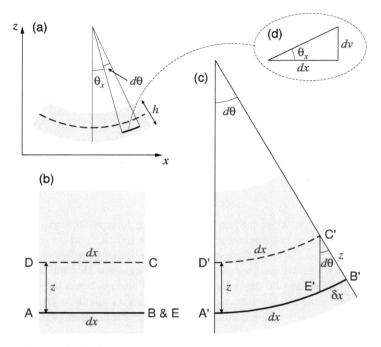

Fig. 7.5 **a** Cross-section in the xz-plane through a thin plate (grey shading), showing the neutral surface (dashed line) and a small fibre within the plate (thick, solid line). Enlargements of the fibre are shown in **b** the undeflected plate, and **c** the deflected plate. **d** An enlargement of a small part of the flexed fibre. Note that all depictions of curvature, angles and dimensions are exaggerated for clarity

surface. Note that points B and E coincide in this unflexed state. When the plate is subject to a downward, transverse force, it flexes and the element ABCD becomes A'B'C'D', shown in Fig. 7.5c.[4]

First, we note that the neutral surface remains unstrained, as explained in Sect. 7.2.4 (Assumption 4), so that its length is the same in both unflexed and flexed states, i.e. $\overline{CD} = \overline{C'D'}$.[5] We also have $\overline{C'D'} = \overline{A'E'}$ because the lines A'D' and E'C' are parallel.

Second, we note that the fibre AB has been extended by an amount $\delta x = \overline{E'B'}$, so that its length is now $\overline{A'B'}$. Hence the strain on the fibre is given by

$$\varepsilon_x = \frac{\text{change in length}}{\text{original length}} = \frac{\delta x}{dx}. \qquad (7.13)$$

---

[4] Note that Fig. 7.5c is drawn with $\theta_x = 0°$ for clarity. The derivations we undertake are largely based on geometry and including an arbitrary, non-zero θ_x just complicates the diagrams; the end results are the same—see Szilard (2004).

[5] The overbar symbol, $\overline{AB}$, means the length of the line AB.

If we now assume that plane cross-sections through the plate that are perpendicular to the undeflected neutral surface remain plane and perpendicular to the deflected neutral surface (Assumption 6), and that the slopes of the deflected neutral surface remain small (Assumption 5), then from triangle B'C'E' in Fig. 7.5c we can see that $\delta x \approx z\, d\theta$ (because $d\theta$ is very small), which when substituted in Eq. 7.13 gives the strain

$$\varepsilon_x = z \frac{\partial \theta_x}{\partial x} , \tag{7.14}$$

where we use the partial derivative symbol (∂) instead of d or δ because these variables are also functions of the y coordinate, and where we add a subscript x to the angle $d\theta$ to distinguish it from a similar angle describing flexure in the y direction. Note that Assumption 5 follows from Assumption 3 (small deflections), and that Assumption 6 is only valid for small deflections of a thin plate; for thick plates and/or large deflections, one must consider lateral, in-plane shear stresses, which complicates matters because lines that were perpendicular to the neutral surface do not remain perpendicular after bending.

We must now relate the angle θ_x to the deflection v. This is achieved by considering a small triangle on the flexed fibre in Fig. 7.5a, which is enlarged in Fig. 7.5d. From the theory of similar triangles, the angles θ_x in Figs. 7.5a and 7.5d are equal, and for small $d\theta$ and small deflections the slope of the triangle does not change over the length of the fibre. Again, for small deflections and small slopes of the neutral surface (Assumptions 3 and 5, respectively), we can see from Fig. 7.5d that the infinitesimal deflection dv is given by $dv \approx \theta_x dx$, or

$$\theta_x = \frac{\partial v}{\partial x} . \tag{7.15}$$

So substituting Eq. 7.15 into Eq. 7.14 gives the strain in the x direction as

$$\varepsilon_x = z \frac{\partial^2 v}{\partial x^2} . \tag{7.16}$$

The analysis in the yz-plane is identical, and we can derive the equation for the y-strain as

$$\varepsilon_y = z \frac{\partial^2 v}{\partial y^2} . \tag{7.17}$$

We can then obtain the normal stresses by substituting Eqs. 7.16 and 7.17 into Eq. 7.12, giving

$$\sigma_x = \frac{Ez}{1 - \nu^2} \left(\frac{\partial^2 v}{\partial x^2} + \nu \frac{\partial^2 v}{\partial y^2} \right) , \tag{7.18}$$

and

$$\sigma_y = \frac{Ez}{1 - \nu^2} \left(\frac{\partial^2 v}{\partial y^2} + \nu \frac{\partial^2 v}{\partial x^2} \right) . \tag{7.19}$$

Now, if the fibre has a thickness in the z direction of dz (and unit width in the y direction), then since stress is force per unit area, the force acting on it in the x direction—by virtue of the in-plane stresses σ_x (Fig. 7.4)—is $\sigma_x \, dz$. This force has a bending moment about the neutral surface of $z\sigma_x dz$, so the total bending moment in the xz-plane, M_x, can be found by integrating these infinitesimal bending moments over the thickness of the plate (h), or

$$M_x = \int_{-h/2}^{h/2} z \, \sigma_x \, dz . \tag{7.20}$$

Substituting Eq. 7.18 into Eq. 7.20, we obtain

$$M_x = \frac{E}{1 - \nu^2} \left(\frac{\partial^2 v}{\partial x^2} + \nu \frac{\partial^2 v}{\partial y^2} \right) \int_{-h/2}^{h/2} z^2 \, dz ,$$

or

$$M_x = \frac{Eh^3}{12(1 - \nu^2)} \left(\frac{\partial^2 v}{\partial x^2} + \nu \frac{\partial^2 v}{\partial y^2} \right) .$$

If we now define the *flexural rigidity* as

$$D \equiv \frac{Eh^3}{12(1 - \nu^2)} , \tag{7.21}$$

we obtain the bending moment in the xz-plane as

$$M_x = D \left(\frac{\partial^2 v}{\partial x^2} + \nu \frac{\partial^2 v}{\partial y^2} \right) . \tag{7.22}$$

In order to obtain the bending moment in the yz-plane, we proceed as above from Eq. 7.20, replacing occurrences of x with y and using Eq. 7.19, and find that the bending moment in the yz-plane is given by

$$M_y = D \left(\frac{\partial^2 v}{\partial y^2} + \nu \frac{\partial^2 v}{\partial x^2} \right) . \tag{7.23}$$

7.2.6   Twisting Moments

The actions of the twisting moments in Fig. 7.1, M_{xy}, M_{yx}, $M_{xy} + dM_{xy}$ and $M_{yx} + dM_{yx}$, serve to shear the plate in the xy-plane, and this enables us to develop a relationship between the twisting moment and the deflection of the plate.

If we take the shear stress and strain components of Eq. 7.7, we obtain

$$\begin{pmatrix} \tau_{xy} \\ \tau_{xz} \\ \tau_{yz} \end{pmatrix} = \begin{pmatrix} \mu & 0 & 0 \\ 0 & \mu & 0 \\ 0 & 0 & \mu \end{pmatrix} \begin{pmatrix} \gamma_{xy} \\ \gamma_{xz} \\ \gamma_{yz} \end{pmatrix} , \tag{7.24}$$

which has no dependence on normal stress or strain. Since we are assuming a state of plane stress (Sect. 7.2.4) we can ignore any stresses—including shear stresses— acting in the z direction (Assumption 7), meaning we have $\tau_{xz} = \tau_{yz} = 0$. Equation 7.24 then becomes

$$\tau_{xy} = \mu \gamma_{xy} , \tag{7.25}$$

where μ is the shear modulus (Sect. 7.2.3), given by

$$\mu = \frac{E}{2(1 + \nu)} , \tag{7.26}$$

which is derived from Eqs. 7.8 and 7.9.

The shear experienced by the plate in the xy-plane is depicted in Fig. 7.6a. Here, a rectangular element ABCD, of side lengths dx and dy, is sheared into the parallelogram A′B′C′D′. As noted in the caption to Fig. 7.2, shear strain is given by the angle $\gamma = \tan \theta$, so from Fig. 7.6a we can write

$$\gamma_1 \approx \frac{du_y}{dx} \quad \text{and} \quad \gamma_2 \approx \frac{du_x}{dy} ,$$

for small angles γ_1 and γ_2. Writing the derivatives as partial derivatives, the total shear strain is given by

$$\gamma_{xy} = \gamma_1 + \gamma_2 = \frac{\partial u_y}{\partial x} + \frac{\partial u_x}{\partial y} = \gamma_{yx} . \tag{7.27}$$

Turning now to the cross-section in Fig. 7.6b, we see that the fibre element AB is deflected downwards to position A′B′, through a distance v in the z direction, and across by a distance u_x in the x direction. If θ_x is a small angle (Assumption 5), then we have

$$u_x = z\theta_x = z\frac{\partial v}{\partial x} , \tag{7.28}$$

where we have used Eq. 7.15 for θ_x. By symmetry, the same is true in the y direction, giving

$$u_y = z\theta_y = z\frac{\partial v}{\partial y} . \tag{7.29}$$

If we substitute Eqs. 7.28 and 7.29 into Eq. 7.27, we obtain an expression for the shear strain in terms of the deflection:

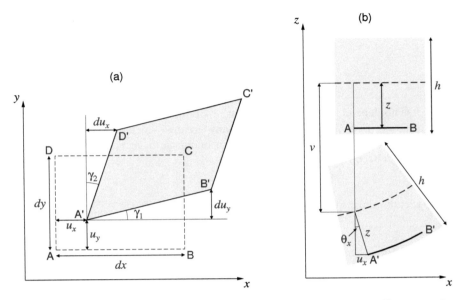

Fig. 7.6 A plate undergoing shear during deflection in **a** plan view (xy-plane), and **b** cross-section (xz-plane). In panel **a**, we let $du_x = (\partial u_x/\partial y)\,dy$, and $du_y = (\partial u_y/\partial x)\,dx$. In panel **b**, the upper grey box represents a part of the undeflected plate, while the lower grey shape represents the corresponding part of the deflected plate, deflected by an amount v. In both, the neutral surface is shown as a dashed line, and the fibre AB as a solid line. The line AB in **b** corresponds to the line AB in **a**, and likewise for A′B′, and the distance u_x in **b** corresponds to the distance u_x in **a**. The angle θ_x is the same as θ_x in Fig. 7.5a

$$\gamma_{xy} = 2z \frac{\partial^2 v}{\partial x\, \partial y} \, . \tag{7.30}$$

Therefore, from Eqs. 7.25, 7.26 and 7.30, the shear stress can be written as

$$\tau_{xy} = \frac{Ez}{1+\nu} \frac{\partial^2 v}{\partial x\, \partial y} \, . \tag{7.31}$$

Proceeding as we did in Sect. 7.2.5 for the justification of Eq. 7.20, the twisting moment in the xy-plane is given by

$$M_{xy} = \int_{-h/2}^{h/2} z\, \tau_{xy}\, dz \, , \tag{7.32}$$

where h is the plate thickness. Substituting Eq. 7.31 into Eq. 7.32, we obtain

$$M_{xy} = \frac{E}{1+\nu} \frac{\partial^2 v}{\partial x\, \partial y} \int_{-h/2}^{h/2} z^2\, dz \, ,$$

or

$$M_{xy} = \frac{Eh^3}{12(1+\nu)} \frac{\partial^2 v}{\partial x \, \partial y} \, .$$

Using Eq. 7.21 for the flexural rigidity, D, we can rewrite this as

$$M_{xy} = \frac{D(1-\nu^2)}{1+\nu} \frac{\partial^2 v}{\partial x \, \partial y} \, ,$$

or

$$M_{xy} = (1-\nu) D \frac{\partial^2 v}{\partial x \, \partial y} \, , \tag{7.33}$$

being the twisting moment in the xy-plane.

7.2.7 Flexural Equations

We are now in a position to express the plate deflection as a partial differential equation in v, rather than in the moments. We do this by simply substituting Eqs. 7.22, 7.23 and 7.33 into the thin plate PDE (Eq. 7.4) and performing the differentiations. This gives the *thin, elastic plate PDE* as

$$D\frac{\partial^4 v}{\partial x^4} + D\frac{\partial^4 v}{\partial y^4} + 2D\frac{\partial^4 v}{\partial x^2 \, \partial y^2} + 2\frac{\partial D}{\partial x}\frac{\partial^3 v}{\partial x^3}$$
$$+ 2\frac{\partial D}{\partial y}\frac{\partial^3 v}{\partial y^3} + 2\frac{\partial D}{\partial x}\frac{\partial^3 v}{\partial x \, \partial y^2} + 2\frac{\partial D}{\partial y}\frac{\partial^3 v}{\partial x^2 \, \partial y}$$
$$+ \left(\frac{\partial^2 D}{\partial x^2} + \nu\frac{\partial^2 D}{\partial y^2}\right)\frac{\partial^2 v}{\partial x^2} + \left(\frac{\partial^2 D}{\partial y^2} + \nu\frac{\partial^2 D}{\partial x^2}\right)\frac{\partial^2 v}{\partial y^2}$$
$$+ 2(1-\nu)\frac{\partial^2 D}{\partial x \, \partial y}\frac{\partial^2 v}{\partial x \, \partial y} = -p_z \tag{7.34}$$

which can be simplified using the *Laplacian operator*

$$\nabla^2 \equiv \frac{\partial^2}{\partial x^2} + \frac{\partial^2}{\partial y^2} \, ,$$

to give

$$\nabla^2 \left(D\nabla^2 v\right) - (1-\nu)\left(\frac{\partial^2 D}{\partial x^2}\frac{\partial^2 v}{\partial y^2} - 2\frac{\partial^2 D}{\partial x \, \partial y}\frac{\partial^2 v}{\partial x \, \partial y} + \frac{\partial^2 D}{\partial y^2}\frac{\partial^2 v}{\partial x^2}\right) = -p_z \, .$$
$$\tag{7.35}$$

Equation 7.35 is a valid PDE when the flexural rigidity is spatially variable, $D(x, y)$. It is a useful equation to solve for deflections when one knows the flexural rigidity distribution and the external load, p_z, such as when one wants to generate synthetic models of plate deflection: in such cases, Eq. 7.35 is usually solved for v using the method of finite differences (Sect. 10.5).

However, an analytic expression for v can be derived if the flexural rigidity is spatially uniform, i.e. not a function of x or y. In this case, Eq. 7.35 may be simplified because any derivative of D will be zero. Setting this condition in Eq. 7.35 gives us

$$D \nabla^2 \nabla^2 v = -p_z ,$$

or

$$D \left(\frac{\partial^4 v}{\partial x^4} + 2 \frac{\partial^4 v}{\partial x^2 \partial y^2} + \frac{\partial^4 v}{\partial y^4} \right) = -p_z , \tag{7.36}$$

which is called the *biharmonic equation*. The opposite signs reflect the fact that p_z is a downward-acting force, while the left-hand side is the *restoring force*, acting upwards to oppose p_z and trying to make the plate flat again.

When applying this theory to the lithosphere, we note that it is not actually an elastic plate of thickness h, but comprises many materials of various rheologies that behave elastically over long timescales (Sect. 1.3.1). Therefore, in geophysics the thickness, h, of the elastic plate shown in Figs. 7.5 and 7.6 is known as the *effective elastic thickness*, T_e, and Eq. 7.21 is written as

$$D \equiv \frac{E \, T_e^3}{12 \, (1 - \nu^2)} . \tag{7.37}$$

7.2.8 Solving the Biharmonic Equation

Since D is a constant, Eq. 7.36 may be solved for v by taking its Fourier transform. Here we use the relation in Eq. 2.49, which is

$$\mathsf{F} \left\{ \frac{\partial^m}{\partial x^m} \frac{\partial^n}{\partial y^n} v(x, y) \right\} = (ik_x)^m (ik_y)^n \, V(k_x, k_y) .$$

Letting $V(\mathbf{k})$ be the Fourier transform of $v(\mathbf{x})$, and $P_z(\mathbf{k})$ be the Fourier transform of the applied load $p_z(\mathbf{x})$, the Fourier transform of Eq. 7.36 is then

$$D \left[(ik_x)^4 V(\mathbf{k}) + 2(ik_x)^2 (ik_y)^2 V(\mathbf{k}) + (ik_y)^4 V(\mathbf{k}) \right] = -P_z(\mathbf{k})$$

or

$$D \left(k_x^4 + 2k_x^2 k_y^2 + k_y^4 \right) V(\mathbf{k}) = -P_z(\mathbf{k}) .$$

Since $k^4 = (k_x^2 + k_y^2)^2$, we finally obtain the equation

$$Dk^4 V(\mathbf{k}) = -P_z(\mathbf{k}) \ . \tag{7.38}$$

The space-domain deflections, $v(\mathbf{x})$, are then just the inverse Fourier transform of $V(\mathbf{k})$.

7.3 Buoyancy

The above theory for plate deflection was developed assuming that the plate is surrounded by air—above it and below it. Since air has an effective density and viscosity of zero (for our purposes), this means there is no resistance to bending provided by the surrounding medium. The lithosphere, though, is underlain by the asthenosphere and can be overlain by water, and the presence of these materials generates extra forces on the plate that have not been considered so far, namely *buoyancy* forces.

In the simplified theories of isostasy (Sect. 1.1), the plate is often referred to as the 'crust' rather than 'lithosphere', and so has a constant density denoted ρ_c. It is taken to lie on a zero-viscosity ('*inviscid*') 'mantle' rather than asthenosphere, with a uniform density ρ_m (with $\rho_m > \rho_c$). And it is overlain by a fluid with lower density ρ_f, which can be air ($\rho_f = \rho_a = 0$) in continental settings, or seawater ($\rho_f = \rho_w = 1030$ kg m^{-3}) in the oceans.

Consider the simple scenario in Fig. 7.7a, which shows an undeflected plate of thickness h whose surface is some depth d underneath a fluid layer: if the fluid is seawater then d will be the ocean depth; if the fluid is air then one would normally take $d = 0$. If some unspecified (for the time being) force acts on the plate's upper surface and depresses it, the plate is deflected downwards by a distance v, and the fluid rushes into the space vacated by the depression, as shown in Fig. 7.7b where we consider just a rectangular column of unit cross-sectional area. We can then find the

Fig. 7.7 A plate of thickness h and density ρ_c is overlain by a fluid of thickness d and density ρ_f and underlain by an inviscid 'mantle' of density ρ_m, with $\rho_f < \rho_c < \rho_m$. **a** The plate is initially flat at all interfaces. **b** It is then deflected by an amount v, and the fluid fills the vacated depression

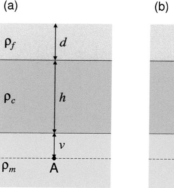
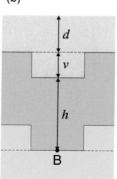

pressure (or force per unit area) generated at the location labelled 'B'—the base of the deflected plate—and compare it with the pressure at the same location but before deflection occurred, which was in the mantle (point 'A'). Using Eq. 1.3, the pressure at point A is the sum of the pressures due to all the layers above it, or

$$p_A = \rho_f g d + \rho_c g h + \rho_m g v \,, \tag{7.39}$$

where g is the gravity acceleration. Similarly, the pressure at point B is

$$p_B = \rho_f g (d + v) + \rho_c g h \,. \tag{7.40}$$

The difference between Eqs. 7.39 and 7.40 gives the change in pressure that occurs during the deflection process, and provides the net force (per unit area) acting on the new base of the plate by virtue of its deflection into a higher-density medium. This is the buoyancy force, $p_b = p_A - p_B$, with

$$p_b = (\rho_m - \rho_f) g v \,. \tag{7.41}$$

This force is directed upwards, with the displaced higher-density mantle trying to recover its old position and pushing the deflected plate back up. Note that Eq. 7.41 does not depend upon the thickness of the plate (h) or its density (ρ_c), and neither does it depend upon the thickness of the overlying layer (d).

We can now write the transverse load specified in Sect. 7.2.1, $p_z(\mathbf{x})$, as the sum of an applied load (which we shall call ℓ) and the buoyancy force, or

$$p_z(\mathbf{x}) = \Delta \rho_{mf} g \, v(\mathbf{x}) + \ell(\mathbf{x}) \,, \tag{7.42}$$

using the notation $\Delta \rho_{ab} \equiv \rho_a - \rho_b$, and assuming that the applied load acts downwards. Substituting Eq. 7.42 into Eq. 7.36, the biharmonic equation now reads

$$D \left(\frac{\partial^4 v}{\partial x^4} + 2 \frac{\partial^4 v}{\partial x^2 \, \partial y^2} + \frac{\partial^4 v}{\partial y^4} \right) + \Delta \rho_{mf} g v = -\ell \,, \tag{7.43}$$

whose Fourier transform is

$$\left(D k^4 + \Delta \rho_{mf} g \right) V(\mathbf{k}) = -L(\mathbf{k}) \tag{7.44}$$

(cf. Eq. 7.38), with solution

$$V(\mathbf{k}) = \frac{-L(\mathbf{k})}{D k^4 + \Delta \rho_{mf} g} \,.$$

The space-domain deflections, $v(\mathbf{x})$, are then just the inverse Fourier transform of $V(\mathbf{k})$.

Note that the density ρ_m in Eqs. 7.43 and 7.44 must be constant. If it were not, then the buoyancy force term in Eq. 7.43 would read $[\rho_m(\mathbf{x}) - \rho_f]g v(\mathbf{x})$, and Eq. 7.44 would not follow from Eq. 7.43 because the Fourier transform of a product of space-domain variables—that is, $\mathsf{F}\{\rho_m(\mathbf{x})v(\mathbf{x})\}$—is actually the convolution of the Fourier transforms of those variables in the wavenumber domain (Sect. 2.2.8). This requirement of uniformity of density applies to all the densities considered in the following flexural models.

The types of applied load, $\ell(\mathbf{x})$, typically used in flexural studies form the subject of the remainder of this chapter. Note that from now on we will drop the argument $(\mathbf{k})$ showing the dependence of the Fourier transforms upon wavenumber, for visual simplicity.

7.4 Surface Loading

In the early days of flexural studies, surface loading was the only type of load considered, with subsurface—or internal—loading not being incorporated in models until later (Kirby 2014). Surface loads comprise an 'initial topography'—created by some tectonic upheaval or volcanism, for example—which is emplaced on the upper surface of the plate and Eq. 7.44 solved for the deflection; from this a 'final topography' after flexure can be derived. Figure 7.8 shows an example of a synthetic, fractal topographic surface acting as the initial load, with the final topography computed for a weak plate ($T_e = 0$ km) and a strong plate ($T_e = 120$ km). The figure shows that the strong plate adequately supports and preserves the short- and medium-wavelength topography without flexing, and that only the long-wavelength loads undergo subsidence and deflect its base. In contrast, the weak plate has no mechanical strength and loads of all sizes must be isostatically compensated (Sect. 1.1); thus the plate base here is a (scaled) mirror-image of the initial topography.

It should also be noted that the plate's geometric thickness (as opposed to its elastic thickness) is not a factor in the computations of flexure, which is why no depth scale is added to Fig. 7.8: the relief at all interfaces in the image is accurate whether the Moho is at 1 km or 100 km depth. This independence will be seen in the ensuing derivations. Indeed, it is only when the gravity field of the system is desired that the geometric thickness of the plate becomes involved (Chap. 8).

7.4.1 Two-Layer Crust Model

Consider an incompressible crust of two layers with density ρ_1 and ρ_2, respectively, shown in Fig. 7.9. The crust is overlain by a fluid of density ρ_f, and itself rests on an inviscid mantle of density ρ_m. Each interface is initially flat and horizontal. Now consider an initial surface load ℓ_T of density ρ_T (where $\rho_f < \rho_T \leq \rho_1$) and geometric amplitude (topography) h_i. We use the subscript 'T' to denote surface (or

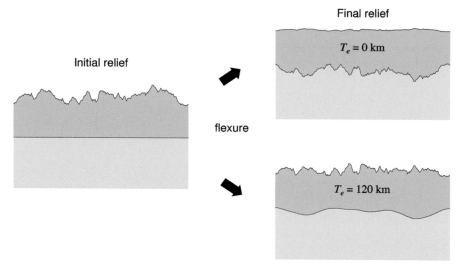

Fig. 7.8 Plate flexure due to surface loading. At left is shown an initial load comprising surface topographic relief emplaced upon a plate with a flat 'Moho' (plate base); at right is shown the final surface topography and Moho relief for two scenarios of effective elastic thickness (T_e)

'top') loading. Working from now on in the Fourier domain (where capital letters denote the Fourier transforms of lower-case variables) the load (in pascals) is given by

$$L_T = \Delta \rho_{Tf} g H_i \ . \tag{7.45}$$

Note that here we are considering the load to be the pressure difference between the load itself and the fluid it displaces. If the surface load were to have the same density as the topmost crustal layer, we would set $\rho_T = \rho_1$, giving $\Delta \rho_{Tf} = \Delta \rho_{1f}$ in Eq. 7.45 and $\Delta \rho_{1T} = 0$ in the following equations.

When the load is emplaced on the surface of the crust it produces a downward deflection of amplitude V_T, and a final surface topography of amplitude H_T, where

$$H_T = H_i + V_T \ , \tag{7.46}$$

shown in Fig. 7.9, where all distances are positive upwards. Thus, the final surface topography, H_T, due to the surface load is the sum of the initial topography, H_i, and the deflection, V_T, which has a negative value in this scenario because it is directed downwards. If we substitute Eq. 7.45 into Eq. 7.46, we get an alternative expression for H_T:

$$H_T = \frac{L_T}{\Delta \rho_{Tf} g} + V_T \ . \tag{7.47}$$

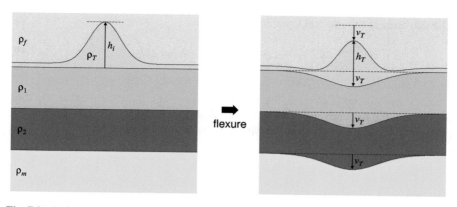

Fig. 7.9 A plate comprising a two-layer crust is subject to an initial surface load of height h_i and density ρ_T (left), which causes the plate (and each interface within the system) to flex by an amount v_T (right). After flexure, the load has elevation h_T

We now derive an expression for the (Fourier transform of the) partial differential equation of plate flexure, Eq. 7.44, in this surface loading scenario. Rather than just changing Eq. 7.44 directly, we will perform a force (or rather, pressure) balance, putting upward forces on the left-hand side of the equation, and downward forces on its right, based on the change in geometry between the left-hand (pre-flexure) side of Fig. 7.9, and its right hand (post-flexure) side. The upward forces are

1. the restoring force from the bending and twisting moments, $Dk^4 V_T$;
2. the buoyancy force from the deflection of the layer 2/mantle interface (layer 2 filling the space vacated by the depressed mantle), $\Delta\rho_{m2} g V_T$;
3. the buoyancy force from the deflection of the layer 1/layer 2 interface (layer 1 filling the space vacated by the depressed layer 2), $\Delta\rho_{21} g V_T$;
4. the buoyancy force from the deflection of the load/layer 1 interface (the load filling the space vacated by the depressed layer 1), $\Delta\rho_{1T} g V_T$;
5. the buoyancy force from the deflection of the fluid/load interface (fluid filling the space vacated by the sinking initial load), $\Delta\rho_{Tf} g V_T$.

The downward forces are only

1. the fluid displaced by the initial surface load (load of higher density replacing fluid of lower density—an 'anti-buoyancy force'), $\Delta\rho_{Tf} g H_i$.

We form our equation by putting upward (positive) forces on the left-hand side and the downward (negative) forces on the right, which gives

$$Dk^4 V_T + \Delta\rho_{m2} g V_T + \Delta\rho_{21} g V_T + \Delta\rho_{1T} g V_T + \Delta\rho_{Tf} g V_T = -\Delta\rho_{Tf} g H_i \quad (7.48)$$

or

$$Dk^4 V_T + \left(\rho_m - \rho_2 + \rho_2 - \rho_1 + \rho_1 - \rho_T + \rho_T - \rho_f\right) g V_T = -\Delta\rho_{Tf} g H_i .$$

The densities of the internal layers cancel, leaving

$$Dk^4 V_T + \Delta \rho_{mf} g V_T = -\Delta \rho_{Tf} g H_i , \tag{7.49}$$

which is the surface loading PDE in the Fourier domain (cf. Eq. 7.44).

Note that we could arrive at the same equation using the total pressure method to compute the net buoyancy force given in Sect. 7.3. Using Fig. 7.9 as a guide, the pressure at a distance v_T below the Moho in the undeflected system is

$$p_A = \rho_f g d + \rho_T g h_i + \rho_1 g t_1 + \rho_2 g t_2 + \rho_m g v_T$$

(where d is the thickness of the fluid layer above the initial load, and t_1 and t_2 are the thicknesses of layers 1 and 2, respectively), while that at the same point in the deflected system is

$$p_B = \rho_f g d + \rho_f g v_T + \rho_T g h_i + \rho_1 g t_1 + \rho_2 g t_2 ,$$

and their difference is the net buoyancy force

$$p_b = \Delta \rho_{mf} g v_T ,$$

as in Eq. 7.49.

When we consider scenarios in which both surface and internal loading feature, it will be useful to distinguish between the deflection experienced by all layers in the fluid/crust/mantle system (V_T) and the deflection of the Moho by a surface load, which we shall denote W_T. As is evident from Fig. 7.9 and the above discussion, we have

$$W_T = V_T , \tag{7.50}$$

i.e. the final Moho topography caused by surface loading is just equal to the deflection of the Moho, because it was initially flat.

We can rewrite the PDE, Eq. 7.49, in a more compact form in terms of V_T and L_T. To that end, using Eq. 7.45, Eq. 7.49 becomes

$$\left(Dk^4 + \Delta \rho_{mf} g \right) V_T = -L_T , \tag{7.51}$$

or

$$W_T = \frac{-L_T}{\Phi g} , \tag{7.52}$$

using Eq. 7.50, where we define

$$\Phi(k) \equiv \frac{Dk^4}{g} + \Delta \rho_{mf} . \tag{7.53}$$

Alternatively, we can rewrite the PDE in terms of W_T and H_T, which will be useful when we consider the gravity field and admittances (Chap. 8). Substituting $L_T = \Delta\rho_{Tf}g(H_T - V_T)$ from Eq. 7.47 into Eq. 7.51 gives

$$\left(Dk^4 + \Delta\rho_{mf}g\right) V_T = -\Delta\rho_{Tf}g(H_T - V_T)$$

which can be rearranged to give

$$W_T = \frac{-\Delta\rho_{Tf}}{\Phi - \Delta\rho_{Tf}} H_T , \qquad (7.54)$$

using Eq. 7.50.

Finally, when we explore the load deconvolution method (Chap. 9) it will be useful to rewrite the PDE in terms of H_T and L_T. If we equate Eqs. 7.52 and 7.54, we obtain

$$\frac{-\Delta\rho_{Tf}}{\Phi - \Delta\rho_{Tf}} H_T = \frac{-L_T}{\Phi g}$$

which can be rearranged to give

$$H_T = \left[\frac{1}{\Delta\rho_{Tf}\, g} - \frac{1}{\Phi g}\right] L_T . \qquad (7.55)$$

Note that some authors include the effects of sediment infill of the flexural depression here (Karner and Watts 1982; Watts 2001; Turcotte and Schubert 2002). However, most do not as it is not an essential part of the model and many regions of the Earth are unsedimented or have only thin sediments. Indeed, one can argue that because sediments are deposited much later than the primary loading event, one should really conduct a new flexural analysis with the sedimentary layer as an initial load upon the post-flexural geometry resulting from the primary loading event. There is also the complication that, in reality, sediments are localised to flexural depressions, while the Fourier transform solution to the biharmonic equation, Eq. 7.44, assumes uniform sediment thickness and density over the whole study area, as noted in Sect. 7.3. In this book, then, it is recommended that sediments be accounted for in the data, rather than in the flexural model. Sediments, and some approaches to model their effect upon flexure, are discussed further in Sect. 11.10.

7.4.2 Multiple-Layer Crust Model

The extension to a model where the crust has many layers is trivial, owing to the cancellation of the densities in Eq. 7.48. Suppose the crust has n layers, with each layer having density ρ_i for $i = 1, \ldots, n$. Again, the crust is overlain by a fluid of density ρ_f, and itself rests on an inviscid mantle of density ρ_m. The initial load is still given by Eq. 7.45, and Eqs. 7.46 and 7.47 still apply. Each interface is initially flat and horizontal, and is deflected by the same amount, v_T with Fourier transform V_T. The force balance equation reads

$$Dk^4 V_T + \Delta\rho_{mn} g V_T + \Delta\rho_{n,n-1} g V_T + \cdots + \Delta\rho_{21} g V_T$$
$$+ \Delta\rho_{1T} g V_T + \Delta\rho_{Tf} g V_T = -\Delta\rho_{Tf} g H_i$$

(cf. Eq. 7.48). This simplifies to

$$Dk^4 V_T + \Delta\rho_{mf} g V_T = -\Delta\rho_{Tf} g H_i ,$$

which is Eq. 7.49, or to

$$V_T = W_T = \frac{-L_T}{\Phi g} , \tag{7.56}$$

using Eqs. 7.45, 7.50 and 7.53; Eq. 7.56 is the same as the equation for a two-layer crust (Eq. 7.52). The procedure to derive expressions for W_T in terms of H_T, and H_T in terms of L_T, is then identical to that outlined in Sect. 7.4.1 because none of the other equations (Eqs. 7.45–7.47 and 7.49–7.52) have changed. Thus we still get Eqs. 7.50 and 7.54, being

$$V_T = W_T = \frac{-\Delta\rho_{Tf}}{\Phi - \Delta\rho_{Tf}} H_T , \tag{7.57}$$

and Eq. 7.55, which is

$$H_T = \left[\frac{1}{\Delta\rho_{Tf} \, g} - \frac{1}{\Phi g} \right] L_T . \tag{7.58}$$

Hence, the plate flexure surface-loading PDE is independent of the number of intra-crustal layers.

7.5 Internal Loading

Loading can act at all depths within the lithosphere, from its surface to its base, and having dealt with surface loading, we now turn to *internal*, or *subsurface*, loading. Examples of different types of internal loads include igneous intrusions or large crustal blocks at the shallower depths, or deeper phenomena such as magmatic under-plating, lithospheric thermal anomalies or large compositional variations.

However, for reasons that will become apparent in Chap. 9, our models can only accommodate two loads. One of these is most often chosen to be a surface load because the topography is easily measurable, whereas internal loads must be deduced from remotely-sensed observations such as seismic, gravity or magnetic data. The other load is typically chosen to lie at the Moho, though Sect. 7.5.2 deals with internal loading at other depths.

7.5.1 Loading at the Moho of a Two-Layer Crust

Consider an incompressible crust of two layers with density ρ_1 and ρ_2, respectively, as in Sect. 7.4. Again, the crust is overlain by a fluid of density ρ_f, and itself rests on an inviscid mantle of density ρ_m, shown in Fig. 7.10. Each interface is initially flat and horizontal. Now consider an initial internal load ℓ_B of density ρ_B (where $\rho_2 \le \rho_B < \rho_m$) and geometric amplitude (relief) w_i. We use the subscript 'B' to denote internal (or 'bottom', 'basal' or 'buried') loading. Working from now on in the Fourier domain (where capital letters denote the Fourier transforms of lower-case variables) the load (in pascals) is given by

$$L_B = \Delta \rho_{mB}\, g\, W_i \ . \tag{7.59}$$

If the internal load were to have the same density as the lower crust, we would set $\rho_B = \rho_2$, giving $\Delta \rho_{mB} = \Delta \rho_{m2}$ in Eq. 7.59, and $\Delta \rho_{B2} = 0$ in the following equations.

When the load is emplaced at the base of the crust (Moho) it produces an upward deflection of amplitude V_B, and a final Moho relief of amplitude W_B, where

$$W_B = W_i + V_B \ , \tag{7.60}$$

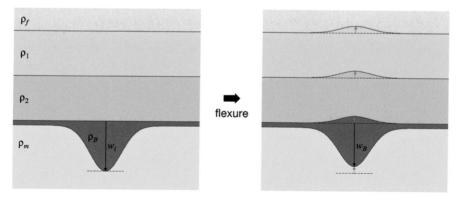

Fig. 7.10 A plate comprising a two-layer crust is subject to an initial subsurface load on the Moho of relief w_i and density ρ_B (left). This causes the plate (and each interface within the system) to flex by an amount v_B, indicated by the small, red arrows (right). After flexure, the load has relief w_B

shown in Fig. 7.10, where all distances are positive upwards. Thus, the final Moho relief, W_B, due to the subsurface load is the sum of the initial Moho relief, W_i, and the deflection, V_B. If we substitute Eq. 7.59 into Eq. 7.60, we get an alternative expression for W_B as

$$W_B = \frac{L_B}{\Delta \rho_{mB} g} + V_B . \qquad (7.61)$$

Proceeding as in Sect. 7.4, we can derive an expression for the Fourier transform of the PDE by performing a force (pressure) balance, putting downward forces on the left-hand side of the equation, and upward forces on its right, based on the change in geometry between the left-hand (pre-flexure) side of Fig. 7.10, and its right-hand (post-flexure) side. Here, we note that in surface loading (Sect. 7.4) the buoyancy forces at each interface arose when material of higher density was displaced by material of lower density, as the latter was deflected downwards into the former (Fig. 7.9). However, with internal loading the deflection is upwards so that at each interface material of higher density is displacing material of lower density, and the forces arising from these displacements are 'negative buoyancy forces'. That is, the force that is generated by the change in state from left-hand to right-hand sides of Fig. 7.10 is the weight of the displacing mass, which acts to return it back to its equilibrium position, i.e. downwards. In terms of force per unit area, the weight of the displacing mass is $\Delta \rho_{ab} g V_B$, for lower-layer a and upper-layer b. Hence, the downward forces are

1. the restoring force from the bending and twisting moments, $Dk^4 V_B$;
2. the negative buoyancy force from the deflection of the load/mantle interface (the mantle occupying the space vacated by the risen load), $\Delta \rho_{mB} g V_B$;
3. the negative buoyancy force from the deflection of the layer 2/load interface (the load occupying the space vacated by the risen layer 2), $\Delta \rho_{B2} g V_B$;
4. the negative buoyancy force from the deflection of the layer 1/layer 2 interface (layer 2 occupying the space vacated by the risen layer 1), $\Delta \rho_{21} g V_B$;
5. the negative buoyancy force from the deflection of the fluid/layer 1 interface (layer 1 occupying the space vacated by the displaced fluid), $\Delta \rho_{1f} g V_B$.

The upward forces are only

1. the mantle displaced by the initial internal load (a buoyancy force arising from a load of lower density replacing the mantle of higher density), $\Delta \rho_{mB} g W_i$.

We form our equation by putting downward (negative) forces on the left-hand side and the upward (positive) forces on the right, which gives

$$-Dk^4 V_B - \Delta \rho_{mB} g V_B - \Delta \rho_{B2} g V_B - \Delta \rho_{21} g V_B - \Delta \rho_{1f} g V_B = \Delta \rho_{mB} g W_i ,$$

or

$$Dk^4 V_B + \left(\rho_m - \rho_B + \rho_B - \rho_2 + \rho_2 - \rho_1 + \rho_1 - \rho_f \right) g V_B = -\Delta \rho_{mB} g W_i .$$

The densities of the internal layers cancel, leaving

$$Dk^4 V_B + \Delta\rho_{mf} g V_B = -\Delta\rho_{mB} g W_i , \qquad (7.62)$$

which is the internal loading PDE in the Fourier domain (see Eq. 7.44).

For completeness, we will show that use of the total pressure method to compute the net buoyancy force given in Sect. 7.3 provides the same answer (for the buoyancy force component). Using Fig. 7.10 as a guide, the pressure at a distance w_i below the Moho in the undeflected system (i.e. at the base of the initial load) is

$$p_A = \rho_f g d + \rho_1 g t_1 + \rho_2 g t_2 + \rho_B g w_i$$

(where d is the thickness of the fluid layer above the surface of crustal layer 1, and t_1 and t_2 are the thicknesses of layers 1 and 2, respectively), while that at the same point in the deflected system is

$$p_B = \rho_f g (d - v_B) + \rho_1 g t_1 + \rho_2 g t_2 + \rho_B g w_i + \rho_m g v_B ,$$

and their difference is the net buoyancy force

$$p_b = \Delta\rho_{mf} g v_B ,$$

as in Eq. 7.62.

Although they have the same value, it will be useful to distinguish between the deflection of the entire system (V_B) and the deflection of the upper surface of the crust (H_B), which forms the visible surface topography post-flexure. As is evident in Fig. 7.10, we have

$$H_B = V_B , \qquad (7.63)$$

i.e. the final surface topography caused by internal loading is just equal to the deflection at the surface of the crust.

It will be useful (in later chapters) to have alternative expressions for Eq. 7.62 available. First, we can express V_B in terms of L_B by substituting Eq. 7.59 into Eq. 7.62, giving

$$\left(Dk^4 + \Delta\rho_{mf} g\right) V_B = -L_B , \qquad (7.64)$$

or, using Eq. 7.63,

$$H_B = \frac{-L_B}{\Phi g} , \qquad (7.65)$$

where Φ is defined by Eq. 7.53.

Next, we can rewrite the PDE in terms of H_B and W_B. Substituting $L_B = \Delta\rho_{mB} g (W_B - V_B)$ from Eq. 7.61 into Eq. 7.64 gives

$$\left(Dk^4 + \Delta\rho_{mf} g\right) V_B = -\Delta\rho_{mB} g (W_B - V_B) ,$$

which can be rearranged to give

$$H_B = \frac{-\Delta\rho_{mB}}{\Phi - \Delta\rho_{mB}} W_B , \qquad (7.66)$$

using Eqs. 7.53 and 7.63.

Finally, we can rewrite the PDE in terms of W_B and L_B by equating Eqs. 7.65 and 7.66, thus:

$$\frac{-\Delta\rho_{mB}}{\Phi - \Delta\rho_{mB}} W_B = \frac{-L_B}{\Phi g} .$$

After some algebra we obtain

$$W_B = \left[\frac{1}{\Delta\rho_{mB} g} - \frac{1}{\Phi g} \right] L_B . \qquad (7.67)$$

7.5.2   Loading Within a Multiple-Layer Crust

The internal load does not need to be placed at the Moho, however, and Fig. 7.11 shows an example of a load placed at the base of the upper crust in a two-layer crust. In order to derive a PDE describing flexure of the system we proceed as in Sect. 7.5.1, but now we will extend the model so that the crust has n layers, with each layer having density ρ_i for $i = 1, \ldots, n$, as in Sect. 7.4.2. As before, the crust is overlain by a fluid of density ρ_f, and itself rests on an inviscid mantle of density ρ_m. Each interface is initially flat and horizontal.

Now consider an initial internal load, ℓ_B, of density ρ_B and geometric amplitude w_i, emplaced at the base of layer j ($j = 1, \ldots, n$), such that $\rho_j \leq \rho_B < \rho_{j+1}$. The Fourier transform of the load is thus given by

$$L_B = \Delta\rho_{j+1,B} g W_i , \qquad (7.68)$$

where $\Delta\rho_{j+1,B} = \rho_{j+1} - \rho_B$. If loading occurs at the base of the lowest crustal layer (i.e. at the Moho, with $j = n$), we define $\rho_{n+1} = \rho_m$, the mantle density. The load produces an upward deflection of amplitude V_B, and a final layer-$(j + 1)$/load interface relief of amplitude W_B, where

$$W_B = W_i + V_B , \qquad (7.69)$$

shown for $n = 2$ in Fig. 7.11. Note that W_B here is not the flexed Moho relief (as it was in Sect. 7.5.1) but the post-flexure relief of the interface at which the load was emplaced. If we substitute Eq. 7.68 into Eq. 7.69, we obtain an alternative expression for W_B as

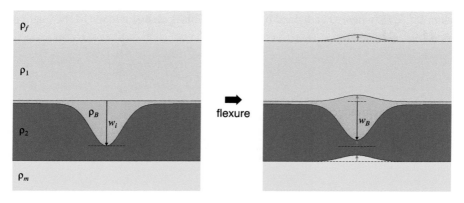

Fig. 7.11 A plate comprising a two-layer crust is subject to an initial subsurface load at the base of layer 1, of relief w_i and density ρ_B (left). This causes the plate (and each interface within the system) to flex by an amount v_B, indicated by the small, red arrows (right). After flexure, the load has relief w_B

$$W_B = \frac{L_B}{\Delta \rho_{j+1,B} g} + V_B \ . \tag{7.70}$$

Proceeding as in Sects. 7.4.2 and 7.5.1, the force balance equation is

$$-Dk^4 V_B - \Delta \rho_{mn} g V_B - \Delta \rho_{n,n-1} g V_B - \cdots - \Delta \rho_{j+1,B} g V_B$$
$$- \Delta \rho_{Bj} g V_B - \cdots - \Delta \rho_{21} g V_B - \Delta \rho_{1f} g V_B \ = \ \Delta \rho_{j+1,B} g W_i \ .$$

The densities of the intra-crustal layers cancel, leaving

$$Dk^4 V_B + \Delta \rho_{mf} g V_B \ = \ -\Delta \rho_{j+1,B} g W_i \ , \tag{7.71}$$

which is the internal loading PDE in the Fourier domain.

As in the Moho-loading case, the final surface topography caused by internal loading is equal to the deflection at the surface, with

$$H_B = V_B \tag{7.72}$$

(see Fig. 7.11). We can rewrite Eq. 7.71 as

$$\left(Dk^4 + \Delta \rho_{mf} g \right) V_B \ = \ -L_B \tag{7.73}$$

using Eq. 7.68, or as

$$H_B \ = \ \frac{-L_B}{\Phi g} \tag{7.74}$$

using Eq. 7.72, where Φ is defined by Eq. 7.53. This is identical to the Moho loading case, Eq. 7.65.

If we now substitute $L_B = \Delta\rho_{j+1,B}g(W_B - V_B)$ from Eq. 7.70 into Eq. 7.73, we obtain

$$\left(Dk^4 + \Delta\rho_{mf}g\right)V_B = -\Delta\rho_{j+1,B}g(W_B - V_B),$$

which can be rearranged to give

$$H_B = \frac{-\Delta\rho_{j+1,B}}{\Phi - \Delta\rho_{j+1,B}}W_B,\tag{7.75}$$

using Eq. 7.72, and which is identical to the Moho-loading case, Eq. 7.66, when $j = n$.

Finally, we can rewrite the PDE in terms of W_B and L_B by equating Eqs. 7.74 and 7.75, thus:

$$\frac{-\Delta\rho_{j+1,B}}{\Phi - \Delta\rho_{j+1,B}}W_B = \frac{-L_B}{\Phi g}.$$

After some algebra we obtain

$$W_B = \left[\frac{1}{\Delta\rho_{j+1,B}g} - \frac{1}{\Phi g}\right]L_B\tag{7.76}$$

(cf. Eq. 7.67). Remember that W_B is the post-flexure relief of the interface at which the load was emplaced, and is not necessarily the flexed Moho relief, as it was called in Sect. 7.5.1.

Thus, it can be seen that the number of intra-crustal layers does not affect the flexure PDE, for either surface or internal loading, though the location of the internal load does.

7.6 Combined Loading

More often than not, both types of load, surface and internal, will be present. Fortunately, because the loading equations (the PDEs in their various forms) are all linear equations, we are allowed to add them in order to ascertain their combined, superposed effects. Thus, the total deflection (V) is the sum of the surface and internal deflections, or

$$V = V_T + V_B,\tag{7.77}$$

shown in Fig. 7.12.

For example, if we want to find the total flexure when surface loading (Sect. 7.4.1) and Moho loading (Sect. 7.5.1) are present, we simply add Eqs. 7.49 and 7.62, giving

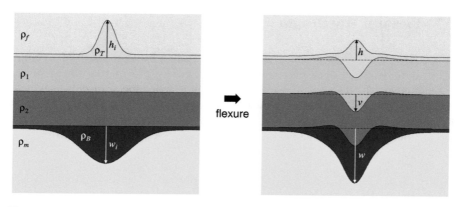

Fig. 7.12 A plate comprising a two-layer crust is subject to an initial surface load of height h_i and density ρ_T, and an initial Moho load of relief w_i and density ρ_B (left). The plate (and each interface within the system) flexes by an amount v, being the sum of the separate deflections due to the top and bottom loads ($v = v_T + v_B$) (right). The final topography (h) and Moho relief (w) are also shown

$$Dk^4(V_T + V_B) + \Delta\rho_{mf}g(V_T + V_B) = -\Delta\rho_{Tf}gH_i - \Delta\rho_{mB}gW_i ,$$

or

$$Dk^4V + \Delta\rho_{mf}gV = -\Delta\rho_{Tf}gH_i - \Delta\rho_{mB}gW_i .$$

If we substitute Eqs. 7.45 and 7.59, we obtain a general PDE for combined surface and Moho loading:

$$\left(Dk^4 + \Delta\rho_{mf}g\right)V = -L_T - L_B ,$$

which could have also been obtained by adding Eqs. 7.51 and 7.64. Indeed, the similarity between Eqs. 7.51, 7.64 and 7.73—which all contain the $Dk^4 + \Delta\rho_{mf}g$ term—suggests that these loading PDEs can readily be summed.

The same applies to the net surface topography (H) and internal-load interface relief (W), with both being the sum of their (linear) surface and internal loading components. Thus we have, for the combined loading surface topography

$$H = H_T + H_B , \tag{7.78}$$

and for the combined loading relief on the internal-load interface

$$W = W_T + W_B , \tag{7.79}$$

shown (in the space domain) in Fig. 7.12.

7.7   Flexural Wavelength

A glance at Fig. 7.9 shows that, under surface loading by a mountain of a certain width, the crust is depressed over a region that extends well beyond the dimensions of the mountain. The distance across the depression is called the *flexural wavelength*, and depends upon T_e. Indeed, one would be correct in thinking that strong plates have a broad but shallow depression, while weak plates have a narrower but deeper depression. Hence, as a final topic in this chapter, we look at the relationship between T_e and flexural wavelength.

We will follow the methods used in Sects. 7.4 and 7.5, but with two differences: we will consider a point load rather than an extended load, and will work in the space domain in one dimension. In using a point load (for example, a delta function), we can ignore the term in the force balance equation that represents the buoyancy force from the deflection of the fluid/load interface (fluid filling the space vacated by the sinking initial load)—which was represented by $\Delta \rho_{Tf} g V_T$ in Eq. 7.48—because a point load essentially has zero width and there is no space being vacated by its sinking. We will also assume that fluid of density ρ_f completely infills the flexural depression, though note that this infill 'fluid' could be air, water, sediment or crust material. Finally, we note that in the 1D space domain the restoring force from the bending moment is written as a fourth-order derivative in the deflection $v(x)$—rather than $Dk^4 V(k)$— and that the delta function load $\ell(x) = 0$, since the load is zero everywhere except at $x = 0$. With all this in mind, the force balance equation reads, for an n-layered crust,

$$D\frac{d^4 v}{dx^4} + \Delta \rho_{mn} g v + \Delta \rho_{n,n-1} g v + \cdots + \Delta \rho_{21} g v + \Delta \rho_{1f} g v = 0 \,,$$

which simplifies to the 1D biharmonic equation

$$D\frac{d^4 v}{dx^4} + \Delta \rho_{mf} g v = 0 \,. \tag{7.80}$$

Differential equations such as these have a solution of form

$$v(x) = e^{\alpha x}(A_1 \cos \alpha x + A_2 \sin \alpha x) + e^{-\alpha x}(A_3 \cos \alpha x + A_4 \sin \alpha x) \,, \tag{7.81}$$

where the A's and α are constants to be determined from specific boundary conditions. First, if the deflection is zero at the (infinitely-distant) edge of the plate, or

$$v(\infty) = 0 \,,$$

then there can be no terms with $e^{+\alpha x}$, meaning that we must set

$$A_1 = A_2 = 0 \,.$$

Therefore Eq. 7.81 becomes

$$v(x) = e^{-\alpha x}(A_3 \cos \alpha x + A_4 \sin \alpha x) . \qquad (7.82)$$

Second, if we require symmetry about the point of application of the load ($x = 0$) then we must have zero gradient there, or

$$\left. \frac{dv}{dx} \right|_{x=0} = 0 .$$

Differentiating Eq. 7.82 gives

$$\frac{dv}{dx} = -\alpha e^{-\alpha x}[A_3(\cos \alpha x + \sin \alpha x) - A_4(\cos \alpha x - \sin \alpha x)] ,$$

and so the boundary condition becomes

$$\left. \frac{dv}{dx} \right|_{x=0} = -\alpha (A_3 - A_4) = 0 ,$$

or

$$A_3 = A_4 \equiv v_0 .$$

Thus Eq. 7.82 becomes

$$v(x) = v_0 e^{-\alpha x}(\cos \alpha x + \sin \alpha x) . \qquad (7.83)$$

We can now tackle Eq. 7.80. Successive derivatives of Eq. 7.83 are

$$\frac{dv}{dx} = -2v_0\alpha e^{-\alpha x} \sin \alpha x ,$$

$$\frac{d^2v}{dx^2} = 2v_0\alpha^2 e^{-\alpha x}(\sin \alpha x - \cos \alpha x) ,$$

$$\frac{d^3v}{dx^3} = 4v_0\alpha^3 e^{-\alpha x} \cos \alpha x ,$$

$$\frac{d^4v}{dx^4} = -4v_0\alpha^4 e^{-\alpha x}(\cos \alpha x + \sin \alpha x) = -4\alpha^4 v .$$

If we substitute the fourth-order derivative into Eq. 7.80, we obtain

$$-4D\alpha^4 v + \Delta\rho_{mf} g v = 0 ,$$

which can be rearranged to give an expression for the constant α,

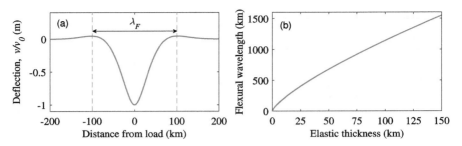

Fig. 7.13 a Surface deformation caused by application of a point load to the surface of a plate of $T_e = 10$ km. The flexural wavelength, $\lambda_F = 203.5$ km. **b** The relationship between elastic thickness (T_e) and flexural wavelength (λ_F), from Eq. 7.85. Parameters for both plots are $\rho_m = 3300$ kg m^{-3} and $\rho_f = 0$ kg m^{-3}

$$\alpha = \left(\frac{\Delta\rho_{mf}g}{4D}\right)^{1/4}, \tag{7.84}$$

which has units of 1/distance and is sometimes referred to as the *flexural parameter*.

Equation 7.83 is plotted in Fig. 7.13a, which shows the typical shape of a flexural depression, that is, with the depression itself flanked by *flexural bulges*. The peaks of these bulges correspond to points of zero gradient, so the distance between them—the *flexural wavelength*—can be found from the first derivative of Eq. 7.83, thus:

$$\frac{dv}{dx} = -2v_0\alpha e^{-\alpha x}\sin\alpha x = 0.$$

The derivative is zero when $\sin\alpha x = 0$, which occurs at $x = 0$ and at

$$x_0 = \pm\frac{\pi}{\alpha}.$$

From Fig. 7.13a the flexural wavelength is twice this distance ($\lambda_F = 2x_0$), or, using Eq. 7.84,

$$\lambda_F = 2\pi\left(\frac{4D}{\Delta\rho_{mf}g}\right)^{1/4}.$$

To express this in terms of T_e we can substitute Eq. 7.37 for D, giving

$$\lambda_F = 2\pi\left[\frac{E\,T_e^3}{3(1-\nu^2)\,\Delta\rho_{mf}g}\right]^{1/4}, \tag{7.85}$$

where E is Young's modulus, and ν is Poisson's ratio. Thus, if we apply a concentrated load to a plate with a given effective elastic thickness then the wavelength of the deformation will be given by λ_F, plotted in Fig. 7.13b for subaerial loading ($\rho_f = 0$ kg m^{-3}).

7.8 Summary

In many tectonic environments, the lithosphere can be modelled as a thin, elastic plate, undergoing flexure from applied loads. Such a plate is constrained to having a thickness no greater than ten per cent of its horizontal extent, and to undergoing deflections that have magnitudes no greater than ten per cent of its thickness. If these conditions are met, then the plate obeys the laws of linear elasticity and solutions for its flexure are readily calculated.

The main types of applied load considered here comprise surface and subsurface loads, and combinations of these. Surface loads could be mountain belts created from tectonic motion or volcanism. Subsurface loads can be emplaced at the base of the plate (such as magmatic underplating), or within it (such as from igneous intrusions). Notably, the geometric thicknesses of the various crustal layers do not feature in the equations of flexure, though the effective elastic thickness (T_e) does, of course, being the thickness of an elastic plate that best models the flexure of the lithosphere.

We also looked at a space-domain solution of the biharmonic equation in one dimension, for the purposes of calculating the flexural wavelength.

7.9 Further Reading

For readers interested in general theories of elasticity then the book by Timoshenko and Goodier (1970) is recommended. However, good books more concerned with plate flexure—though from an engineering viewpoint—are those by Timoshenko and Woinowsky-Krieger (1959), Ghali et al. (2003), and Szilard (2004). Note though that most engineering texts use a positive-downwards convention, in contrast to our positive-upwards convention.

The books by Watts (2001) and Turcotte and Schubert (2002) are essential reading as not only do they cover flexure, but they also have chapters on rheology, and investigate non-elastic materials. Note, though, that these books discuss the flexure of 1D beams and the cylindrical bending of plates, rather than full 2D plate flexure.

Finally, the many papers that deal with flexure of thin, elastic plates (and other types) are summarised in the review by Kirby (2014). Of note are those that 'started it all off': Banks et al. (1977), who were the first to model the continental lithosphere as a thin, elastic plate subject to surface loading; Louden and Forsyth (1982), who developed internal loading in the oceans; and McNutt (1983), who invoked internal loading in a continental setting. The first time that combined surface and internal loading appeared was in the seminal paper by Forsyth (1985).

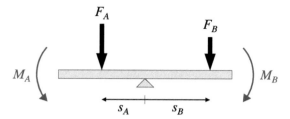

Fig. 7.14 A seesaw, illustrating static equilibrium. Force F_A is applied a distance s_A to the left of the fulcrum, generating anticlockwise moment M_A. Another force F_B is applied a distance s_B to the right of the fulcrum, generating clockwise moment M_B

Appendix

Moments

A moment (M) is a turning force about a fixed axis that will cause an object to rotate about that axis. It is measured as the product of the applied force (F) and distance from the axis (s), or

$$M = Fs ,$$

and is thus measured in units of newton metres. If an object is in static equilibrium, the clockwise moments acting on it must cancel the anticlockwise moments, or $\sum M = 0$. Consider the seesaw in Fig. 7.14, where a force F_A is applied a distance s_A to the left of the fulcrum, while another force F_B is applied a distance s_B to the right of the fulcrum. F_A thus creates an anticlockwise moment, while F_B generates a clockwise moment. If the seesaw is to remain motionless then the sum of the moments must be zero, or

$$F_A s_A + F_B s_B = 0 ,$$

using the convention that distances to the right of the fulcrum and upward forces are positive, while distances to the left and downward forces are negative.

References

Banks RJ, Parker RL, Huestis SP (1977) Isostatic compensation on a continental scale: local versus regional mechanisms. Geophys J R Astron Soc 51:431–452

Forsyth DW (1985) Subsurface loading and estimates of the flexural rigidity of continental lithosphere. J Geophys Res 90(B14):12,623–12,632

Ghali A, Neville AM, Brown TG (2003) Structural analysis: a unified classical and matrix approach, 5th edn. Spon Press, London

Karner GD, Watts AB (1982) On isostasy at Atlantic-type continental margins. J Geophys Res 87(B4):2923–2948

Kirby JF (2014) Estimation of the effective elastic thickness of the lithosphere using inverse spectral methods: the state of the art. Tectonophys 631:87–116

Louden KE, Forsyth DW (1982) Crustal structure and isostatic compensation near the Kane fracture zone from topography and gravity measurements: 1. Spectral analysis approach. Geophys J R Astron Soc 68:725–750

McNutt MK (1983) Influence of plate subduction on isostatic compensation in northern California. Tectonics 2:399–415

Szilard R (2004) Theories and applications of plate analysis. Wiley, Hoboken

Timoshenko SP, Goodier JN (1970) Theory of elasticity, 3rd edn. McGraw-Hill, New York

Timoshenko SP, Woinowsky-Krieger S (1959) Theory of plates and shells, 2nd edn. McGraw-Hill, New York

Turcotte DL, Schubert G (2002) Geodynamics, 2nd edn. Cambridge University Press, Cambridge

Watts AB (2001) Isostasy and flexure of the lithosphere. Cambridge University Press, Cambridge

Chapter 8
Gravity and Admittance of a Flexed Plate

8.1 Introduction

Figure 7.8 and the discussion in Sect. 7.4 showed that knowledge of the surface topography is not enough on its own to provide an estimate of the effective elastic thickness, T_e: one also needs to know the relief of the flexed Moho so that the two surfaces can be compared (using the admittance or coherency). But while the topographic surface is easy to observe, the Moho relief is not. Seismic methods can sometimes yield reasonably good estimates of Moho depth, but such data are expensive and difficult to acquire, and existing data sets do not provide global coverage at a sufficient resolution.

Another way to estimate the Moho relief is to invert gravity data. While gravity observations are not a direct measurement of the Moho, they are plentiful and provide global coverage at a high spatial resolution (Sect. 11.2.3). Gravity[1] is an attractive force that is generated by matter, and although Newton's equations of gravitation are framed in terms of mass, they are more usefully rewritten—for our purposes—in terms of the mass per unit volume, or density. Doing this tells us that changes in density within a region will generate changes in the gravity field. Hence, as two of the largest density contrasts within the lithosphere/asthenosphere system are those occurring at its upper surface (the crust-air or crust-water interface) and at the crust-mantle boundary (the Moho), the gravity field observed at the surface of the Earth (or reasonably close to it) will be dominated by the effect of these density contrasts. In principle, then, the structure of the observed gravity field will resemble the structure of the relief at these two interfaces. For example, at thinned crust, where high-density mantle displaces lower-density crust, there will be a gravity high; at an ocean trench, where low-density water takes the place of higher-density crust, there will be a gravity low (Fig. 8.1). If there is no relief, and the interface is flat and smooth, then there will be gravity (there will always be gravity) but no gravity anomaly or change in gravity.

---

[1] Geodesists distinguish between 'gravitation' and 'gravity' by defining the former to be the force generated by mass, and the latter to be the vector sum of gravitation and the centrifugal forces experienced by an object on a spinning body, such as the Earth. This is elaborated upon in Sect. 8.2.

© Springer Nature Switzerland AG 2022
J. Kirby, *Spectral Methods for the Estimation of the Effective Elastic Thickness of the Lithosphere*, Advances in Geophysical and Environmental Mechanics and Mathematics, https://doi.org/10.1007/978-3-031-10861-7_8

Fig. 8.1 Cross-section
through oceanic lithosphere,
showing an ocean trench
(left) and upwelling of the
mantle (right), and the
corresponding free-air
gravity anomaly (red line)

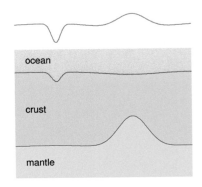

Elastic thickness estimation methods, therefore, use the gravity anomaly—
specifically the Bouguer anomaly—to estimate the relief on the Moho. This approach
is not perfect, and there are assumptions that must be made, but it does overcome the
comparative lack of global seismic data. The methods used to determine the gravity
anomaly due to plate flexure form the subject of this chapter.

8.2   Gravity Anomalies

In order to understand gravity anomalies, we first need to recap Newton's laws of
gravity, specifically when they apply to the gravity field of a planet and its effects on
objects within that field.

8.2.1   Gravity and Gravitation

When applied to celestial mechanics, *Newton's law of universal gravitation* states
that the gravitational force exerted by a planet of mass M upon a body of mass m,
where the centres of mass of the two objects are separated by a distance l, is given
by the vector equation

$$\mathbf{F} = -\frac{\mathcal{G}Mm}{l^2}\,\hat{\mathbf{l}},$$

where $\mathcal{G}(= 6.6743 \times 10^{-11}\ \mathrm{m^3kg^{-1}s^{-2}})$ is the Newtonian gravitational constant, $\mathbf{l}$ is
the 3D vector in the direction from the centre of mass of the planet to the centre of
mass of the body, $\hat{\mathbf{l}}$ is its unit vector, and $l = |\mathbf{l}|$ is its magnitude. The minus sign
reflects the fact that the gravitational force is directed towards the planet, opposite
to the direction of $\mathbf{l}$. The gravitational force is measured in newtons and, as can be
seen, varies with both M and m.

This latter dependency (upon the mass of the smaller body) is inconvenient as one would like to describe a gravitational field that is dependent only upon the mass of the planet and one's distance from it, and not upon one's own mass as well. So, from Newton's second law of motion (force equals mass times acceleration, or $\mathbf{F} = m\mathbf{a}$) we can represent the field as the *acceleration* one experiences towards the planet due to its gravitational pull. Thus we have

$$m\mathbf{a} = -\frac{\mathcal{G}Mm}{l^2}\,\hat{\mathbf{l}}\,,$$

or[2]

$$\mathbf{b} = -\frac{\mathcal{G}M}{l^2}\,\hat{\mathbf{l}}\,, \tag{8.1}$$

where we give the *gravitational acceleration*, or *gravitational field*, its own symbol, $\mathbf{b}$.

There is, however, another force at play. All objects within or on the Earth or within its atmosphere will rotate with the planet as it spins upon its axis, once every 23 h and 56 min. This rotation imparts a centrifugal force which acts to oppose gravitation, though its effect decreases towards the poles where there is no rotation with respect to the north-south pole axis. The *rotational acceleration*, $\mathbf{f}$, of a body rotating about the planet's axis is given by the vector equation

$$\mathbf{f} = \omega^2 \mathbf{p}\,,$$

where ω is the body's angular velocity (usually equal to the angular velocity of the planet if the object is fixed on or under its surface), and $\mathbf{p}$ is the vector representing the perpendicular distance to the body from the rotation axis (i.e. parallel to the equatorial plane), having magnitude $p = l\cos\phi'$, where ϕ' is the geocentric latitude of the body. The rotational acceleration has a maximum value on the equatorial plane, approximately $0.034\ \mathrm{ms}^{-2}$ or 0.3% of the gravitational acceleration, decreasing to zero at the poles.

The net force, gravitation plus rotation, is called *gravity*, and the corresponding acceleration is the vector sum of the gravitational and rotational accelerations, called the *gravity acceleration* or *gravity field*, $\mathbf{g}$, where

$$\mathbf{g} = \mathbf{b} + \mathbf{f}\,.$$

The gravity acceleration—or actually its scalar value g—is the quantity that is measured with a gravity meter, and is the quantity that appears in many physical formulae

---

[2] The cancellation of m either side of this equation assumes that they are the same thing. Fortunately, experiments have shown that, as best as we can measure them, *inertial mass* (from Newton's laws of motion) is indeed equal to gravitational mass (from Newton's laws of gravitation). This equality, known as the *weak equivalence principle*, is a cornerstone of Einstein's laws of relativity.

(rather than the gravitational acceleration, b, which does not include the rotational contribution).

8.2.2   Gravity Potential and the Geoid

As well as an acceleration, it is also useful to be able to express gravity as a *potential*. Potential is a scalar quantity and is defined as the work[3] done in bringing a unit mass from infinity to a point closer to the gravitating body, or

$$V = \frac{1}{m} \int_{\infty}^{l} \mathbf{F} \cdot d\mathbf{l} = \int_{\infty}^{l} \mathbf{b} \cdot d\mathbf{l} \ . \tag{8.2}$$

If we use Eq. 8.1 for $\mathbf{b}$ and perform the integration, we find the *gravitational potential*, V, at a distance l from the source is given by

$$V = \frac{\mathcal{G}M}{l} \ . \tag{8.3}$$

Potential is a measure of the energy in a system, though it is measured in joules per kilogram, rather than joules.

In addition to the integral equation relationship in Eq. 8.2, it is also useful to express the gravitational field as a differential equation in the potential. Expanding the unit vector in Eq. 8.1 in 3D Cartesian coordinates[4] gives the vector components of the gravitational acceleration, $\mathbf{b} = (b_x, b_y, b_z)$, as

$$b_x = -\frac{\mathcal{G}M}{l^2}\frac{x-x'}{l} = -\frac{\mathcal{G}M(x-x')}{l^3} \ , \tag{8.4a}$$

$$b_y = -\frac{\mathcal{G}M}{l^2}\frac{y-y'}{l} = -\frac{\mathcal{G}M(y-y')}{l^3} \ , \tag{8.4b}$$

$$b_z = -\frac{\mathcal{G}M}{l^2}\frac{z-z'}{l} = -\frac{\mathcal{G}M(z-z')}{l^3} \tag{8.4c}$$

(Hofmann-Wellenhof and Moritz 2006), where the coordinates of the centre of mass of the gravitating source are (x', y', z')[5] and those of the distant body are (x, y, z), and where

---

[3] Work equals force times distance, and is measured in joules.

[4] Here we use 3D geocentric Cartesian coordinates, whose origin is at the centre of mass of the planet, where the z-axis is aligned with the planet's mean rotational axis and points north, and the x- and y-axes are in the equatorial plane. Other origins and orientations may be used, however, since the laws of physics are the same in all reference frames.

[5] Usually chosen to be $(0, 0, 0)$.

$$l = \sqrt{(x - x')^2 + (y - y')^2 + (z - z')^2} \tag{8.5}$$

is the distance between them. So if V is given by Eq. 8.3 and l by Eq. 8.5, then it should now be easy to recognise that

$$b_x = \frac{\partial V}{\partial x}, \quad b_y = \frac{\partial V}{\partial y}, \quad b_z = \frac{\partial V}{\partial z}.$$

Therefore, using the gradient operator, which in 3D Cartesian coordinates is

$$\nabla \equiv \left(\frac{\partial}{\partial x}, \frac{\partial}{\partial y}, \frac{\partial}{\partial z} \right),$$

we can write

$$\mathbf{b} = \nabla V.$$

Thus, the gravitational field is the gradient of the gravitational potential.

Correspondingly, we have that the gravity acceleration is the gradient of the *gravity potential*, W, or

$$\mathbf{g} = \nabla W.$$

Hence, if the gravity potential is represented by a series of contours throughout space, $W(x, y, z)$, then the gravity field, $\mathbf{g}$, is stronger when these contours are closer together, and weaker when they are further apart, like the proximity of contours on a topographic map indicating steep or gentle terrain. And just as one can plot out lines of equal height contour on a topographic map, one can map out surfaces of equal gravity potential through space. Such surfaces are called *equipotential surfaces*, and there are an infinite number of them surrounding any gravitating body. On the Earth, that particular equipotential surface that most closely approximates mean sea level is called the *geoid*, and as such it is often used as a datum for elevations, with heights referenced to it called *orthometric heights* (Hofmann-Wellenhof and Moritz 2006).

Equation 8.3 gives the expression for the gravitational potential of a point source, while Eq. 8.4 give the expression for the gravitational field components of a point source. What if we want to find the potential or field at some measurement point P due to a gravitating body with an extended shape? First, as will be explained in Sect. 8.2.4, we require that P is not inside the gravitating body, although it can be on its surface. We can then divide the body into many tiny elements of mass dm and volume dv, where the element at any location Q inside the body has density $\rho(Q)$; thus we have $dm = \rho(Q)\, dv$. Next, consider the potential felt at P due to just one of the elements, say at Q. If P is a distance l away from Q, then the gravitational potential, dV, felt at P due to the mass element at Q is given by Eq. 8.3 as

$$dV(P) = \frac{\mathcal{G}\, dm}{l} = \frac{\mathcal{G}\, \rho(Q)\, dv}{l}.$$

Now, from the principle of superposition, the total potential due to the whole gravitating body is just the sum of the potentials due to all the tiny mass elements contained within the body, or

$$V(P) = \sum_{body} dV(P) \;\rightarrow\; \int_{v} dV(P) \,,$$

where the rules of calculus enable the sum of infinitesimal elements to be represented as an integral over a volume. Thus we have

$$V(P) = \mathcal{G} \int_{v} \frac{\rho(Q)}{l} \, dv \,.$$

If the gravitating body has a uniform (constant) density, ρ, then we can write

$$V(P) = \mathcal{G}\rho \int_{v} \frac{1}{l} \, dv \,.$$

The choice of coordinate system will dictate the form of the volume integral. For example, in 3D geocentric Cartesian coordinates, (x, y, z), the volume element is $dv = dx\,dy\,dz$, while in spherical polar coordinates, (r, ϕ, λ), we would have $dv = r^2 \cos\phi\,dr\,d\phi\,d\lambda$.

Similarly, we can derive expressions for the gravitational field components of an extended source. For example, following the above method, the b_z component in 3D geocentric Cartesian coordinates is

$$b_z(P) = -\mathcal{G} \iiint_{v} \rho(Q) \frac{z - z'}{l^3} \, dx'\,dy'\,dz' \qquad (8.6)$$

(from Eq. 8.4c), where the measurement point P has coordinates (x, y, z), and the points Q inside the body have coordinates (x', y', z'). The distance between them is given by Eq. 8.5.

8.2.3 Normal Gravity

Another effect that it is important to account for when modelling the Earth's gravity field is the fact that the planet is not a sphere but is an *oblate ellipsoid of revolution*, as noted in Sect. 6.1. Indeed, the reason the Earth is ellipsoidal is because its rotation causes equatorial regions to bulge outwards more than polar regions, such that the equator is about 21 km further away from the centre of the planet than are the poles. This oblateness, coupled with the effect of the Earth's rotation, means that the gravity acceleration is approximately 0.052 ms^{-2} weaker on the equator compared to the poles.

Table 8.1 Constants of the GRS80 reference ellipsoid (Moritz 2000)

Quantity	Symbol	Value
Semi-major axis	a	6,378,137 m
Semi-minor axis	b	6,356,752.3141 m
Geocentric constant	$\mathcal{G}M$	$3.986\ 005 \times 10^{14}$ m^{3}s^{-2}
Angular velocity	ω	$7.292\ 115 \times 10^{-5}$ rad s^{-1}
Equatorial normal gravity	γ_a	9.780 326 771 5 ms^{-2}
Polar normal gravity	γ_b	9.832 186 368 5 ms^{-2}
Flattening	f	0.003 352 810 681 18
Geodetic parameter	m	0.003 449 786 003 08

The ellipsoid model that best fits the geoid is called a *reference ellipsoid*, and is defined by four parameters. Two of these govern its geometric shape and size (its equatorial semi-major axis and polar semi-minor axis[6]), and the other two describe its physical properties, being its mass and angular velocity. The two reference ellipsoids in common use today are GRS80 (Moritz 2000) and WGS84 (NGA 2014). Table 8.1 gives the values of some ellipsoidal constants for the GRS80 ellipsoid.

A reference ellipsoid has two purposes. First, its size and shape provide a geometric surface upon which to 'lay' a geodetic datum, enabling the definition of latitudes and longitudes (see Chap. 6). Second, being a massive and rotating body, it provides a first-order approximation to the actual gravity field of the Earth (the zeroth-order approximation being the sphere), accounting for more than 99% of the observed gravitational acceleration, and all of the observed rotational acceleration. The gravity field due to the reference ellipsoid is known as *normal gravity*, γ. Its value on the surface of the ellipsoid is calculated using *Somigliana's equation*

$$\gamma = \frac{a\gamma_a \cos^2 \phi + b\gamma_b \sin^2 \phi}{\sqrt{a^2 \cos^2 \phi + b^2 \sin^2 \phi}} \tag{8.7}$$

(Hofmann-Wellenhof and Moritz 2006), where ϕ is geodetic latitude, and the values of the GRS80 ellipsoidal constants are given in Table 8.1.

If the ellipsoid accounts for more than 99% of the observed gravitational acceleration and all of the observed rotational acceleration, what then of the remaining one per cent? The reference ellipsoid is a smooth, rotating body with homogeneous density and cannot exactly model the real Earth, with its undulating topography and lateral and vertical variations in density. But if one subtracts the gravity effect of the ellipsoid from the actual gravity field of the Earth, one is left with just the gravitational effect of the density and topography variations (and no rotation). This is the *gravity anomaly*, Δg, where

---

[6] Actually, the two defining geometric parameters are the semi-major axis and dynamical form factor, J_2, from which the semi-minor axis can be derived.

$$\Delta g = g - \gamma . \tag{8.8}$$

As it stands, Eq. 8.8 is not the finished product as we need to be more specific about where g and γ are measured/estimated. We will return to this in Sect. 8.2.4, but for now we note that, because 99% of the gravity field has been subtracted, the residual gravity anomalies have very small values compared to the gross gravity acceleration of the Earth. As mentioned, the variation in gravity acceleration from poles to equator is roughly 0.052 ms^{-2}, but on more local scales gravity anomalies vary by up to only ±0.001 ms^{-2}. Furthermore, modern gravity meters are capable of measuring variations in gravity to a precision of 10^{-8} ms^{-2}. For these reasons, geophysicists and geodesists use another unit of measurement instead of ms^{-2}, the *milliGal*, where

$$1 \text{ mGal} = 10^{-5} \text{ ms}^{-2} .$$

Hence, the equatorial-polar gravity variation is approximately 5200 mGal, local gravity variations due to density changes are of the order of ±100 mGal or less, and modern instruments can measure to 1 μGal.

8.2.4   Free-Air Anomalies

Equation 8.8 as it stands is not enough to define a consistent gravity anomaly for use in geophysics or geodesy, because we need to define exactly where g and γ are measured, computed or estimated. The historical definition of the gravity anomaly, which persists to this day, is that g should be the gravity acceleration as would be measured on the geoid (at point P in Fig. 8.2)—if that were possible—and that γ should be evaluated on the ellipsoid (at point Q) using Eq. 8.7, or

$$\Delta g = g_P - \gamma_Q . \tag{8.9}$$

Except at sea—where all gravity measurements are made on the geoid, approximately—it is usually not possible to measure g_P directly, and one must estimate it using observations made on the topographic surface and approximations about the behaviour of the gravity field inside matter. This is not as easy as it might sound, even considering that one does not need to solve this problem for regions below the geoid. First, compositional variations and faulting cause the density of rocks to change quite rapidly, both horizontally and vertically, and more often than not the geology is poorly known. Therefore a mean density for topographic rocks must be assumed.

Second, the equations governing the behaviour of the gravity field in free space or air are very different to those governing its behaviour inside matter. In a vacuum (or the Earth's atmosphere at a push) the gravitational potential obeys *Laplace's equation*, being

$$\nabla^2 V = 0 , \tag{8.10}$$

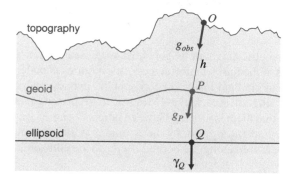

Fig. 8.2 Gravity measured at a point O on the topographic surface, g_{obs}, must be reduced to a value g_P on the geoid at a point P below O along the geoidal normal, using the free-air correction over the distance $PO = h$. The gravity anomaly is then $\Delta g = g_P - \gamma_Q$, where Q is the point on the ellipsoid below P along the ellipsoidal normal

where ∇^2 is the *Laplacian operator*, which in 3D Cartesian coordinates is

$$\nabla^2 \equiv \frac{\partial^2}{\partial x^2} + \frac{\partial^2}{\partial y^2} + \frac{\partial^2}{\partial z^2} \, .$$

Solutions to Laplace's equation are called *harmonic functions* and they are very well-behaved and predictable. The gravitational potential is harmonic because $\nabla^2(l^{-1}) = 0$.[7] In contrast, within matter of density ρ the gravitational potential obeys *Poisson's equation*, being

$$\nabla^2 V = -4\pi \mathcal{G} \rho \, ,$$

which reverts to Laplace's equation outside matter (when $\rho = 0$). However, there are no analytic solutions to Poisson's equation, which presents a problem when one wishes to understand how the gravity field or potential behave inside matter.

As an aside, note that we are now referring to the gravitational potential (V), rather than gravity potential (W), even though the gravity meter is measuring the effects of both gravitation and rotation. We can do this because the subtraction of the normal gravity field from the observation in Eq. 8.9 accounts for and removes all the rotational acceleration, leaving Δg just a function of gravitation only. This is fortunate because Poisson's equation for the gravity potential is $\nabla^2 W = -4\pi \mathcal{G} \rho + 2\omega^2$, which does not revert to Laplace's equation when $\rho = 0$, meaning W is not harmonic even outside matter.

So, given gravity observations, g_{obs}, made on the topographic surface (at point O in Fig. 8.2), how does one estimate g_P on the geoid without digging a hole? One approach is to begin with a Taylor series expansion from point P to point O along the normal vector to the geoid, with

---

[7] If one is comfortable with calculus, this can be proved quite easily using Eq. 8.5.

$$g_{\text{obs}} = g_P + \frac{\partial g}{\partial h} h + \frac{1}{2} \frac{\partial^2 g}{\partial h^2} h^2 + \cdots ,$$

where h, the orthometric height or height above mean sea level, must be small compared to the radius of the Earth. Unfortunately, this requires knowledge of the gravity field and its vertical gradient inside the topographic masses, and as we established above this is very difficult to achieve. However, if we pretend that the topography does not exist and that there is just air[8] between points O and P, we can approximate the actual gravity gradient, $\partial g / \partial h$, by the normal gravity gradient, $\partial \gamma / \partial h_e$, where h_e is the height measured along the normal vector to the ellipsoid, giving[9]

$$g_{\text{obs}} = g_P + \frac{\partial \gamma}{\partial h_e} h + \frac{1}{2} \frac{\partial^2 \gamma}{\partial h_e^2} h^2 + \cdots .$$

Rearranging this we find that the gravity on the geoid is given by

$$g_P = g_{\text{obs}} - \frac{\partial \gamma}{\partial h_e} h - \frac{1}{2} \frac{\partial^2 \gamma}{\partial h_e^2} h^2 - \cdots ,$$

more commonly written as

$$g_P = g_{\text{obs}} + \delta g_F ,$$

where the *free-air correction* is

$$\delta g_F = -\frac{\partial \gamma}{\partial h_e} h - \frac{1}{2} \frac{\partial^2 \gamma}{\partial h_e^2} h^2 - \cdots .$$

Hence the *free-air anomaly* is just

$$\Delta g_F = g_{\text{obs}} + \delta g_F - \gamma , \tag{8.11}$$

from Eq. 8.9, where γ is evaluated on the ellipsoid (point Q) using Eq. 8.7 (see Fig. 8.2). The derivatives of the normal gravity are given by

$$\frac{\partial \gamma}{\partial h_e} = -\frac{2\gamma}{a} \left(1 + f + m - 2f \sin^2 \phi \right) ,$$

and

$$\frac{\partial^2 \gamma}{\partial h_e^2} = \frac{6\gamma}{a^2}$$

---

[8] Hence the name 'free-air': we pretend that the observation point is hanging in empty space with no topography between it and the geoid.

[9] Note, even though we replace the gradients along $h = PO$ with the gradients along $h_e = QP$, we still only wish to evaluate these gradients over PO, so we do not replace h with h_e, which is a height above the ellipsoid.

(Hofmann-Wellenhof and Moritz 2006), where γ is normal gravity evaluated on the ellipsoid, a is the semi-major axis length, ϕ is geodetic latitude, f is the *flattening* of the ellipsoid, and m is the *geodetic parameter* (Table 8.1). Hence, the free-air correction is

$$\delta g_F = 2\gamma \left(1 + f + m - 2f \sin^2 \phi\right) \frac{h}{a} - 3\gamma \frac{h^2}{a^2}, \tag{8.12}$$

truncated to second order. In practice the second-order term can be neglected unless one is conducting surveys at high altitude: it reaches a value of 1 mGal at elevations of around 3700 m. The first-order term in Eq. 8.12 is latitude-dependent, but only varies from $0.30834\,h$ at the poles to $0.30877\,h$ on the equator, so frequently the approximation

$$\delta g_F \approx 0.3086\,h$$

is used, for h in metres and δg_F in mGal.

Even though a correction has been made for the elevation of the gravity meter, free-air anomalies are still strongly correlated with topography. This is because the free-air correction does not account for the gravitational attraction of the topographic mass, and of course there is more mass under high topography and less mass under low. Nevertheless, free-air anomalies provide a good reflection of the actual gravitational field of the Earth, containing the signature of all density variations from the surface of the planet down to its core.

8.2.5 *Bouguer Anomalies*

The retention of the topographic gravity signal by free-air anomalies does not suit all purposes, though. In physical geodesy, geoid determination requires that Stokes' theorem is satisfied: that all the mass of the Earth is contained within the geoid (Hofmann-Wellenhof and Moritz 2006). Therefore geodesists often make a further correction to free-air anomalies called Helmert condensation, which essentially moves the topographic masses below the geoid while preserving the mass of the Earth and the gravitational field due to these masses.

Geophysicists, also, are often less interested in the gravitational signal due to the topographic masses above the geoid than they are in the signals from deeper sources. But rather than condense the topographic masses—which still retains their signal— geophysicists prefer to remove them altogether, by the combined application of the *simple Bouguer correction* and *terrain correction*, yielding the *Bouguer anomaly*. In contrast, geodesists rarely use the Bouguer anomaly because the Bouguer correction decreases the mass of the Earth: a 'geoid' determined from Bouguer anomalies would correspond to a planet of smaller mass than the Earth.

8.2.5.1 The Bouguer Correction

An expression for the simple Bouguer correction at a gravity station on some topography of height h can be derived by modelling the topography at the station as a flat cylinder centred on the station, of uniform density ρ, height h, and infinite radius (Fig. 8.3a). Consider for a moment a 3D Cartesian coordinate system, (x, y, z), with its origin on the geoid vertically below the gravity station, its z-axis pointing vertically upwards, and its x- and y-axes in the plane of the geoid (Fig. 8.3a). The vertical component of the gravitational field due to the mass of the cylinder can then be evaluated using the principle of superposition of many tiny mass elements within the cylinder, as outlined in Sect. 8.2.2. If we desire the gravity at a point P on the surface of the cylinder at Cartesian coordinates $(0, 0, h)$, due to tiny mass elements at points $Q(x', y', z')$ within the cylinder, then Eq. 8.6 in this Cartesian coordinate system is

$$b_z(P) \;=\; -\mathcal{G}\rho \int_0^h \int_{-\infty}^{\infty} \int_{-\infty}^{\infty} \frac{h - z'}{l^3}\, dx'\, dy'\, dz' \,, \tag{8.13}$$

where the distance between points P and Q is

$$l \;=\; \sqrt{x'^2 + y'^2 + (h - z')^2} \,.$$

Noting that a cylinder has rotational symmetry, we transform to a 3D cylindrical coordinate system, (r, θ, z), with the same origin and z-axis orientation as the 3D Cartesian system, and its r- and θ-axes in the plane of the geoid (Fig. 8.3a). The relationship between the Cartesian and cylindrical coordinate systems is given by the transformation equations

$$x = r \cos\theta \,, \qquad y = r \sin\theta \,, \qquad z = z \,.$$

The volume element in the integral transforms as

$$dx'\, dy'\, dz' \;\longrightarrow\; r'\, dr'\, d\theta'\, dz' \,,$$

while the distance between points P and Q becomes

$$l \;=\; \sqrt{r'^2 + (h - z')^2} \,,$$

where

$$r' \;=\; \sqrt{x'^2 + y'^2} \,.$$

We can now write Eq. 8.13 in the cylindrical coordinate system as

$$b_z(P) \;=\; -\mathcal{G}\rho \int_0^h \int_0^{2\pi} \int_0^{\infty} \frac{h - z'}{[r'^2 + (h - z')^2]^{3/2}}\, r'\, dr'\, d\theta'\, dz' \,.$$

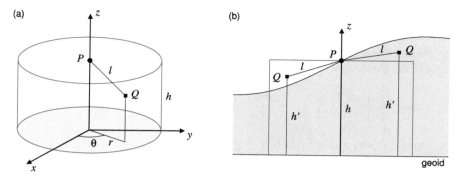

Fig. 8.3 a The Bouguer plate correction, showing the Cartesian and cylindrical coordinate systems. **b** The terrain correction, also showing the Bouguer plate in red (side view)

Performing the θ integral first gives

$$b_z(P) = -2\pi\mathcal{G}\rho \int_0^h \int_0^\infty \frac{r'\,(h - z')}{[r'^2 + (h - z')^2]^{3/2}}\,dr'\,dz'$$

$$= -2\pi\mathcal{G}\rho \int_0^h \left[\int_0^\infty \frac{r'}{[r'^2 + (h - z')^2]^{3/2}}\,dr' \right] (h - z')\,dz' \ .$$

Then, using the method of integration by substitution, it can be shown that the r' integral in square brackets is

$$\int_0^\infty \frac{r'}{[r'^2 + (h - z')^2]^{3/2}}\,dr' = \frac{1}{h - z'} \ ,$$

giving

$$b_z(P) = -2\pi\mathcal{G}\rho \int_0^h dz' \ .$$

This trivial integration then gives the attraction due to the cylinder as

$$\delta g_B = -2\pi\mathcal{G}\rho h \ ,$$

where we have now given it the symbol δg_B, the *simple Bouguer correction* or sometimes the *Bouguer plate correction*. The attraction is negative because it acts in a direction opposite to the positive z-axis.

In the general case where the topography is overlain by a fluid of density ρ_f (for example at sea, where h is the bathymetry and the fluid is seawater), the simple Bouguer correction is

$$\delta g_B = -2\pi\mathcal{G}\left(\rho_0 - \rho_f\right) h \ , \tag{8.14}$$

where we now write the density of the cylinder as the *Bouguer reduction density*, ρ_0. Historically, it was common to set the value of the reduction density to 2670 kg m^{-3}, a practice that continues today. If we use this value for the reduction density then the simple Bouguer correction has values, on land and at sea, of

$$\delta g_B \approx \begin{cases} -0.1120\,h\,, & \text{land}: \quad h \geq 0\,, \quad \rho_f = 0 \\ +0.0688\,|h|\,, & \text{sea}: \quad h < 0\,, \quad \rho_f = \rho_w = 1030 \text{ kg m}^{-3} \end{cases}$$

for h in metres and δg_B in mGal.

The *simple Bouguer anomaly* is just the simple Bouguer correction applied to the free-air anomaly, or

$$\Delta g_{SB} = \Delta g_F + \delta g_B\,,$$

where Δg_F is given by Eq. 8.11. It can be seen, then, that application of the simple Bouguer correction on land represents the removal of (the gravitational attraction of) rock of density ρ_0 and elevation h, replacing it with air of density zero. The remaining gravity anomaly (the Bouguer anomaly) is hence free from the complex signature of the topography and thus reveals deeper density contrasts. At sea, because it is of opposite sign, the correction represents the removal of seawater of density ρ_w and depth h, replacing it with rock of density ρ_0 and thickness h. This action removes the density contrast at the sea floor, and again, the Bouguer anomaly will only contain the signature of deeper density contrasts within the Earth. Thus, Bouguer anomalies on land are generally large and negative, while at sea they are generally large and positive (as shown in Fig. 1.3).

One can extend this reasoning to the calculation of Bouguer anomalies in other environments. Scenarios A and D in Fig. 8.4 and Table 8.4 have already been dis-

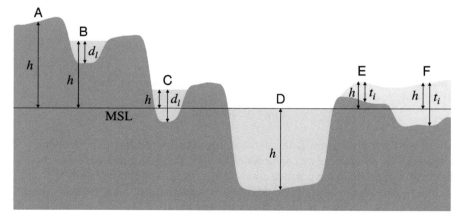

Fig. 8.4 Topographic profile showing the parameters of the simple Bouguer corrections for the scenarios listed in Table 8.2. Bedrock is shown in brown shading, ocean and lakes in blue, and ice in cyan; MSL is mean sea level (i.e. the geoid)

Table 8.2 Simple Bouguer corrections (δg_B) in a variety of environments (NGS 1994). The densities are: ρ_0 is the Bouguer reduction density (2670 kg m^{-3}), ρ_w is the density of seawater (1030 kg m^{-3}), ρ_l is the density of fresh lake water (1000 kg m^{-3}), and ρ_i is the density of ice (917 kg m^{-3}). The depths and thicknesses are: h is the topographic surface (i.e. the distance from mean sea level (MSL) to either the surface of the rocky topography, the sea floor, the lake surface or the ice sheet surface), d_l is the depth of the lake, and t_i is the thickness of the ice sheet/cap

Environment	Bouguer correction	Conditions
A: Land	$-2\pi\mathcal{G}\rho_0 h$	$h > 0$
B: Lake (bottom above MSL)	$-2\pi\mathcal{G}\left[\rho_l d_l + \rho_0(h - d_l)\right]$	$h > d_l > 0$
C: Lake (bottom below MSL)	$-2\pi\mathcal{G}\left[\rho_l d_l + \rho_0(h - d_l)\right]$	$d_l > h > 0$
D: Ocean	$-2\pi\mathcal{G}\left(\rho_0 - \rho_w\right)h$	$h < 0$
E: Ice (bottom above MSL)	$-2\pi\mathcal{G}\left[\rho_i t_i + \rho_0(h - t_i)\right]$	$h > t_i > 0$
F: Ice (bottom below MSL)	$-2\pi\mathcal{G}\left[\rho_i t_i + \rho_0(h - t_i)\right]$	$t_i > h > 0$

cussed, above, but the figure also shows how Bouguer corrections are evaluated for the cases of lakes and ice caps with their bases above or below mean sea level (MSL). For example, in scenario C, the surface of a lake of depth d_l is at elevation h above MSL, while its bottom is below sea level, at (negative) elevation $h - d_l$. To form the Bouguer anomaly here one must: (1) remove the gravity effect of lake water above MSL, of density ρ_l and thickness h, via a Bouguer correction of

$$\delta g_{B,1} = -2\pi\mathcal{G}\rho_l h ,$$

where $h > 0$, and (2) replace the lake water below MSL, of density ρ_l and thickness $h - d_l$, with the same thickness $(h - d_l)$ of rock of density ρ_0, via a further Bouguer correction of

$$\delta g_{B,2} = +2\pi\mathcal{G}\rho_l(h - d_l) - 2\pi\mathcal{G}\rho_0(h - d_l) ,$$

where $h - d_l < 0$ (i.e. remove water, add rock). The total Bouguer correction, $\delta g_{B,1} + \delta g_{B,2}$, for scenario C is thus

$$\delta g_B = -2\pi\mathcal{G}\left[\rho_l d_l + \rho_0(h - d_l)\right] ,$$

as noted in Table 8.4. Note, this derivation also applies to scenario F (an ice sheet with its bottom below MSL), with t_i and ρ_i replacing d_l and ρ_l, respectively.

The simple Bouguer correction will be revisited in Sects. 11.3 and 11.10.

8.2.5.2   The Terrain Correction

Using a flat-topped cylinder of infinite radius is generally a crude approximation to the actual topography surrounding a gravity station. For that reason the simple Bouguer correction can be refined using a *terrain correction*, which adjusts for the slope of the topography in the vicinity of the measurement point. An expression for the terrain correction can be derived using the same procedure that gave us the Bouguer plate correction, though with different limits in the z' integral.

Figure 8.3b shows a tiny mass element at point Q and height h'. The gravitational attraction at the gravity station, P, due to all the mass elements within the topography but above the Bouguer plate (to the right of P in Fig. 8.3b), and all the negative-mass elements in empty space below the plate (to the left of P), is given by

$$\delta g_T = -\mathcal{G}\rho \int_h^{h'(x',y')} \int_{-\infty}^{\infty} \int_{-\infty}^{\infty} \frac{h - z'}{l^3} \, dx' \, dy' \, dz' \tag{8.15}$$

instead of Eq. 8.13 (Moritz 1968). Equation 8.15 is the formula to calculate the terrain correction (in 3D Cartesian coordinates) and it cannot much be simplified due to the topography-dependent term, $h'(x', y')$, in the upper limit of the z' integral. Its triple integral can be converted to a more computationally-efficient form, however, and Li and Sideris (1994), for example, rewrote Eq. 8.15 as a convolution integral and made use of the fast Fourier transform to implement it quickly and efficiently using digital elevation models.

Thus, the *refined* or *complete Bouguer anomaly* is the terrain-corrected simple Bouguer anomaly, or

$$\Delta g_B = \Delta g_F + \delta g_B + \delta g_T \, .$$

8.3   Upward/Downward Continuation of Gravity

The development of models of plate flexure, specifically the derivation of theoretical admittance and coherence formulae, require that the gravitational attraction of the topography and of any density contrasts within the lithospheric model are known. And since we are using the gravity anomaly, we need to know these values on the geoid. As one might suspect, we proceed using the principle of superposition of tiny mass elements.

Consider a layer of material of variable density $\rho(x', y')$, bounded by an upper surface $h_1(x', y')$ and a lower surface $h_2(x', y')$, as shown in Fig. 8.5a. Now consider an observation point $P(x, y, z_0)$ above the layer where P is confined to the plane $z = z_0$, and we require $z_0 > \max(h_1)$ and $\min(h_1) > \max(h_2)$ for all (x, y) and (x', y') (frequently the lower surface is chosen to be flat, but we will proceed assuming it has relief for a while). Examples where this scenario applies are a ship on the ocean surface (z_0) with h_1 being the seafloor bathymetry, or when z_0 is the geoid and h_1 is the Moho.

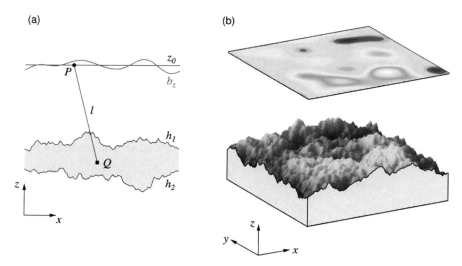

Fig. 8.5 **a** Cross-section through a layer of material bounded by surfaces $h_1(x, y)$ and $h_2(x, y)$, and its gravity field, $b_z(x, y)$, as measured on the plane z_0. **b** The scenario in perspective view, though now the layer has a flat lower surface

If we consider many tiny mass elements at source points $Q(x', y', z')$ within the layer, then the vertical component of the gravitational field due to the layer as measured at P is evaluated using the principle of superposition outlined in Sect. 8.2.2. So Eq. 8.6 can be written as

$$b_z(P) = -\mathcal{G} \int_{h_2(x',y')}^{h_1(x',y')} \int_{-\infty}^{\infty} \int_{-\infty}^{\infty} \rho(x', y') \frac{z_0 - z'}{l^3} \, dx' \, dy' \, dz' , \qquad (8.16)$$

where the distance between P and Q is

$$l = \sqrt{(x - x')^2 + (y - y')^2 + (z_0 - z')^2} . \qquad (8.17)$$

Eq. 8.16 is solved by taking its 2D Fourier transform [over (x, y) coordinates rather than (x', y') coordinates, because b_z is a function of the former rather than the latter]. This gives

$$\mathsf{F}\{b_z(x, y, z_0)\} = -\mathcal{G} \int_{-\infty}^{\infty} \int_{-\infty}^{\infty} \rho(x', y')$$
$$\times \left[\int_{h_2(x',y')}^{h_1(x',y')} (z_0 - z') \, \mathsf{F}\left\{\frac{1}{l^3}\right\} dz' \right] dx' \, dy' .$$

Now, $\mathsf{F}\{l^{-3}\}$ was solved in Sect. 2.6.5 using the Hankel transform. So substitution of Eq. 2.57 gives

$$\mathsf{F}\{b_z(x, y, z_0)\} = -\mathcal{G} \int_{-\infty}^{\infty} \int_{-\infty}^{\infty} \rho(x', y') \left[\int_{h_2(x',y')}^{h_1(x',y')} (z_0 - z') \right.$$

$$\left. \times 2\pi\, e^{-i(k_x x' + k_y y')}\, \frac{e^{-k(z_0 - z')}}{z_0 - z'}\, dz' \right] dx'\, dy' \; ,$$

where we have replaced the z in Eq. 2.57 with z_0 to align with our notation in Eq. 8.17, and where the substitution is only valid under the condition that $z_0 > z'$ and $k > 0$, where $k = (k_x^2 + k_y^2)^{1/2}$. The $(z_0 - z')$ terms cancel, so this can be rewritten as

$$\mathsf{F}\{b_z(\mathbf{x}, z_0)\} = -2\pi\mathcal{G} \int_{-\infty}^{\infty} \int_{-\infty}^{\infty} \rho(\mathbf{x}')\, I_{z'}(\mathbf{x}')\, e^{-i\mathbf{k}\cdot\mathbf{x}'}\, d^2\mathbf{x}' \; , \qquad (8.18)$$

where we now write $\mathbf{x} = (x, y)$ and $\mathbf{x}' = (x', y')$, and where we have let

$$I_{z'}(\mathbf{x}') = \int_{h_2(\mathbf{x}')}^{h_1(\mathbf{x}')} e^{-k(z_0 - z')}\, dz' \; . \qquad (8.19)$$

Before proceeding with the z' integral, we note that Eq. 8.18 actually has the form of a 2D Fourier transform, or

$$\mathsf{F}\{b_z(\mathbf{x}, z_0)\} = -2\pi\mathcal{G}\, \mathsf{F}\{\rho(\mathbf{x}')\, I_{z'}(\mathbf{x}')\} \; . \qquad (8.20)$$

Returning to Eq. 8.19, the z' integral has a trivial solution:

$$I_{z'}(\mathbf{x}') = \left[\frac{e^{-k(z_0 - z')}}{k} \right]_{h_2(\mathbf{x}')}^{h_1(\mathbf{x}')}$$

$$= k^{-1} e^{-kz_0} \left(e^{kh_1(\mathbf{x}')} - e^{kh_2(\mathbf{x}')} \right) \; .$$

We could leave this in its current form, but Parker (1972) chose to expand the exponential terms in square brackets as a power series, using the well-known identity

$$e^x = \sum_{n=0}^{\infty} \frac{x^n}{n!} \; .$$

Doing this gives

$$I_{z'}(\mathbf{x}') = k^{-1} e^{-kz_0} \left(\sum_{n=0}^{\infty} \frac{k^n h_1^n(\mathbf{x}')}{n!} - \sum_{n=0}^{\infty} \frac{k^n h_2^n(\mathbf{x}')}{n!} \right)$$

$$= e^{-kz_0} \sum_{n=1}^{\infty} \frac{k^{n-1}}{n!} \left[h_1^n(\mathbf{x}') - h_2^n(\mathbf{x}') \right] \; , \qquad (8.21)$$

where we commence the summation at $n = 1$ because the $n = 0$ term is zero. Finally, if we substitute Eq. 8.21 into Eq. 8.20, we obtain

$$\mathsf{F}\{b_z(\mathbf{x}, z_0)\} = -2\pi\mathcal{G}\, e^{-kz_0}\, \mathsf{F}\left\{\rho(\mathbf{x}) \sum_{n=1}^{\infty} \frac{k^{n-1}}{n!}\left[h_1^n(\mathbf{x}) - h_2^n(\mathbf{x})\right]\right\}$$

or

$$\mathsf{F}\{b_z(\mathbf{x}, z_0)\} = -2\pi\mathcal{G}\, e^{-kz_0} \sum_{n=1}^{\infty} \frac{k^{n-1}}{n!}\, \mathsf{F}\left\{\rho(\mathbf{x})\left[h_1^n(\mathbf{x}) - h_2^n(\mathbf{x})\right]\right\}, \qquad (8.22)$$

where we have replaced $\mathbf{x}'$ with $\mathbf{x}$ since the grids are coincident.

If one is concerned only with the attraction due to a single interface, $h(\mathbf{x})$, that separates two materials that have a constant density contrast of $\Delta\rho$ (Fig. 8.5b), then one can set $h_2(\mathbf{x}) = 0$ and $\rho(\mathbf{x}) = \Delta\rho$ in Eq. 8.22, giving

$$\mathsf{F}\{b_z(\mathbf{x}, z_0)\} = -2\pi\mathcal{G}\Delta\rho\, e^{-kz_0} \sum_{n=1}^{\infty} \frac{k^{n-1}}{n!}\, \mathsf{F}\{h^n(\mathbf{x})\}, \qquad (8.23)$$

which is known informally as *Parker's formula*. Equation 8.23 contains a summation of Fourier transforms, and for most data sets successive terms become small enough to be negligible after several iterations (Parker 1972). It is a very useful equation for many applications, but here—especially with the load deconvolution method—the summation presents a problem because Eq. 8.23 must be inverted for $h(\mathbf{x})$ given $b_z(\mathbf{x}, z_0)$. While this is possible, it does not make the deconvolution equations particularly elegant, so it is common to take the linear approximation.[10] Writing $G(\mathbf{k})$ for the Fourier transform of $b_z(\mathbf{x}, z_0)$ and $H(\mathbf{k})$ for the Fourier transform of $h(\mathbf{x})$, Eq. 8.23 becomes, for $n = 1$,

$$G(\mathbf{k}) = 2\pi\mathcal{G}\Delta\rho\, e^{-kz_0}\, H(\mathbf{k}), \qquad (8.24)$$

where we have ignored the minus sign (which merely tells us that gravity acts in opposition to our choice of z-axis direction). Equation 8.24 is now invertible for the Fourier transform of the interface relief:

$$H(\mathbf{k}) = \frac{G(\mathbf{k})\, e^{kz_0}}{2\pi\mathcal{G}\Delta\rho}. \qquad (8.25)$$

Equations 8.24 and 8.25 are often referred to as the *upward* and *downward continuation* formulae, respectively. The presence of the $\exp(-kz_0)$ term in the former produces a smooth surface from a rough one. The presence of the $\exp(+kz_0)$ term

---

[10] Several studies have analysed the effect of including higher-order terms, summarised in Kirby (2014).

in the latter, however, yields a rough surface from a smooth, but at the expense of amplifying any high-frequency noise present in $G(\mathbf{k})$, sometimes prohibitively. Care must be taken, therefore, when performing downward continuation.

8.4 Gravity from Surface Loading

Flexure of a thin, elastic plate by surface loading causes all the previously-flat interfaces within the plate to bend by the same amount, being the deflection v_T as explained in Sect. 7.4. As noted in Sect. 8.1, the bending of the interfaces generates relief on those interfaces, manifesting as lateral density variations that cause a change in the gravitational field, giving a gravity anomaly.

Here, and in Sect. 8.5, we will derive expressions for the gravity anomaly generated as a result of plate flexure, using the linear form of the upward continuation formula, Eq. 8.24. Since the depth to various interfaces appears in this equation, we will now need to specify this quantity where we did not in the analysis of flexure (Sects. 7.4 and 7.5).

We also need to distinguish between the two types of gravity anomaly—free-air and Bouguer (Sects. 8.2.4 and 8.2.5)—the difference between them being that the free-air anomaly includes the gravitational effect of the topography (or bathymetry at sea), while the Bouguer anomaly has had this removed via application of the Bouguer plate and terrain corrections. Since these two types of anomaly differ by only the topographic signal, in the ensuing derivations we will use a symbol β_g which takes the values

$$
\beta_g \equiv \begin{cases} 0, & \text{Bouguer}, \\ 1, & \text{free-air}. \end{cases} \tag{8.26}
$$

As will be seen, β_g will multiply the topographic effect, removing it when we are dealing with Bouguer anomalies.

8.4.1 Two-Layer Crust Model

Following the scenario of Sect. 7.4.1, consider a crust of two layers with density ρ_1 and ρ_2, respectively, where the crust is overlain by a fluid of density ρ_f, and rests on an inviscid mantle of density ρ_m (Fig. 8.6). If the mean thicknesses of the two layers are t_1 and t_2, respectively ($t_1, t_2 > 0$), and the mean depth (below some datum) to the top of layer 1 is d ($d \geq 0$), then the mean depth to the layer 1/layer 2 interface is $z_1 = d + t_1$, and the mean depth to the layer 2/mantle interface is $z_2 = d + t_1 + t_2$ (so $z_1, z_2 > 0$).

After the crust is subjected to an initial surface load, ℓ_T, of density ρ_T (where $\rho_f < \rho_T \leq \rho_1$) it flexes and settles into the geometry shown in Fig. 8.6. Each flexed interface acts as a source that generates a gravitational field, shown as the red hatched

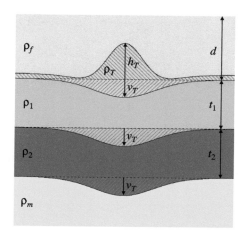

Fig. 8.6 Surface loading on a two-layer crust as in Fig. 7.9, but showing the regions that generate a gravity anomaly in red hatched shading. Note that the depths to the interfaces are given by $z_1 = d + t_1$, and $z_2 = d + t_1 + t_2$. On land we would set $d = 0$ and $\rho_f = 0$

shading in Fig. 8.6, with a magnitude dependent upon its depth and density contrast. Using Eq. 8.24, the gravitational attraction of each deflected interface is:

1. the displacement of the fluid by the topography at mean depth d, $2\pi\mathcal{G}\Delta\rho_{Tf}e^{-kd}H_T$;
2. the deflection of the load/layer 1 interface by an amount V_T at mean depth d, $2\pi\mathcal{G}\Delta\rho_{1T}e^{-kd}V_T$;
3. the deflection of the layer 1/layer 2 interface by an amount V_T at mean depth z_1, $2\pi\mathcal{G}\Delta\rho_{21}e^{-kz_1}V_T$;
4. the deflection of the layer 2/mantle interface by an amount V_T at mean depth z_2, $2\pi\mathcal{G}\Delta\rho_{m2}e^{-kz_2}V_T$.

If the surface load were to have the same density as the topmost crustal layer, we would set $\rho_T = \rho_1$, giving $\Delta\rho_{Tf} = \Delta\rho_{1f}$ and $\Delta\rho_{1T} = 0$ in the following equations. Also note that on land the mean 'depth' to the topography will be zero, while at sea it will be the mean ocean depth in the region.

As mentioned above, the free-air and Bouguer gravity anomalies differ with respect to the inclusion of the topographic gravitational effect, i.e. the term in item 1, above. The free-air anomaly will include this term while the Bouguer anomaly will not, so we multiply this term by the factor β_g (Eq. 8.26). The total gravity anomaly is now just the sum of the gravitational attractions due to the flexed interfaces, thus:

$$G_T = 2\pi\mathcal{G}\beta_g\Delta\rho_{Tf}e^{-kd}H_T + 2\pi\mathcal{G}\Delta\rho_{1T}e^{-kd}V_T + 2\pi\mathcal{G}\Delta\rho_{21}e^{-kz_1}V_T$$
$$+ 2\pi\mathcal{G}\Delta\rho_{m2}e^{-kz_2}V_T .\qquad(8.27)$$

In order to derive an expression for the admittance (Sect. 8.6.1), we need to rewrite Eq. 8.27 in terms of the final topography, H_T. We do this by substituting V_T from Eqs. 7.50 and 7.54 into Eq. 8.27, giving, after some algebra,

$$G_T = \frac{2\pi \mathcal{G} \Delta \rho_{Tf}}{\Phi - \Delta \rho_{Tf}} \left\{ \left[\beta_g (\Phi - \Delta \rho_{Tf}) - \Delta \rho_{1T} \right] e^{-kd} - \Delta \rho_{21} e^{-kz_1} \right.$$
$$\left. - \Delta \rho_{m2} e^{-kz_2} \right\} H_T \,, \tag{8.28}$$

where Φ is defined by Eq. 7.53, being

$$\Phi \equiv \frac{Dk^4}{g} + \Delta \rho_{mf} \,.$$

If we need an expression for G_T in terms of the applied surface load, L_T (which we will when we investigate the load deconvolution method in Chap. 9), we can substitute H_T from Eq. 7.55 into Eq. 8.28, giving, after plenty of algebra,

$$G_T = \frac{2\pi \mathcal{G}}{\Phi g} \left\{ \left[\beta_g (\Phi - \Delta \rho_{Tf}) - \Delta \rho_{1T} \right] e^{-kd} - \Delta \rho_{21} e^{-kz_1} \right.$$
$$\left. - \Delta \rho_{m2} e^{-kz_2} \right\} L_T \,. \tag{8.29}$$

8.4.2   Multiple-Layer Crust Model

Just as we did in Sect. 7.4.2, the two-layer model of Sect. 8.4.1 can be extended to a crust comprising many distinct layers. Remembering that each deflected interface generates a gravitational attraction, and that the gravity anomaly is just the linear sum of these attractions, we can write

$$G_T = 2\pi \mathcal{G} \, \beta_g \Delta \rho_{Tf} e^{-kd} H_T + 2\pi \mathcal{G} \Delta \rho_{1T} e^{-kd} V_T + 2\pi \mathcal{G} \sum_{i=1}^{n} \Delta \rho_{i+1,i} e^{-kz_i} V_T$$

(cf. Eq. 8.27), where n is the number of intra-crustal layers, z_i is the depth to the base of crustal layer i, and $\Delta \rho_{i+1,i}$ is the density contrast at the base of layer i (with $\rho_{n+1} = \rho_m$, the mantle density).

If we want to express the gravity in terms of the topography, we can substitute V_T from Eq. 7.57 into the above equation for G_T, giving, after some algebra,

$$G_T = \frac{2\pi \mathcal{G} \Delta \rho_{Tf}}{\Phi - \Delta \rho_{Tf}} \left\{ \left[\beta_g (\Phi - \Delta \rho_{Tf}) - \Delta \rho_{1T} \right] e^{-kd} - \sum_{i=1}^{n} \Delta \rho_{i+1,i} \, e^{-kz_i} \right\} H_T$$
$$\tag{8.30}$$

(cf. Eq. 8.28). Similarly, an expression for G_T in terms of the applied surface load, L_T, can be derived by substituting H_T from Eq. 7.58 into Eq. 8.30, giving

$$G_T = \frac{2\pi\mathcal{G}}{\Phi g}\left\{\left[\beta_g(\Phi - \Delta\rho_{Tf}) - \Delta\rho_{1T}\right]e^{-kd} - \sum_{i=1}^{n}\Delta\rho_{i+1,i}\,e^{-kz_i}\right\}L_T \quad (8.31)$$

(cf. Eq. 8.29).

8.5 Gravity from Internal Loading

As with surface loading (Sect. 8.4), the flexure of interfaces where there is a density contrast generates a gravitational attraction. In this section we derive expressions for the gravity anomalies due to the internal loading scenarios discussed in Sect. 7.5.

8.5.1 Loading at the Moho of a Two-Layer Crust

Following the scenario of Sect. 7.5.1, consider a crust of two layers with density ρ_1 and ρ_2, respectively, where the crust is overlain by a fluid of density ρ_f, and rests on an inviscid mantle of density ρ_m (Fig. 8.7). As in Sect. 8.4.1, if the mean thicknesses of the two layers are t_1 and t_2, respectively ($t_1, t_2 > 0$), and the mean depth (below some datum) to the top of layer 1 is d ($d \geq 0$), then the depth to the layer 1/layer 2 interface is $z_1 = d + t_1$, and the depth to the layer 2/mantle interface is $z_2 = d + t_1 + t_2$ (so $z_1, z_2 > 0$).

After the crust is subjected to an initial surface load, ℓ_B, of density ρ_B (where $\rho_2 \leq \rho_B < \rho_m$) it flexes and settles into the geometry shown in Fig. 8.7. Each flexed interface acts as a source that generates a gravitational field—shown as the red

Fig. 8.7 Internal loading at the Moho of a two-layer crust as in Fig. 7.10, but showing the regions that generate a gravity anomaly in red hatched shading. The small, upward arrows denote the deflection, v_B. The depths to the interfaces are given by $z_1 = d + t_1$, and $z_2 = d + t_1 + t_2$. On land we would set $d = 0$ and $\rho_f = 0$

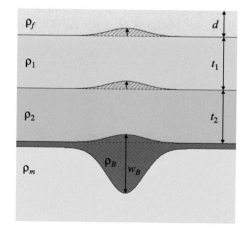

hatched shading in Fig. 8.7—with a magnitude dependent upon its depth and density contrast. Using Eq. 8.24, the gravitational attraction of each deflected interface is:

1. the deflection of the fluid/layer 1 interface by an amount V_B at mean depth d,
   $2\pi \mathcal{G} \Delta \rho_{1f} e^{-kd} V_B$;
2. the deflection of the layer 1/layer 2 interface by an amount V_B at mean depth z_1,
   $2\pi \mathcal{G} \Delta \rho_{21} e^{-kz_1} V_B$;
3. the deflection of the layer 2/load interface by an amount V_B at mean depth z_2,
   $2\pi \mathcal{G} \Delta \rho_{B2} e^{-kz_2} V_B$;
4. the displacement of the mantle by the final Moho load at mean depth z_2,
   $2\pi \mathcal{G} \Delta \rho_{mB} e^{-kz_2} W_B$.

If the internal load were to have the same density as the lower crust, we would set $\rho_B = \rho_2$, giving $\Delta \rho_{mB} = \Delta \rho_{m2}$ and $\Delta \rho_{B2} = 0$ in the following equations. Also note that on land the mean 'depth' to the topography will be zero, while at sea it will be the mean ocean depth in the region.

As before (Sect. 8.4.1), we multiply the topographic gravitational effect (the term in item 1, above) by the factor β_g (Eq. 8.26) to distinguish between free-air and Bouguer anomalies. The total gravity anomaly is then the sum of the gravitational attractions due to the flexed interfaces, thus:

$$G_B = 2\pi \mathcal{G} \beta_g \Delta \rho_{1f} e^{-kd} V_B \; + \; 2\pi \mathcal{G} \Delta \rho_{21} e^{-kz_1} V_B \; + \; 2\pi \mathcal{G} \Delta \rho_{B2} e^{-kz_2} V_B$$
$$+ \; 2\pi \mathcal{G} \Delta \rho_{mB} e^{-kz_2} W_B \; . \tag{8.32}$$

If we want an expression for the admittance (Sect. 8.6.2), we need to rewrite Eq. 8.32 in terms of the final topography, H_B. We do this by substituting V_B from Eq. 7.63 and W_B from Eq. 7.66 into Eq. 8.32, giving, after some algebra,

$$G_B \; = \; 2\pi \mathcal{G} \left[\beta_g \Delta \rho_{1f} e^{-kd} \; + \; \Delta \rho_{21} e^{-kz_1} \; - \; (\Phi - \Delta \rho_{m2}) \, e^{-kz_2} \right] H_B \; , \tag{8.33}$$

where Φ is defined by Eq. 7.53. Finally, to arrive at an expression for G_B in terms of L_B, we substitute H_B from Eq. 7.65 into Eq. 8.33, giving

$$G_B = \frac{2\pi \mathcal{G}}{\Phi_g} \left[-\beta_g \Delta \rho_{1f} e^{-kd} \; - \; \Delta \rho_{21} e^{-kz_1} \; + \; (\Phi - \Delta \rho_{m2}) \, e^{-kz_2} \right] L_B \; . \tag{8.34}$$

Note how both Eqs. 8.33 and 8.34 are independent of the load density, ρ_B.

8.5.2 Loading within a Multiple-Layer Crust

For a more general case of internal loading, we now find the gravity field due to the loading scenario presented in Sect. 7.5.2. Recall, this described a crust of n layers, with each layer having density ρ_i for $i = 1, \cdots, n$. As before, the crust is overlain by a fluid of density ρ_f, and rests on an inviscid mantle of density ρ_m. The initial

Fig. 8.8 Internal loading of a two-layer crust at the base of layer 1, as in Fig. 7.11, showing the regions that generate a gravity anomaly in red hatched shading. The small, upward arrows denote the deflection, v_B. The depths to the interfaces are given by $z_1 = d + t_1$, and $z_2 = d + t_1 + t_2$. On land we would set $d = 0$ and $\rho_f = 0$

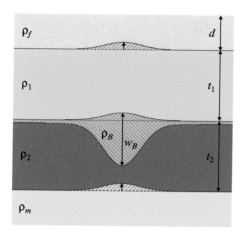

internal load has density ρ_B and is emplaced at the base of layer j ($j = 1, \cdots, n$), such that $\rho_j \leq \rho_B < \rho_{j+1}$ (with $\rho_{n+1} = \rho_m$, the mantle density). An example of the geometry for $n = 2$ is shown in Fig. 8.8.

The total gravity anomaly is then the sum of the gravitational attractions due to the flexed interfaces, or

$$
\begin{aligned}
G_B = {} & 2\pi\mathcal{G}\beta_g \Delta\rho_{1f} e^{-kd} V_B \; + \; 2\pi\mathcal{G}\Delta\rho_{21} e^{-kz_1} V_B \; + \; 2\pi\mathcal{G}\Delta\rho_{32} e^{-kz_2} V_B \; + \; \cdots \\
& + 2\pi\mathcal{G}\Delta\rho_{j,j-1} e^{-kz_{j-1}} V_B \; + \; 2\pi\mathcal{G}\Delta\rho_{Bj} e^{-kz_j} V_B \\
& + 2\pi\mathcal{G}\Delta\rho_{j+1,B} e^{-kz_j} W_B \; + \; 2\pi\mathcal{G}\Delta\rho_{j+2,j+1} e^{-kz_{j+1}} V_B \; + \; \cdots \\
& + 2\pi\mathcal{G}\Delta\rho_{n,n-1} e^{-kz_{n-1}} V_B \; + \; 2\pi\mathcal{G}\Delta\rho_{mn} e^{-kz_n} V_B \; .
\end{aligned} \tag{8.35}
$$

While part of Eq. 8.35 can be rewritten as a summation over all interfaces, the summation will create a term that does not exist in reality: $2\pi\mathcal{G}\Delta\rho_{j+1,j} e^{-kz_j} V_B$. This term gives the gravitational attraction of the interface between layers j and $j + 1$, whereas in reality these two layers are separated by the load and do not touch one another (see Fig. 8.8). Therefore we must subtract that term from the summation, leaving Eq. 8.35 as

$$
G_B = 2\pi\mathcal{G} \left[\beta_g \Delta\rho_{1f} e^{-kd} V_B \; + \; \sum_{i=1}^{n} \Delta\rho_{i+1,i} e^{-kz_i} V_B \; - \; \Delta\rho_{j+1,j} e^{-kz_j} V_B \right.
$$
$$
\left. + \; \Delta\rho_{Bj} e^{-kz_j} V_B \; + \; \Delta\rho_{j+1,B} e^{-kz_j} W_B \right],
$$

which simplifies to

$$G_B = 2\pi\mathcal{G}\left[\left(\beta_g \Delta\rho_{1f}e^{-kd} + \sum_{i=1}^{n}\Delta\rho_{i+1,i}e^{-kz_i} - \Delta\rho_{j+1,B}e^{-kz_j}\right)V_B\right.$$

$$\left. + \Delta\rho_{j+1,B}e^{-kz_j}W_B\right]. \tag{8.36}$$

In order to rewrite Eq. 8.36 in terms of H_B, we substitute V_B from Eq. 7.72 and W_B from Eq. 7.75, giving

$$G_B = 2\pi\mathcal{G}\left[\left(\beta_g \Delta\rho_{1f}e^{-kd} + \sum_{i=1}^{n}\Delta\rho_{i+1,i}e^{-kz_i} - \Delta\rho_{j+1,B}e^{-kz_j}\right)H_B\right.$$

$$\left. - (\Phi - \Delta\rho_{j+1,B})\,e^{-kz_j}\,H_B\right],$$

or

$$G_B = 2\pi\mathcal{G}\left[\beta_g \Delta\rho_{1f}e^{-kd} + \sum_{i=1}^{n}\Delta\rho_{i+1,i}e^{-kz_i} - \Phi\,e^{-kz_j}\right]H_B, \tag{8.37}$$

where Φ is defined by Eq. 7.53. Using the Kronecker delta

$$\delta_{ij} = \begin{cases} 1, & i = j \\ 0, & i \neq j \end{cases},$$

Eq. 8.37 can be written in a more compact form as

$$G_B = 2\pi\mathcal{G}\left[\beta_g \Delta\rho_{1f}e^{-kd} + \sum_{i=1}^{n}(\Delta\rho_{i+1,i} - \delta_{ij}\Phi)\,e^{-kz_i}\right]H_B. \tag{8.38}$$

For example, we can recover Eq. 8.33 by choosing $n = j = 2$ in Eq. 8.38:

$$G_B = 2\pi\mathcal{G}\left[\beta_g \Delta\rho_{1f}e^{-kd} + \Delta\rho_{21}e^{-kz_1} + (\Delta\rho_{m2} - \Phi)\,e^{-kz_2}\right]H_B.$$

If we then substitute H_B from Eq. 7.74 into Eq. 8.38, we get an expression for G_B in terms of L_B, or

$$G_B = \frac{2\pi\mathcal{G}}{\Phi g}\left[-\beta_g \Delta\rho_{1f}e^{-kd} - \sum_{i=1}^{n}(\Delta\rho_{i+1,i} - \delta_{ij}\Phi)\,e^{-kz_i}\right]L_B. \tag{8.39}$$

Remember, $\rho_{n+1} = \rho_m$, the mantle density. Note how both Eqs. 8.38 and 8.39 are independent of the load density, ρ_B.

8.6   The Admittance of Theoretical Models

We saw in Sect. 5.2 how the admittance, Q, is a transfer function or filter from topography to gravity, containing information about lithospheric parameters such as elastic thickness, crust and mantle densities, and the depths to various layers within the crust, for example. While the main method of T_e estimation nowadays is load deconvolution with the coherence (Chap. 9), some studies still use the admittance so it is worthwhile exploring this quantity further.

When written as a transfer function relating the Earth's actual gravity and topography, the admittance is best represented by Eq. 5.2 due to the presence of non-flexural 'noise' (Sect. 5.2.1). However when developing theoretical models of the admittance—as we are doing here—we can assume a noise-free system, in which case Eq. 5.6 is more relevant, being

$$G(\mathbf{k}) \;=\; Q(\mathbf{k})\,H(\mathbf{k})\,,$$

where $G(\mathbf{k})$ and $H(\mathbf{k})$ are the Fourier transforms of the gravity anomaly and topography, respectively.

Up to now we have considered the flexure and gravity field of single-loading models only (surface-only and internal-only loading). For such models, Eqs. 8.28, 8.33 and 8.38 can be written in the form of

$$G(\mathbf{k}) \;=\; Q_R(\mathbf{k})\,H(\mathbf{k})\,, \tag{8.40}$$

where $Q_R(\mathbf{k})$ is a real-valued, wavenumber-dependent function: the (real part of the) admittance. But because the function $Q_R(\mathbf{k})$ is real-valued and not complex, it does not alter the phase of the topography under the multiplication in Eq. 8.40, meaning that the topography and gravity have the same phase (Sects. 5.2.2 and 5.2.3). The wavenumber-dependence of $Q_R(\mathbf{k})$ merely results in the scaling of the topography transform, albeit by different values at different wavenumbers, and does not affect the phase. Thus we can write

$$Q_R(\mathbf{k}) \;=\; \frac{G(\mathbf{k})}{H(\mathbf{k})}\,, \tag{8.41}$$

which means it is a simple task to divide the expressions we derived for gravity by H. This is why we made a point of rewriting the gravity in terms of the topography in Sects. 8.4 and 8.5.

Just as there are two types of gravity anomaly, there are also two types of admittance. If free-air anomalies are used as $G(\mathbf{k})$ in Eq. 8.41 then one obtains the *free-air admittance*; if Bouguer anomalies are used then one obtains the *Bouguer admittance*.[11] Using the parameter β_g from Eq. 8.26, setting $\beta_g = 1$ gives an expression for the free-air admittance, while setting $\beta_g = 0$ gives the Bouguer admittance.

---

[11] Note that the free-air admittance is given the symbol Z in some texts, rather than Q.

8.6.1   Surface Loading

A general equation for the admittance of surface-only loading (Q_T) is obtained by dividing Eq. 8.30 by H_T, giving

$$Q_T = \frac{2\pi \mathcal{G} \Delta \rho_{Tf}}{\Phi - \Delta \rho_{Tf}} \left\{ [\beta_g (\Phi - \Delta \rho_{Tf}) - \Delta \rho_{1T}] e^{-kd} - \sum_{i=1}^{n} \Delta \rho_{i+1,i} e^{-kz_i} \right\},$$
$$(8.42)$$

where Φ is given by Eq. 7.53, n is the number of crustal layers, z_i (> 0) is the mean depth to the base of crustal layer i, and $\Delta \rho_{i+1,i}$ is the density contrast at the base of layer i (with $\rho_{n+1} = \rho_m$, the mantle density). To obtain the admittance for continental environments one sets $d = 0$; at sea d is the (positive) mean seafloor depth in the region.

For example, in the case of surface loading of a single-layer crust ($n = 1$) of density $\rho_1 = \rho_c$, where the load also has density $\rho_T = \rho_c$, Eq. 8.42 becomes

$$Q_T = 2\pi \mathcal{G} \Delta \rho_{cf} \left(\beta_g e^{-kd} - \frac{\Delta \rho_{mc}}{\Phi - \Delta \rho_{cf}} e^{-kz_m} \right),$$

or

$$Q_T = 2\pi \mathcal{G} \Delta \rho_{cf} \left[\beta_g e^{-kd} - e^{-kz_m} \left(1 + \frac{Dk^4}{\Delta \rho_{mc} g} \right)^{-1} \right], \qquad (8.43)$$

where z_m is the depth to the base of the crust (the Moho). Figure 8.9 shows examples of this admittance in continental and oceanic environments. Historically, the Bouguer admittance has been used on land, while the free-air admittance is typically used at sea, though the two differ only in the term containing β_g (i.e. by $2\pi \mathcal{G} \Delta \rho_{cf} e^{-kd}$, which becomes $2\pi \mathcal{G} \rho_c$ on land). The marine free-air admittance is dominated by the e^{-kd} term at mid- to short wavelengths, causing it to decrease back to values close to zero.

Figure 8.9 shows that the admittance is highly dependent upon T_e, with the change from low to high admittance occurring at longer wavelengths as T_e increases. In this fashion, comparison of such theoretical curves against observed admittances can yield T_e for a study area. However, the admittance is also strongly sensitive to the thickness of the crust, with thicker crust also giving a transition from low to high admittance at longer wavelengths (Fig. 8.10a), mimicking a stronger plate. The admittance is relatively insensitive to variations in crustal density, except at the shorter wavelengths[12] (Fig. 8.10b), and very insensitive to variations in mantle density at all wavelengths (Fig. 8.10c). Therefore the sensitivity of the admittance to its various parameters somewhat impedes its utility for T_e estimation.

---

[12] This is the case for the free-air admittance, which tends to zero at long wavelengths, and to $2\pi \mathcal{G} \rho_c$ at short wavelengths (on land). With the Bouguer admittance, the sensitivity to variations in ρ_c is at the long wavelengths, since the Bouguer admittance tends to $-2\pi \mathcal{G} \rho_c$ at long wavelengths (on land), and to zero at short wavelengths.

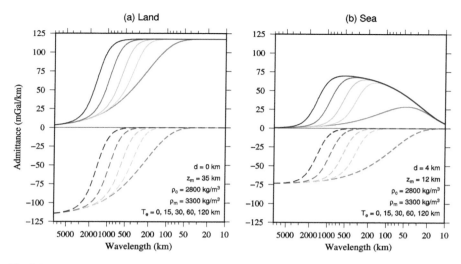

Fig. 8.9 Free-air admittances (solid curves) and Bouguer admittances (dashed curves) for a surface-loading model with a single-layer crust, given by Eq. 8.43 with the parameter values shown in the figure. The curves are colour-coded for various T_e values: red curves show Airy isostasy ($T_e = 0$ km), while the other curves are (from lighter to darker tones) $T_e = 15$ km, 30 km, 60 km and 120 km. Wavelength (λ) and wavenumber (k) are related by $\lambda = 2\pi/k$

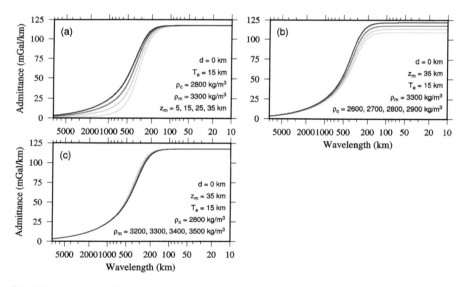

Fig. 8.10 Surface-loading, free-air admittances for a single-layer continental crust (Eq. 8.43) showing the variation with: **a** depth to Moho (z_m), **b** crust density (ρ_c), and **c** mantle density (ρ_m). Lighter curves show smaller values of the parameter, darker curves show higher values, indicated in the panels

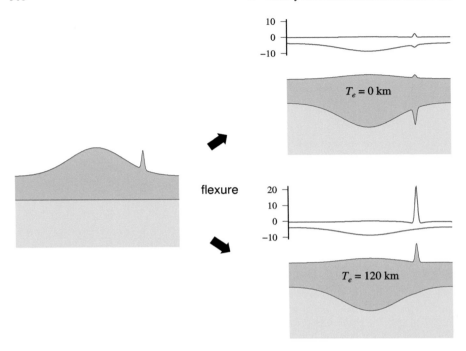

Fig. 8.11 At left is shown an initial load comprising surface topographic relief emplaced upon a plate with a flat 'Moho' (plate base). At right is shown the final surface topography and Moho relief for two scenarios of plate strength, together with their respective free-air anomalies (blue) and Bouguer anomalies (red) (mGal)

To understand the physical significance of the admittance, recall that it is a transfer function, or filter, from topography to gravity, as in Eq. 5.6 ($G = QH$). Thus a high (absolute) value of the admittance at a certain wavelength will amplify the effect of the topography, while an admittance close to zero will suppress those harmonics in the topography. Figure 8.11 shows four loading scenarios (large and small loads on weak and strong plates) and their free-air and Bouguer anomalies. Consider the very long wavelength end of the spectrum first. A very large load on a weak plate (and even a very large load on a strong plate) will be compensated by local (Airy) isostasy, rather than being supported by the mechanical strength of the plate. Therefore, because the load is hydrostatically compensated by deflections of interfaces alone (i.e. by density contrasts), the gravitational attractions of the topography and deflected interfaces should approximately cancel, leaving long-wavelength free-air anomalies with very small values. So in order to produce very small free-air anomalies from the equation $G = QH$, the very long wavelength free-air admittance must be approximately zero, as seen in Fig. 8.9.

At the other end of the spectrum, short-wavelength loads generate free-air anomalies commensurate with their post-flexure topographic relief, having a free-air admittance value of $2\pi \mathcal{G} \rho_c$ on land (Fig. 8.9 and Eq. 8.43). That is, 1 km of topographic elevation will produce a free-air anomaly of 112 mGal,[13] for a topographic density of 2670 kg m^{-3}. At these short wavelengths, the compensation of the load (at the Moho in this model) does not contribute to the free-air anomaly, as explained following. On the one hand, if the plate is strong then the load is supported by the plate strength and there is no compensation by interface deflection; because there is no interface deflection no gravity anomaly is generated (Fig. 8.11). On the other hand, if the plate is weak, then even though the Moho is deflected and the load is isostatically compensated, the gravitational effect of the deflected Moho is strongly attenuated[14] and does not cancel the gravitational attraction of the topography. Note though, for shallow interfaces the attenuation is less marked and the gravity effect of the deflected interface is numerically larger, but negative; this reduces the free-air anomaly and hence free-air admittance (Fig. 8.10a).

Finally, we will look at the case where the plate has zero rigidity, and loading is governed by Airy isostasy. When $T_e = 0$ km in the single-layer crust example considered above, Eq. 8.43 becomes the Airy admittance,

$$Q_A = 2\pi \mathcal{G} \Delta \rho_{cf} \left(\beta_g e^{-kd} - e^{-kz_m} \right) , \tag{8.44}$$

shown in Fig. 8.9.

8.6.2 Internal Loading

If Eq. 8.38 is divided by H_B then we obtain an equation for the internal loading admittance,

$$Q_B = 2\pi \mathcal{G} \left[\beta_g \Delta \rho_{1f} e^{-kd} + \sum_{i=1}^{n} (\Delta \rho_{i+1,i} - \delta_{ij} \Phi) e^{-kz_i} \right] , \tag{8.45}$$

where Φ is given by Eq. 7.53, and

$$\delta_{ij} = \begin{cases} 1, & i = j \\ 0, & i \neq j \end{cases}$$

is the Kronecker delta. There are n crustal layers ($n \geq 1$), z_i (> 0) is the mean depth to the base of crustal layer i, and $\Delta \rho_{i+1,i}$ is the density contrast at the base of layer i (with $\rho_{n+1} = \rho_m$, the mantle density). To obtain the admittance for continental environments set $d = 0$; at sea d is the (positive) mean seafloor depth in the region.

---

[13] Note the similarity with the Bouguer plate correction for unit topography, Eq. 8.14, i.e. the gravitational effect of the topography.
[14] This is embodied in the e^{-kz} terms in Eqs. 8.30 and 8.42. When the product kz is large this exponential is very small: a large kz can result from short-wavelength loads (large k) and/or deep interfaces (large z).

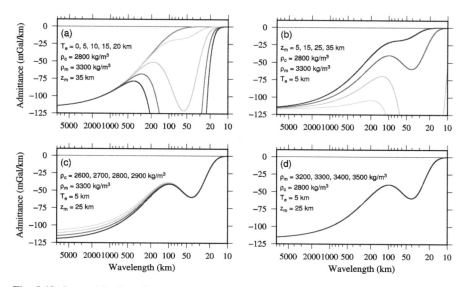

Fig. 8.12 Internal-loading, Bouguer admittances for a single-layer continental crust (Eq. 8.46) showing the variation with: **a** T_e, **b** depth to Moho (z_m), **c** crust density (ρ_c), **d** mantle density (ρ_m); lighter curves show smaller values of the parameter, darker curves show higher values, indicated in the panels. The red curve in **a** shows Airy isostasy ($T_e = 0$ km)

The internal load occurs at the base of layer j ($j = 1, \ldots, n$); its density does not need to be specified (Sect. 8.5).

For a simple example of a single-layer crust of density ρ_c, Eq. 8.45 gives the Bouguer admittance (with $n = j = 1$) as

$$Q_B = 2\pi \mathcal{G} \left(\Delta \rho_{mc} - \Phi \right) e^{-k z_m} ,$$

or

$$Q_B = -2\pi \mathcal{G} \Delta \rho_{cf} \, e^{-k z_m} \left(1 + \frac{D k^4}{\Delta \rho_{cf} g} \right) , \qquad (8.46)$$

where z_m is the depth to the base of the crust (the Moho). Equation 8.46 is plotted in Fig. 8.12a for different values of T_e (holding z_m fixed), and in Fig. 8.12b for different values of z_m (holding T_e fixed). Note how the two sets of curves are quite similar, showing a large (negative) increase in admittance in the mid- to short wavelengths. As for surface loading, such a similarity makes it difficult to reliably estimate T_e from the admittance, unless the depths to intra-crustal interfaces are reliably constrained. The sensitivity of the internal-loading admittance to the model densities is shown in Fig. 8.12c for crustal density and Fig. 8.12d for mantle density. As in the surface-loading case, the Bouguer admittance is only sensitive to crustal density at the long wavelengths (or at short wavelengths for free-air admittance), and highly insensitive to mantle density.

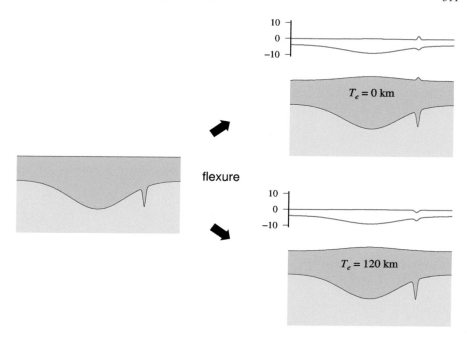

Fig. 8.13 At left is shown an initial load comprising relief emplaced at the base ('Moho') of a plate with a flat upper surface. At right is shown the final surface topography and Moho relief for two scenarios of plate strength, together with their respective free-air anomalies (blue) and Bouguer anomalies (red) (mGal)

Figure 8.13 seeks to explain the shape of the internal loading admittance curves. Very long-wavelength internal loads behave in the same way as surface loads, giving similar results (Figs. 8.9 and 8.11), so the discussion for such loads in Sect. 8.6.1 applies here, too. The same goes for loads of any size applied to a very weak plate (Airy isostasy). However, when a medium- or short-wavelength load is applied to a strong plate, while a (negative) Bouguer anomaly is generated (by virtue of the Moho relief displaced by the post-flexure load), the load cannot significantly flex the plate and only a very small surface topography is generated. Thus, at such wavelengths, the Bouguer admittance must be very large (and negative, Fig. 8.12a) so that when it multiplies the very small topography (in $G = QH$) it produces a non-zero (negative) Bouguer anomaly.

Finally, when $T_e = 0$ km in the single-layer crust model, Eq. 8.46 becomes the Airy Bouguer admittance,

$$Q_A = -2\pi \mathcal{G} \Delta \rho_{cf} \, e^{-k z_m} \, , \tag{8.47}$$

shown in Fig. 8.12a (cf. Eq. 8.44).

8.7   Combined Loading

As in Sect. 7.6, we must consider the reality that both surface and internal loading will be present. And just as we have added the gravity effects of various density contrasts to form gravity anomalies, we can simply add the gravity anomalies due to surface and internal loading to give a total, combined gravity anomaly,

$$G = G_T + G_B \ . \tag{8.48}$$

However, we cannot add surface and internal admittances to give a combined-loading admittance, meaning that

$$Q \neq Q_T + Q_B \ ,$$

because such an equation does not consider the fact that surface and internal loading processes are independent and will generally have random phases (remember, both Q_T and Q_B are real valued). Rather, we must note that the combined admittance is the ratio of the combined gravity (Eq. 8.48) to the combined topography (Eq. 7.78), or

$$Q \ = \ \frac{G}{H} \ = \ \frac{G_T + G_B}{H_T + H_B} \ = \ \frac{Q_T H_T + Q_T H_B}{H_T + H_B} \ .$$

We will discuss and explore combined loading in Chap. 9.

8.8   Summary

T_e estimation relies upon accurate knowledge of both the surface topography and Moho relief. While seismic data, for example, can provide estimates of the latter, they are generally of low accuracy and resolution, with a relatively poor global coverage. In contrast, gravity data generally have good accuracy and sufficient resolution for our purposes and furthermore have global coverage; using potential theory, gravity data can be used as a proxy for the depth to the Moho or any other subsurface interface. To this end, in this chapter we investigated the relevant aspects of potential theory with the twin aims of defining the gravity anomaly and its variants, and finding a mathematical relationship between the gravity anomaly and subsurface mass distributions. The latter led to the derivation of the expression known as Parker's formula.

We then derived expressions for the gravity field of the surface and internal loading scenarios introduced in Chap. 7, specifically their free-air and Bouguer anomalies. These then led to theoretical equations for the admittance between the gravity anomalies and topography. From plots of these expressions, we saw how the admittance can be very sensitive to relatively small variations in its geophysical parameters—

particularly elastic thickness and crustal layer thicknesses—suggesting caution when using it for T_e estimation, as, for example, a change in depth to Moho can masquerade as a change in T_e.

8.9 Further Reading

Gravity potential theory and the definition of gravity anomalies are dealt with in textbooks classed under the subject of physical geodesy, the prime examples being Heiskanen and Moritz (1967) (or its successor, Hofmann-Wellenhof and Moritz (2006), which is more readily available), Torge (2012), or Vaníček and Krakiwsky (1986). However, geodesists tend not to use the flat Earth approximation and the theory is typically framed in spherical or ellipsoidal coordinates. Therefore the reader is also encouraged to look at the books by Blakely (1996), Grant and West (1965), Turcotte and Schubert (2002), and Watts (2001), and the journal article by Parker (1972), for the planar treatment.

Regarding theoretical plate models—and their gravity and admittance expressions—the articles cited in Sect. 7.9 are again recommended: Banks et al. (1977), Louden and Forsyth (1982), McNutt (1983), and Forsyth (1985), plus the article by Karner (1982), the book by Watts (2001), and the review by Kirby (2014).

References

Banks RJ, Parker RL, Huestis SP (1977) Isostatic compensation on a continental scale: local versus regional mechanisms. Geophys J R Astron Soc 51:431–452

Blakely RJ (1996) Potential theory in gravity and magnetic applications. Cambridge University Press, Cambridge

Forsyth DW (1985) Subsurface loading and estimates of the flexural rigidity of continental lithosphere. J Geophys Res 90(B14):12,623–12,632

Grant FS, West GF (1965) Interpretation Theory in Applied Geophysics. McGraw-Hill, New York

Heiskanen WA, Moritz H (1967) Physical Geodesy. WH Freeman, San Francisco

Hofmann-Wellenhof B, Moritz H (2006) Physical Geodesy, 2nd edn. Springer, Vienna

Karner GD (1982) Spectral representation of isostatic models. BMR J Aust Geol Geophys 7:55–62

Kirby JF (2014) Estimation of the effective elastic thickness of the lithosphere using inverse spectral methods: the state of the art. Tectonophys 631:87–116

Louden KE, Forsyth DW (1982) Crustal structure and isostatic compensation near the Kane fracture zone from topography and gravity measurements: 1. Spectral analysis approach. Geophys J R Astron Soc 68:725–750

Li YC, Sideris MG (1994) Improved gravimetric terrain corrections. Geophys J Int 119:740–752

McNutt MK (1983) Influence of plate subduction on isostatic compensation in northern California. Tectonics 2:399–415

Moritz H (1968) On the use of the terrain correction in solving Molodensky's problem. Ohio State University Report 108, Department of Geodetic Science and Surveying, Ohio State University, Columbus

Moritz H (2000) Geodetic Reference System 1980. J Geod 74:128–133

NGA (2014) Department of Defense World Geodetic System 1984: Its Definition and Relationships with Local Geodetic Systems. National Geospatial-Intelligence Agency, Arnold, Missouri. Available at https://earth-info.nga.mil/GandG/publications/NGA_STND_0036_1_0_0_WGS84/NGA.STND.0036_1.0.0_WGS84.pdf. Accessed 3 Jun 2020

NGS (1994) Annex O: Gravity Control Formulas. The Blue Book, National Geodetic Survey, National Oceanic and Atmospheric Administration. Available at https://geodesy.noaa.gov/FGCS/BlueBook Accessed 9 August 2021

Parker RL (1972) The rapid calculation of potential anomalies. Geophys J R Astron Soc 31:447–455

Torge W (2012) Geodesy, 4th edn. de Gruyter, Berlin

Turcotte DL, Schubert G (2002) Geodynamics, 2nd edn. Cambridge University Press, Cambridge

Vaníček P, Krakiwsky EJ (1986) Geodesy: The Concepts, 2nd edn. Elsevier, Amsterdam

Watts AB (2001) Isostasy and Flexure of the Lithosphere. Cambridge University Press, Cambridge

Chapter 9
The Load Deconvolution Method

9.1 Introduction

In Sect. 8.6 we derived expressions for the theoretical admittance of surface and internal loading of the lithosphere. Clearly, one might expect that these equations could be compared against actual, observed admittance data from gravity and topography measurements, so that by varying parameters such as effective elastic thickness (T_e) and the thicknesses and densities of crustal layers in the equations, one would find an admittance curve that best matched the data. Indeed, this was the primary method of T_e estimation for a ten-year period in the 1970s and 1980s. Although the admittance between gravity and topography had been estimated and interpreted in terms of the density structure of the lithosphere in the early 1970s (Dorman and Lewis 1970, 1972; Lewis and Dorman 1970), it was not until the study by McKenzie and Bowin (1976) that it was used for T_e estimation. In the following decade many studies used the method, or variants on it, in which theoretical curves—such as those given by Eq. 8.43—were fitted to observed admittances, yielding parameters such as T_e, depth to Moho, or crustal density (Banks et al. 1977; Banks and Swain 1978; Watts 1978; Cochran 1979; Detrick and Watts 1979; Karner and Watts 1982, 1983)—see Kirby (2014) for a more detailed summary.

These studies, however, all used surface-only loading models (Sects. 7.4, 8.4 and 8.6.1), which led Forsyth (1985) to question their accuracy. He suggested that the surface-loading, curve-fitting method would underestimate the true elastic thickness because the method does not account for the possible existence of internal loads (shown in Fig. 9.1). Furthermore, such studies would need to window a large area in order to capture all possible flexural wavelengths (Sect. 7.7), which runs the risk of merging tectonic provinces with a variety of elastic thicknesses. And because the formula used to estimate observed admittance (Eq. 5.4) preferentially weights regions with high topography—such as mountain ranges, where T_e is likely to be lower because the rocks are younger—the resulting estimate for the study area would

© Springer Nature Switzerland AG 2022
J. Kirby, *Spectral Methods for the Estimation of the Effective Elastic Thickness of the Lithosphere*, Advances in Geophysical and Environmental Mechanics and Mathematics, https://doi.org/10.1007/978-3-031-10861-7_9

Fig. 9.1 Observed free-air
admittance of a synthetic
model with $T_e = 45$ km and
equal surface and internal
loading (black dots). The
blue line shows the
best-fitting predicted
admittance when a
surface-loading-only model
is forced, having $T_e = 20$
km. The red line shows the
best-fitting predicted
admittance using load
deconvolution, having
$T_e = 44$ km

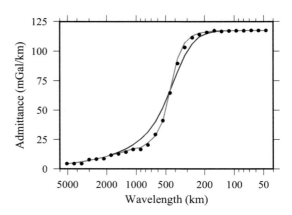

be biased downwards. To show all this, he demonstrated that the admittance of a
single surface-loading-only model with low elastic thickness could be mimicked by
a weighted average of many internal-loading-only admittances whose T_e values were
allowed to vary from zero up to 100 km, and where the weights were determined by
the post-flexure surface topography.

The problem was how to distinguish between these two very different scenarios
that gave very similar predictions, and Forsyth (1985) suggested two new methods.
First, using the internal loading model of McNutt (1983) together with Banks et al.
(1977)'s surface loading model, Forsyth developed a combined loading model that
sought to better emulate actual loading of the lithosphere. Second, he proposed that
surface and internal loading could be distinguished by using the coherence between
Bouguer anomalies and topography instead of their admittance. The coherence also
has the advantage that it weights gravity and topography equally whereas the admittance is biased towards topography (compare Eqs. 5.4 and 5.14). And furthermore,
as we shall see, the coherence is much less sensitive than the admittance to the depth
to the internal load.

The method became known as 'the Bouguer coherence method', or just 'the
coherence method', or even sometimes 'Forsyth's method'. However, a better name,
given by Lowry and Smith (1994), is *load deconvolution* as it is more descriptive of
the actual procedure, and it acknowledges that the method can actually also be applied
to the (Bouguer or free-air) admittance. Instead of merely fitting a theoretical curve
to observed data, the method 'deconvolves' the initial surface and internal loads on
the plate from the present-day gravity and topography, based on an assumed elastic
thickness. In this fashion, the precise structure of the loads can be accounted for at
different wavelengths, something the curve-fitting method does not do.

In this chapter we will explore the load deconvolution method, but will present
it in a slightly different way to the original by Forsyth (1985), for reasons given in
Sect. 9.5.

9.2 Combined Loading

The concept of combined surface and internal loading was introduced in Sects. 7.6 and 8.7, where we saw that the total gravity anomaly and topography after flexure are just linear sums of their surface (T) and internal (B) components. Thus Eqs. 8.48 and 7.78 were

$$G(\mathbf{k}) = G_T(\mathbf{k}) + G_B(\mathbf{k}) \tag{9.1}$$

and

$$H(\mathbf{k}) = H_T(\mathbf{k}) + H_B(\mathbf{k}) , \tag{9.2}$$

respectively, where we are still dealing with the Fourier transforms of these quantities (hence the dependence on 2D wavenumber, $\mathbf{k}$). You might also recall that in these sections we spent a little time rearranging the equations so that we could write the topography and gravity in terms of the initial loads, $L_B(\mathbf{k})$ and $L_T(\mathbf{k})$; the purpose of this endeavour will be revealed here.

A glance at Eq. 7.55, for example, shows that it can be written in the form $H_T(\mathbf{k}) = \kappa_T(k)L_T(\mathbf{k})$, where the *deconvolution coefficient* κ_T is a wavenumber-dependent function.[1] This is also true for the other surface loading equations, the internal loading equations, and the equations relating gravity to loads in Chap. 8. Thus, in general the *predicted gravity anomalies and topography* of surface and internal loading can be written

$$G_T(\mathbf{k}) = \mu_T(k) L_T(\mathbf{k}) , \tag{9.3a}$$
$$G_B(\mathbf{k}) = \mu_B(k) L_B(\mathbf{k}) , \tag{9.3b}$$
$$H_T(\mathbf{k}) = \kappa_T(k) L_T(\mathbf{k}) , \tag{9.3c}$$
$$H_B(\mathbf{k}) = \kappa_B(k) L_B(\mathbf{k}) , \tag{9.3d}$$

where the μs are called the *gravity deconvolution coefficients* and the κs are the *flexural deconvolution coefficients*, and are dependent on T_e. If we substitute Eq. 9.3 into Eqs. 9.1 and 9.2, we get

$$G(\mathbf{k}) = \mu_T(k)L_T(\mathbf{k}) + \mu_B(k)L_B(\mathbf{k})$$

and

$$H(\mathbf{k}) = \kappa_T(k)L_T(\mathbf{k}) + \kappa_B(k)L_B(\mathbf{k}) ,$$

which, following Banks et al. (2001), can be written in matrix form as

---

[1] For the isotropic plates considered in this book the deconvolution coefficients are isotropic and are functions of 1D (radially-averaged) wavenumber, $k \equiv |\mathbf{k}|$. For plates with an orthotropic or anisotropic response to loading the deconvolution coefficients are not isotropic and are functions of 2D wavenumber—see Swain and Kirby (2003).

$$\begin{pmatrix} G \\ H \end{pmatrix} = \begin{pmatrix} \mu_T & \mu_B \\ \kappa_T & \kappa_B \end{pmatrix} \begin{pmatrix} L_T \\ L_B \end{pmatrix} , \tag{9.4}$$

dropping the dependence on wavenumber.

The exact form of the deconvolution coefficients will depend upon the loading scenario. For the general case of an n-layer crust, the two flexural deconvolution coefficients are

$$\kappa_T(k) = \frac{1}{\Delta\rho_{Tf}\, g} - \frac{1}{\Phi(k)g} \tag{9.5}$$

from Eq. 7.58, and

$$\kappa_B(k) = \frac{-1}{\Phi(k)\, g} \tag{9.6}$$

from Eq. 7.74. The two gravity deconvolution coefficients are

$$\mu_T(k) = \frac{2\pi\mathcal{G}}{\Phi(k)g} \left\{ \left[\beta_g(\Phi(k) - \Delta\rho_{Tf}) - \Delta\rho_{1T} \right] e^{-kd} - \sum_{i=1}^{n} \Delta\rho_{i+1,i}\, e^{-kz_i} \right\} \tag{9.7}$$

from Eq. 8.31, and

$$\mu_B(k) = \frac{2\pi\mathcal{G}}{\Phi(k)\, g} \left\{ -\beta_g \Delta\rho_{1f} e^{-kd} - \sum_{i=1}^{n} \left[\Delta\rho_{i+1,i} - \delta_{ij}\, \Phi(k) \right] e^{-kz_i} \right\} \tag{9.8}$$

from Eq. 8.39. To remind ourselves, there are n crustal layers ($n \geq 1$), z_i is the (positive) mean depth to the base of crustal layer i, and $\Delta\rho_{i+1,i}$ is the density contrast at the base of layer i (with $\rho_{n+1} = \rho_m$, the mantle density). For continental environments one sets $d = 0$ and $\rho_f = 0$; at sea d is the (positive) mean seafloor depth in the region, and $\rho_f = \rho_w$, the seawater density. The surface load has density ρ_T, and the internal load occurs at the base of layer j ($j = 1, \cdots, n$) though its density does not need to be specified; δ_{ij} is the Kronecker delta. If the gravity anomaly is chosen to be the free-air anomaly then $\beta_g = 1$; for Bouguer anomalies set $\beta_g = 0$ (Eq. 8.26). Finally, Φ is defined by Eq. 7.53, being

$$\Phi(k) \equiv \frac{Dk^4}{g} + \Delta\rho_{mf} . $$

9.3   Combined-Loading Coherency, Coherence and Admittance

Before pursuing the details of the load deconvolution method, it will be instructive to understand the coherence between gravity and topography of a flexed plate under combined loading, and how it relates to the plate's geophysical parameters. Hence, we need to derive theoretical expressions for the coherence, much like we did for the admittance in Sect. 8.6. Here, however, we will also derive expressions for the coherency as, even though Forsyth (1985) framed his method in terms of the coherence, the coherency is the more fundamental quantity (Sect. 5.3). To recap, Eq. 5.12 gives the coherency as

$$\Gamma(k) \; = \; \frac{\langle G(\mathbf{k}) \, H^*(\mathbf{k}) \rangle}{\langle G(\mathbf{k}) \, G^*(\mathbf{k}) \rangle^{1/2} \, \langle H(\mathbf{k}) \, H^*(\mathbf{k}) \rangle^{1/2}} \; , \tag{9.9}$$

which we write as a function of 1D wavenumber (k) because we will be dealing exclusively with isotropic plates. The coherence can then be found from

$$\gamma^2(k) \; = \; |\Gamma(k)|^2 \tag{9.10}$$

(Sect. 5.3.2). Additionally, since the load deconvolution method can also be applied to the admittance, we recall Eq. 5.4:

$$Q(k) \; = \; \frac{\langle G(\mathbf{k}) \, H^*(\mathbf{k}) \rangle}{\langle H(\mathbf{k}) \, H^*(\mathbf{k}) \rangle} \; . \tag{9.11}$$

The angular brackets indicate the averaging procedures explained in Sects. 9.4.2 (multitaper method) and 9.4.3 (wavelet method).

9.3.1 Predicted Coherency, Coherence and Admittance

Theoretical equations for the coherency and admittance can be derived by substituting Eq. 9.4 into Eqs. 9.9 and 9.11. But since these latter two equations have terms in common, we will compute the cross- and auto-spectra, GH^*, GG^* and HH^*, separately. Making the afore-mentioned substitutions we see that

$$\begin{aligned}
\langle GH^* \rangle \; &= \; \langle (\mu_T L_T + \mu_B L_B) (\kappa_T L_T + \kappa_B L_B)^* \rangle \\
&= \; \langle \mu_T \kappa_T |L_T|^2 + \mu_B \kappa_B |L_B|^2 + \mu_T \kappa_B L_T L_B^* + \mu_B \kappa_T L_B L_T^* \rangle .
\end{aligned}$$

Now, the load deconvolution method hinges upon the two loading processes—surface and internal—being independent of one another (Forsyth 1985). That is, the two initial loads are statistically uncorrelated. One way of expressing this is that, upon averaging, the cross-spectral terms vanish, or

$$\langle L_T L_B^* \rangle \; = \; \langle L_B L_T^* \rangle \; = \; 0 \tag{9.12}$$

(though see Sect. 9.7). With this condition we thus have

$$\langle GH^* \rangle = \langle \mu_T \kappa_T |L_T|^2 + \mu_B \kappa_B |L_B|^2 \rangle,   \tag{9.13}$$

which is real-valued. Similarly, we can show that

$$\langle GG^* \rangle = \langle \mu_T^2 |L_T|^2 + \mu_B^2 |L_B|^2 \rangle,   \tag{9.14}$$

and

$$\langle HH^* \rangle = \langle \kappa_T^2 |L_T|^2 + \kappa_B^2 |L_B|^2 \rangle,   \tag{9.15}$$

which are also real-valued. We can now substitute Eqs. 9.13–9.15 into Eq. 9.9, giving the *predicted coherency* as

$$\Gamma_{pr}(k) = \frac{\langle \mu_T \kappa_T |L_T|^2 + \mu_B \kappa_B |L_B|^2 \rangle}{\langle \mu_T^2 |L_T|^2 + \mu_B^2 |L_B|^2 \rangle^{1/2} \langle \kappa_T^2 |L_T|^2 + \kappa_B^2 |L_B|^2 \rangle^{1/2}},   \tag{9.16}$$

while the *predicted coherence* is, from Eqs. 9.10 and 9.16,

$$\gamma_{pr}^2(k) = \frac{\langle \mu_T \kappa_T |L_T|^2 + \mu_B \kappa_B |L_B|^2 \rangle^2}{\langle \mu_T^2 |L_T|^2 + \mu_B^2 |L_B|^2 \rangle \langle \kappa_T^2 |L_T|^2 + \kappa_B^2 |L_B|^2 \rangle}.   \tag{9.17}$$

Note how the predicted coherency is a real-valued variable, and not complex as it ought to be (Chap. 5). The lack of an imaginary part arose under the assumption of independent initial loads (Eq. 9.12) and will be discussed further in Sect. 9.7. The *predicted admittance* is also real-valued, and is found by substituting Eqs. 9.13–9.15 into Eq. 9.11, thus:

$$Q_{pr}(k) = \frac{\langle \mu_T \kappa_T |L_T|^2 + \mu_B \kappa_B |L_B|^2 \rangle}{\langle \kappa_T^2 |L_T|^2 + \kappa_B^2 |L_B|^2 \rangle}.   \tag{9.18}$$

An alternative expression for the predicted coherency can be found by substituting Eq. 9.3 into Eq. 9.16, giving

$$\Gamma_{pr}(k) = \frac{\langle G_T H_T^* + G_B H_B^* \rangle}{\langle |G_T|^2 + |G_B|^2 \rangle^{1/2} \langle |H_T|^2 + |H_B|^2 \rangle^{1/2}},   \tag{9.19}$$

which tells us that the predicted coherency measures the degree of correlation between predicted gravity and topography (assuming independence of surface and internal loading processes). Similarly, substitution of Eqs. 9.3 into Eq. 9.18 gives an alternative expression for the predicted admittance, thus:

$$Q_{pr}(k) = \frac{\langle G_T H_T^* + G_B H_B^* \rangle}{\langle |H_T|^2 + |H_B|^2 \rangle}.   \tag{9.20}$$

Even though Eqs. 9.19 and 9.20 look like they are complex-valued (because of the cross-spectral terms in the numerator), Eqs. 9.16 and 9.19 give identical, real-valued results in practice, as do Eqs. 9.18 and 9.20.

It is worthwhile stressing the significance of this result, that the predicted admittance and coherency are real-valued, as it will prove useful when we come to implementing the load deconvolution method. Following on from the discussion in Sects. 5.2.2 and 5.3.3, it tells us that any gravity anomaly predicted from the topography using the predicted admittance—as in $G = Q_{pr} H$—will be in phase with the topography. This is the reason why it is very difficult, if not impossible, to construct a useful isostatic anomaly, at least using spectral methods (Kirby 2019).

9.3.2 The Load Ratio

Forsyth (1985)'s next initiative was to express the initial loads as a ratio. Although he wrote it slightly differently, the ratio of initial internal-to-surface load amplitudes, or *load ratio*, $f(k)$, is found from

$$f^2(k) = \frac{\langle |L_B(\mathbf{k})|^2 \rangle}{\langle |L_T(\mathbf{k})|^2 \rangle} \qquad (9.21)$$

(Banks et al. 2001), where, again, we use isotropic averaging with isotropic plates so that the load ratio is a function of 1D wavenumber. If surface-loading processes are operating exclusively, then $f(k) = 0$ at all wavenumbers[2]; if there is only internal loading then $f(k) \to \infty, \forall k$. Another way of representing the load ratio, f, is through the *internal load fraction*, F, where

$$F(k) = \frac{f(k)}{1 + f(k)}, \qquad (9.22)$$

being the ratio of the internal load to the total load. The internal load fraction is slightly easier to plot than is the load ratio as it is bounded between values of zero and one, whereas f is unbounded. Surface-only loading corresponds to $f = F = 0$, while internal-only loading has $f = \infty$ but $F = 1$; equal combined loading has $f = 1$ and $F = 0.5$. If one needs to retrieve f from F, then the inverse of Eq. 9.22 is

$$f(k) = \frac{F(k)}{1 - F(k)}.$$

---

[2] This is usually written as $f(k) = 0, \forall k$, where the symbol $\forall$ means 'for all', as in $f(k) = 0$ at all values of k.

9.3.3   Theoretical Coherency, Coherence and Admittance

Calculation of the load ratio is not essential for T_e estimation using load deconvolution: rather, the load ratio emerges as an additional piece of information, as will be seen in Sect. 9.4. As well as providing geological information, the load ratio is useful for deriving theoretical equations for the coherency and admittance of combined loading regimes. If we assume that

$$\left\langle \mu_T(k)\kappa_T(k)\,|L_T(\mathbf{k})|^2 \right\rangle \;=\; \mu_T(k)\kappa_T(k)\left\langle |L_T(\mathbf{k})|^2 \right\rangle \tag{9.23}$$

etc., then we can divide numerator and denominator of Eq. 9.18 by $\left\langle |L_T|^2 \right\rangle$, and use Eq. 9.21 to get the *theoretical admittance* as

$$Q_{th}(k) \;=\; \frac{\mu_T\kappa_T + \mu_B\kappa_B f^2}{\kappa_T^2 + \kappa_B^2 f^2}\;. \tag{9.24}$$

Remember, the deconvolution coefficients and load ratio, f, are functions of (1D) wavenumber. In the same fashion we can derive the *theoretical coherency* from Eq. 9.16 as

$$\varGamma_{th}(k) \;=\; \frac{\mu_T\kappa_T + \mu_B\kappa_B f^2}{\left(\mu_T^2 + \mu_B^2 f^2\right)^{1/2}\left(\kappa_T^2 + \kappa_B^2 f^2\right)^{1/2}}\;, \tag{9.25}$$

while the *theoretical coherence* is found from Eq. 9.25 using Eq. 9.10, giving

$$\gamma_{th}^2(k) \;=\; \frac{\left(\mu_T\kappa_T + \mu_B\kappa_B f^2\right)^2}{\left(\mu_T^2 + \mu_B^2 f^2\right)\left(\kappa_T^2 + \kappa_B^2 f^2\right)}\;. \tag{9.26}$$

Again, the theoretical admittance, coherency and coherence are real-valued variables (though see Sect. 9.7). Note, though, that the load deconvolution method does not use Eqs. 9.24–9.26 to estimate T_e—see Sect. 9.4.

We now have theoretical equations with which to explore the properties of the coherency/coherence and admittance. First, in the limit of surface-only loading ($f(k) = 0,\ \forall k$) Eq. 9.24 becomes

$$Q_{th}(k) \;=\; \frac{\mu_T}{\kappa_T} \;=\; \frac{G_T}{H_T} \;=\; Q_T\;,$$

using Eqs. 9.3a and 9.3c, while in the limit of internal-only loading ($f(k) \to \infty,\ \forall k$) Eq. 9.24 gives

$$Q_{th}(k) \;=\; \frac{\mu_B}{\kappa_B} \;=\; \frac{G_B}{H_B} \;=\; Q_B\;,$$

using Eqs. 9.3b and 9.3d. Both of these results are expected, and merely state that under surface-loading conditions the combined admittance formula reverts to the surface-loading admittance, and under internal-loading conditions the combined admittance formula reverts to the internal-loading admittance. With the coherency, however, substitution of $f = 0$ into Eq. 9.25 gives

$$\Gamma_{th}(k) = \frac{\mu_T \kappa_T}{\sqrt{\mu_T^2} \sqrt{\kappa_T^2}} = \pm 1 \,,$$

depending on the signs of the square roots, while substitution of $f \to \infty$ gives

$$\Gamma_{th}(k) = \frac{\mu_B \kappa_B f^2}{\sqrt{\mu_B^2 f^2} \sqrt{\kappa_B^2 f^2}} = \pm 1 \,.$$

These results tell us that, under surface-only loading or internal-only loading, the gravity and topography will always be coherent at all wavenumbers, no matter what the strength of the plate. This is because the coherency is a measure of the phase (rather than amplitude) relationship between two signals (Sect. 5.3.1), and under a single type of loading—and in the absence of noise—the gravity and topography will always be in phase (Sect. 5.2.2). Thus, the coherency and coherence are only meaningful concepts in the presence of combined loading.

The classical load deconvolution method uses the observed and predicted Bouguer coherence to estimate T_e (though it can use the Bouguer or free-air admittance, but never the free-air coherence—see Sects. 9.4 and 11.6). So to study the properties of the Bouguer coherence we will use the simple example of a single-layer crust ($n = 1$) of density $\rho_1 = \rho_c$ overlying a mantle of density $\rho_2 = \rho_m$, where the surface load has density $\rho_T = \rho_c$. The internal load is emplaced at the base of the crust ($j = 1$), which is at depth z_m. Because we are focussing on the Bouguer coherence we set $\beta_g = 0$, which means that we do not need to specify an ocean depth, d. With these parameter values, Eqs. 9.5–9.8 become

$$\kappa_T = \frac{1}{\Delta \rho_{cf} g} - \frac{1}{\Phi g} \,, \tag{9.27a}$$

$$\kappa_B = \frac{-1}{\Phi g} \,, \tag{9.27b}$$

$$\mu_T = \frac{-2\pi \mathcal{G} \Delta \rho_{mc} \, e^{-kz_m}}{\Phi g} \,, \tag{9.27c}$$

$$\mu_B = \frac{-2\pi \mathcal{G} \, (\Delta \rho_{mc} - \Phi) \, e^{-kz_m}}{\Phi g} \,. \tag{9.27d}$$

Substitution of Eq. 9.27 into Eq. 9.25 gives a rather cumbersome equation that can be made more elegant by using the two dimensionless, wavenumber-dependent variables used by Forsyth (1985)[3]:

---

[3] Forsyth (1985) used a continental model with $\rho_f = 0$.

$$\xi(k) \; = \; 1 + \frac{Dk^4}{\Delta\rho_{mc}g} \; , \qquad \phi(k) \; = \; 1 + \frac{Dk^4}{\Delta\rho_{cf}g} \; . \qquad (9.28)$$

After some algebra, and noting that

$$\xi(k) \; = \; \frac{\Phi(k) - \Delta\rho_{cf}}{\Delta\rho_{mc}} \; , \qquad \phi(k) \; = \; \frac{\Phi(k) - \Delta\rho_{mc}}{\Delta\rho_{cf}} \; ,$$

we obtain the theoretical Bouguer coherency for this plate model as

$$\Gamma_{th}(k) \; = \; \frac{-\xi - \phi f^2 r^2}{\left(1 + \phi^2 f^2 r^2\right)^{1/2} \left(\xi^2 + f^2 r^2\right)^{1/2}} \; , \qquad (9.29)$$

where we define the density ratio

$$r \; \equiv \; \frac{\Delta\rho_{cf}}{\Delta\rho_{mc}} \; . \qquad (9.30)$$

The theoretical coherence is just the square of Eq. 9.29, or

$$\gamma_{th}^2(k) \; = \; \frac{\left(-\xi - \phi f^2 r^2\right)^2}{\left(1 + \phi^2 f^2 r^2\right) \left(\xi^2 + f^2 r^2\right)} \; . \qquad (9.31)$$

Note how, for this single-layer model, the coherency and coherence are independent of crustal thickness (internal load depth), though the same is not true for multiple-layer crusts.

It is now easy to see that if $T_e = 0$ km then $\xi(k) = \phi(k) = 1$ from Eq. 9.28, and that $\Gamma_{th}(k) = -1$ and $\gamma_{th}^2(k) = 1$ at all wavenumbers. Thus, the coherency and coherence are only meaningful concepts when the plate has non-zero rigidity. We can see this in the simple model shown in Fig. 9.2. When the plate has zero strength (Airy isostasy) the initial surface and internal loads achieve an identical state of isostatic equilibrium, producing identical surface topography and Bouguer anomalies. This means that it is not possible to distinguish between surface and internal loading when $T_e = 0$ km, and the coherence, shown in Fig. 9.3a, is one at all wavenumbers. At the other extreme, when the plate has an infinite rigidity the Bouguer coherence is zero at all wavenumbers because the Bouguer anomaly and topography are not coherent (Figs. 9.2d and 9.3c). When the plate has a non-zero rigidity, though, the relationship between the topography and Bouguer anomalies from the surface and internal load components is (almost) unique, and depends upon the plate strength. So not only can the effects of surface and internal loading be separated, but also the strength of the plate can be estimated via the coherence (Figs. 9.2c and 9.3b).

The theoretical Bouguer coherency and coherence (Eqs. 9.29 and 9.31) are plotted in Fig. 9.4 for various values of T_e and load ratio, f, where the load ratio is taken to be uniform, i.e. having the same value at all wavenumbers. First note that

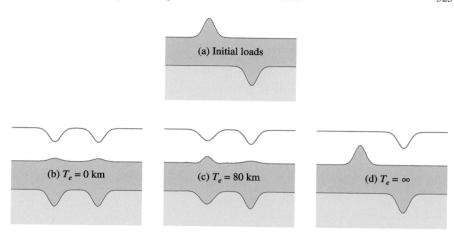

Fig. 9.2 **a** Two initial loads, one comprising surface topographic relief, the other being relief emplaced at the Moho. **b–d** The final surface topography and Moho relief for the indicated elastic thickness, together with the respective Bouguer anomalies (red)

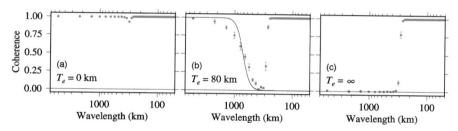

Fig. 9.3 The observed coherence of the three sets of Bouguer anomalies and topography in Fig. 9.2b–d, shown as red circles with error bars. The blue curve in **b** is the theoretical Bouguer coherence for a plate with $T_e = 80$ km and $f = 1$. The coherence is one at short wavelengths in all models owing to the lack of signal at these harmonics in the loads. Multitaper method with NW $= 3$, $K = 5$

the coherency has exclusively negative values (though see Sect. 9.7), reflecting the anti-correlation between Bouguer anomaly and topography (shown in Fig. 9.2 for example), while the coherence is positive (being a squared quantity). Both, though, are almost bimodal, with the coherence taking values of either one or zero at most wavelengths. A coherence of one indicates that the Bouguer anomaly and topography are coherent at those wavelengths, as shown in Fig. 9.2a; as noted above, this tells us that all loads of such wavelengths, whether surface or internal, are compensated isostatically. At wavelengths where the coherence is zero the Bouguer anomaly and topography are incoherent; this tells us that loads of these wavelengths are supported by the mechanical strength of the plate. Alternatively, as noted in Sect. 5.3.2, the coherence is a measure of the fraction of energy in the gravity field that can be predicted from the topography using the admittance: when it is one, then one hundred per cent of the energy in the Bouguer anomaly arises from the topography (Airy

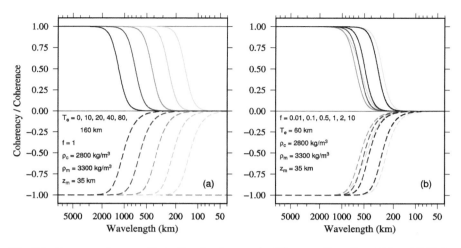

Fig. 9.4 Bouguer coherency (dashed) and coherence (solid) curves from Eqs. 9.29 and 9.31 for a single-layer continental crust showing the variation with: **a** T_e, **b** load ratio (f); lighter curves show smaller values of the parameter, darker curves show higher values, indicated in the panels. The red curve in **a** shows Airy isostasy ($T_e = 0$ km). Note that f is assumed to be uniform, i.e. wavenumber-independent

isostasy); when it is zero, then none of the energy in the Bouguer anomaly is due to the topography (and if there is a non-zero Bouguer anomaly it will have arisen from another source, such as a flexed Moho which does not generate surface topography).

Second, Fig. 9.4a shows that the coherency and coherence are very sensitive to variations in T_e. When the plate is very rigid the change from one to zero coherence (the *rollover*) occurs at long wavelengths because a rigid plate can mechanically support even large loads (i.e. there are lots of harmonics with zero coherence). When the plate is very weak the rollover occurs at short wavelengths because most loads cannot be mechanically supported and must be isostatically compensated (i.e. there are lots of harmonics with a coherence of one). The third observation of note is that the coherency and coherence are relatively insensitive to variations in load ratio as long as f is not too high or too low (see the curves for $f = 0.01$ and 10 in Fig. 9.4b). These extreme values of load ratio tell us that either surface or internal loading is dominant and, as we saw above (and substitution of $f = 0$ or $f = \infty$ into Eq. 9.29 shows us), the coherency tends to a uniform value of -1, and the coherence to 1, in such cases.

Indeed, Simons and Olhede (2013) derived an equation for the *transition wavenumber* (the rollover) of the coherence, by solving Eq. 9.31 for wavenumber. The transition wavenumber is the wavenumber at which the Bouguer coherence has a value of 0.5, and provides the wavelength separating coherent from incoherent gravity and topography (or compensated from supported topography). And because the coherence rollover is generally very steep, the separation between these two regimes is usually very clear. Using a slightly different notation from Simons and Olhede (2013),

Fig. 9.5 The coherence transition wavelength, λ_t, as a function of T_e from Eq. 9.32 for $f = 1$ (red line). Also shown is the flexural wavelength (λ_F), from Eq. 7.85 with $\rho_f = 0$ kg m^{-3} (dashed blue line). Other parameters are $\rho_m = 3300$ kg m^{-3} and $\rho_c = 2800$ kg m^{-3}

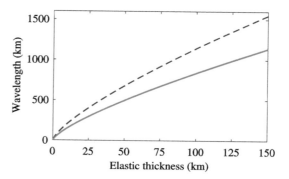

Fig. 9.6 The coherence transition wavelength, λ_t in km, as a function of T_e and load ratio (expressed in terms of the internal load fraction, F, from Eq. 9.22). Thus the red curve in Fig. 9.5 is the locus of $F = 0.5$ here. Note that, for a given T_e value, λ_t has a maximum value at approximately $F = 0.22$ ($f = 0.28$)

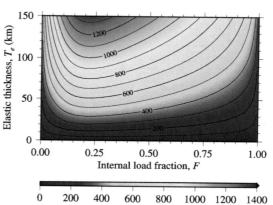

the equation for the transition wavenumber is

$$k_t = \left[\frac{\Delta\rho_{mc}\, g}{2Df} \left(1 - f(1+r) + f^2 r + \sqrt{\beta} \right) \right]^{1/4}, \qquad (9.32)$$

where

$$\beta = 1 + 2f(1-r) + f^2(1+4r+r^2) - 2f^3 r(1-r) + f^4 r^2$$

and where r is defined by Eq. 9.30. The transition wavelength, obtained from $\lambda_t = 2\pi/k_t$, is plotted in Fig. 9.5 (for equal surface and internal loads) which shows that it is somewhat smaller than the flexural wavelength for the same elastic thickness (Sect. 7.7). But since the transition wavelength depends on both T_e and load ratio, Fig. 9.6 shows how λ_t varies with both these quantities. As noted above, if f is not very small or very large, then the transition wavelength is relatively insensitive to f, the more so for lower elastic thicknesses.

The dependence of the Bouguer coherence on density and load depth is shown in Fig. 9.7. The curves in Fig. 9.7c–9.7f were generated using Eq. 9.31, but to show the

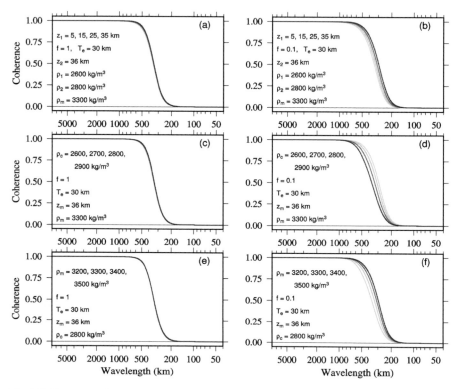

Fig. 9.7 Bouguer coherence curves for continental crust of **a** and **b** two layers, and **c–f** a single layer, showing the variation with: **a** and **b** depth to internal load (z_1), **c** and **d** crustal density (ρ_c), **e** and **f** mantle density (ρ_m); lighter curves show smaller values of the parameter, darker curves show higher values, indicated in the panels. The left-hand panels **a**, **c** and **e** show models with $f = 1$, while the right-hand panels **b**, **d** and **f** show models with $f = 0.1$

variation with load depth (Fig. 9.7a, b) a two-layer model was used instead because the single-layer model is not dependent upon any depth at all—see Eq. 9.31. The left-hand panels of Fig. 9.7 show that the Bouguer coherence is almost completely independent of crust and mantle densities and depth to internal load, though when surface loading becomes dominant ($f < 0.1$, approximately) a minor dependency reveals itself (the right-hand panels of Fig. 9.7). It can be seen, therefore, that the high sensitivity to T_e, low sensitivity to load ratio, and lack of sensitivity to density and depth of loading, is what makes the Bouguer coherence such a good diagnostic tool for effective elastic thickness estimation.

In contrast, we saw in Sect. 8.6 how the surface-loading and internal-loading admittances did not always lend themselves to T_e estimation as they were rather sensitive to depth of loading and certain densities, as well as elastic thickness. But what of the combined admittance? Following the procedure to derive a theoretical coherency/coherence shown above, we can derive an expression for the theoretical

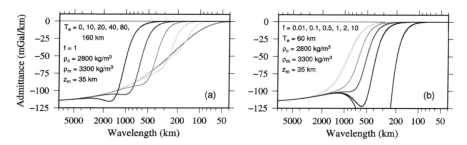

Fig. 9.8 As Fig. 9.4, but for the Bouguer admittance, using Eq. 9.33

admittance of a single-layer crust model by substituting Eqs. 9.27 into Eq. 9.24. If we also use Eqs. 9.28 and 9.30 we can derive an equation for the theoretical Bouguer admittance of a single-layer crust as

$$Q_{th}(k) = -2\pi G \Delta\rho_{cf} e^{-kz_m} \left(\frac{\xi + \phi f^2 r^2}{\xi^2 + f^2 r^2} \right).$$ (9.33)

Note that Eq. 9.33 reverts to the admittance for Airy isostasy (Eqs. 8.44 or 8.47) when $T_e = 0$ km, to the surface-loading admittance (Eq. 8.43) when $f = 0$, and to the internal-loading admittance (Eq. 8.46) when $f = \infty$. Eq. 9.33 is plotted in Fig. 9.8 for varying values of T_e and f. As seen, not only is the admittance sensitive to T_e variations, it is also very sensitive to variations in the load ratio. These observations, coupled with the high sensitivity to depth of loading and, to a lesser extent, crustal density (Fig. 9.9), can make T_e estimation using the admittance particularly frustrating and prone to error.

9.4 T_e Estimation with Load Deconvolution

Having now understood the properties of the Bouguer coherence and the nature of combined loading, we can proceed with our investigation of the load deconvolution method.

9.4.1 Overview of Load Deconvolution

9.4.1.1 Uniform-f Inversion

First though, let us for a moment consider what would occur if T_e were to be estimated using a curve-fitting method, as was the normal procedure before Forsyth's paper in

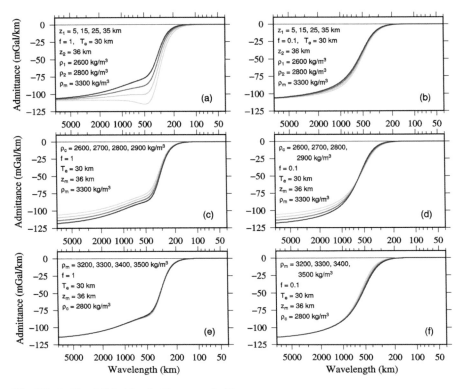

Fig. 9.9 As Fig. 9.7, but for the Bouguer admittance

1985, though with the admittance. In such an endeavour, one would take a theoretical equation for the Bouguer coherence such as Eq. 9.31 and adjust its parameters (mainly T_e and f, but also layer thicknesses and densities) until one found a curve that best fit the observed coherence, thus giving the best-fit T_e. In doing this, though, one would need to assume that the load ratio is uniform; i.e. assume that $f(k) = f_0$, $\forall k$, for some constant f_0, and treat f_0 as one of the inversion parameters.[4] Fitting a theoretical coherence curve to the observed coherence has been called *uniform-f inversion* (Kirby and Swain 2006) because of this assumption (see also Sect. 11.7).

---

[4] Actually, this assumption of uniform f in theoretical equations for the coherence does not have to be made, and one could assign some kind of analytic equation to $f(k)$ in Eq. 9.26, in which f is wavenumber-dependent, such as $f(k) = a + bf + cf^2$ for example. However, one would then need to treat a, b and c as inversion parameters to be estimated, thus increasing the complexity of the procedure.

9.4.1.2 Load Reconstruction, I

Forsyth (1985)'s load deconvolution method was designed to sidestep this need to invert for f, or to have to assume its value, and he did this by reconstructing the initial loads from the observed, present-day gravity and topography. This is achieved by assuming a T_e value and inverting Eq. 9.4, giving the so-called *deconvolution equations*,

$$\begin{pmatrix} L_T \\ L_B \end{pmatrix} = \frac{1}{\Delta} \begin{pmatrix} \kappa_B & -\mu_B \\ -\kappa_T & \mu_T \end{pmatrix} \begin{pmatrix} G \\ H \end{pmatrix} , \tag{9.34}$$

where the determinant of the matrix is

$$\Delta(k) = \mu_T \kappa_B - \mu_B \kappa_T . \tag{9.35}$$

One must take care when evaluating Eq. 9.35 as, in certain circumstances, it can take a value of zero. For example, if we use the deconvolution coefficients for a single-layer crust (Eqs. 9.27), then the determinant is given by

$$\Delta(k) = \frac{-2\pi \mathcal{G} e^{-kz_m}}{\Delta\rho_{cf} g^2} \frac{Dk^4}{Dk^4 + \Delta\rho_{mf} g} ,$$

which is zero when $T_e = 0$ or $k = 0$.

Note that Forsyth (1985) expressed the deconvolution equations in terms of relief on the initial loads rather than the loads themselves. He also used the Moho relief rather than the gravity anomaly, and while this should make no difference to the eventual formula for the predicted coherence, Kirby and Swain (2011) found that a computational difference existed, favouring use of gravity rather than Moho relief, as will be explained in Sect. 9.5.

9.4.1.3 Coherence Versus SRC

Forsyth (1985) utilised the coherence as the diagnostic tool, but more recently Kirby and Swain (2009) instead recommended use of the squared real coherency (SRC, Eq. 5.17), for the following reason. We saw in Sect. 9.3.1 that the assumption of uncorrelated initial loads led to a real-valued predicted coherency between gravity and topography. In contrast, the observed coherency is complex-valued (Sect. 5.3.3). If one then converts the observed coherency to observed coherence and predicted coherency to predicted coherence, through the operation $\gamma^2 = |\Gamma|^2$ (Eq. 9.10), the predicted coherence will just be the square of the real part of the predicted coherency (because there is no imaginary part), while the observed coherence will be the sum of the squares of the real and imaginary parts of the observed coherency [or, coherence = SRC + SIC (Eq. 5.18), where SIC is the squared imaginary coherency]. That is, the latter contains parts which the former does not, meaning the one cannot properly be compared with the other. One should actually compare real part with real part, and imaginary part with imaginary part (should they exist) separately.[5]

The effect upon T_e estimation is shown in Fig. 9.10, where the synthetic gravity and topography were generated from flexure of an elliptical T_e model (see Sect. 10.5). Here it can be seen that the observed coherence between the synthetic data has much more variability than the SRC, due, of course, to the presence of the SIC embedded in the coherence. The SIC increases the overall coherence relative to the SRC, causing T_e to be underestimated in many places, and giving an overall 'noisier' feel to the T_e map.

9.4.1.4   Predicted SRC

Once the initial loads have been found, one then computes the *predicted SRC* (defined by Eq. 5.17), being

$$\Gamma_{pr,R}^2(k) = \frac{\left\langle \mu_T \kappa_T |L_T|^2 + \mu_B \kappa_B |L_B|^2 \right\rangle^2}{\left\langle \mu_T^2 |L_T|^2 + \mu_B^2 |L_B|^2 \right\rangle \left\langle \kappa_T^2 |L_T|^2 + \kappa_B^2 |L_B|^2 \right\rangle}, \qquad (9.36)$$

from Eq. 9.16. Note how the predicted SRC (Eq. 9.36) is identical to the predicted coherence (Eq. 9.17), because the latter contains no imaginary part. As usual, the angular brackets indicate some averaging procedure which will depend upon whether one is using multitapers or wavelets (or some other technique) as the analysis method (Sects. 9.4.2 and 9.4.3). Note that an alternative expression for the predicted SRC—giving identical results—can be found from Eq. 9.19 as

$$\Gamma_{pr,R}^2(k) = \frac{\left\langle G_T H_T^* + G_B H_B^* \right\rangle^2}{\left\langle |G_T|^2 + |G_B|^2 \right\rangle \left\langle |H_T|^2 + |H_B|^2 \right\rangle}. \qquad (9.37)$$

So, while Eq. 9.36 is not an equation for the amount of correlation between L_T and L_B,[6] it does give the predicted SRC, which, as shown in Eq. 9.37, measures the amount of correlation between predicted gravity (G_T and G_B) and topography (H_T and H_B). This is why the predicted SRC can be meaningfully compared with the SRC between observed gravity and topography.

9.4.1.5   Load Reconstruction, II

The predicted gravity and topography are obtained from the reconstructed initial loads, which are obtained from the observed gravity and topography. For example,

---

[5] This will be elaborated upon in Sect. 9.6.

[6] Such a formula would be $\mathrm{Re}\langle L_T L_B^* \rangle^2 / (\langle |L_T|^2 \rangle \langle |L_B|^2 \rangle)$, which should be random because the loads are uncorrelated.

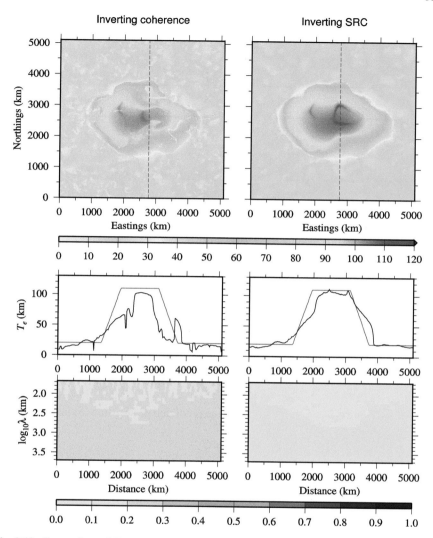

Fig. 9.10 Comparison of T_e recovered by inverting the coherence (left-hand panels) against that recovered by inverting the squared real coherency (SRC; right-hand panels). The top panels show the maps of recovered T_e, while the middle panels show cross-sections—along the dashed lines in the T_e maps—through the model T_e (red line) and the recovered T_e (black line). The bottom panels show slices in the ys- plane—along the same dashed lines—through the 3D observed coherence and SRC. Fan wavelet method with $|\mathbf{k}_0| = 5.336$

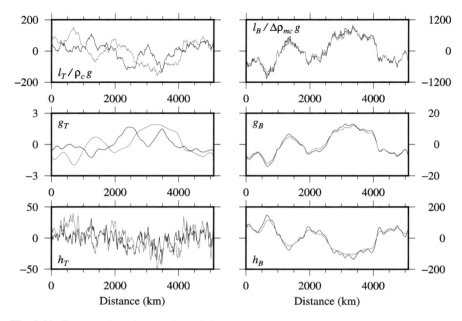

Fig. 9.11 Cross-sections through the relief of the reconstructed initial surface loads and corresponding predicted gravity and topography components in the space domain, using the correct T_e (10 km; black lines) and an incorrect T_e (30 km; red lines). The scale for load relief and topography is in metres, that for the gravity anomaly is in mGal. Multitaper method with NW = 1, $K = 1$

from Eqs. 9.3 and 9.34, the component of the predicted gravity due to surface loading is

$$G_T = \frac{\mu_T}{\Delta} (\kappa_B G - \mu_B H) \ ,$$

while the component of the predicted topography due to internal loading is

$$H_B = \frac{\kappa_B}{\Delta} (-\kappa_T G + \mu_T H) \ .$$

But since the observed gravity and topography have 'real-world' amplitudes and phases, one would expect the predicted gravity and topography components to have similar amplitudes and phases, though weighted by the deconvolution coefficients. And since the deconvolution coefficients depend upon elastic thickness, any change in T_e will change both the amplitudes and phases of the reconstructed initial loads and predicted gravity and topography.

An example is shown in Fig. 9.11. Here, two random, fractal initial loads flexed a plate of $T_e = 10$ km, generating a Bouguer anomaly and topography (see Chap. 10), which were then used to reconstruct the initial loads and the predicted gravity and topography components using a second elastic thickness, T_e', in Eqs. 9.3 and 9.34. When $T_e' = T_e$, the initial loads were recovered perfectly (black lines in Fig. 9.11);

Fig. 9.12 Observed
Bouguer SRC (black circles
with error bars) of a
synthetic plate with $T_e = 50$
km and $f = 1$ ($F = 0.5$),
with the predicted SRC (red
line) and internal load
fraction ($F(k)$, green line)
from load deconvolution.
Also shown is the theoretical
SRC (blue line) and the line
$F = 0.5$ (dashed green line).
Multitaper method with NW
$= 1$, $K = 1$

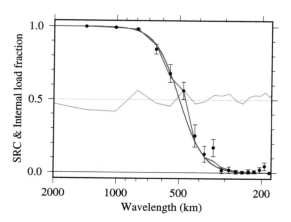

but when any other value was used for T_e' (30 km in Fig. 9.11), the reconstructed
initial loads differed from the known initial loads in both amplitude and phase.

Hence, the variability in the amplitudes and phases of the predicted gravity and
topography cause the predicted SRC to possess a variability not seen in the smooth
curves of the theoretical coherence, as shown in Fig. 9.12. And the same is true for the
load ratio. Indeed, the variability in the predicted SRC almost mimics the variability
in both the observed SRC and the load ratio (at least in the harmonics around the
rollover). So when T_e is adjusted in order to find the best fit between predicted and
observed SRCs, it is actually the spectra of the loads that is being adjusted, which
then affects the predicted gravity and topography, which then affects the predicted
SRC. In such a fashion it is possible to mimic the observed SRC very well, and
merely by adjusting one number (T_e). Contrast this with the uniform-f, curve-fitting
method where the coherence curve possesses no variability and two numbers (T_e and
f) must be adjusted.

9.4.1.6 Inversion for T_e

As noted, elastic thickness is found by adjusting its value such that the predicted
SRC is a best fit to the observed SRC ($\Gamma_{obs,R}^2$), given the errors on the observed SRC
($\sigma_{\Gamma_{obs,R}^2}$). That is, the best-fitting T_e is the one that minimises the *chi-squared statistic*

$$\chi^2(T_e) = \sum_k \left(\frac{\Gamma_{obs,R}^2(k) - \Gamma_{pr,R}^2(k; T_e)}{\sigma_{\Gamma_{obs,R}^2}(k)} \right)^2 , \qquad (9.38)$$

where we write the predicted SRC as a function of T_e. The crude approach would be
to evaluate χ^2 for many T_e values over a range, and select that T_e value where χ^2 is
a minimum, as shown in Fig. 9.13. Such a method would also provide errors on the
T_e estimate, being the intersections of the chi-squared curve with the line given by

Table 9.1 $\Delta\chi^2$ values corresponding to the indicated confidence level (CL) for a single inversion parameter (Press et al. 1992)

CL:	68.27%	90%	95.45%	99%	99.73%	99.99%
$\Delta\chi^2$:	1.00	2.71	4.00	6.63	9.00	15.10

Fig. 9.13 The chi-squared statistic (red curve) for a range of T_e values. The best-fit T_e (128.4 km) occurs at the minimum chi-squared value, and has a 95% confidence interval of $\Delta T_e =$ 120.6–135.8 km ($\Delta\chi^2 = 4$)

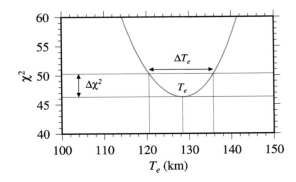

$\chi^2_{\min} + \Delta\chi^2$, where $\Delta\chi^2$ depends upon the desired confidence level and the number of inversion parameters, and takes values as given in Table 9.1. For example, Fig. 9.13 shows that the best-fitting T_e is 128.4 km, with a 95% confidence interval of 120.6–135.8 km (see also Sect. 11.8).

There do exist, fortunately, many methods for the minimisation of functions and inverse problem theory in general (Tarantola 2005), but it is not the intention of this book to explore these. Nevertheless, a fast method that has proved faithful to this author at least, is the root-finding algorithm known as *Brent's method*, which can also be used to find the errors on the parameter (T_e) (Press et al. 1992).

Regarding the errors on the observed SRC in Eq. 9.38, one can use either the analytic formulae (Sect. 5.5.2) or jackknife error estimates (Sect. 5.5.3). However, it is also acceptable to weight the difference between observed and predicted SRC by the inverse of the wavenumber, as in

$$\chi^2(T_e) = \sum_k \left(\frac{\Gamma^2_{obs,R}(k) - \Gamma^2_{pr,R}(k; T_e)}{k} \right)^2 .$$

This method, known as *inverse wavenumber weighting*, downweights the noisier SRC estimates that sometimes occur at high wavenumbers. It also generally results in smoother T_e maps since there is none of the spatial variability that is present in analytic or jackknife errors.

9.4.2 Load Deconvolution with Multitapers

The procedure used to create a map of elastic thickness over a study area using the moving-window multitaper method is given here as a series of steps to follow. For the explanation of symbols, see the identified equation or section number.

1. Choose a plate model; i.e. specify the number of crustal layers, the thicknesses and densities of the layers, the mantle density, and the location of the internal load. These parameters can be fixed beforehand because the Bouguer coherence is so insensitive to them (Fig. 9.7).
2. Choose values of NW and K for the Slepian tapers (Sects. 3.3.4 and 3.3.5), and the size of the moving windows (Sect. 3.5).
3. Extract gravity and topography data in the moving window.
4. Compute the observed, radially-averaged, squared real coherency (SRC) of the data in the window by following the steps in Sects. 5.4.1.1 and 5.4.1.2 (which we include here for completeness). First, take the Bouguer anomaly and topography data, taper them, and take their Fourier transforms (Eq. 5.20),

$$\check{G}_{ij}(\mathbf{k}) = \mathsf{F}\{g(\mathbf{x}) \, w_{ij}(\mathbf{x})\} \ ,$$
$$\check{H}_{ij}(\mathbf{k}) = \mathsf{F}\{h(\mathbf{x}) \, w_{ij}(\mathbf{x})\} \ ,$$

where we now write $g(\mathbf{x})$ for $g_{\xi\eta}$, and $w_{ij}(\mathbf{x})$ for $w_{ij\xi\eta}$. Calculate the (i, j)th cross- and auto-eigenspectra (Eq. 5.21),

$$\hat{S}_{ij}^{(gh)}(\mathbf{k}) = \check{G}_{ij}(\mathbf{k}) \, \check{H}_{ij}^*(\mathbf{k}) \ ,$$
$$\hat{S}_{ij}^{(gg)}(\mathbf{k}) = \check{G}_{ij}(\mathbf{k}) \, \check{G}_{ij}^*(\mathbf{k}) \ ,$$
$$\hat{S}_{ij}^{(hh)}(\mathbf{k}) = \check{H}_{ij}(\mathbf{k}) \, \check{H}_{ij}^*(\mathbf{k}) \ ,$$

and then radially average these (Eq. 5.23):

$$\bar{S}_{ij}^{(gh)}(k) = \frac{1}{N_m} \sum_{l=1}^{N_m} \hat{S}_{ij}^{(gh)}(\mathbf{k}_l) \ ,$$
$$\bar{S}_{ij}^{(gg)}(k) = \frac{1}{N_m} \sum_{l=1}^{N_m} \hat{S}_{ij}^{(gg)}(\mathbf{k}_l) \ ,$$
$$\bar{S}_{ij}^{(hh)}(k) = \frac{1}{N_m} \sum_{l=1}^{N_m} \hat{S}_{ij}^{(hh)}(\mathbf{k}_l) \ ,$$

where we write k for $|\mathbf{k}|$. Next, average these radially-averaged spectra over the tapers (Eq. 5.24):

$$\bar{S}_{gh}(k) \;=\; \frac{1}{K^2} \sum_{i=1}^{K} \sum_{j=1}^{K} \bar{S}_{ij}^{(gh)}(k) \,,$$

$$\bar{S}_{gg}(k) \;=\; \frac{1}{K^2} \sum_{i=1}^{K} \sum_{j=1}^{K} \bar{S}_{ij}^{(gg)}(k) \,,$$

$$\bar{S}_{hh}(k) \;=\; \frac{1}{K^2} \sum_{i=1}^{K} \sum_{j=1}^{K} \bar{S}_{ij}^{(hh)}(k) \,.$$

Finally, and based on Eq. 5.25, calculate the observed, radially-averaged SRC for the window:

$$\Gamma_{obs,R}^{2}(k) \;=\; \frac{\left\{ \mathrm{Re}\!\left[\bar{S}_{gh}(k)\right]\right\}^{2}}{\bar{S}_{gg}(k)\,\bar{S}_{hh}(k)} \,.$$

Note, the errors on the SRC must also be computed, using either analytic formulae or jackknife error estimates (Sect. 5.5).

5. Take the Fourier transforms of the windowed Bouguer anomaly and topography data, giving $G(\mathbf{k})$ and $H(\mathbf{k})$.
6. Assume a starting value of T_e.
7. Calculate the deconvolution coefficients, $\mu_T(k)$, $\mu_B(k)$, $\kappa_T(k)$ and $\kappa_B(k)$ (e.g., Eqs. 9.5–9.8), for the plate model and T_e value.
8. Recreate the two initial surface and internal loads through load deconvolution using Eq. 9.34, which is, in this context,

$$\begin{pmatrix} L_T(\mathbf{k}) \\ L_B(\mathbf{k}) \end{pmatrix} \;=\; \frac{1}{\Delta(k)} \begin{pmatrix} \kappa_B(k) & -\mu_B(k) \\ -\kappa_T(k) & \mu_T(k) \end{pmatrix} \begin{pmatrix} G(\mathbf{k}) \\ H(\mathbf{k}) \end{pmatrix} \,,$$

where the determinant is

$$\Delta(k) \;=\; \mu_T(k)\,\kappa_B(k) \;-\; \mu_B(k)\,\kappa_T(k) \,.$$

Care must be taken not to evaluate the determinant—and therefore neither the loads—at zero wavenumber. As the zero wavenumber represents the 'DC value' (Sect. 2.4.2) which is not important in flexural studies, the two initial loads can be assigned values of zero here.

9. Calculate the surface and internal components of the predicted gravity and topography (Eq. 9.3):

$$G_T(\mathbf{k}) \;=\; \mu_T(k)\,L_T(\mathbf{k}) \,,$$
$$G_B(\mathbf{k}) \;=\; \mu_B(k)\,L_B(\mathbf{k}) \,,$$
$$H_T(\mathbf{k}) \;=\; \kappa_T(k)\,L_T(\mathbf{k}) \,,$$
$$H_B(\mathbf{k}) \;=\; \kappa_B(k)\,L_B(\mathbf{k}) \,.$$

10. Take the inverse Fourier transform of the surface and internal components of the predicted gravity and topography, giving $g_T(\mathbf{x})$, $g_B(\mathbf{x})$, $h_T(\mathbf{x})$ and $h_B(\mathbf{x})$ [and, if a load ratio is desired, inverse Fourier transform $L_T(\mathbf{k})$ and $L_B(\mathbf{k})$ to get $\ell_T(\mathbf{x})$ and $\ell_B(\mathbf{x})$].

11. Using the same NW and K values that were used to calculate the observed SRC, taper the space-domain predicted gravity and topography components, and take their Fourier transforms (Eq. 5.20):

$$
\begin{aligned}
\check{G}_{T,ij}(\mathbf{k}) &= \mathsf{F}\{g_T(\mathbf{x})\,w_{ij}(\mathbf{x})\} \ , \\
\check{G}_{B,ij}(\mathbf{k}) &= \mathsf{F}\{g_B(\mathbf{x})\,w_{ij}(\mathbf{x})\} \ , \\
\check{H}_{T,ij}(\mathbf{k}) &= \mathsf{F}\{h_T(\mathbf{x})\,w_{ij}(\mathbf{x})\} \ , \\
\check{H}_{B,ij}(\mathbf{k}) &= \mathsf{F}\{h_B(\mathbf{x})\,w_{ij}(\mathbf{x})\} \ .
\end{aligned}
$$

If a load ratio is desired, do the same to the two initial load components:

$$
\begin{aligned}
\check{L}_{T,ij}(\mathbf{k}) &= \mathsf{F}\{\ell_T(\mathbf{x})\,w_{ij}(\mathbf{x})\} \ , \\
\check{L}_{B,ij}(\mathbf{k}) &= \mathsf{F}\{\ell_B(\mathbf{x})\,w_{ij}(\mathbf{x})\} \ .
\end{aligned}
$$

12. Next, calculate the (i,j)th cross- and auto-eigenspectra from Eq. 5.21, but do not include $T{-}B$ cross-spectral terms, as in Eq. 9.37:

$$
\begin{aligned}
\hat{S}_{ij}^{(gh)}(\mathbf{k}) &= \check{G}_{T,ij}(\mathbf{k})\,\check{H}_{T,ij}^{*}(\mathbf{k}) + \check{G}_{B,ij}(\mathbf{k})\,\check{H}_{B,ij}^{*}(\mathbf{k}) \ , \\
\hat{S}_{ij}^{(gg)}(\mathbf{k}) &= \check{G}_{T,ij}(\mathbf{k})\,\check{G}_{T,ij}^{*}(\mathbf{k}) + \check{G}_{B,ij}(\mathbf{k})\,\check{G}_{B,ij}^{*}(\mathbf{k}) \ , \\
\hat{S}_{ij}^{(hh)}(\mathbf{k}) &= \check{H}_{T,ij}(\mathbf{k})\,\check{H}_{T,ij}^{*}(\mathbf{k}) + \check{H}_{B,ij}(\mathbf{k})\,\check{H}_{B,ij}^{*}(\mathbf{k}) \ .
\end{aligned}
$$

Do the same to get the load auto-eigenspectra if a load ratio is desired:

$$
\begin{aligned}
\hat{S}_{ij}^{(LT)}(\mathbf{k}) &= \check{L}_{T,ij}(\mathbf{k})\check{L}_{T,ij}^{*}(\mathbf{k}) \ , \\
\hat{S}_{ij}^{(LB)}(\mathbf{k}) &= \check{L}_{B,ij}(\mathbf{k})\check{L}_{B,ij}^{*}(\mathbf{k}) \ .
\end{aligned}
$$

13. Then radially average the cross- and auto-eigenspectra (Eq. 5.23):

$$
\bar{S}_{ij}^{(gh)}(k) = \frac{1}{N_m}\sum_{l=1}^{N_m}\hat{S}_{ij}^{(gh)}(\mathbf{k}_l) \ ,
$$

$$
\bar{S}_{ij}^{(gg)}(k) = \frac{1}{N_m}\sum_{l=1}^{N_m}\hat{S}_{ij}^{(gg)}(\mathbf{k}_l) \ ,
$$

$$
\bar{S}_{ij}^{(hh)}(k) = \frac{1}{N_m}\sum_{l=1}^{N_m}\hat{S}_{ij}^{(hh)}(\mathbf{k}_l) \ .
$$

Radially average the load auto-eigenspectra if a load ratio is desired:

$$\bar{S}_{ij}^{(LT)}(k) = \frac{1}{N_m} \sum_{l=1}^{N_m} \hat{S}_{ij}^{(LT)}(\mathbf{k}_l) \, ,$$

$$\bar{S}_{ij}^{(LB)}(k) = \frac{1}{N_m} \sum_{l=1}^{N_m} \hat{S}_{ij}^{(LB)}(\mathbf{k}_l) \, .$$

14. Next, average these radially-averaged spectra over the tapers (Eq. 5.24):

$$\bar{S}_{gh}(k) = \frac{1}{K^2} \sum_{i=1}^{K} \sum_{j=1}^{K} \bar{S}_{ij}^{(gh)}(k) \, ,$$

$$\bar{S}_{gg}(k) = \frac{1}{K^2} \sum_{i=1}^{K} \sum_{j=1}^{K} \bar{S}_{ij}^{(gg)}(k) \, ,$$

$$\bar{S}_{hh}(k) = \frac{1}{K^2} \sum_{i=1}^{K} \sum_{j=1}^{K} \bar{S}_{ij}^{(hh)}(k) \, .$$

And, if a load ratio is desired, do likewise:

$$\bar{S}_{LT}(k) = \frac{1}{K^2} \sum_{i=1}^{K} \sum_{j=1}^{K} \bar{S}_{ij}^{(LT)}(k) \, ,$$

$$\bar{S}_{LB}(k) = \frac{1}{K^2} \sum_{i=1}^{K} \sum_{j=1}^{K} \bar{S}_{ij}^{(LB)}(k) \, .$$

15. Finally, and based on Eq. 5.25, calculate the predicted, radially-averaged SRC for the window:

$$\Gamma_{pr,R}^2(k) = \frac{\left\{ \mathrm{Re}\left[\bar{S}_{gh}(k) \right] \right\}^2}{\bar{S}_{gg}(k)\, \bar{S}_{hh}(k)} \, .$$

If a load ratio is desired, perform the following operation (Eq. 9.21):

$$f(k) = \sqrt{\frac{\bar{S}_{LB}(k)}{\bar{S}_{LT}(k)}} \, .$$

16. Form the chi-squared statistic (Eq. 9.38) between observed and predicted SRCs, using the errors on the observed SRC as weights (Section 5.5), thus:

$$\chi^2(T_e) = \sum_{n=1}^{N_k} \left(\frac{\Gamma^2_{obs,R}(k_n) - \Gamma^2_{pr,R}(k_n)}{\sigma_{\Gamma^2_{obs,R}}(k_n)} \right)^2 ,$$

where N_k is the number of wavenumber annuli, and k_n is the nth annulus.

17. If χ^2 is a minimum proceed to the next step; otherwise choose another T_e value and go back to step 7.
18. Calculate the errors on the best-fitting T_e using the chi-squared curve or Brent's method.
19. Allocate the best-fitting T_e value and its errors [and $f(k)$] to the coordinates of the window centre, proceed to the next moving window, and go back to step 3.

If one is calculating a single estimate of elastic thickness for the whole study area (i.e. not creating a map of spatial T_e variations), then obviously the 'window' mentioned at step 2 and thereafter is the entire study area, and one terminates the loop at the last step. One then also has the option of inverse Fourier transforming the initial loads and topography components corresponding to the best-fitting T_e to get $\ell_T(\mathbf{x})$, $\ell_B(\mathbf{x})$, $g_T(\mathbf{x})$, $g_B(\mathbf{x})$, $h_T(\mathbf{x})$ and $h_B(\mathbf{x})$ in the space domain, as did Bechtel et al. (1987) and Zuber et al. (1989), as shown in Fig. 9.11.

In steps 5–10 of the above procedure, the surface and internal components of the predicted gravity and topography ($g_T(\mathbf{x})$, $g_B(\mathbf{x})$, $h_T(\mathbf{x})$ and $h_B(\mathbf{x})$) are reconstructed in the same window used to compute the observed SRC. An alternative procedure instead reconstructs these components over the entire study area, but then windows and tapers them only over the area of the current window in order to compute the predicted SRC (steps 11–15). While this alternative would undoubtedly increase computation time, synthetic tests found that it reduced the bias and variance of recovered T_e estimates (Pérez-Gussinyé et al. 2004, 2007).

9.4.3 Load Deconvolution with Wavelets

The procedure used to create a map of elastic thickness over a study area using the fan wavelet method is given here as a series of steps to follow. For the explanation of symbols, see the identified equation or section number.

1. Choose a plate model; i.e. specify the number of crustal layers, the thicknesses and densities of the layers, the mantle density, and the location of the internal load. These parameters can be fixed beforehand because the Bouguer coherence is so insensitive to them (Fig. 9.7).
2. Choose a value of the central wavenumber, $|\mathbf{k}_0|$, for the Morlet wavelet (Sects. 4.3.3 and 4.11.1).
3. Compute the observed, local wavelet squared real coherency (SRC) of the Bouguer anomaly and topography data over the entire study area by following the steps in Sect. 5.4.2.1 (which we include here for completeness). First, compute the Morlet wavelet coefficients at each azimuth of the fan (Eqs. 5.26):

$$\widetilde{G}(s, \mathbf{x}, \alpha) = \mathsf{F}^{-1}\{G(\mathbf{k})\, \Psi_{s,\alpha}^*(\mathbf{k})\} \ ,$$

$$\widetilde{H}(s, \mathbf{x}, \alpha) = \mathsf{F}^{-1}\{H(\mathbf{k})\, \Psi_{s,\alpha}^*(\mathbf{k})\} \ .$$

Alternatively, one could use space-domain convolution if the very long wavelengths are of particular interest[7] (Eq. 5.27). Next, form the Morlet wavelet cross- and auto-scalograms at each azimuth (Eq. 5.28):

$$S_{gh}(k_e, \mathbf{x}, \alpha) = \widetilde{G}(k_e, \mathbf{x}, \alpha)\, \widetilde{H}^*(k_e, \mathbf{x}, \alpha) \ ,$$

$$S_{gg}(k_e, \mathbf{x}, \alpha) = \widetilde{G}(k_e, \mathbf{x}, \alpha)\, \widetilde{G}^*(k_e, \mathbf{x}, \alpha) \ ,$$

$$S_{hh}(k_e, \mathbf{x}, \alpha) = \widetilde{H}(k_e, \mathbf{x}, \alpha)\, \widetilde{H}^*(k_e, \mathbf{x}, \alpha) \ ,$$

where we convert wavelet scale to the equivalent Fourier wavenumber (Eq. 4.41). Then compute the fan wavelet cross- and auto-scalograms by azimuthal averaging (Eq. 5.29):

$$S_{gh}(k_e, \mathbf{x}) = \frac{1}{N_\alpha} \sum_{j=1}^{N_\alpha} S_{gh}(k_e, \mathbf{x}, \alpha_j) \ ,$$

$$S_{gg}(k_e, \mathbf{x}) = \frac{1}{N_\alpha} \sum_{j=1}^{N_\alpha} S_{gg}(k_e, \mathbf{x}, \alpha_j) \ ,$$

$$S_{hh}(k_e, \mathbf{x}) = \frac{1}{N_\alpha} \sum_{j=1}^{N_\alpha} S_{hh}(k_e, \mathbf{x}, \alpha_j) \ .$$

Finally, and based on Eq. 5.30, calculate the observed, local wavelet SRC over the entire study area:

$$\Gamma_{obs,R}^2(k_e, \mathbf{x}) = \frac{\{\mathsf{Re}\left[S_{gh}(k_e, \mathbf{x})\right]\}^2}{S_{gg}(k_e, \mathbf{x})\, S_{hh}(k_e, \mathbf{x})} \ .$$

Note, the errors on the SRC must also be computed, using either analytic formulae or jackknife error estimates (Sect. 5.5).

4. Select grid node $\mathbf{x}_i$, where $i = 1, \cdots, N_x N_y$.
5. Assume a starting value of T_e.
6. Calculate the deconvolution coefficients as a function of equivalent Fourier wavenumber, $\mu_T(k_e)$, $\mu_B(k_e)$, $\kappa_T(k_e)$ and $\kappa_B(k_e)$ (e.g., Eqs. 9.5–9.8), for the plate model and T_e value.

---

[7] Having very high-accuracy wavelet coefficients at very long wavelengths is only important when one needs to invert the free-air admittance in a study area that is potentially subject to mantle convection forces (Kirby and Swain 2014). The long-wavelength coherence, however, has a value of one whether or not such forces are at play, meaning space-domain convolution is not essential.

7. Through load deconvolution, recreate the two initial surface and internal loads from the wavelet coefficients computed in step 4, using Eq. 9.34, which is, in this context,

$$\begin{pmatrix} \widetilde{L_T}(k_e, \mathbf{x}_i, \alpha) \\ \widetilde{L_B}(k_e, \mathbf{x}_i, \alpha) \end{pmatrix} = \frac{1}{\Delta(k_e)} \begin{pmatrix} \kappa_B(k_e) & -\mu_B(k_e) \\ -\kappa_T(k_e) & \mu_T(k_e) \end{pmatrix} \begin{pmatrix} \widetilde{G}(k_e, \mathbf{x}_i, \alpha) \\ \widetilde{H}(k_e, \mathbf{x}_i, \alpha) \end{pmatrix},$$

(9.39)

where the determinant is

$$\Delta(k_e) = \mu_T(k_e)\kappa_B(k_e) - \mu_B(k_e)\kappa_T(k_e).$$

Since the wavelet transform is never computed at zero scale, we will always have $k_e \neq 0$, and the determinant and loads can be evaluated at all equivalent Fourier wavenumbers.

8. Form the Morlet wavelet auto-scalograms of the initial loads at each azimuth (Eq. 5.28):

$$S_{LT}(k_e, \mathbf{x}_i, \alpha) = \widetilde{L_T}(k_e, \mathbf{x}_i, \alpha)\, \widetilde{L_T}^*(k_e, \mathbf{x}_i, \alpha),$$
$$S_{LB}(k_e, \mathbf{x}_i, \alpha) = \widetilde{L_B}(k_e, \mathbf{x}_i, \alpha)\, \widetilde{L_B}^*(k_e, \mathbf{x}_i, \alpha).$$

9. Azimuth-average these to get the fan wavelet auto-scalograms of the initial loads (Eq. 5.29):

$$S_{LT}(k_e, \mathbf{x}_i) = \frac{1}{N_\alpha} \sum_{j=1}^{N_\alpha} S_{LT}(k_e, \mathbf{x}_i, \alpha_j),$$

$$S_{LB}(k_e, \mathbf{x}_i) = \frac{1}{N_\alpha} \sum_{j=1}^{N_\alpha} S_{LB}(k_e, \mathbf{x}_i, \alpha_j).$$

10. Calculate the square of the wavelet load ratio (Eq. 9.21):

$$f^2(k_e, \mathbf{x}_i) = \frac{S_{LB}(k_e, \mathbf{x}_i)}{S_{LT}(k_e, \mathbf{x}_i)}.$$

If one desires the load ratio, take the square root.

11. Finally, based on Eq. 9.25, calculate the predicted, local wavelet SRC at location $\mathbf{x}_i$:

$$\Gamma^2_{pr,R}(k_e, \mathbf{x}_i) = \frac{\left[\mu_T(k_e)\kappa_T(k_e) + \mu_B(k_e)\kappa_B(k_e)f^2(k_e, \mathbf{x}_i)\right]^2}{\left[\mu_T^2(k_e) + \mu_B^2(k_e)f^2(k_e, \mathbf{x}_i)\right]\left[\kappa_T^2(k_e) + \kappa_B^2(k_e)f^2(k_e, \mathbf{x}_i)\right]}.$$

(9.40)

We will look at the origin of this equation at the end of these steps.

12. Form the chi-squared statistic (Eq. 9.38) between observed and predicted SRCs, using the errors on the observed SRC as weights (Sect. 5.5), thus:

$$\chi^2(\mathbf{x}_i; T_e) = \sum_{n=1}^{N_k} \left(\frac{\Gamma_{obs,R}^2(k_{e,n}, \mathbf{x}_i) - \Gamma_{pr,R}^2(k_{e,n}, \mathbf{x}_i)}{\sigma_{\Gamma_{obs,R}^2}(k_{e,n}, \mathbf{x}_i)} \right)^2 ,$$

where N_k is the number of equivalent Fourier wavenumbers (i.e. the number of scales), and $k_{e,n}$ is the nth equivalent Fourier wavenumber.

13. If χ^2 is a minimum proceed to the next step; otherwise choose another T_e value and go back to step 6.

14. Calculate the errors on the best-fitting T_e using the chi-squared curve or Brent's method.

15. Allocate the best-fitting T_e value and its errors [and $f(k_e, \mathbf{x}_i)$] to grid node $\mathbf{x}_i$, choose the next adjacent grid node, $\mathbf{x}_{i+1}$, and go back to step 5.

Equation 9.40 resembles the equation for the theoretical coherency (Eq. 9.25), but actually evaluates the predicted, rather than theoretical, SRC, because it uses the observed gravity and topography to reconstruct the initial loads, and thence load ratio. If we begin with Eq. 9.36, there is the term $\langle \mu_T^2 |L_T|^2 \rangle$, for example, where, in the wavelet method, the averaging is performed over Morlet wavelet azimuth, and the $|L_T|^2$ term is given by $S_{LT}(k_e, \mathbf{x}_i, \alpha)$ (see step 8, above). Since the deconvolution coefficients are independent of azimuth, they can be taken outside the angular brackets, giving

$$\langle \mu_T^2(k_e) S_{LT}(k_e, \mathbf{x}_i, \alpha) \rangle = \mu_T^2(k_e) \langle S_{LT}(k_e, \mathbf{x}_i, \alpha) \rangle = \mu_T^2(k_e) S_{LT}(k_e, \mathbf{x}_i) ,$$

where the last equality is from step 9, above. Thus, Eq. 9.36 can be written as

$$\Gamma_{pr,R}^2(k_e, \mathbf{x}_i) = \frac{\left[\mu_T(k_e)\kappa_T(k_e)S_{LT}(k_e, \mathbf{x}_i) + \mu_B(k_e)\kappa_B(k_e)S_{LB}(k_e, \mathbf{x}_i) \right]^2}{\left\{ \begin{array}{c} \left[\mu_T^2(k_e)S_{LT}(k_e, \mathbf{x}_i) + \mu_B^2(k_e)S_{LB}(k_e, \mathbf{x}_i) \right] \\ \times \left[\kappa_T^2(k_e)S_{LT}(k_e, \mathbf{x}_i) + \kappa_B^2(k_e)S_{LB}(k_e, \mathbf{x}_i) \right] \end{array} \right\}} .$$

We can now divide numerator and denominator by $[S_{LT}(k_e, \mathbf{x}_i)]^2$ to get Eq. 9.40.

If one is calculating a single estimate of elastic thickness for the whole study area (i.e. not creating a map of spatial T_e variations), then one must compute the global SRC at step 3 by spatially averaging the fan wavelet cross- and auto-scalograms, using Eqs. 5.31, to get the global cross- and auto-scalograms $S_{gh}(k_e)$, $S_{gg}(k_e)$ and $S_{hh}(k_e)$. From these, one can then compute the observed, global wavelet SRC,

$$\Gamma_{obs,R}^2(k_e) = \frac{\{\text{Re}\left[S_{gh}(k_e) \right]\}^2}{S_{gg}(k_e)\, S_{hh}(k_e)}$$

(cf. Eq. 5.32). Then, to compute the predicted global SRC, after step 9 one must spatially average the fan wavelet auto-scalograms of the initial loads, using Eq. 5.31, to get $S_{LT}(k_e)$ and $S_{LB}(k_e)$, and then form the global wavelet load ratio, $f^2(k_e)$. The predicted, global wavelet SRC is then given by

$$\Gamma^2_{pr,R}(k_e) = \frac{\left[\mu_T(k_e)\kappa_T(k_e) + \mu_B(k_e)\kappa_B(k_e)f^2(k_e)\right]^2}{\left[\mu_T^2(k_e) + \mu_B^2(k_e)f^2(k_e)\right]\left[\kappa_T^2(k_e) + \kappa_B^2(k_e)f^2(k_e)\right]} .$$

It can be seen that with the wavelet method the cross- and auto scalograms and SRC of the observed data need only be computed once, in contrast to the multitaper method where the cross- and auto-eigenspectra and SRC of the observed data must be evaluated in each window. In addition, when computing the predicted SRC the multitaper method involves further inverse and forward Fourier transforms of the predicted gravity and topography at each taper and in each window, whereas the wavelet method requires no further wavelet transform. Hence, the wavelet method is computationally faster, at least compared to moving-window multitaper techniques.

9.5 Load Versus Gravity Deconvolution

As noted in Sect. 9.4.1, Forsyth (1985) expressed the initial loads in terms of the topography and Moho relief, and computed the predicted coherence from their surface and internal components. The methods presented above, however, follow Banks et al. (2001) in that they express the initial loads in terms of the topography and gravity anomaly. From a theoretical standpoint there should be no difference between results from the two methods, as gravity and Moho relief are simply related by Parker's formula, Eq. 8.24 (ignoring higher-order terms). Nevertheless, Kirby and Swain (2011) did detect differences between T_e estimates from the two approaches, such that use of gravity gave more accurate results (when tested on synthetic flexural models with known elastic thickness). In this section we will reproduce that work, showing the differences between deconvolution using Moho relief and gravity anomaly.

9.5.1 Using Loads

Instead of Eq. 9.4, which expresses the present-day gravity and topography in terms of the initial loads, we can write

$$\begin{pmatrix} W \\ H \end{pmatrix} = \begin{pmatrix} \nu_T & \nu_B \\ \kappa_T & \kappa_B \end{pmatrix} \begin{pmatrix} L_T \\ L_B \end{pmatrix} , \tag{9.41}$$

which expresses the topography and relief of the internal-load interface[8] in terms of the initial loads, where the predicted internal-load interface relief and topography components can be written

---

[8] If the internal load occurs at the Moho, then W is the Moho relief.

$$W_T(\mathbf{k}) = \nu_T(k) L_T(\mathbf{k}) , \tag{9.42a}$$

$$W_B(\mathbf{k}) = \nu_B(k) L_B(\mathbf{k}) , \tag{9.42b}$$

$$H_T(\mathbf{k}) = \kappa_T(k) L_T(\mathbf{k}) , \tag{9.42c}$$

$$H_B(\mathbf{k}) = \kappa_B(k) L_B(\mathbf{k}) \tag{9.42d}$$

(cf. Eq. 9.3). The κs are the flexural deconvolution coefficients (unchanged from Eq. 9.3) while the νs are called the *load deconvolution coefficients*.

As noted in Sect. 9.2, the exact form of the deconvolution coefficients will depend upon the loading scenario. For the general case of an n-layer crust, the two flexural deconvolution coefficients are given by Eqs. 9.5 and 9.6, while the two load deconvolution coefficients are

$$\nu_T(k) = \frac{-1}{\Phi(k)\,g} , \tag{9.43}$$

from Eq. 7.56, and

$$\nu_B(k) = \frac{1}{\Delta\rho_{j+1,B}g} - \frac{1}{\Phi(k)\,g} , \tag{9.44}$$

from Eq. 7.76. As described in Sect. 9.2, there are n crustal layers ($n \geq 1$), and $\Delta\rho_{i+1,i}$ is the density contrast at the base of layer i (with $\rho_{n+1} = \rho_m$, the mantle density). The surface load has density ρ_T, and the internal load with density ρ_B occurs at the base of layer j ($j = 1, \cdots, n$). Finally, Φ is defined by Eq. 7.53, where one sets $\rho_f = 0$ on land, and $\rho_f = \rho_w$, the seawater density, at sea.

Following the procedure in Sect. 9.3.1, substituting Eq. 9.41 into Eq. 9.9 and then taking the squared real component gives the predicted SRC using load deconvolution equations as

$$\Gamma^2_{pr,R,W}(k) = \frac{\left\langle \nu_T \kappa_T |L_T|^2 + \nu_B \kappa_B |L_B|^2 \right\rangle^2}{\left\langle \nu_T^2 |L_T|^2 + \nu_B^2 |L_B|^2 \right\rangle \left\langle \kappa_T^2 |L_T|^2 + \kappa_B^2 |L_B|^2 \right\rangle} \tag{9.45}$$

(cf. Eq. 9.17). Note that if Eq. 9.42 is substituted into Eq. 9.45 we obtain

$$\Gamma^2_{pr,R,W}(k) = \frac{\left\langle W_T H_T^* + W_B H_B^* \right\rangle^2}{\left\langle |W_T|^2 + |W_B|^2 \right\rangle \left\langle |H_T|^2 + |H_B|^2 \right\rangle} ,$$

which is the equation used by Forsyth (1985) to calculate the predicted coherence.

As the initial loads are derived from the observed gravity and topography it is instructive to substitute them back into Eq. 9.45 so that we can fully understand how the coherence (or SRC) is being computed. The initial loads are obtained by inverting Eq. 9.41, giving

$$\begin{pmatrix} L_T \\ L_B \end{pmatrix} = \frac{1}{\Delta_W} \begin{pmatrix} \kappa_B & -\nu_B \\ -\kappa_T & \nu_T \end{pmatrix} \begin{pmatrix} W \\ H \end{pmatrix} , \tag{9.46}$$

where the determinant of the matrix is

$$\Delta_W = \nu_T \kappa_B - \nu_B \kappa_T . \qquad (9.47)$$

Now, Forsyth (1985) used a plate model comprising a single-layer crust so that the internal load was emplaced at the Moho. He also calculated the Moho relief, W, from the Bouguer anomaly, G, using the linear approximation of Parker's formula, Eq. 8.25, or

$$W = \frac{G}{2\pi \mathcal{G} \Delta \rho_{mc} e^{-kz_m}} . \qquad (9.48)$$

Thus, Eq. 9.46 gives the two initial loads as

$$L_T = \Delta_W^{-1} (\kappa_B W - \nu_B H) ,$$
$$L_B = \Delta_W^{-1} (-\kappa_T W + \nu_T H) ,$$

but it will prove more convenient later if we use Eq. 9.48 and write them as

$$L_T = \Delta_W^{-1} \left(2\pi \mathcal{G} \Delta \rho_{mc} e^{-kz_m}\right)^{-1} J_B , \qquad (9.49a)$$
$$L_B = \Delta_W^{-1} \left(2\pi \mathcal{G} \Delta \rho_{mc} e^{-kz_m}\right)^{-1} J_T , \qquad (9.49b)$$

where we define

$$J_B = \kappa_B G - 2\pi \mathcal{G} \Delta \rho_{mc} e^{-kz_m} \nu_B H , \qquad (9.50a)$$
$$J_T = -\kappa_T G + 2\pi \mathcal{G} \Delta \rho_{mc} e^{-kz_m} \nu_T H . \qquad (9.50b)$$

If we now substitute Eq. 9.49 into Eq. 9.45 we obtain, after some algebra, the predicted coherence as computed by deconvolution using Moho relief as

$$\Gamma^2_{pr,R,W}(k) =$$

$$\frac{\left\langle e^{2kz_m} \Delta_W^{-2} \left(\nu_T \kappa_T |J_B|^2 + \nu_B \kappa_B |J_T|^2\right)\right\rangle^2}{\left\langle e^{2kz_m} \Delta_W^{-2} \left(\nu_T^2 |J_B|^2 + \nu_B^2 |J_T|^2\right)\right\rangle \left\langle e^{2kz_m} \Delta_W^{-2} \left(\kappa_T^2 |J_B|^2 + \kappa_B^2 |J_T|^2\right)\right\rangle} . \qquad (9.51)$$

9.5.2 Using Gravity

Although we have already derived formulae for the predicted SRC using the gravity deconvolution coefficients (Eq. 9.36), we need to express it in a similar form to

Eq. 9.51 in order to compare the two approaches. Again, since the initial loads are derived from the observed gravity and topography, we will substitute them back into Eq. 9.36, but first we will express them in a form similar to Eq. 9.49. From Eq. 9.34 we have

$$L_T = \Delta_G^{-1} (\kappa_B G - \mu_B H) \,, \tag{9.52a}$$

$$L_B = \Delta_G^{-1} (-\kappa_T G + \mu_T H) \,, \tag{9.52b}$$

where the determinant of the matrix is

$$\Delta_G = \mu_T \kappa_B - \mu_B \kappa_T \,. \tag{9.53}$$

Staying with the single-layer crust model, the gravity deconvolution equations were given in Eq. 9.27 as

$$\mu_T = \frac{-2\pi \mathcal{G}}{\Phi g} \Delta\rho_{mc} \, e^{-kz_m} \,, \tag{9.54a}$$

$$\mu_B = \frac{-2\pi \mathcal{G}}{\Phi g} (\Delta\rho_{mc} - \Phi) \, e^{-kz_m} \,, \tag{9.54b}$$

while, for a single-layer crust, the load deconvolution equations, Eqs. 9.43 and 9.44, are

$$\nu_T = \frac{-1}{\Phi g} \,, \tag{9.55a}$$

$$\nu_B = \frac{1}{\Delta\rho_{mc} g} - \frac{1}{\Phi g} \,. \tag{9.55b}$$

In comparing Eqs. 9.54 and 9.55 it can thus be seen that

$$\mu_T = 2\pi \mathcal{G} \Delta\rho_{mc} \, e^{-kz_m} \nu_T \,, \tag{9.56a}$$

$$\mu_B = 2\pi \mathcal{G} \Delta\rho_{mc} \, e^{-kz_m} \nu_B \,. \tag{9.56b}$$

Furthermore, using Eq. 9.56, we can therefore also derive the relationship

$$\Delta_G = 2\pi \mathcal{G} \Delta\rho_{mc} \, e^{-kz_m} \Delta_W \tag{9.57}$$

between the matrix determinants, Eqs. 9.47 and 9.53. Thus, from Eqs. 9.52, 9.56 and 9.57 we can derive

$$L_T = (2\pi \mathcal{G} \Delta \rho_{mc})^{-1} e^{kz_m} \Delta_W^{-1} J_B \,, \tag{9.58a}$$

$$L_B = (2\pi \mathcal{G} \Delta \rho_{mc})^{-1} e^{kz_m} \Delta_W^{-1} J_T \,, \tag{9.58b}$$

where the J_T and J_B are given by Eq. 9.50.

If we now substitute Eqs. 9.56 and 9.58 into Eq. 9.36 we obtain, after some algebra, the predicted SRC as computed by deconvolution using the Bouguer anomaly, or

$$\Gamma^2_{pr,R,G}(k) =$$

$$\frac{\left\langle e^{kz_m} \Delta_W^{-2} \left(\nu_T \kappa_T \,|J_B|^2 + \nu_B \kappa_B \,|J_T|^2 \right) \right\rangle^2}{\left\langle \Delta_W^{-2} \left(\nu_T^2 \,|J_B|^2 + \nu_B^2 \,|J_T|^2 \right) \right\rangle \left\langle e^{2kz_m} \Delta_W^{-2} \left(\kappa_T^2 \,|J_B|^2 + \kappa_B^2 \,|J_T|^2 \right) \right\rangle} \,. \tag{9.59}$$

In order to compare Eqs. 9.51 and 9.59, we let

$$X(k) = \Delta_W^{-2} \left(\nu_T \kappa_T \,|J_B|^2 + \nu_B \kappa_B \,|J_T|^2 \right) \,,$$

$$Y(k) = \Delta_W^{-2} \left(\nu_T^2 \,|J_B|^2 + \nu_B^2 \,|J_T|^2 \right) \,,$$

$$Z(k) = \Delta_W^{-2} \left(\kappa_T^2 \,|J_B|^2 + \kappa_B^2 \,|J_T|^2 \right) \,,$$

in which case Eq. 9.51 can be written as

$$\Gamma^2_{pr,R,W}(k) = \frac{\left\langle e^{2kz_m} X \right\rangle^2}{\left\langle e^{2kz_m} Y \right\rangle \left\langle e^{2kz_m} Z \right\rangle} \,, \tag{9.60}$$

while Eq. 9.59 can be written as

$$\Gamma^2_{pr,R,G}(k) = \frac{\left\langle e^{kz_m} X \right\rangle^2}{\left\langle Y \right\rangle \left\langle e^{2kz_m} Z \right\rangle} \,. \tag{9.61}$$

We can now see that use of the load deconvolution equations give a predicted SRC with terms of e^{2kz_m} within each of the averaging brackets (Eq. 9.60). In contrast the gravity deconvolution equations give a predicted SRC with only one e^{kz_m} term and one e^{2kz_m} term (Eq. 9.61). If the annuli were narrow enough to assume but one value of k then all these exponential terms could be taken out of the averaging brackets and would cancel, making Eqs. 9.60 and 9.61 identical. However, this does not occur in most practical implementations, and the exponential terms become prohibitively large, biasing the predicted SRC and hence recovered T_e. It is therefore recommended that the gravity deconvolution equations are used, because the corresponding predicted SRC contains fewer exponential terms.

9.6   Model Noise

9.6.1   Categorising Noise

When a predicted SRC is computed during load deconvolution, it is a trivial procedure to also compute the predicted admittance, from Eqs. 9.18 or 9.20.[9] Figure 9.14 shows the predicted admittance corresponding to the predicted SRC in Fig. 9.12, and it can be seen that there is a very good fit between observed and predicted admittances. Indeed, as Forsyth wrote when describing load deconvolution,

> in solving for the amplitudes of the surface and subsurface loads ... we find solutions which will exactly reproduce the gravity and topographic signals, thus automatically matching the admittance

(Forsyth 1985, p. 12,629). This good fit occurs because the load deconvolution procedure tries to assign every feature and harmonic present in the observed gravity and topography to the predicted gravity and topography. That is, every signal is modelled as being part of the flexural model. As Bechtel et al. (1987, p. 448) put it:

> by considering both surface and subsurface loading the entire gravity signal can be accounted for, leaving no isostatic anomalies.

However, this appropriation of potentially non-flexural signals—such as mantle convection or observational errors—by a flexural model can potentially pose problems when it comes to reconstructing the loads and estimating T_e. Such signals have been discussed in Sect. 5.2.1 and were termed 'noise', in that they are not represented by an isostatic response function. In the context of load deconvolution we can go further, and define 'noise' as being those signals in the observed gravity and topography that cannot (and should not) be modelled by thin elastic plate flexure as implemented by load deconvolution. It is thus 'model noise' rather than instrument noise (although that exists too).

Identification—and ideally removal—of such noise is difficult, of course. We can, however, make use of the fact that load deconvolution predicts that the post-flexure gravity and topography are in phase, and that their predicted coherency and admittance are real-valued (Sect. 9.3.1). Thus, using the technique noted in Sect. 9.4.1.3—of inverting only the real part of the observed coherency (or its square, the SRC)—will avoid including any out-of-phase signals (or *uncorrelated noise*) in the flexural model. As explained in Sect. 5.3.3, the uncorrelated, or out-of-phase harmonics in the observed gravity and topography lurk in the imaginary coherency.

While trivial to implement, this technique will not remove *correlated noise* though, being non-flexural signals that are in phase with the flexural signals. And one can safely assume that the real world contains just as much correlated as uncorrelated noise, though the noise may not always be of a sufficient amplitude or bandwidth as to affect load deconvolution. Unfortunately, such noise is much more difficult to

---

[9] Note, we are not talking about inverting the admittance for T_e here; that will be dealt with in Sect. 11.6.

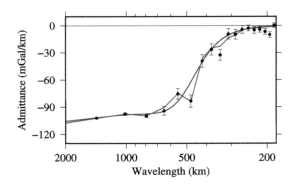

Fig. 9.14 Observed Bouguer admittance (black circles with error bars), with the predicted admittance (red line) generated from the same computation that created Fig. 9.12. Also shown is the theoretical admittance (blue line)

identify, let alone sidestep or remove. One tactic we can employ, though, is to use the information in the imaginary part of the observed coherency. Even though this reflects the amount of uncorrelated noise, if one assumes that non-flexural processes contain, on average, equivalent amounts of correlated and uncorrelated noise, then pinpointing the location (in space and wavenumber) of uncorrelated noise via the imaginary coherency may also hint at the location of the partnering correlated noise.

Noise is detrimental to T_e estimation via load deconvolution because the addition of noise to one of two coherent signals reduces their coherence, as has been discussed in Section 5.3.2. We show this again in Fig. 9.15a—this time using a flexural model rather than an analytic relationship—where the coherence (SRC) between grids of synthetic Bouguer anomaly and topography is greatly reduced by the addition of band-limited noise to the Bouguer anomaly. The noise field was a random, fractal surface, bandpass filtered between 1000 and 2000 km wavelength, with a standard deviation 1.4 times that of the Bouguer anomaly to which it was added. As will be elaborated upon later, it is important to note that the coherence is only reduced at those wavelengths where the noise has similar power to (or greater power than) the Bouguer anomaly (Fig. 9.15b).

9.6.2 Unexpressed Loading

A potential source of noise proposed fairly recently (McKenzie 2003) was a type of loading that produced a gravity anomaly but no signal in the surface topography, referred to as *unexpressed loading*. Since load deconvolution expects all flexural signals to be expressed in the observed gravity and topography to varying degrees, this type of loading—if it exists—has the potential to bias T_e estimates made using classical load deconvolution.

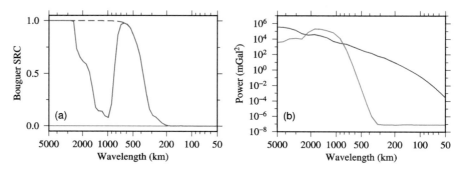

Fig. 9.15 a The dashed blue line shows the observed SRC between synthetic Bouguer anomaly and topography grids, while the red line shows the SRC when band-limited fractal noise was added to the Bouguer anomaly. **b** Power spectra of the Bouguer anomaly grid (magenta line) and the band-limited fractal noise (green line). (See also Fig. 5.6a.) The fan wavelet method in global mode with $|\mathbf{k}_0| = 7.547$ was used to generate all SRCs and power spectra

A mathematical model[10] for unexpressed loading can be devised by setting $H = 0$ in Eq. 9.4, or

$$\begin{pmatrix} G' \\ 0 \end{pmatrix} = \begin{pmatrix} \mu_T & \mu_B \\ \kappa_T & \kappa_B \end{pmatrix} \begin{pmatrix} L'_T \\ L'_B \end{pmatrix} , \tag{9.62}$$

from which we can obtain the condition

$$\kappa_T L'_T + \kappa_B L'_B = 0 ,$$

or

$$L'_T = -\frac{\kappa_B}{\kappa_T} L'_B . \tag{9.63}$$

This tells us that the initial surface and internal loads that create unexpressed loading must be in phase (because the ratio κ_B / κ_T is a real number). Clearly this violates the condition that the initial loads must be statistically uncorrelated (Eq. 9.12), immediately suggesting that unexpressed loading cannot be adequately modelled by load deconvolution.

If we now substitute Eq. 9.63 into Eq. 9.62 we find the post-flexure gravity anomaly to be given by

$$G' = \eta_0 L'_B , \tag{9.64}$$

where η_0 is the 'zero topography transfer function', given by[11]

---

[10] The equations given in McKenzie (2003) contain errors, which were corrected by Kirby and Swain (2009).

[11] Note, Kirby and Swain (2009) expressed η_0 as a transfer function from W'_i, the initial internal load relief, to G'. Eq. 7.59 relates W'_i to L'_B.

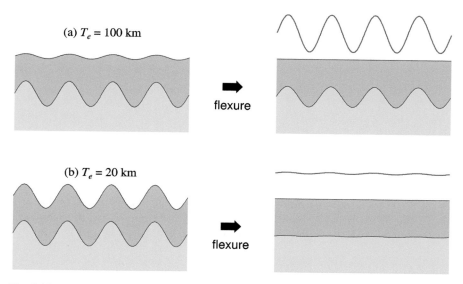

Fig. 9.16 Unexpressed loading for $T_e = $ **a** 100 km, and **b** 20 km. When two initial loads with relief shown at left flex a plate, the resulting upper surface is flat, as shown at right. Here, the initial internal load is a sinusoidal surface; the corresponding initial surface load is calculated from Eq. 9.63, while the final Moho relief is obtained from Eq. 9.67. The red curves show the final Bouguer anomaly, from Eq. 9.64

$$\eta_0(k) = \mu_B - \frac{\mu_T \kappa_B}{\kappa_T} . \tag{9.65}$$

For example, if we substitute the deconvolution coefficients corresponding to a single-layer crust (Eq. 9.27) into Eq. 9.65, the zero topography transfer function becomes

$$\eta_0 = \frac{2\pi \mathcal{G} e^{-k z_m}}{g} \left(1 + \frac{\Delta \rho_{mc} g}{D k^4} \right)^{-1} . \tag{9.66}$$

Alternatively, we can find the post-flexure Moho undulation (assuming the internal load is applied here) by applying the same treatment to Eq. 9.41, giving

$$W' = \left(\nu_B - \frac{\nu_T \kappa_B}{\kappa_T} \right) L'_B , \tag{9.67}$$

where the νs are given by Eqs. 9.43 and 9.44. The principle of unexpressed loading is illustrated in Fig. 9.16, for strong and weak plates.

Unexpressed loading was proposed as an explanation for large gravity anomalies that are not associated with any obvious topography. Such anomalies often occur in ancient cratons where the topography has been heavily eroded, leaving a very subdued—and often almost flat—landscape. Of course the most simple explanation

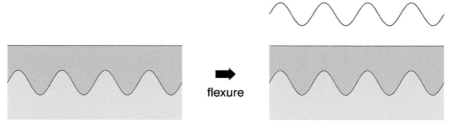

Fig. 9.17 Internal-only initial (expressed) loading (left) of a plate of high T_e can generate large Bouguer anomalies (red curve) and flat topography (right)

for the coexistence of subdued topography and large gravity anomalies is that of a strong plate able to resist the buoyancy force exerted by the internal loads, as shown in Fig. 9.17; if the plate were any weaker then the internal loads would generate surface topography (see, for example, Fig. 9.2). Load deconvolution[12] can adequately explain such a scenario.

The unexpressed loading model was actually devised in an attempt to rationalise the belief of some researchers (McKenzie and Fairhead 1997; Maggi et al. 2000; Jackson 2002) that continental T_e should rarely exceed 25 km, even in cratons. In order to explain the very high T_e estimates (>100 km) returned by load deconvolution in some continental interiors, these researchers proposed that they were a consequence of 'noise' contaminating the Bouguer coherence. The unexpressed loading model was presented as a mechanism for the noise, providing an explanation that allowed for subdued topography, large gravity anomalies and low T_e.

Unfortunately, though, this model does not actually fulfil the purpose for which it was invented, namely that one can have subdued topography and large gravity anomalies yet still have low T_e. This is hinted at in Fig. 9.16b, which shows that when T_e is low the gravity anomalies arising from unexpressed loading will be small, and is confirmed in Fig. 9.18a, which shows the RMS amplitude of such gravity anomalies (in green) compared to those from expressed loading (in red). If global T_e is indeed less than 25 km, then initial, unexpressed, internal loads would need to be at least eight times larger than their expressed cousins in order to produce the same gravity anomaly. This seems unrealistic. The unexpressed loading model thus cannot explain how the large gravity anomalies of continental interiors can be associated with low T_e.

A plot of the zero topography transfer function, η_0, verifies this (Fig. 9.18b). When T_e is large the transfer function has a larger amplitude at most wavelengths, and, for a given initial internal load, the final post-flexure gravity anomalies will be large, from Eq. 9.64. When T_e is low, however, the opposite is true, and the gravity anomalies will be small. This is thus a mathematical justification of the objection raised by Pérez-Gussinyé et al. (2004), that as erosion and sedimentation acted to reduce the topographic amplitude, a weak plate would rebound, reducing the amplitude of the

---

[12] And Occam's razor.

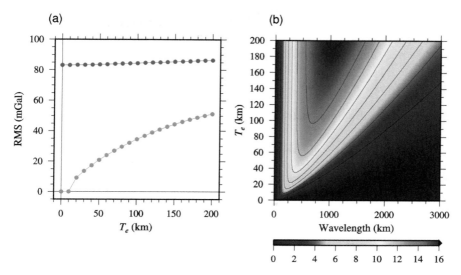

Fig. 9.18 **a** RMS values of post-flexure, synthetic Bouguer anomalies from expressed (red) and unexpressed (green) loading, for a range of T_e values from 0 to 200 km. The same initial load was used in both cases. **b** The amplitude of the zero topography transfer function, η_0 (mGal/km), as a function of wavelength, for a range of T_e values from 0 to 200 km. The plate model here has a single-layer crust, with η_0 taken from Eq. 9.66

internal loads and hence Bouguer anomalies. Conversely, a strong plate would resist rebound and the Bouguer anomalies would be largely preserved. Therefore, it seems that unexpressed loading is merely an interesting—but unrealistic—curiosity.

9.7 Correlated Initial Loads

9.7.1 Correlated-Load Theory

In the derivation of the predicted coherency and admittance (Sect. 9.3.1) we needed to assume that the two initial loads, L_T and L_B, were statistically uncorrelated, expressed via Eq. 9.12, or

$$\langle L_T L_B^* \rangle = \langle L_B L_T^* \rangle = 0 . \tag{9.68}$$

As will be shown, this restriction is made to ensure that the predicted coherence has a rollover that is diagnostic of T_e, and is not 1 at all wavelengths. Then in Sect. 9.6.2 we saw how perfectly-correlated initial loads (i.e. initial loads that are completely in phase) can create a scenario that cannot adequately be modelled by load deconvolution—so-called unexpressed loading. In unexpressed loading the initial load correlation is complete (100%), but one can also explore the effect of smaller degrees of correlation (Kirby and Swain 2009).

This can be done by expressing the Fourier transforms of the loads in polar form, as

$$L_T(\mathbf{k}) = |L_T(\mathbf{k})|\, e^{i\theta_T(\mathbf{k})}\,, \qquad L_B(\mathbf{k}) = |L_B(\mathbf{k})|\, e^{i\theta_B(\mathbf{k})}\,,$$

where the phase difference between the two initial loads is

$$\delta(\mathbf{k}) = \theta_B(\mathbf{k}) - \theta_T(\mathbf{k})\,.$$

Now we follow the same procedure as in Sect. 9.3.1, but instead of assuming that the initial loads are uncorrelated, as in Eq. 9.68, we write

$$L_B L_T^* = |L_T||L_B|\, e^{i\delta}\,, \tag{9.69a}$$

$$L_T L_B^* = |L_T||L_B|\, e^{-i\delta}\,. \tag{9.69b}$$

We can then show that Eq. 9.13 becomes

$$
\begin{aligned}
\langle GH^* \rangle &= \left\langle (\mu_T L_T + \mu_B L_B)(\kappa_T L_T + \kappa_B L_B)^* \right\rangle \\
&= \left\langle \mu_T \kappa_T |L_T|^2 + \mu_B \kappa_B |L_B|^2 + (\mu_T \kappa_B + \mu_B \kappa_T)|L_T||L_B|\cos\delta \right. \\
&\quad \left. + i\,(\mu_B \kappa_T - \mu_T \kappa_B)|L_T||L_B|\sin\delta \right\rangle\,,
\end{aligned}
\tag{9.70}
$$

Eq. 9.14 becomes

$$
\begin{aligned}
\langle GG^* \rangle &= \left\langle (\mu_T L_T + \mu_B L_B)(\mu_T L_T + \mu_B L_B)^* \right\rangle \\
&= \left\langle \mu_T^2 |L_T|^2 + \mu_B^2 |L_B|^2 + 2\mu_T \mu_B |L_T||L_B|\cos\delta \right\rangle\,,
\end{aligned}
\tag{9.71}
$$

and Eq. 9.15 becomes

$$
\begin{aligned}
\langle HH^* \rangle &= \left\langle (\kappa_T L_T + \kappa_B L_B)(\kappa_T L_T + \kappa_B L_B)^* \right\rangle \\
&= \left\langle \kappa_T^2 |L_T|^2 + \kappa_B^2 |L_B|^2 + 2\kappa_T \kappa_B |L_T||L_B|\cos\delta \right\rangle\,.
\end{aligned}
\tag{9.72}
$$

One could now construct correlated-load versions of the predicted coherency and admittance, similar to Eqs. 9.16 and 9.18. However, in order to estimate T_e from these, one would need to somehow determine the initial load phase difference, $\delta(\mathbf{k})$. One approach would be to use the initial loads that are already computed as part of the load deconvolution procedure. Then, from Eqs. 2.7 and 9.69a, we can estimate their phase difference from

$$\delta(\mathbf{k}) = \tan^{-1}\left(\frac{\mathrm{Im}[L_B L_T^*]}{\mathrm{Re}[L_B L_T^*]} \right)\,.$$

To the best of this author's knowledge such an endeavour has not yet been attempted. Perhaps one reason for this is that a predicted coherency computed using Eqs. 9.70–9.72 exactly reproduces the observed coherency for any value of T_e, which is not that useful. Thus, some kind of assumption about the loads—like that given by Eq. 9.68—is essential for T_e estimation.

Nevertheless, our general understanding of the flexural process can be improved by pursuing this investigation into correlated initial loading, and we shall begin by developing theoretical equations such as we did in Sect. 9.3.3. These were derived using assumptions such as that in Eq. 9.23, or

$$\langle \mu_T \kappa_T |L_T|^2 \rangle = \mu_T \kappa_T \langle |L_T|^2 \rangle ,$$

etc. That is, wavenumber-dependent terms not containing data, such as the deconvolution coefficients, can be taken out of the averaging brackets because they are constants at a given wavenumber. Second, we shall assume that, for a given load, its amplitudes and phases are independent, giving

$$\langle |L_T||L_B| \cos \delta \rangle = \langle |L_T| \rangle \langle |L_B| \rangle \langle \cos \delta \rangle .$$

Third, if the phase difference is independent of azimuth, then we can write

$$\langle \cos \delta \rangle = \cos \delta , \qquad \langle \sin \delta \rangle = \sin \delta .$$

Finally, from Eq. 9.21 we will assume that

$$f(k) = \frac{\langle |L_B| \rangle}{\langle |L_T| \rangle} .$$

Applying these assumptions to Eqs. 9.70–9.72, and then substituting the results into Eqs. 9.9 and 9.11, we follow the procedure in Sect. 9.3.3 and obtain the theoretical admittance as

$$Q_{th}(k) =$$
$$\frac{\mu_T \kappa_T + \mu_B \kappa_B f^2 + (\mu_T \kappa_B + \mu_B \kappa_T) f \cos \delta + i (\mu_B \kappa_T - \mu_T \kappa_B) f \sin \delta}{\kappa_T^2 + \kappa_B^2 f^2 + 2 \kappa_T \kappa_B f \cos \delta}$$

$$(9.73)$$

(cf. Eq. 9.24), and the theoretical coherency as

$$\Gamma_{th}(k) =$$
$$\frac{\mu_T \kappa_T + \mu_B \kappa_B f^2 + (\mu_T \kappa_B + \mu_B \kappa_T) f \cos \delta + i (\mu_B \kappa_T - \mu_T \kappa_B) f \sin \delta}{\left(\mu_T^2 + \mu_B^2 f^2 + 2 \mu_T \mu_B f \cos \delta \right)^{1/2} \left(\kappa_T^2 + \kappa_B^2 f^2 + 2 \kappa_T \kappa_B f \cos \delta \right)^{1/2}}$$

$$(9.74)$$

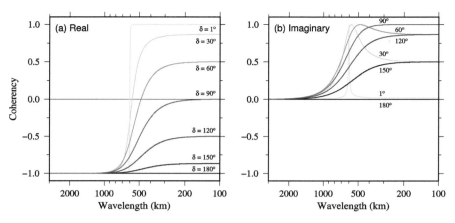

Fig. 9.19 **a** Real and **b** imaginary theoretical Bouguer coherency curves from Eq. 9.75. The red curve corresponds to an initial load phase difference of $\delta(k) = 90°$, $\forall k$; the brown curves correspond to $\delta(k) = 1°$, $30°$, $60°$, $120°$, $150°$, $180°$, $\forall k$, with lighter curves showing smaller values of δ, and darker curves showing higher values. Other parameters are: $T_e = 60$ km, $f = 1$, $z_m = 35$ km, $\rho_c = 2800$ kg m^{-3} and $\rho_m = 3300$ kg m^{-3}

(cf. Eq. 9.25). In the simple example of a single-layer crust used elsewhere in this chapter we can derive an expression for the theoretical Bouguer coherency as

$$\Gamma_{th}(k) = \frac{-\xi - \phi f^2 r^2 + (\phi\xi + 1) fr \cos\delta + i (\phi\xi - 1) fr \sin\delta}{\left(1 + \phi^2 f^2 r^2 - 2\phi fr \cos\delta\right)^{1/2} \left(\xi^2 + f^2 r^2 - 2\xi fr \cos\delta\right)^{1/2}}$$

(9.75)

(cf. Eq. 9.29), where ξ, ϕ and r are given by Eqs. 9.28 and 9.30. The real and imaginary components of Eq. 9.75 are plotted in Fig. 9.19 for several values of initial load phase difference, δ.

There are many interesting facets to these correlated-loads equations. First, note how both admittance and coherency are now complex variables, whereas their sister equations derived assuming uncorrelated initial loads are real-valued. This brings them more in line with the observed admittance and coherency, which are also complex. Second, note how the real coherency can assume positive values (at the shorter-wavelength end of the spectrum, and for $\delta < 90°$, in Fig. 9.19a), whereas classical models predict only negative or zero coherency (e.g. Fig. 9.4). Pérez-Gussinyé et al. (2009) made use of this theory in their study of African T_e, identifying regions of initial load correlation around the East African Rift.

Figure 9.19a also reveals that, for a given T_e, the coherency rollover migrates to longer wavelengths as δ decreases. This means that if initial-load correlation is present in a region, then T_e could be overestimated if the observed SRC is inverted using the classical (uncorrelated-loads) load deconvolution method.

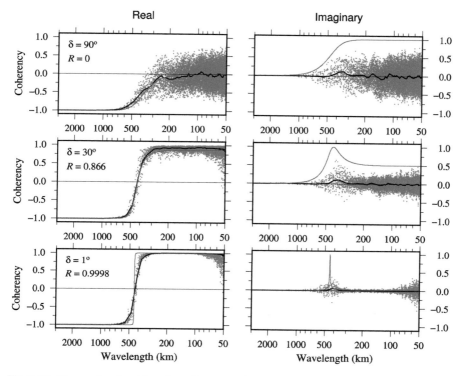

Fig. 9.20 Observed real and imaginary Bouguer coherency between synthetic Bouguer anomaly and topography pairs generated from initial loads with three different correlation coefficients, of R = 0, 0.866 and 0.9998 (T_e = 40 km, f = 1). The grey dots show the actual 2D coherency estimates in the upper two quadrants of the wavenumber domain ($k_y \geq 0$) plotted as a function of wavelength (cf. Fig. 5.4), while the black lines show the radially-averaged (1D) coherency (multitaper method with NW = 3, K = 3). The red lines show the theoretical Bouguer coherency curves for the noted δ values, from Eq. 9.75

9.7.2 Simulation with Synthetic Models

We can also simulate initial load correlation in synthetic models to test this theory. Figure 9.20 shows both 1D (radially-averaged) and 2D Bouguer coherency estimates between three synthetic gravity/topography pairs, each generated from initial loads with different correlation coefficients between them (R = 0, 0.866 and 0.9998) (see Sect. 10.3 and Eq. 10.3). While there is scatter[13] in the raw 2D real coherency estimates, they follow the theoretical curve at wavelengths longer than the rollover wavelength, and the radially-averaged real coherency follows the theoretical curve over

---

[13] The amount of scatter is dependent not only upon the random correlations noted in Sects. 10.2 and 10.4, but also upon the number of tapers (K) used in the estimation: the more tapers the smaller the scatter, as expected since variance is inversely proportional to K (Eq. 3.28).

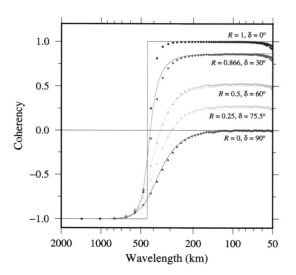

Fig. 9.21 The brown circles show 100-mean, radially-averaged multitaper estimates (NW = 3, $K = 3$) of the observed real Bouguer coherency between synthetic Bouguer anomaly and topography pairs generated from initial loads with a known correlation coefficient (R, indicated). The brown lines show the theoretical real Bouguer coherency curves with different δ values, from Eq. 9.75

the whole spectrum. Note, though, how the 2D estimates of the imaginary component are scattered about zero, and how the 1D radially-averaged imaginary coherency oscillates about zero and does not follow the theoretical curve at wavelengths shorter than the rollover. While this could be viewed as a failure of the model, it can be seen that the theoretical imaginary coherency curve provides an envelope that contains the 2D estimates, squeezing them closer to the abscissa (zero-coherency axis) as the correlation coefficient, R, increases. So this behaviour is actually not that surprising: when the two initial loads are uncorrelated ($R = 0$) there are still unavoidable correlations at random locations/wavenumbers, and the initial load coherency has a large scatter (see Fig. 10.5); but as the correlation is increased (e.g. $R = 0.866$ in Fig. 9.20), these random correlations are 'ironed out' until they are almost completely removed ($R = 0.9998$). This demonstrates again that the imaginary coherency is a measure of the amount of uncorrelated harmonics in two signals (Sect. 5.3.3).

Figure 9.20 also hints at a relationship between the initial-load correlation coefficient (R), the initial-load phase difference (δ), and the short-wavelength final-load, real coherency. We can investigate this using synthetic models, but since any two fractal models are susceptible to the effects of random correlations, clearer results are obtained by averaging the coherencies between 100 model pairs (see Sect. 10.4). Some such averaged real coherencies—for five different initial-load correlation coefficients—are shown in Fig. 9.21. First, we can see that if the initial loads have a correlation coefficient of R (as given in Eq. 10.3), the resulting real observed coherency is faithfully described by a theoretical real coherency curve with an initial-load phase difference given by

$$\delta = \cos^{-1} R,$$

suggesting that

$$R = \cos \delta .$$

This result quasi-validates the correlated-loads theory presented here, since the assignment of R to an initial-load pair is independent of the flexural equations used to generate the post-flexure gravity and topography.

This finding also suggests that the case of $\delta = 90°$ in the above theory describes randomly-correlated (or statistically independent) initial loads ($R = 0$). This is further reinforced by noting that when we set $\delta = 90°$ in Eqs. 9.73 and 9.74, their real parts revert to the classical equations derived under the assumption of uncorrelated initial loads, Eqs. 9.24 and 9.25.

9.7.3 Phase Relationships

We can also see in Fig. 9.21 that the short-wavelength asymptote of the real Bouguer coherency is equal to the initial-load correlation coefficient, or

$$\Gamma_R(k \to \infty) = R .$$

This behaviour is also exhibited in the phase on the Bouguer coherency (and admittance, since they are equal; Eq. 5.13). For a single-layer crust, Eqs. 2.7 and 9.75 give the theoretical Bouguer coherency phase as

$$\varphi_{th}(k) = \tan^{-1}\left[\frac{(\phi\xi - 1)\, fr \sin \delta}{-\xi - \phi f^2 r^2 + (\phi\xi + 1)\, fr \cos \delta}\right], \tag{9.76}$$

plotted in Fig. 9.22. Here we can see that the short-wavelength asymptote of the Bouguer coherency phase is equal to the initial-load phase, or

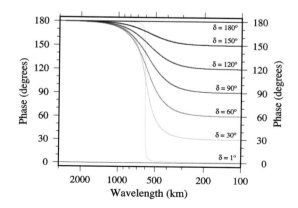

Fig. 9.22 As Fig. 9.19, but for the Bouguer phase from Eq. 9.76

$$\varphi_{th}(k \to \infty) = \delta.$$

These observations tell us that initial loads that have too short a wavelength to flex the plate are preserved during the flexural event, with whatever phase relationship they had before flexure being invariant under flexure. This theory thus represents an improvement upon the classical uncorrelated-loads theory, which, as has been noted several times in this book (Sects. 5.2.2, 5.3.3, 9.3.1 and 9.6.1), predicts a real-valued admittance, implying that the post-flexure gravity and topography will always be in phase at all wavelengths.[14] Instead, Fig. 9.22 shows us that at long wavelengths the Bouguer anomaly and topography will be in anti-phase (180°), while at short wavelengths their phase can take any value.

9.8   Some Theoretical Considerations

The deconvolution coefficients (for example Eqs. 9.5–9.8) are derived using (1) the solution of the biharmonic equation in the wavenumber domain, Eq. 7.44, and (2) Parker's formula, Eq. 8.23. Both of these equations are themselves derived under the assumption of continuous Fourier theory, that is, the representation of a continuous and infinitely-long signal using infinitely-repeating sines and cosines (Chap. 2). Therefore, it is legitimate to ask whether the deconvolution coefficients evaluated as functions of equivalent Fourier wavenumber—and therefore wavelet scale—are compatible with those evaluated using the actual Fourier wavenumber.

At first thought, the answer to this question is no: in general, the wavelet transform cannot replace the Fourier transform when solving differential or integral equations. Certainly, Eqs. 7.44 and Eq. 8.23 cannot be derived using any wavelet transform. However, an empirical study by Kirby (2005) showed that, out of six commonly-used wavelets, only the Morlet wavelet was able to reproduce the Fourier power spectrum correctly. This is because this wavelet uses (Gaussian-modulated) complex exponentials as its basis, similar—in frequency at least—to the basis functions of the Fourier transform. So rewriting Eq. 9.34 as Eq. 9.39—by simply replacing Fourier wavenumber with equivalent Fourier wavenumber, and Fourier transform with wavelet transform—works, as borne out by synthetic testing (Kirby and Swain 2008). And perhaps more importantly, when T_e is estimated by comparing the observed and predicted SRCs, both of these are derived from the same wavelet transforms (of gravity and topography data); in other words, like is being compared with like. Contrast this with the uniform-f method (Sects. 9.4.1.1 and 11.7),

---

[14] Specifically, the classical uncorrelated-loads theory predicts that the Bouguer anomaly and topography will be in anti-phase (180°) at all wavelengths because the real part of the Bouguer admittance is always negative and its imaginary part always zero, and $\tan^{-1}(\text{zero/negative}) = 180°$. With free-air anomalies, the classical theory predicts that these will sometimes have a 0° and sometimes a 180° phase relationship with the topography, because the (real) free-air admittance can take positive or negative values.

where—essentially—a wavelet transform (the observed SRC) is being compared with a Fourier transform (the theoretical SRC).

This point was actually made by Pérez-Gussinyé et al. (2004), but in the context of the multitaper method. They were concerned that small windows would bias T_e estimates because of spectral leakage and a loss of resolution. Consider a spectral estimate at wavenumber k. This estimate actually contains information over a wavenumber range $k \pm W$, where W is the bandwidth of the Slepian tapers; the smaller the bandwidth, the less the leakage and the better resolved the estimate. But bandwidth is proportional to the time-bandwidth product, NW, and inversely proportional to the window size (Eqs. 3.25 and 5.38, Sects. 3.3.4 and 5.6). So using small windows with high-NW tapers will increase spectral leakage (see also Sect. 11.11). Therefore, one can reasonably ask whether or not the predicted SRC is a true representation of the theoretical SRC. Pérez-Gussinyé et al. (2004)'s response was that it didn't matter, as long as both observed and predicted SRC were computed using the same tapers over the same window: any bias caused by spectral leakage would affect both equally, and thus not affect the T_e estimate during minimisation of the chi-squared statistic. But as noted above, if multitapered SRCs are compared with theoretical curves, then the T_e estimate is likely to be biased.

9.9 Summary

In this chapter we introduced the concept of combined surface and internal loading, and developed equations to represent the observed gravity anomaly and topography in terms of the applied, initial loads on a plate. Then, assuming that the two loading processes were independent, we saw how one could derive a predicted coherence between gravity and topography, whose form was governed by both T_e and the initial loads themselves. After exploring the theoretical properties of the Bouguer coherence and admittance under combined loading, we investigated Forsyth's load deconvolution method, whereby T_e is estimated by comparing the observed Bouguer squared real coherency (SRC) with the predicted SRC, the latter being derived from a load reconstruction based on observed gravity and topography data. We then studied the differences between implementation of the method using multitapers and wavelets, and saw how load deconvolution from observed gravity data, rather than derived Moho relief, can provide more robust estimates of the coherence/SRC. Next, a discussion about the role of model noise showed how this is unlikely to be relevant as it does not satisfactorily explain how large gravity anomalies in eroded cratons could be associated with anything other than a very rigid plate. Finally, we looked at how the classical model, which assumes independent initial loads, can be improved by allowing initial load correlation, although we noted that such a model has not yet been used to directly estimate T_e using a load deconvolution-like procedure.

9.10   Further Reading

The load deconvolution method is best explained in journal articles rather than any book, the most obvious being Forsyth (1985). Fortunately, since this paper there have been many descriptions and summaries of the method, though often with diverse notation. Besides Forsyth's paper, good places to start are the papers written not long after the method was first proposed, by Bechtel et al. (1987), Ebinger et al. (1989) and Zuber et al. (1989); the article by Lowry and Smith (1994) is also helpful. The relatively long-lived controversy surrounding model noise and cratonic T_e estimates appears in many articles, but the paper that kicked it off (McKenzie and Fairhead 1997) is perhaps not the best place to start as it sowed much confusion; McKenzie (2003) does a better job. Responses to these two papers were numerous, best summarised in Kirby (2014) and Watts (2021). The topic of correlated initial loads is discussed in Wieczorek (2007, 2015), Kirby and Swain (2009) and Audet (2014), and seems to be a feature of planetary, rather than Earthly, T_e studies (Ding et al. 2019). Finally, other perspectives of load deconvolution appear in the reviews by Simons and Olhede (2013), Audet (2014) and Kirby (2014), and the book by Watts (2001).

References

Audet P (2014) Toward mapping the effective elastic thickness of planetary lithospheres from a spherical wavelet analysis of gravity and topography. Phys Earth Planet Inter 226:48–82

Banks RJ, Swain CJ (1978) The isostatic compensation of East Africa. Proc R Soc Lond A 364:331–352

Banks RJ, Parker RL, Huestis SP (1977) Isostatic compensation on a continental scale: local versus regional mechanisms. Geophys J R Astron Soc 51:431–452

Banks RJ, Francis SC, Hipkin RG (2001) Effects of loads in the upper crust on estimates of the elastic thickness of the lithosphere. Geophys J Int 145:291–299

Bechtel TD, Forsyth DW, Swain CJ (1987) Mechanisms of isostatic compensation in the vicinity of the East African Rift, Kenya. Geophys J R Astron Soc 90:445–465

Cochran JR (1979) An analysis of isostasy in the world's oceans: 2. Midocean ridge crests. J Geophys Res 84(B9):4713–4729

Detrick RS, Watts AB (1979) An analysis of isostasy in the world's oceans: 3. Aseismic ridges. J Geophys Res 84(B7):3637–3653

Ding M, Lin J, Gu C, Huang Q, Zuber MT (2019) Variations in Martian lithospheric strength based on gravity/topography analysis. J Geophys Res Planets 124:3095–3118

Dorman LM, Lewis BTR (1970) Experimental isostasy, 1: Theory of the determination of the Earth's isostatic response to a concentrated load. J Geophys Res 75:3357–3365

Dorman LM, Lewis BTR (1972) Experimental isostasy, 3: Inversion of the isostatic Green function and lateral density changes. J Geophys Res 77:3068–3077

Ebinger CJ, Bechtel TD, Forsyth DW, Bowin CO (1989) Effective elastic plate thickness beneath the East African and Afar plateaus and dynamic compensation of the uplifts. J Geophys Res 94(B3):2883–2901

Forsyth DW (1985) Subsurface loading and estimates of the flexural rigidity of continental lithosphere. J Geophys Res 90(B14):12,623–12,632

Jackson J (2002) Strength of the continental lithosphere: time to abandon the jelly sandwich? GSA Today 12:4–10

Karner GD, Watts AB (1982) On isostasy at Atlantic-type continental margins. J Geophys Res 87(B4):2923–2948

Karner GD, Watts AB (1983) Gravity anomalies and flexure of the lithosphere at mountain ranges. J Geophys Res 88(B12):10,449–10,477

Kirby JF (2005) Which wavelet best reproduces the Fourier power spectrum? Comput Geosci 31:846–864

Kirby JF (2014) Estimation of the effective elastic thickness of the lithosphere using inverse spectral methods: the state of the art. Tectonophys 631:87–116

Kirby JF (2019) On the pitfalls of Airy isostasy and the isostatic gravity anomaly in general. Geophys J Int 216:103–122

Kirby JF, Swain CJ (2006) Mapping the mechanical anisotropy of the lithosphere using a 2D wavelet coherence, and its application to Australia. Phys Earth Planet Inter 158:122–138

Kirby JF, Swain CJ (2008) An accuracy assessment of the fan wavelet coherence method for elastic thickness estimation. Geochem Geophys Geosyst 9:Q03022. https://doi.org/10.1029/2007GC001773, (Correction, Geochem Geophys Geosyst 9:Q05021. https://doi.org/10.1029/2008GC002071, 2008)

Kirby JF, Swain CJ (2009) A reassessment of spectral T_e estimation in continental interiors: the case of North America. J Geophys Res 114(B8):B08401. https://doi.org/10.1029/2009JB006356

Kirby JF, Swain CJ (2011) Improving the spatial resolution of effective elastic thickness estimation with the fan wavelet transform. Comput Geosci 37:1345–1354

Kirby JF, Swain CJ (2014) The long wavelength admittance and effective elastic thickness of the Canadian shield. J Geophys Res Solid Earth 119:5187–5214

Lewis BTR, Dorman LM (1970) Experimental isostasy, 2: an isostatic model for the USA derived from gravity and topography data. J Geophys Res 75:3367–3386

Lowry AR, Smith RB (1994) Flexural rigidity of the Basin and Range–Colorado Plateau–Rocky Mountain transition from coherence analysis of gravity and topography. J Geophys Res 99(B10):20,123–20,140

Maggi A, Jackson JA, McKenzie D, Priestley K (2000) Earthquake focal depths, effective elastic thickness, and the strength of the continental lithosphere. Geology 28:495–498

McKenzie D (2003) Estimating T_e in the presence of internal loads. J Geophys Res 108(B9):2438. https://doi.org/10.1029/2002JB001766

McKenzie DP, Bowin C (1976) The relationship between bathymetry and gravity in the Atlantic Ocean. J Geophys Res 81:1903–1915

McKenzie D, Fairhead JD (1997) Estimates of the effective elastic thickness of the continental lithosphere from Bouguer and free air gravity anomalies. J Geophys Res 102(B12):27,523–27,552

McNutt MK (1983) Influence of plate subduction on isostatic compensation in Northern California. Tectonics 2:399–415

Pérez-Gussinyé M, Lowry AR, Watts AB, Velicogna I (2004) On the recovery of effective elastic thickness using spectral methods: examples from synthetic data and from the Fennoscandian Shield. J Geophys Res 109(B10):B10409. https://doi.org/10.1029/2003JB002788

Pérez-Gussinyé M, Lowry AR, Watts AB (2007) Effective elastic thickness of South America and its implications for intracontinental deformation. Geochem Geophys Geosyst 8:Q05009. https://doi.org/10.1029/2006GC001511

Pérez-Gussinyé M, Metois M, Fernández M, Vergés J, Fullea J, Lowry AR (2009) Effective elastic thickness of Africa and its relationship to other proxies for lithospheric structure and surface tectonics. Earth Planet Sci Lett 287:152–167

Press WH, Teukolsky SA, Vetterling WT, Flannery BP (1992) Numerical Recipes in Fortran 77, 2nd edn. Cambridge University Press, Cambridge

Simons FJ, Olhede SC (2013) Maximum-likelihood estimation of lithospheric flexural rigidity, initial-loading fraction and load correlation, under isotropy. Geophys J Int 193:1300–1342

Swain CJ, Kirby JF (2003) The coherence method using a thin anisotropic elastic plate model. Geophys Res Lett 30:2014. https://doi.org/10.1029/2003GL018350

Tarantola A (2005) Inverse Problem Theory and Methods for Model Parameter Estimation. SIAM, Philadelphia

Watts AB (1978) An analysis of isostasy in the world's oceans: 1. Hawaiian-Emperor seamount chain. J Geophys Res 83(B12):5989–6004

Watts AB (2001) Isostasy and Flexure of the Lithosphere. Cambridge University Press, Cambridge

Watts AB (2021) Isostasy. In: Gupta HK (ed) Encyclopedia of Solid Earth Geophysics, 2nd edn. Springer, Cham, pp 831–847

Wieczorek MA (2007) Gravity and topography of the terrestrial planets. In: Schubert G (ed) Treatise on Geophysics, vol 10. Elsevier, Amsterdam, pp 165–206

Wieczorek MA (2015) Gravity and topography of the terrestrial planets. In: Schubert G (ed) Treatise on Geophysics, vol 10, 2nd edn. Elsevier, Amsterdam, pp 153–193

Zuber MT, Bechtel TD, Forsyth DW (1989) Effective elastic thickness of the lithosphere and the mechanisms of isostatic compensation in Australia. J Geophys Res 94(B7):9353–9367

Chapter 10
Synthetic Testing

10.1 Introduction

When faced with a 'new' method or technique, it is usually a good idea to test its accuracy, or calibrate it. In flexural studies, this was first done by Macario et al. (1995), who tested the load deconvolution method of Forsyth (1985) using Monte Carlo simulations. They first generated synthetic Bouguer anomalies and topography from flexure of a plate of known elastic thickness, then using the load deconvolution method they inverted the Bouguer coherence between the synthetic gravity and topography to see how closely they retrieved the known T_e. Doing this many times and for several T_e values, they found that, on average, elastic thickness was recovered well for weaker plates, but tended to be underestimated when the plates were more rigid.

Many subsequent studies undertook the same accuracy assessment, especially when new methods or plate models were being introduced. Notable articles are provided in Sect. 10.7. Most of these studies followed Macario et al. (1995) by using two random, fractal surfaces as the initial loads. The advantage in using fractal surfaces is that, to a good approximation, both the gravity field and topography of the Earth have a fractal distribution, so the models used are not unrealistic. Furthermore, since there are a potentially infinite number of such surfaces if they are created using a random number generator, each pair of synthetic gravity anomaly and topography data sets will be independent from other pairs, allowing for meaningful statistical analyses.

10.2 Fractal Surfaces

A fractal structure—whether a 1D line or 2D surface, or indeed any other construction—is characterised by a parameter called the *Hurst exponent* (H). The Hurst exponent measures the rate at which the autocorrelation of a signal decreases

J. Kirby, *Spectral Methods for the Estimation of the Effective Elastic Thickness of the Lithosphere*, Advances in Geophysical and Environmental Mechanics and Mathematics, https://doi.org/10.1007/978-3-031-10861-7_10

as the lag between pairs of values increases, and takes values between zero and one. To illustrate the concept of the Hurst exponent, consider a time series signal. If the time series has a Hurst exponent in the range $0 < H < 0.5$, a high value of the signal at a certain time is likely to be immediately followed by a low signal value, which is likely to be followed by a high value, and so on, far into the future; that is, the increments of the signal are negatively correlated. The lower the value of H, the more frequent and extreme this reversal of values. Thus, such signals tend to oscillate about their mean value and are known as *mean-reverting* or *anti-persistent* signals. In contrast, in signals with $0.5 < H < 1$, a high signal value is likely to be immediately followed by another high value, and yet another high value at the next time; future values are likely to remain high, and the increments of the signal are positively correlated. The higher the value of H, the more persistent the value of the signal, and the less likely it is to change considerably. Hence, such signals are known as *persistent* signals.

If a signal has $H = 0.5$, then it is said to exhibit *Brownian motion*.[1] Falling between the negative autocorrelation in signals with $H < 0.5$, and the positive autocorrelation in signals with $H > 0.5$, the values in Brownian motion signals are uncorrelated: it is equally likely that a high value or a low value will follow any given value in the time series, meaning future values are hard to predict. Hence, signals with other values of the Hurst exponent are said to exhibit *fractional Brownian motion* (fBm). All such signals are also said to be *self-affine*, meaning that they look the same regardless of the scale of magnification, or alternatively, have scale-invariant statistical properties.

Another metric for the characterisation of fBm processes is the *fractal dimension* (D_F). Focussing now on 2D surfaces (as opposed to 1D lines or time series), the relationship between the Hurst exponent and fractal dimension is given by

$$D_F = 3 - H \, ,$$

meaning that the fractal dimension of a surface takes values limited by $2 < D_F < 3$. Thus, from the above discussion, persistent surfaces are characterised by $2 < D_F < 2.5$, while anti-persistent surfaces are characterised by $2.5 < D_F < 3$. As shown in Fig. 10.1, persistent surfaces are smoother and less variable, with relatively low power at high wavenumbers. In contrast, anti-persistent surfaces are much rougher, with comparatively high power at high wavenumbers (while still maintaining a red power spectrum). Indeed, self-affine surfaces have a power spectrum that obeys an inverse-power law. That is, for wavenumber $\mathbf{k} = (k_x, k_y)$, the power spectrum has the form

$$S(k_x, k_y) = \left(k_x^2 + k_y^2\right)^{-\beta/2} = |\mathbf{k}|^{-\beta} \, ,$$

---

[1] Or to possess 'Brown noise', named after the botanist Robert Brown, and nothing to do with the colour [although Brown noise is often called *red noise* because the low frequencies (red end of the spectrum) have higher power than the high frequencies (blue end of the spectrum)]. It is also interesting to note that Brownian motion is the integral of uncorrelated white noise.

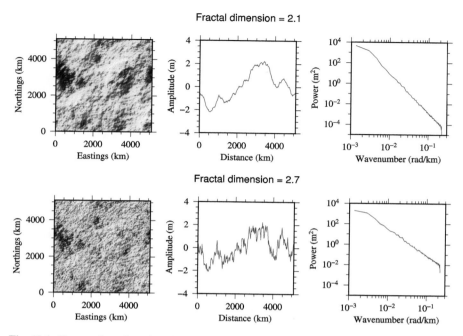

Fig. 10.1 Two random, fractal surfaces generated using the spectral synthesis method of Saupe (1988), with the same initial seeds but different fractal dimensions (left panels). The middle panels show cross-sections through the surfaces at 2540 km northing. The right panels show the radially-averaged, multitapered power spectra of the surfaces (NW = 1, $K = 1$)

where β is called the *spectral exponent*. For 2D surfaces, the relationship between fractal dimension and spectral exponent is

$$\beta = 8 - 2D_F \,, \tag{10.1}$$

while that between the Hurst and spectral exponents is

$$\beta = 2(H+1) \,.$$

Thus, for 2D surfaces[2] we have $2 < \beta < 4$, with anti-persistent surfaces having $2 < \beta < 3$, and persistent surfaces having $3 < \beta < 4$.

There are several methods available for the generation of random, fractal surfaces on a computer, but the one used by most T_e studies is the spectral synthesis method (Saupe 1988). Each surface generated using this method has a uniformly-distributed random phase, and a Gaussian-distributed random amplitude, which follows a power-law curve with a spectral exponent determined by the choice of fractal dimension

---

[2] In n dimensions the relationships are: $0 < H < 1$ always; $D_F = (n+1) - H$, meaning $n < D_F < n+1$; $\beta = (3n+2) - 2D_F$, and $\beta = 2H + n$, meaning $n < \beta < n+2$.

Fig. 10.2 Slices in the
ys-plane through the 3D
coherency between two
random, fractal surfaces of
fractal dimension 2.5; real
and imaginary components.
Fan wavelet method with
$|\mathbf{k}_0| = 5.336$

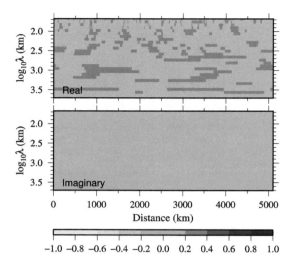

(Eq. 10.1). Owing to the random nature of the process, each surface should therefore be uncorrelated—on average—with other such surfaces. However, while this lack of average correlation is true, Fig. 10.2 shows that there are random pockets of high coherency, both positive and negative and real and imaginary, where the two random surfaces momentarily correlate in space or in wavelength. The effect of such fleeting correlations has an effect on T_e estimation, meaning that the recovered T_e values from such synthetic data will, in general, be different.

Fractal Surface Generation Code

Since the reference Saupe (1988) is not commonly available, the spectral synthesis code is provided here, as a subroutine written in the Fortran90 language. It should therefore be compiled using a Fortran90 or 95 compiler. It calls a subroutine `fft` written by Singleton (1968), though any fast Fourier transform routine can be supplemented. It also calls the functions `ran2` and `gasdev` from Press et al. (1992) (not included).

```
!*****************************************************************
! subroutine specsynth
!
! Fortran77/90 code to compute a periodic, random, fractal
!   surface, of zero mean and unit variance, based on the
!   SpectralSynthesisFM2D algorithm in:
!   Saupe D (1988) Algorithms for random fractals. In:
!     Peitgen H-O, Saupe D (eds) The Science of Fractal Images.
!     Springer, New York, pp 71-136.
!
```

```
! Input:
!  fd = fractal dimension (2 < fd < 3).
!
!  nx = number of grid nodes in x-direction;
!  ny = number of grid nodes in y-direction;
!  (nx should equal ny).
!
!  idump, idumr = initial seeds for random number generator:
!    idumr for amplitude, idump for phase.
!    These can be any, but different, integers.
!
! Output:
!  zr = fractal surface array.
!
! Calls functions ran2 and gasdev, and subroutine fft.
!*************************************************************

subroutine specsynth(zr,nx,ny,fd,idump,idumr)

implicit none

! arrays :
real(kind=8),intent(out):: zr(ny,nx)
real(kind=8):: zi(ny,nx)
complex(kind=8):: z(0:ny,0:nx)

! variables :
integer,intent(in):: nx,ny
integer,intent(inout):: idump,idumr
integer:: nxy,i,j,i0,j0
real(kind=8),intent(in):: fd
real(kind=8):: twopi,beta,rad,phase,zmean,zstd
real(kind=8):: gasdev,ran2

!-------------------------------------------------------------
twopi = 8 * atan(1.d0)

! spectral exponent :
beta = 8 - 2*fd


!-------------------------------------------------------------
! generate surface in wavenumber domain :

z = (0.d0,0.d0)
do i=0,ny/2
```

```
do j=0,nx/2
 phase = twopi * ran2(idump)
 if (i/=0 .or. j/=0) then
   rad = gasdev(idumr) * real(i*i + j*j)**(-beta/4)
 else
   rad = 0.d0
 end if
 z(i,j) = cmplx( rad*cos(phase) , rad*sin(phase) )
 if (i==0) then
   i0 = 0
 else
   i0 = ny - i
 end if
 if (j==0) then
   j0 = 0
 else
   j0 = nx - j
 end if
 z(i0,j0) = cmplx( rad*cos(phase) , -rad*sin(phase) )
 end do
end do

z(ny/2,0) = cmplx( real(z(ny/2,0)) , 0.d0)
z(0,nx/2) = cmplx( real(z(0,nx/2)) , 0.d0)
z(ny/2,nx/2) = cmplx( real(z(ny/2,nx/2)) , 0.d0)

do i=1,ny/2-1
 do j=1,nx/2-1
  phase = twopi * ran2(idump)
  rad = gasdev(idumr) * real(i*i + j*j)**(-beta/4)
  z(i,nx-j) = cmplx( rad*cos(phase) , rad*sin(phase) )
  z(ny-i,j) = cmplx( rad*cos(phase) , -rad*sin(phase) )
 end do
end do

!------------------------------------------------------------
! inverse Fourier transform :

zr = real(z(0:ny-1,0:nx-1))
zi = aimag(z(0:ny-1,0:nx-1))

nxy = nx * ny
call fft(zr,zi,nxy,ny,ny,-1)
call fft(zr,zi,nxy,nx,nxy,-1)
zr = zr / nxy
```

```
    !-----------------------------------------------------------
    ! assign zero mean and unit variance :

    zmean = sum(zr) / nxy
    zstd = sqrt( sum(zr**2)/nxy - zmean**2)

    zr = (zr - zmean) / zstd

    !-----------------------------------------------------------
    return
    end subroutine specsynth
```

10.3 The Initial Loads

The method outlined in Sect. 10.2 is used to generate two periodic, random, fractal surfaces with zero mean and unit variance, which we will call $s_T(\mathbf{x})$ and $s_B(\mathbf{x})$. If these are rescaled by multiplying them by a factor of 100, say, they can be used to represent the amplitudes of the relief of the two initial loads in metres, or

$$h_i(\mathbf{x}) = 100\, s_T(\mathbf{x}), \tag{10.2a}$$
$$w_i(\mathbf{x}) = 100\, s_B(\mathbf{x}). \tag{10.2b}$$

The initial surface load is then computed using a space-domain version of Eq. 7.45, or

$$\ell_T(\mathbf{x}) = \Delta\rho_{Tf}\, g\, h_i(\mathbf{x}),$$

where ρ_T is the density of the load (see Sect. 7.4.1). The initial internal load is computed using a space-domain version of Eq. 7.68, or

$$\ell_B(\mathbf{x}) = \Delta\rho_{j+1,B}\, g\, w_i(\mathbf{x}),$$

where the load (of density ρ_B) is emplaced at the base of crustal layer j (see Sect. 7.5.2). However, in order for the loads to have a load ratio f, where f is a single real number chosen by the user, the internal load needs to be rescaled as

$$\ell_B(\mathbf{x}) \longrightarrow f\, \frac{\Delta\rho_{Tf}}{\Delta\rho_{j+1,B}}\, \ell_B(\mathbf{x}),$$

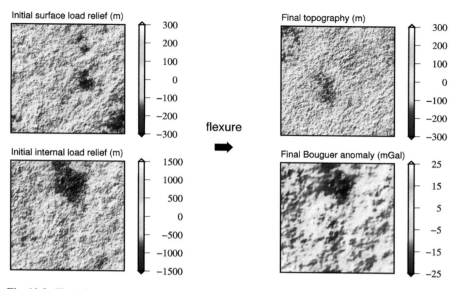

Fig. 10.3 The left-hand panels show two random, fractal surfaces used as relief on the initial loads, with a load ratio of $f = 1$. The right-hand panels show the topography and Bouguer anomaly after flexure of a plate with uniform $T_e = 40$ km. All grids have been assigned a zero mean. Grid dimensions are 5100 km × 5100 km

where $0 < f < \infty$. Thus, since the surfaces s_T and s_B both have unit variance, the variances of the initial loads have the following relationship:

$$\text{var}\{\ell_B(\mathbf{x})\} = f^2 \text{var}\{\ell_T(\mathbf{x})\} .$$

The left hand panels of Fig. 10.3 show an example of the fractal relief of two initial loads. Note that the `SpectralSynthesisFM2D` algorithm given in Sect. 10.2 produces periodic surfaces, with no discontinuities in the value or gradient of the surface between opposite edges (Fig. 10.4). Real-world data, however, are not periodic, and any manipulation of them using Fourier—or space-domain convolution—methods will give rise to the Gibbs phenomenon (Sect. 2.5.4), potentially causing errors in derived results. One must bear this in mind when testing or calibrating a method using synthetic modelling, as periodic surfaces will provide ideal results (see also Sect. 11.11).

Finally, if one wishes to introduce a degree of correlation between the two loads (Sect. 9.7), as did Macario et al. (1995) in their synthetic testing, this can be done in the following way. From the two random, fractal surfaces used as foundation for the initial surface and internal loads, $s_T(\mathbf{x})$ and $s_B(\mathbf{x})$, create a third surface, $s'_B(\mathbf{x})$, from them using

$$s'_B(\mathbf{x}) = R\, s_T(\mathbf{x}) + s_B(\mathbf{x})\sqrt{1 - R^2} , \qquad (10.3)$$

Fig. 10.4 Four identical
synthetic surfaces laid
side-by-side to show the
effect of periodic boundary
conditions. There are no
discontinuities in the value
or gradient of the surface at
its edges

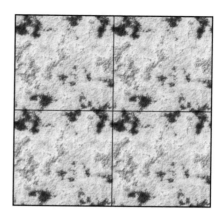

(cf. Eq. 5.15) where R is the chosen correlation coefficient ($-1 \le R \le 1$) (Macario
et al. 1995). The new surface, $s'_B(\mathbf{x})$ can then be used as the foundation for the initial
internal load in Eq. 10.2b.

10.4   Uniform-T_e Plates

To obtain gravity anomalies and topography from flexure of a uniform-T_e plate, one
can use the Fourier-domain solution of the biharmonic equation (Sect. 7.2.8). For
combined loading, the solutions for gravity and topography are given by Eq. 9.4,
reproduced here as

$$G(\mathbf{k}) = \mu_T(k)L_T(\mathbf{k}) + \mu_B(k)L_B(\mathbf{k}) \, ,$$
$$H(\mathbf{k}) = \kappa_T(k)L_T(\mathbf{k}) + \kappa_B(k)L_B(\mathbf{k}) \, ,$$

where $L_T(\mathbf{k})$ and $L_B(\mathbf{k})$ are the Fourier transforms of the loads computed in
Sect. 10.3, and for the general case of an n-layer crust, the deconvolution coefficients are given by Eqs. 9.5–9.8, evaluated using assumed values of T_e, crust and
mantle densities, and crustal layer thicknesses. One then takes the inverse Fourier
transform of $G(\mathbf{k})$ and $H(\mathbf{k})$ to obtain the gravity anomalies, $\Delta g(\mathbf{x})$, and topography,
$h(\mathbf{x})$, in the space domain. Figure 10.3 shows an example of two initial loads and
the corresponding post-flexure topography and Bouguer anomaly, from flexure of a
plate with a uniform T_e of 40 km.

While Fig. 10.2 showed how the random nature of the synthetic, fractal surfaces
can give rise to regions (in space and wavenumber) of high coherency between
them, Fig. 10.5 shows how these random correlations propagate into the coherency
between the final gravity and topography. First, note how the 2D estimates of the
initial-load coherency (the grey dots) can attain relatively large values ($|\Gamma| > 0.5$),

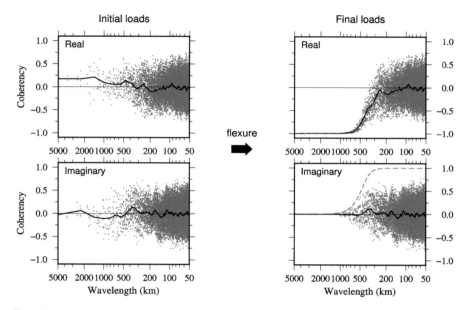

Fig. 10.5 The coherency between the surfaces in Fig. 10.3. At left is the coherency between the two initial loads, at right is that between the final gravity and topography (multitaper method with NW $= 3$, $K = 3$). The black lines show the radially-averaged (1D) coherency; the grey dots show the actual 2D coherency estimates in the upper two quadrants of the wavenumber domain (cf. Fig. 5.4). The dashed red lines show the theoretical Bouguer coherency curves from Eq. 9.75, for $T_e = 40$ km, $f = 1$, and $\delta = 90°$

while retaining an average (the black lines) close to zero—as one would hope, given that the initial loads should be statistically uncorrelated (Sects. 9.3.1 and 9.7). Under the flexural operation, however, the long-wavelength[3] estimates are molded into a form dictated by theory, exhibiting the classic rollover in the real part, while the long-wavelength imaginary parts become exactly zero.[4] Figure 10.5 also shows that the short-wavelength, post-flexure coherency estimates, in contrast, retain their pre-flexure structure because such short-wavelength loads cannot flex the plate and their spatial structure is unaltered. This observation could potentially be useful with real data, as it means that the observed coherency between gravity and topography reveals information about the short-wavelength initial loads without having to recreate them using load deconvolution (Sects. 9.7.2 and 9.7.3).

Synthetic modelling has many uses, though. Macario et al. (1995) used it to check the accuracy of load deconvolution using the Bouguer coherence, while Kirby and Swain (2008) and Pérez-Gussinyé et al. (2009a), for example, have used it to calibrate

---

[3] 'Long wavelength' and 'short wavelength' here mean relative to the coherency rollover, dictated of course by the model T_e value.

[4] See Sect. 9.7 for a discussion on the relationship between the imaginary components of the observed and theoretical coherencies.

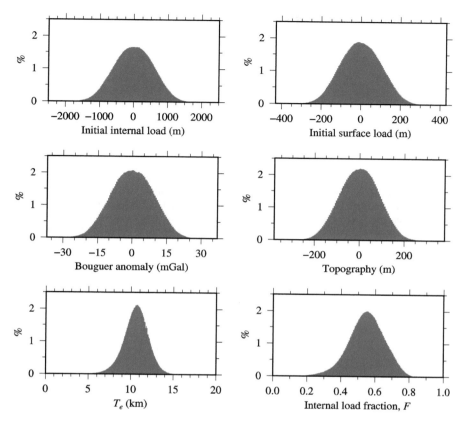

Fig. 10.6 Histograms of the two initial load reliefs, the post-flexure Bouguer anomaly and topography for a plate of $T_e = 10$ km and $f = 1$, and the T_e and F recovered using the fan wavelet method with $|\mathbf{k}_0| = 5.336$. In all cases the grids are 256×256 nodes and there were 100 models, giving 6,553,600 observations in each histogram

the wavelet and multitaper methods, respectively. In such studies, one hundred pairs of synthetic gravity and topography grids are generated at a certain T_e value, and the chosen analysis method then used to recover the model T_e from each pair; an average and standard deviation over the one hundred results then provides a measure of the accuracy of the analysis method at that T_e value. Figure 10.6 shows some results for a plate with $T_e = 10$ km. First, it can be seen that the random, fractal initial loads have a Gaussian distribution (as expected), and that this distribution translates through the flexural equations so that the post-flexure gravity and topography also have a Gaussian distribution. Furthermore, it can be seen that the recovered T_e grids have Gaussian distributions as well. The wavelet transform in local mode was used to generate Fig. 10.6, so each record in the histogram represents T_e at one grid node within a 256×256 grid, where there are one hundred such grids.

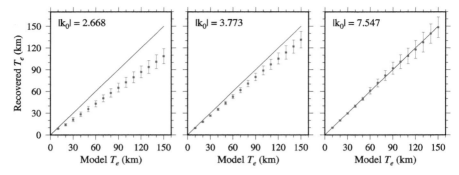

Fig. 10.7 T_e recovery capability of the global fan wavelet method of load deconvolution with the Bouguer SRC, for three $|\mathbf{k}_0|$ values, with gravity and topography generated on 5100×5100 km grids at 20 km grid spacing, from uniform-T_e plate models. Red squares show mean recovered T_e values after averaging one hundred results, with error bars showing one standard deviation

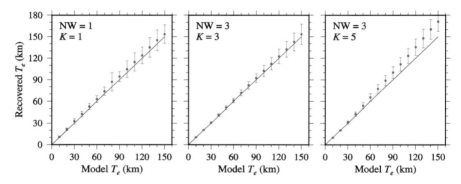

Fig. 10.8 As Fig. 10.7, but for the multitaper method at the indicated NW and K values

Figure 10.7 shows the results of some tests on varying the central wavenumber in the global fan wavelet method, in which a single T_e estimate is recovered for a gravity-topography pair. Here, the analysis is performed at many T_e values. As reported in Kirby and Swain (2011), low values of $|\mathbf{k}_0|$ provide underestimates of the true T_e, when T_e is uniform over a large area. And Fig. 10.8 shows the effect of varying the time-bandwidth product (NW) and the number of tapers (K) in the multitaper method when that is used to obtain a single T_e estimate over the grid. It is seen that, if one wishes to reduce variance in the SRC by using higher-NW tapers, then one should not use the full complement of tapers allowed at that NW value, otherwise one risks overestimating T_e.

The examples given in Figs. 10.7 and 10.8 show just a few objectives of how synthetic modelling can help assess the accuracy of a method—in this case the variation of $|\mathbf{k}_0|$ in the wavelet method and of NW and K in the multitaper method. One can, of course, check many things, such as the effect that the size of the area, the grid spacing of the data, or the wrong assumption of a parameter (e.g. internal load depth) has upon the recovered T_e. But while synthetic modelling with uniform-T_e plates can provide an improved understanding of a T_e-estimation method, it does suffer from being too 'pure'. Elastic thickness is highly unlikely to be constant over very large areas, and, as noted in Sect. 10.3, the loads and derived gravity and topography grids are periodic, unlike real data. It is useful, then, to try and mimic the real world by using aperiodic data and spatially-variable elastic thicknesses in the models. The former can be achieved by, for example, generating periodic gravity and topography over a large grid (512×512 grid nodes, say) and then performing the T_e-recovery method on an extracted—and thus aperiodic—subset (256×256 grid nodes, say), as long as the model T_e is uniform. However, the latter—using spatially-variable T_e models—cannot be undertaken using the outlined Fourier-based method, and one must use the methods given in Sect. 10.5.

10.5 Variable-T_e Plates

If one wishes to generate a synthetic model where T_e is spatially variable, one cannot solve Eq. 7.35 by taking its Fourier transform, as we did in Sects. 7.2.8 and 10.4. This is because both the flexural rigidity and the deflections are functions of the spatial vector, and the Fourier relation in Eq. 2.49 does not apply to products of spatially-variable functions or their derivatives. Instead one must solve this PDE for the deflections using a numerical method; that is, one that does not generate an analytic solution such as Eq. 7.38, but instead provides a scenario-dependent numerical solution that must be recalculated every time a parameter of the system changes. Many numerical methods exist for the solution of such differential equations, but the one we will consider here is the method of *finite differences*, which approximates derivatives of a function by the linear gradient of the function over fixed intervals. For the reader unfamiliar with finite differences, a brief introduction is provided in the Appendix to this chapter. That Appendix also provides the difference equations for each partial derivative, for instance the difference equations representing $\partial v/\partial x$, or $\partial^3 v/\partial x^2 \partial y$, etc.

The PDE we are seeking to convert into a finite difference equation (FDE) is Eq. 7.35, the thin, elastic plate PDE, or more specifically, Eqs. 7.34 and 7.42, which are

$$
\begin{aligned}
D\frac{\partial^4 v}{\partial x^4} &+ D\frac{\partial^4 v}{\partial y^4} + 2D\frac{\partial^4 v}{\partial x^2 \partial y^2} + 2\frac{\partial D}{\partial x}\frac{\partial^3 v}{\partial x^3} \\
&+ 2\frac{\partial D}{\partial y}\frac{\partial^3 v}{\partial y^3} + 2\frac{\partial D}{\partial x}\frac{\partial^3 v}{\partial x \partial y^2} + 2\frac{\partial D}{\partial y}\frac{\partial^3 v}{\partial x^2 \partial y} \\
+ \left(\frac{\partial^2 D}{\partial x^2} + v\frac{\partial^2 D}{\partial y^2}\right)\frac{\partial^2 v}{\partial x^2} &+ \left(\frac{\partial^2 D}{\partial y^2} + v\frac{\partial^2 D}{\partial x^2}\right)\frac{\partial^2 v}{\partial y^2} \\
+ 2(1-v)\frac{\partial^2 D}{\partial x \partial y}\frac{\partial^2 v}{\partial x \partial y} &+ \Delta\rho_{mf} g\, v = -\ell \quad (10.4)
\end{aligned}
$$

where $v(\mathbf{x})$ is the deflection of the plate, $D(\mathbf{x})$ is its flexural rigidity, ℓ is the applied initial load, and $\Delta\rho_{mf}gv$ is the buoyancy force generated by displacement of the mantle. As noted in the Appendix to this chapter, we must assume that both $v(\mathbf{x})$ and $D(\mathbf{x})$ are given on complete and regular grids. Furthermore, when converting a PDE into a FDE it is commonplace to assume—where possible—equal grid intervals in each direction, or $\Delta x = \Delta y = \Delta$.

The goal of the finite difference method is to represent Eq. 10.4 by the matrix equation

$$\mathbf{D}\mathbf{v} = \mathbf{l}, \tag{10.5}$$

where $\mathbf{D}$ is a matrix of the derivatives of the known flexural rigidity, $\mathbf{v}$ is a matrix of the unknown deflections, and $\mathbf{l}$ is a matrix of the known loads. The deflection matrix is then found by matrix inversion, or

$$\mathbf{v} = \mathbf{D}^{-1}\mathbf{l}. \tag{10.6}$$

The $\mathbf{D}$ matrix is populated with finite difference representations of each derivative of D that appears in Eq. 10.4 using the difference equations given in the Appendix to this chapter. Thus, the first term in Eq. 10.4 is written as

$$D\frac{\partial^4 v}{\partial x^4} = D_{i,j}\left(v_{i+2,j} - 4v_{i+1,j} + 6v_{i,j} - 4v_{i-1,j} + v_{i-2,j}\right),$$

for a point on the grid with registration in the x direction of i, and registration in the y direction of j. The second term is written as

$$D\frac{\partial^4 v}{\partial y^4} = D_{i,j}\left(v_{i,j+2} - 4v_{i,j+1} + 6v_{i,j} - 4v_{i,j-1} + v_{i,j-2}\right),$$

and the third term as

$$2D\frac{\partial^4 v}{\partial x^2 \partial y^2} = 2D_{i,j}\left(4v_{i,j} - 2v_{i+1,j} - 2v_{i-1,j} - 2v_{i,j+1} - 2v_{i,j-1} + v_{i+1,j+1}\right.$$
$$\left. + v_{i-1,j+1} + v_{i+1,j-1} + v_{i-1,j-1}\right).$$

These are all terms for the fourth-order partial derivatives, which, because they are of even order, can be written in a form that is symmetric to the grid node (i, j). In contrast, the odd-order derivatives are referenced to a point that is not on the grid [for example the point $(i + \frac{1}{2}, j + \frac{1}{2})$]. Fortunately, in the thin, elastic plate PDE all the odd-order derivatives (first- and third-order) multiply one another, so when we transcribe these derivatives we preserve their anti-symmetry but split them into two parts, with the leading parts multiplying each other, and the trailing parts multiplying one another. Thus the fourth term in Eq. 10.4 is written as

$$2\frac{\partial D}{\partial x}\frac{\partial^3 v}{\partial x^3} = \left(D_{i+1,j} - D_{i,j}\right)\left(v_{i+2,j} - 3v_{i+1,j} + 3v_{i,j} - v_{i-1,j}\right)$$
$$+ \left(D_{i,j} - D_{i-1,j}\right)\left(v_{i+1,j} - 3v_{i,j} + 3v_{i-1,j} - v_{i-2,j}\right) .$$

The other odd-order derivatives are transcribed as

$$2\frac{\partial D}{\partial y}\frac{\partial^3 v}{\partial y^3} = \left(D_{i,j+1} - D_{i,j}\right)\left(v_{i,j+2} - 3v_{i,j+1} + 3v_{i,j} - v_{i,j-1}\right)$$
$$+ \left(D_{i,j} - D_{i,j-1}\right)\left(v_{i,j+1} - 3v_{i,j} + 3v_{i,j-1} - v_{i,j-2}\right) ,$$

$$2\frac{\partial D}{\partial x}\frac{\partial^3 v}{\partial x \partial y^2} = \left(D_{i+1,j} - D_{i,j}\right)\left(2v_{i,j} - 2v_{i+1,j} - v_{i,j+1} - v_{i,j-1}\right.$$
$$\left. + v_{i+1,j-1} + v_{i+1,j+1}\right)$$
$$+ \left(D_{i,j} - D_{i-1,j}\right)\left(-2v_{i,j} + 2v_{i-1,j} + v_{i,j+1} + v_{i,j-1}\right.$$
$$\left. - v_{i-1,j-1} - v_{i-1,j+1}\right) ,$$

and

$$2\frac{\partial D}{\partial y}\frac{\partial^3 v}{\partial x^2 \partial y} = \left(D_{i,j+1} - D_{i,j}\right)\left(2v_{i,j} - 2v_{i,j+1} - v_{i+1,j} - v_{i-1,j}\right.$$
$$\left. + v_{i-1,j+1} + v_{i+1,j+1}\right)$$
$$+ \left(D_{i,j} - D_{i,j-1}\right)\left(-2v_{i,j} + 2v_{i,j-1} + v_{i+1,j} + v_{i-1,j}\right.$$
$$\left. - v_{i-1,j-1} - v_{i+1,j-1}\right) .$$

The remaining terms are all of second order, which are transcribed as

$$\frac{\partial^2 D}{\partial x^2}\frac{\partial^2 v}{\partial x^2} = \left(D_{i+1,j} - 2D_{i,j} + D_{i-1,j}\right)\left(v_{i+1,j} - 2v_{i,j} + v_{i-1,j}\right) ,$$

$$v\frac{\partial^2 D}{\partial y^2}\frac{\partial^2 v}{\partial x^2} = v\left(D_{i,j+1} - 2D_{i,j} + D_{i,j-1}\right)\left(v_{i+1,j} - 2v_{i,j} + v_{i-1,j}\right) ,$$

$$\frac{\partial^2 D}{\partial y^2}\frac{\partial^2 v}{\partial y^2} = \left(D_{i,j+1} - 2D_{i,j} + D_{i,j-1}\right)\left(v_{i,j+1} - 2v_{i,j} + v_{i,j-1}\right) ,$$

$$v\frac{\partial^2 D}{\partial x^2}\frac{\partial^2 v}{\partial y^2} = v\left(D_{i+1,j} - 2D_{i,j} + D_{i-1,j}\right)\left(v_{i,j+1} - 2v_{i,j} + v_{i,j-1}\right) ,$$

and

$$2(1 - v)\frac{\partial^2 D}{\partial x\, \partial y}\frac{\partial^2 v}{\partial x\, \partial y} = \frac{1 - v}{8}\left(D_{i+1,j+1} - D_{i-1,j+1} - D_{i+1,j-1}\right.$$
$$+ D_{i-1,j-1}\right)\left(v_{i+1,j+1} - v_{i-1,j+1} - v_{i+1,j-1}\right.$$
$$\left.+ v_{i-1,j-1}\right).$$

Note that we also have the buoyancy force term, $\Delta\rho_{mf}\,g\,v_{i,j}$, on the left-hand side of Eq. 10.4, as it is also a function of the deflection. Note also that the right-hand sides of all of these finite difference equations must be divided by Δ^4.

If we now replace each term in Eq. 10.4 with its finite difference equivalent shown above, we can write Eq. 10.4 as the sum

$$a_1 v_{i,j-2} + a_2 v_{i-1,j-1} + a_3 v_{i,j-1} + a_4 v_{i+1,j-1} + a_5 v_{i-2,j}$$
$$+ a_6 v_{i-1,j} + a_7 v_{i,j} + a_8 v_{i+1,j} + a_9 v_{i+2,j}$$
$$+ a_{10} v_{i-1,j+1} + a_{11} v_{i,j+1} + a_{12} v_{i+1,j+1} + a_{13} v_{i,j+2} = -\ell_{i,j}\Delta^4,\quad (10.7)$$

for the point (i, j), where the coefficients of the vs can be shown to be, after some algebra,

$$a_7 = 8(1 + v)D_{i,j} + (3 - 2v)\left(D_{i+1,j} + D_{i-1,j} + D_{i,j+1} + D_{i,j-1}\right)$$
$$+ \Delta\rho_{mf}\,g\,\Delta^4$$
$$a_3 = -2(1 + v)D_{i,j} - 4D_{i,j-1} - (1 - v)\left(D_{i+1,j} + D_{i-1,j}\right)$$
$$a_6 = -2(1 + v)D_{i,j} - 4D_{i-1,j} - (1 - v)\left(D_{i,j+1} + D_{i,j-1}\right)$$
$$a_8 = -2(1 + v)D_{i,j} - 4D_{i+1,j} - (1 - v)\left(D_{i,j+1} + D_{i,j-1}\right)$$
$$a_{11} = -2(1 + v)D_{i,j} - 4D_{i,j+1} - (1 - v)\left(D_{i+1,j} + D_{i-1,j}\right)$$
$$a_2 = D_{i-1,j} + D_{i,j-1} + \mathcal{D}$$
$$a_4 = D_{i+1,j} + D_{i,j-1} - \mathcal{D}$$
$$a_{10} = D_{i-1,j} + D_{i,j+1} - \mathcal{D}$$
$$a_{12} = D_{i+1,j} + D_{i,j+1} + \mathcal{D}$$
$$a_1 = D_{i,j-1}$$
$$a_5 = D_{i-1,j}$$
$$a_9 = D_{i+1,j}$$
$$a_{13} = D_{i,j+1}\qquad\qquad (10.8)$$

for the point (i, j), where

$$\mathcal{D} = \frac{1}{8}(1 - v)\left(D_{i+1,j+1} - D_{i-1,j+1} - D_{i+1,j-1} + D_{i-1,j-1}\right).$$

Fig. 10.9 Schematic
illustration of Eq. 10.7 (and
therefore also Eq. 10.9),
showing the deflections
experienced by the
surrounding grid nodes when
a load is applied at grid node
(i, j). The values of the
deflection coefficients (the
as) are given by Eq. 10.8

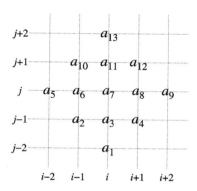

Figure 10.9 shows how the deflection coefficients relate to the grid nodes. Note that
Eq. 10.7 can also be written as the matrix equation

$$\frac{1}{\Delta^4} \begin{pmatrix} a_1 & a_2 & a_3 & a_4 & a_5 & a_6 & a_7 & a_8 & a_9 & a_{10} & a_{11} & a_{12} & a_{13} \end{pmatrix} \begin{pmatrix} v_{i,j-2} \\ v_{i-1,j-1} \\ v_{i,j-1} \\ v_{i+1,j-1} \\ v_{i-2,j} \\ v_{i-1,j} \\ v_{i,j} \\ v_{i+1,j} \\ v_{i+2,j} \\ v_{i-1,j+1} \\ v_{i,j+1} \\ v_{i+1,j+1} \\ v_{i,j+2} \end{pmatrix} = -\ell_{i,j} . \quad (10.9)$$

Now, if the space-domain grid has N_x nodes in the x direction and N_y nodes in
the y direction, and thus a total of $M = N_x N_y$ nodes, then there will be a total of M
versions of Eq. 10.9. These versions can be collected together and represented by
the matrix equation Eq. 10.5, being

$$\mathbf{D} \mathbf{v} = \mathbf{l} .$$

The load vector, $\mathbf{l}$, is

$$\mathbf{l} = \begin{pmatrix} -\ell_{1,1} \\ -\ell_{2,1} \\ \vdots \\ -\ell_{N_x,1} \\ -\ell_{1,2} \\ \vdots \\ -\ell_{N_x,2} \\ -\ell_{1,3} \\ \vdots \\ -\ell_{N_x,N_y} \end{pmatrix}.$$

Note the ordering. While the load itself is represented by a 2D grid, $\ell(x, y)$, the load vector is a 1D column vector (or a $M \times 1$ matrix) with its first element in the bottom-left (south-west) corner of the 2D grid, and with its subsequent elements proceeding from left to right along the rows of the grid, stepping 'up' a row, and eventually terminating in the top-right (north-east) corner. We also require the deflection vector, $\mathbf{v}$, to have the same ordering and the same size, with

$$\mathbf{v} = \begin{pmatrix} v_{1,1} \\ v_{2,1} \\ \vdots \\ v_{N_x,1} \\ v_{1,2} \\ \vdots \\ v_{N_x,2} \\ v_{1,3} \\ \vdots \\ v_{N_x,N_y} \end{pmatrix}.$$

The rigidity matrix, $\mathbf{D}$, is an accumulation of the deflection coefficients. Since both $\mathbf{v}$ and $\mathbf{l}$ are of size $M \times 1$, $\mathbf{D}$ must be a $M \times M$ matrix. A typical row of $\mathbf{D}$ looks like this:

$$(\cdots 0\ 0\ a_1\ 0\ 0\ \cdots\ 0\ a_2\ a_3\ a_4\ 0\ \cdots\ a_5\ a_6\ a_7\ a_8\ a_9\ \cdots\ 0\ a_{10}\ a_{11}\ a_{12}\ 0\ \cdots\ 0\ 0\ a_{13}\ 0\ 0\ \cdots)$$

where the as are evaluated at the (i, j) values corresponding to the chosen element $\ell_{i,j}$ in the load vector. The dots represent zeros, and since each row in $\mathbf{D}$ has M elements there will be $M - 13$ zeros in each row. Note the correspondence between the structure of this row and the structure of the deflection coefficients in Fig. 10.9. Furthermore, there are M rows in the rigidity matrix, and each row resembles the one above it except the elements are moved one place to the right. This gives **D** a *block-diagonal* structure, in general.

However, proper construction of the rigidity matrix—and then solution of Eq. 10.5—requires *boundary conditions*. These dictate the behaviour of the plate at its edges, and allow us to determine expressions for those terms of a finite difference equation that lie 'outside' the plate. For instance, in civil engineering the 'fixed edge' boundary condition is sometimes stipulated, whereby the deflection and slope of the plate are both zero at one or more of its edges. When generating synthetic flexural models, though, it is convenient to assume *periodic* boundary conditions, whereby the behaviour of the plate at one edge is identical to that at its opposite edge.[5] In terms of finite difference equations this would mean that, for $i = 1, \ldots, N_x$, and $j = 1, \ldots, N_y$, we would have

$$
\begin{aligned}
v_{0,j} &= v_{N_x,j} \\
v_{-1,j} &= v_{N_x-1,j} \\
v_{N_x+1,j} &= v_{1,j} \\
v_{N_x+2,j} &= v_{2,j} \; ,
\end{aligned}
$$

for all values of j, and

$$
\begin{aligned}
v_{i,0} &= v_{i,N_y} \\
v_{i,-1} &= v_{i,N_y-1} \\
v_{i,N_y+1} &= v_{i,1} \\
v_{i,N_y+2} &= v_{i,2} \; ,
\end{aligned}
$$

for all values of i (and similarly for $D_{i,j}$). For example, if $N_x = N_y = 5$, the $(i, j) = (1, 1)$ version of Eq. 10.9 with periodic boundary conditions is

$$
\frac{1}{\Delta^4} \begin{pmatrix} a_1 \ a_2 \ a_3 \ a_4 \ a_5 \ a_6 \ a_7 \ a_8 \ a_9 \ a_{10} \ a_{11} \ a_{12} \ a_{13} \end{pmatrix}
\begin{pmatrix}
v_{1,4} \\
v_{5,5} \\
v_{1,5} \\
v_{2,5} \\
v_{4,1} \\
v_{5,1} \\
v_{1,1} \\
v_{2,1} \\
v_{3,1} \\
v_{5,2} \\
v_{1,2} \\
v_{2,2} \\
v_{1,3}
\end{pmatrix}
= -\ell_{1,1} \; ,
$$

[5] This periodicity occurs with the Fourier domain solutions of the biharmonic equation for uniform flexural rigidity (Sect. 10.4).

and the first (top) row of the rigidity matrix would be

$$\left(a_7 \ a_8 \ a_9 \ a_5 \ a_6 \ a_{11} \ a_{12} \ 0 \ 0 \ a_{10} \ a_{13} \ 0 \ 0 \ 0 \ 0 \ a_1 \ 0 \ 0 \ 0 \ 0 \ a_3 \ a_4 \ 0 \ 0 \ a_2 \right) .$$

Once the rigidity matrix has been formed, the deflection matrix is then found from Eq. 10.6, or

$$\mathbf{v} \ = \ \mathbf{D}^{-1}\mathbf{l} \, ,$$

using a suitable method to invert the rigidity matrix. The elements of $\mathbf{v}$ form the space-domain deflections, $v(\mathbf{x})$. For example, if both surface and internal loading are present, and the crust has a single layer, then the initial load is the sum of Eqs. 7.45 and 7.59 (in the space domain), or

$$\ell(\mathbf{x}) \ = \ \Delta\rho_{cf} g h_i(\mathbf{x}) \ + \ \Delta\rho_{mc} g w_i(\mathbf{x}) \, .$$

As before, random, fractal surfaces can be used as the initial load reliefs, h_i and w_i. We then obtain the final surface topography by substituting the space-domain versions of Eqs. 7.46 and 7.63 into Eq. 7.78, and using Eq. 7.77, giving

$$h(\mathbf{x}) \ = \ h_i(\mathbf{x}) \ + \ v(\mathbf{x}) \, .$$

We can also obtain the final Moho relief by substituting the space-domain versions of Eqs. 7.50 and 7.60 into Eq. 7.79, and using Eq. 7.77 to get

$$w(\mathbf{x}) \ = \ w_i(\mathbf{x}) \ + \ v(\mathbf{x}) \, .$$

The Bouguer anomaly can then be found from w using Parker's formula, Eq. 8.24 (or its full version, Eq. 8.23).

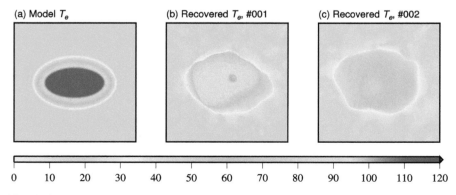

(a) Model T_e (b) Recovered T_e, #001   (c) Recovered T_e, #002

0   10   20   30   40   50   60   70   80   90   100   110   120

Fig. 10.10 a A model T_e distribution used to generate synthetic gravity anomalies and topography. **b** and **c** T_e recovered from two different random, fractal models, using the fan wavelet method with $|\mathbf{k}_0| = 7.547$

Figure 10.10a shows a spatially-variable T_e distribution from which synthetic Bouguer anomalies and topography were generated. Figure 10.10b, c shows the recovered T_e from two such random, fractal models.

10.6 Summary

The accuracy of T_e-estimation methods may be tested using synthetic modelling. The particular approach described here is one that has been used extensively in flexural studies, and uses random, fractal surfaces as the initial surface and internal loads emplaced upon a plate of known elastic thickness. This chapter provides details of a method that is used to compute such fractal surfaces. The equations of plate flexure provided in Chaps. 7, 8 and 9 then allow for the computation of the gravity anomaly and topography after flexure, given a known T_e distribution. If T_e is uniform across the study area, then the post-flexure surface topography and Moho relief may be obtained using the Fourier transform solution of the biharmonic equation. However, if the T_e distribution possesses a spatial variability, then the thin, elastic plate PDE must be solved using alternative methods. In this chapter the method of finite differences is used, and some background theory on the finite difference method is also provided. In both cases, Parker's method is used to obtain the post-flexure Bouguer anomaly from the Moho relief.

Once the final gravity and topography have been computed, the chosen analysis method—e.g. load deconvolution using wavelets—is used to try and recover the known, input T_e distribution. If the method used to generate the initial loads produces independent loads (which the one presented here does), then the procedure can be repeated many times, providing a statistical assessment of the accuracy of the analysis method. This is recommended, because the random nature of the initial loads means that the recovered T_e values are seldom identical.

10.7 Further Reading

As noted in Sect. 10.1, the synthetic testing procedure was initiated by Macario et al. (1995), so this paper is essential reading. Other T_e-oriented articles that have used Macario's method (or similar) are Lowry and Smith (1994), Simons et al. (2000), Swain and Kirby (2003), Stark et al. (2003), Simons et al. (2003), Pérez-Gussinyé et al. (2004), Kirby and Swain (2004), Swain and Kirby (2006), Audet and Mareschal (2007), Crosby (2007), Kirby and Swain (2008, 2009), Pérez-Gussinyé et al. (2009a, b), Kirby and Swain (2011), and Simons and Olhede (2013). The review article by Kirby (2014) contains a summary of synthetic testing in the discipline.

Most of these studies use initial loads with a fractal geometry. However, the topic of fractal geometries can be highly mathematical, going into a level of detail far beyond the knowledge required in this book. Therefore 'good' texts covering fractals to our level are the books by Peitgen and Saupe (1988) (which contains the chapter by Saupe (1988), cited in Sect. 10.2) and Turcotte (1997), and the articles by Huang and Turcotte (1989) and Turcotte (1989). Of particular interest, however, is the article by Simons and Olhede (2013) who use surfaces generated from the Matérn spectral class. Such surfaces do not have a strict straight-line log power spectrum plot, thus offering slightly improved flexibility when attempting to mimic the actual power spectra of topography or gravity.

While the finite difference method has been introduced here, it is a vast topic, and by no means the only method available to solve differential equations numerically. Useful textbooks covering finite differences (and other methods) to a greater degree than is presented here—and in the context of plate flexure too—are those by Ghali and Neville (1997) (and its updated version by Ghali et al. (2003)), and Szilard (2004). In addition to synthetic testing of T_e-estimation methods, finite differences have also been used to model lithospheric flexure in other contexts, notably by Sheffels and McNutt (1986), van Wees and Cloetingh (1994), Stewart and Watts (1997), Jordan and Watts (2005), Wyer and Watts (2006), Braun et al. (2013), and Zhang et al. (2018).

Appendix

Finite Differences

The method of finite differences is one of many methods available to solve differential equations numerically, in this case by representing the derivatives of the function as algebraic equations. It requires the discretisation of the space domain so that values of the function exist on a complete and regular grid, and calculates the derivatives by considering rates of change over fixed intervals.

Working in the one-dimensional space domain, the first-order derivative of a continuous function $v(x)$ at the point x is given by the well-known difference quotient of calculus as

$$\frac{dv}{dx}\bigg|_x = \lim_{\delta \to 0} \frac{v(x + \delta) - v(x)}{\delta} \, ,$$

shown as the dashed, red line in Fig. 10.11. If the function v is instead given as a discrete sequence, v_i ($i = 0, 1, \ldots, N - 1$), where the interval between elements (grid spacing) is a constant that cannot approach zero (Δ), then an approximation of the first-order derivative is given by

$$\frac{dv}{dx}\bigg|_{i+\frac{1}{2}} \approx \frac{v_{i+1} - v_i}{\Delta} \, .$$

Fig. 10.11 The red curve is a continuous function $v(x)$, while the dashed, red line shows its gradient at $x = i$. The black circles are the discrete points of the sampled sequence, v_i. The gradient of the line labelled 'a' is $dv/dx|_{i-1/2}$, that of the line labelled 'b' is $dv/dx|_{i+1/2}$, while that of the line labelled 'c' is $dv/dx|_i$

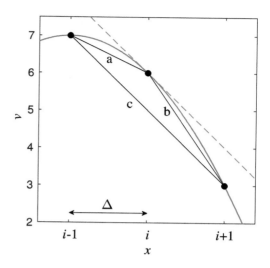

Strictly speaking, this derivative (shown by the gradient of the line labelled 'b' in Fig. 10.11) is calculated at a point halfway between grid node i and grid node $i + 1$, that is, at a point that does not exist in the original sequence because it is 'off-grid'. Thus, a better approach, and one that gives the derivative at the node i exactly, would be to take the mean of the slopes either side of node i, or

$$\frac{dv}{dx}\bigg|_i = \frac{1}{2}\left(\frac{dv}{dx}\bigg|_{i+\frac{1}{2}} + \frac{dv}{dx}\bigg|_{i-\frac{1}{2}}\right)$$
$$= \frac{1}{2}\left(\frac{v_{i+1} - v_i}{\Delta} + \frac{v_i - v_{i-1}}{\Delta}\right), \tag{10.10}$$

where we now replace the 'approximately equal to' sign with an equality. As shown in Fig. 10.11, the mean derivative (shown by the gradient of the line labelled 'c' in Fig. 10.11) is a more accurate representation of the true derivative than are either of the adjacent derivatives (the lines labelled 'a' and 'b'). This approach, of assigning derivatives to the point between the two end grid nodes rather than the end nodes themselves, is called the method of *central differences*.

Note that Eq. 10.10 can be simplified to

$$\frac{dv}{dx}\bigg|_i = \frac{v_{i+1} - v_{i-1}}{2\Delta}.$$

While this representation is common in finite difference solutions of differential equations, we will not make use of it here, instead preferring the form of Eq. 10.10, for reasons given in Sect. 10.5.

For a function of two independent variables, $v(x, y)$, discretised to the 2D sequence $v_{i,j}$, we must find finite difference equations that represent its partial derivatives in each direction. Fortunately this does not present extra work and is a logical extension of the formulation for ordinary derivatives introduced above. As in Eq. 10.10, the first-order partial derivative of v in the x-direction is given by

$$\frac{\partial v}{\partial x}\bigg|_{i,j} = \frac{1}{2}\left(\frac{\partial v}{\partial x}\bigg|_{i+\frac{1}{2},j} + \frac{\partial v}{\partial x}\bigg|_{i-\frac{1}{2},j}\right)$$

$$= \frac{v_{i+1,j} - v_{i,j}}{2\Delta x} + \frac{v_{i,j} - v_{i-1,j}}{2\Delta x},$$

where the indices i and j denote grid nodes in the x and y coordinates, respectively. We can represent this equation semi-graphically by the *stencil*

$$2\frac{\partial v}{\partial x} = \frac{1}{\Delta x}\begin{bmatrix} \cdot & \cdot & \cdot & \cdot & \cdot \\ \cdot & \cdot & \cdot & \cdot & \cdot \\ \cdot & \cdot & -1 & +1 & \cdot \\ \cdot & \cdot & \cdot & \cdot & \cdot \\ \cdot & \cdot & \cdot & \cdot & \cdot \end{bmatrix} + \frac{1}{\Delta x}\begin{bmatrix} \cdot & \cdot & \cdot & \cdot & \cdot \\ \cdot & \cdot & \cdot & \cdot & \cdot \\ \cdot & -1 & +1 & \cdot & \cdot \\ \cdot & \cdot & \cdot & \cdot & \cdot \\ \cdot & \cdot & \cdot & \cdot & \cdot \end{bmatrix},$$

where the dots ($\cdot$) indicate a value of zero. The central element of the stencil represents the node (i, j), with surrounding elements representing $i + 1$, $j - 1$, etc. Similarly, the first-order partial derivative of v in the y-direction is given by

$$\frac{\partial v}{\partial y}\bigg|_{i,j} = \frac{1}{2}\left(\frac{\partial v}{\partial y}\bigg|_{i,j+\frac{1}{2}} + \frac{\partial v}{\partial y}\bigg|_{i,j-\frac{1}{2}}\right)$$

$$= \frac{v_{i,j+1} - v_{i,j}}{2\Delta y} + \frac{v_{i,j} - v_{i,j-1}}{2\Delta y},$$

which is illustrated by the stencil

$$2\frac{\partial v}{\partial y} = \frac{1}{\Delta y}\begin{bmatrix} \cdot & \cdot & \cdot & \cdot & \cdot \\ \cdot & \cdot & +1 & \cdot & \cdot \\ \cdot & \cdot & -1 & \cdot & \cdot \\ \cdot & \cdot & \cdot & \cdot & \cdot \\ \cdot & \cdot & \cdot & \cdot & \cdot \end{bmatrix} + \frac{1}{\Delta y}\begin{bmatrix} \cdot & \cdot & \cdot & \cdot & \cdot \\ \cdot & \cdot & \cdot & \cdot & \cdot \\ \cdot & \cdot & +1 & \cdot & \cdot \\ \cdot & \cdot & -1 & \cdot & \cdot \\ \cdot & \cdot & \cdot & \cdot & \cdot \end{bmatrix}.$$

To compute the finite difference equations for the second-order derivatives, we find the gradient of the first-order derivative using central differences. Thus, the second-order partial derivative in the x direction is the gradient of the first-order x-derivative between nodes $i - \frac{1}{2}$ and $i + \frac{1}{2}$, or

$$\frac{\partial^2 v}{\partial x^2}\bigg|_{i,j} = \frac{1}{\Delta x}\left(\frac{\partial v}{\partial x}\bigg|_{i+\frac{1}{2},j} - \frac{\partial v}{\partial x}\bigg|_{i-\frac{1}{2},j}\right)$$

$$= \frac{1}{\Delta x}\left(\frac{v_{i+1,j} - v_{i,j}}{\Delta x} - \frac{v_{i,j} - v_{i-1,j}}{\Delta x}\right)$$

$$= \frac{v_{i+1,j} - 2v_{i,j} + v_{i-1,j}}{\Delta x^2},$$

which is represented by the stencil

$$\frac{\partial^2 v}{\partial x^2} = \frac{1}{\Delta x^2}\begin{bmatrix} \cdot & \cdot & \cdot & \cdot & \cdot \\ \cdot & \cdot & \cdot & \cdot & \cdot \\ \cdot & +1 & -2 & +1 & \cdot \\ \cdot & \cdot & \cdot & \cdot & \cdot \\ \cdot & \cdot & \cdot & \cdot & \cdot \end{bmatrix}.$$

Similarly, in the y direction the second-order partial derivative is given by

$$\frac{\partial^2 v}{\partial y^2}\bigg|_{i,j} = \frac{1}{\Delta y}\left(\frac{\partial v}{\partial y}\bigg|_{i,j+\frac{1}{2}} - \frac{\partial v}{\partial y}\bigg|_{i,j-\frac{1}{2}}\right)$$

$$= \frac{1}{\Delta y}\left(\frac{v_{i,j+1} - v_{i,j}}{\Delta y} - \frac{v_{i,j} - v_{i,j-1}}{\Delta y}\right)$$

$$= \frac{v_{i,j+1} - 2v_{i,j} + v_{i,j-1}}{\Delta y^2},$$

represented by the stencil

$$\frac{\partial^2 v}{\partial y^2} = \frac{1}{\Delta y^2}\begin{bmatrix} \cdot & \cdot & \cdot & \cdot & \cdot \\ \cdot & \cdot & +1 & \cdot & \cdot \\ \cdot & \cdot & -2 & \cdot & \cdot \\ \cdot & \cdot & +1 & \cdot & \cdot \\ \cdot & \cdot & \cdot & \cdot & \cdot \end{bmatrix}.$$

Finally, the cross-derivative is found by taking the gradient of the x-derivative in the y direction, or

$$\frac{\partial^2 v}{\partial x \partial y}\bigg|_{i,j} = \frac{1}{2\Delta y}\left(\frac{\partial v}{\partial x}\bigg|_{i,j+1} - \frac{\partial v}{\partial x}\bigg|_{i,j-1}\right)$$

$$= \frac{1}{2\Delta y}\left(\frac{v_{i+1,j+1} - v_{i-1,j+1}}{2\Delta x} - \frac{v_{i+1,j-1} - v_{i-1,j-1}}{2\Delta x}\right)$$

$$= \frac{v_{i+1,j+1} - v_{i-1,j+1} - v_{i+1,j-1} + v_{i-1,j-1}}{4\Delta x \Delta y}.$$

This partial derivative is illustrated by the stencil

$$4 \frac{\partial^2 v}{\partial x\, \partial y} = \frac{1}{\Delta x\, \Delta y} \begin{bmatrix} \cdot & \cdot & \cdot & \cdot & \cdot \\ \cdot & -1 & \cdot & +1 & \cdot \\ \cdot & \cdot & \cdot & \cdot & \cdot \\ \cdot & +1 & \cdot & -1 & \cdot \\ \cdot & \cdot & \cdot & \cdot & \cdot \end{bmatrix}.$$

When computing the third-order derivatives, as for the first-order derivatives, we split them into two parts, taking the mean of the third-order derivatives either side of node (i, j). Thus, for the third-order partial x-derivative we have

$$\left.\frac{\partial^3 v}{\partial x^3}\right|_{i,j} = \frac{1}{2}\left(\left.\frac{\partial^3 v}{\partial x^3}\right|_{i+\frac{1}{2},j} + \left.\frac{\partial^3 v}{\partial x^3}\right|_{i-\frac{1}{2},j} \right).$$

Then, representing the third-order derivatives as gradients of the second-order derivatives, using central differences we can write

$$\left.\frac{\partial^3 v}{\partial x^3}\right|_{i,j} = \frac{1}{2}\left[\frac{1}{\Delta x}\left(\left.\frac{\partial^2 v}{\partial x^2}\right|_{i+1,j} - \left.\frac{\partial^2 v}{\partial x^2}\right|_{i,j} \right) + \frac{1}{\Delta x}\left(\left.\frac{\partial^2 v}{\partial x^2}\right|_{i,j} - \left.\frac{\partial^2 v}{\partial x^2}\right|_{i-1,j} \right) \right]$$

$$= \frac{1}{2}\left[\frac{1}{\Delta x}\left(\frac{v_{i+2,j} - 2v_{i+1,j} + v_{i,j}}{\Delta x^2} - \frac{v_{i+1,j} - 2v_{i,j} + v_{i-1,j}}{\Delta x^2} \right) \right.$$
$$\left. + \frac{1}{\Delta x}\left(\frac{v_{i+1,j} - 2v_{i,j} + v_{i-1,j}}{\Delta x^2} - \frac{v_{i,j} - 2v_{i-1,j} + v_{i-2,j}}{\Delta x^2} \right) \right],$$

where we have used the finite difference representations of the second-order derivatives. Collecting terms within parentheses we obtain

$$\left.\frac{\partial^3 v}{\partial x^3}\right|_{i,j} = \frac{v_{i+2,j} - 3v_{i+1,j} + 3v_{i,j} - v_{i-1,j}}{2\Delta x^3}$$
$$+ \frac{v_{i+1,j} - 3v_{i,j} + 3v_{i-1,j} - v_{i-2,j}}{2\Delta x^3},$$

illustrated by the stencil

$$2 \frac{\partial^3 v}{\partial x^3} = \frac{1}{\Delta x^3} \begin{bmatrix} \cdot & \cdot & \cdot & \cdot & \cdot \\ \cdot & \cdot & \cdot & \cdot & \cdot \\ \cdot & -1 & +3 & -3 & +1 \\ \cdot & \cdot & \cdot & \cdot & \cdot \\ \cdot & \cdot & \cdot & \cdot & \cdot \end{bmatrix} + \frac{1}{\Delta x^3} \begin{bmatrix} \cdot & \cdot & \cdot & \cdot & \cdot \\ \cdot & \cdot & \cdot & \cdot & \cdot \\ -1 & +3 & -3 & +1 & \cdot \\ \cdot & \cdot & \cdot & \cdot & \cdot \\ \cdot & \cdot & \cdot & \cdot & \cdot \end{bmatrix}.$$

For the third-order partial y-derivative we proceed as for the x-derivative, starting with

$$\left.\frac{\partial^3 v}{\partial y^3}\right|_{i,j} = \frac{1}{2}\left(\left.\frac{\partial^3 v}{\partial y^3}\right|_{i,j+\frac{1}{2}} + \left.\frac{\partial^3 v}{\partial y^3}\right|_{i,j-\frac{1}{2}}\right),$$

and obtaining

$$\left.\frac{\partial^3 v}{\partial y^3}\right|_{i,j} = \frac{v_{i,j+2} - 3v_{i,j+1} + 3v_{i,j} - v_{i,j-1}}{2\Delta y^3}$$

$$+ \frac{v_{i,j+1} - 3v_{i,j} + 3v_{i,j-1} - v_{i,j-2}}{2\Delta y^3},$$

which is represented by the stencil

$$2\frac{\partial^3 v}{\partial y^3} = \frac{1}{\Delta y^3}\begin{bmatrix} \cdot & \cdot & +1 & \cdot & \cdot \\ \cdot & \cdot & -3 & \cdot & \cdot \\ \cdot & \cdot & +3 & \cdot & \cdot \\ \cdot & \cdot & -1 & \cdot & \cdot \\ \cdot & \cdot & \cdot & \cdot & \cdot \end{bmatrix} + \frac{1}{\Delta y^3}\begin{bmatrix} \cdot & \cdot & \cdot & \cdot & \cdot \\ \cdot & \cdot & +1 & \cdot & \cdot \\ \cdot & \cdot & -3 & \cdot & \cdot \\ \cdot & \cdot & +3 & \cdot & \cdot \\ \cdot & \cdot & -1 & \cdot & \cdot \end{bmatrix}.$$

The third-order partial cross-derivatives are derived in a similar manner. For the y-derivative of the second-order partial x-derivative we begin with its mean value either side of the node (i, j), or

$$\left.\frac{\partial^3 v}{\partial x^2\,\partial y}\right|_{i,j} = \frac{1}{2}\left(\left.\frac{\partial^3 v}{\partial x^2\,\partial y}\right|_{i,j+\frac{1}{2}} + \left.\frac{\partial^3 v}{\partial x^2\,\partial y}\right|_{i,j-\frac{1}{2}}\right),$$

which can then be written as

$$\left.\frac{\partial^3 v}{\partial x^2\,\partial y}\right|_{i,j} = \frac{1}{2}\left[\frac{1}{\Delta y}\left(\left.\frac{\partial^2 v}{\partial x^2}\right|_{i,j+1} - \left.\frac{\partial^2 v}{\partial x^2}\right|_{i,j}\right)\right.$$

$$\left.+ \frac{1}{\Delta y}\left(\left.\frac{\partial^2 v}{\partial x^2}\right|_{i,j} - \left.\frac{\partial^2 v}{\partial x^2}\right|_{i,j-1}\right)\right],$$

where we write the y-derivative of the second-order partial x-derivative as the y-gradient of the second-order partial x-derivative. We then replace the second-order partial x-derivatives with their finite difference equations, giving, after some rearrangement

$$\left.\frac{\partial^3 v}{\partial x^2\,\partial y}\right|_{i,j} = \frac{v_{i+1,j+1} - 2v_{i,j+1} + v_{i-1,j+1} - v_{i+1,j} + 2v_{i,j} - v_{i-1,j}}{2\Delta x^2\Delta y}$$

$$+ \frac{v_{i+1,j} - 2v_{i,j} + v_{i-1,j} - v_{i+1,j-1} + 2v_{i,j-1} - v_{i-1,j-1}}{2\Delta x^2\Delta y}.$$

Again, we could simplify this equation further but choose not to, as explained in Sect. 10.5. This partial derivative is represented by the stencil

$$
2\frac{\partial^3 v}{\partial x^2 \partial y} = \frac{1}{\Delta x^2 \Delta y}
\begin{bmatrix}
\cdot & \cdot & \cdot & \cdot & \cdot \\
\cdot & +1 & -2 & +1 & \cdot \\
\cdot & -1 & +2 & -1 & \cdot \\
\cdot & \cdot & \cdot & \cdot & \cdot \\
\cdot & \cdot & \cdot & \cdot & \cdot
\end{bmatrix}
$$

$$
+ \frac{1}{\Delta x^2 \Delta y}
\begin{bmatrix}
\cdot & \cdot & \cdot & \cdot & \cdot \\
\cdot & \cdot & \cdot & \cdot & \cdot \\
\cdot & +1 & -2 & +1 & \cdot \\
\cdot & -1 & +2 & -1 & \cdot \\
\cdot & \cdot & \cdot & \cdot & \cdot
\end{bmatrix} .
$$

Similarly, for the x-derivative of the second-order partial y-derivative we begin with

$$
\left. \frac{\partial^3 v}{\partial x \partial y^2} \right|_{i,j} = \frac{1}{2}\left(\left. \frac{\partial^3 v}{\partial x \partial y^2} \right|_{i+\frac{1}{2},j} + \left. \frac{\partial^3 v}{\partial x \partial y^2} \right|_{i-\frac{1}{2},j} \right) ,
$$

and arrive at

$$
\left. \frac{\partial^3 v}{\partial x \partial y^2} \right|_{i,j} = \frac{v_{i+1,j+1} - 2v_{i+1,j} + v_{i+1,j-1} - v_{i,j+1} + 2v_{i,j} - v_{i,j-1}}{2\Delta x \Delta y^2}
$$

$$
+ \frac{v_{i,j+1} - 2v_{i,j} + v_{i,j-1} - v_{i-1,j+1} + 2v_{i-1,j} - v_{i-1,j-1}}{2\Delta x \Delta y^2} ,
$$

which has stencil

$$
2\frac{\partial^3 v}{\partial x \partial y^2} = \frac{1}{\Delta x \Delta y^2}
\begin{bmatrix}
\cdot & \cdot & \cdot & \cdot & \cdot \\
\cdot & \cdot & -1 & +1 & \cdot \\
\cdot & \cdot & +2 & -2 & \cdot \\
\cdot & \cdot & -1 & +1 & \cdot \\
\cdot & \cdot & \cdot & \cdot & \cdot
\end{bmatrix}
$$

$$
+ \frac{1}{\Delta x \Delta y^2}
\begin{bmatrix}
\cdot & \cdot & \cdot & \cdot & \cdot \\
\cdot & -1 & +1 & \cdot & \cdot \\
\cdot & +2 & -2 & \cdot & \cdot \\
\cdot & -1 & +1 & \cdot & \cdot \\
\cdot & \cdot & \cdot & \cdot & \cdot
\end{bmatrix} .
$$

Finally, finite difference expressions for the three fourth-order derivatives are obtained from the gradients of the third-order derivatives. For the fourth-order partial x derivative we have

$$\frac{\partial^4 v}{\partial x^4}\bigg|_{i,j} = \frac{1}{\Delta x}\left(\frac{\partial^3 v}{\partial x^3}\bigg|_{i+\frac{1}{2},j} - \frac{\partial^3 v}{\partial x^3}\bigg|_{i-\frac{1}{2},j}\right),$$

where the third-order derivatives can be written as the gradients of the second-order derivatives, or

$$\frac{\partial^4 v}{\partial x^4}\bigg|_{i,j} = \frac{1}{\Delta x}\left[\frac{1}{\Delta x}\left(\frac{\partial^2 v}{\partial x^2}\bigg|_{i+1,j} - \frac{\partial^2 v}{\partial x^2}\bigg|_{i,j}\right)\right.$$
$$\left. - \frac{1}{\Delta x}\left(\frac{\partial^2 v}{\partial x^2}\bigg|_{i,j} - \frac{\partial^2 v}{\partial x^2}\bigg|_{i-1,j}\right)\right].$$

Using the difference equations for the second-order derivatives we obtain

$$\frac{\partial^4 v}{\partial x^4}\bigg|_{i,j} = \frac{1}{\Delta x^2}\left(\frac{v_{i+2,j} - 2v_{i+1,j} + v_{i,j}}{\Delta x^2} - \frac{v_{i+1,j} - 2v_{i,j} + v_{i-1,j}}{\Delta x^2}\right)$$
$$- \frac{1}{\Delta x^2}\left(\frac{v_{i+1,j} - 2v_{i,j} + v_{i-1,j}}{\Delta x^2} - \frac{v_{i,j} - 2v_{i-1,j} + v_{i-2,j}}{\Delta x^2}\right),$$

or

$$\frac{\partial^4 v}{\partial x^4}\bigg|_{i,j} = \frac{v_{i+2,j} - 4v_{i+1,j} + 6v_{i,j} - 4v_{i-1,j} + v_{i-2,j}}{\Delta x^4},$$

with stencil

$$\frac{\partial^4 v}{\partial x^4} = \frac{1}{\Delta x^4}\begin{bmatrix} \cdot & \cdot & \cdot & \cdot & \cdot \\ \cdot & \cdot & \cdot & \cdot & \cdot \\ +1 & -4 & +6 & -4 & +1 \\ \cdot & \cdot & \cdot & \cdot & \cdot \\ \cdot & \cdot & \cdot & \cdot & \cdot \end{bmatrix}.$$

In a similar fashion, the fourth-order partial y derivative begins with

$$\frac{\partial^4 v}{\partial y^4}\bigg|_{i,j} = \frac{1}{\Delta y}\left(\frac{\partial^3 v}{\partial y^3}\bigg|_{i,j+\frac{1}{2}} - \frac{\partial^3 v}{\partial y^3}\bigg|_{i,j-\frac{1}{2}}\right),$$

and ends with

$$\frac{\partial^4 v}{\partial y^4}\bigg|_{i,j} = \frac{v_{i,j+2} - 4v_{i,j+1} + 6v_{i,j} - 4v_{i,j-1} + v_{i,j-2}}{\Delta y^4}$$

which has stencil

$$\frac{\partial^4 v}{\partial y^4} = \frac{1}{\Delta y^4} \begin{bmatrix} \cdot & \cdot & +1 & \cdot & \cdot \\ \cdot & \cdot & -4 & \cdot & \cdot \\ \cdot & \cdot & +6 & \cdot & \cdot \\ \cdot & \cdot & -4 & \cdot & \cdot \\ \cdot & \cdot & +1 & \cdot & \cdot \end{bmatrix} .$$

The fourth-order cross-derivative is derived in a similar manner, starting with the x-gradient of the third-order partial cross-derivative:

$$\frac{\partial^4 v}{\partial x^2 \, \partial y^2}\bigg|_{i,j} = \frac{1}{\Delta x} \left(\frac{\partial^3 v}{\partial x \, \partial y^2}\bigg|_{i+\frac{1}{2},j} - \frac{\partial^3 v}{\partial x \, \partial y^2}\bigg|_{i-\frac{1}{2},j} \right) .$$

Replacing the third-order derivatives with the x-gradients of the second-order partial y-derivatives we obtain

$$\frac{\partial^4 v}{\partial x^2 \, \partial y^2}\bigg|_{i,j} = \frac{1}{\Delta x} \left[\frac{1}{\Delta x} \left(\frac{\partial^2 v}{\partial y^2}\bigg|_{i+1,j} - \frac{\partial^2 v}{\partial y^2}\bigg|_{i,j} \right) \right.$$
$$\left. - \frac{1}{\Delta x} \left(\frac{\partial^2 v}{\partial y^2}\bigg|_{i,j} - \frac{\partial^2 v}{\partial y^2}\bigg|_{i-1,j} \right) \right] ,$$

which gives, when replacing the second-order derivatives with their finite difference equations and collecting like terms,

$$\frac{\partial^4 v}{\partial x^2 \, \partial y^2}\bigg|_{i,j} = \frac{1}{\Delta x^2 \Delta y^2} \left[4v_{i,j} - 2\left(v_{i+1,j} + v_{i-1,j} + v_{i,j+1} + v_{i,j-1}\right) \right.$$
$$\left. + \left(v_{i+1,j+1} + v_{i-1,j+1} + v_{i+1,j-1} + v_{i-1,j-1}\right) \right] .$$

The stencil for this derivative is

$$\frac{\partial^4 v}{\partial x^2 \, \partial y^2} = \frac{1}{\Delta x^2 \Delta y^2} \begin{bmatrix} \cdot & \cdot & \cdot & \cdot & \cdot \\ \cdot & +1 & -2 & +1 & \cdot \\ \cdot & -2 & +4 & -2 & \cdot \\ \cdot & +1 & -2 & +1 & \cdot \\ \cdot & \cdot & \cdot & \cdot & \cdot \end{bmatrix} .$$

References

Audet P, Mareschal J-C (2007) Wavelet analysis of the coherence between Bouguer gravity and topography: application to the elastic thickness anisotropy in the Canadian Shield. Geophys J Int 168:287–298

Braun J, Deschamps F, Rouby D, Dauteuil O (2013) Flexure of the lithosphere and the geodynamical evolution of non-cylindrical rifted passive margins: results from a numerical model incorporating variable elastic thickness, surface processes and 3D thermal subsidence. Tectonophys 604:72–82

Crosby AG (2007) An assessment of the accuracy of admittance and coherence estimates using synthetic data. Geophys J Int 171:25–54

Forsyth DW (1985) Subsurface loading and estimates of the flexural rigidity of continental lithosphere. J Geophys Res 90(B14):12,623–12,632

Ghali A, Neville AM (1997) Structural analysis: a unified classical and matrix approach, 4th edn. E & FN Spon, London

Ghali A, Neville AM, Brown TG (2003) Structural analysis: a unified classical and matrix approach, 5th edn. Spon Press, London

Huang J, Turcotte DL (1989) Fractal mapping of digitized images: application to the topography of Arizona and comparisons with synthetic images. J Geophys Res 94(B6):7491–7495

Jordan TA, Watts AB (2005) Gravity anomalies, flexure and the elastic thickness structure of the India-Eurasia collisional system. Earth Planet Sci Lett 236:732–750

Kirby JF (2014) Estimation of the effective elastic thickness of the lithosphere using inverse spectral methods: the state of the art. Tectonophys 631:87–116

Kirby JF, Swain CJ (2004) Global and local isostatic coherence from the wavelet transform. Geophys Res Lett 31:L24608. https://doi.org/10.1029/2004GL021569

Kirby JF, Swain CJ (2008) An accuracy assessment of the fan wavelet coherence method for elastic thickness estimation. Geochem Geophys Geosyst 9:Q03022 https://doi.org/10.1029/2007GC001773, (Correction, Geochem Geophys Geosyst 9:Q05021. https://doi.org/10.1029/2008GC002071, 2008)

Kirby JF, Swain CJ (2009) A reassessment of spectral T_e estimation in continental interiors: the case of North America. J Geophys Res 114(B8):B08401. https://doi.org/10.1029/2009JB006356

Kirby JF, Swain CJ (2011) Improving the spatial resolution of effective elastic thickness estimation with the fan wavelet transform. Comput Geosci 37:1345–1354

Lowry AR, Smith RB (1994) Flexural rigidity of the Basin and Range–Colorado Plateau–Rocky Mountain transition from coherence analysis of gravity and topography. J Geophys Res 99(B10):20,123–20,140

Macario A, Malinverno A, Haxby WF (1995) On the robustness of elastic thickness estimates obtained using the coherence method. J Geophys Res 100(B8):15,163–15,172

Pérez-Gussinyé M, Lowry AR, Watts AB, Velicogna I (2004) On the recovery of effective elastic thickness using spectral methods: examples from synthetic data and from the Fennoscandian Shield. J Geophys Res 109(B10):B10409. https://doi.org/10.1029/2003JB002788

Pérez-Gussinyé M, Swain CJ, Kirby JF, Lowry AR (2009a) Spatial variations of the effective elastic thickness, T_e, using multitaper spectral estimation and wavelet methods: examples from synthetic data and application to South America. Geochem Geophys Geosyst 10:Q04005. https://doi.org/10.1029/2008GC002229

Pérez-Gussinyé M, Metois M, Fernández M, Vergés J, Fullea J, Lowry AR (2009b) Effective elastic thickness of Africa and its relationship to other proxies for lithospheric structure and surface tectonics. Earth Planet Sci Lett 287:152–167

Peitgen H-O, Saupe D (1988) The science of fractal images. Springer, New York

Press WH, Teukolsky SA, Vetterling WT, Flannery BP (1992) Numerical recipes in Fortran 77, 2nd edn. Cambridge University Press, Cambridge

Saupe D (1988) Algorithms for random fractals. In: Peitgen H-O, Saupe D (eds) The science of fractal images. Springer, New York, pp 71–136

Sheffels B, McNutt M (1986) Role of subsurface loads and regional compensation in the isostatic balance of the Transverse Ranges, California: evidence for intracontinental subduction. J Geophys Res 91(B6):6419–6431

Simons FJ, Olhede SC (2013) Maximum-likelihood estimation of lithospheric flexural rigidity, initial-loading fraction and load correlation, under isotropy. Geophys J Int 193:1300–1342

Simons FJ, Zuber MT, Korenaga J (2000) Isostatic response of the Australian lithosphere: estimation of effective elastic thickness and anisotropy using multitaper spectral analysis. J Geophys Res 105(B8):19,163–19,184

Simons FJ, van der Hilst RD, Zuber MT (2003) Spatiospectral localization of isostatic coherence anisotropy in Australia and its relation to seismic anisotropy: implications for lithospheric deformation. J Geophys Res 108(B5):2250. https://doi.org/10.1029/2001JB000704

Singleton RC (1968) An algorithm for computing the mixed radix fast Fourier transform. IEEE Trans Audio Electroacoust AU-17(2):93–102

Stark CP, Stewart J, Ebinger CJ (2003) Wavelet transform mapping of effective elastic thickness and plate loading: validation using synthetic data and application to the study of southern African tectonics. J Geophys Res 108(B12):2558. https://doi.org/10.1029/2001JB000609

Stewart J, Watts AB (1997) Gravity anomalies and spatial variations of flexural rigidity at mountain ranges. J Geophys Res 102(B3):5327–5352

Swain CJ, Kirby JF (2003) The effect of 'noise' on estimates of the elastic thickness of the continental lithosphere by the coherence method. Geophys Res Lett 30:1574. https://doi.org/10.1029/2003GL017070

Swain CJ, Kirby JF (2006) An effective elastic thickness map of Australia from wavelet transforms of gravity and topography using Forsyth's method. Geophys Res Lett 33:L02314. https://doi.org/10.1029/2005GL025090

Szilard R (2004) Theories and applications of plate analysis. Wiley, Hoboken

Turcotte DL (1989) Fractals in geology and geophysics. Pure Appl Geophys 131:171–196

Turcotte DL (1997) Fractals and chaos in geology and geophysics, 2nd edn. Cambridge University Press, Cambridge

van Wees JD, Cloetingh S (1994) A finite-difference technique to incorporate spatial variations in rigidity and planar faults into 3-D models for lithospheric flexure. Geophys J Int 117:179–195

Wyer P, Watts AB (2006) Gravity anomalies and segmentation at the East Coast, USA continental margin. Geophys J Int 166:1015–1038

Zhang J, Sun Z, Xu M, Yang H, Zhang Y, Li F (2018) Lithospheric 3-D flexural modelling of subducted oceanic plate with variable effective elastic thickness along the Manila Trench. Geophys J Int 215:2071–2092

Chapter 11
Practical T_e Estimation

11.1 Introduction

Having described in detail the theory behind T_e estimation using load deconvolution, and introduced a method to test its efficacy (synthetic modelling), all that remains now is to show how the method is applied in the real world. This naturally starts with a discussion about gravity and topography data as these are all that is required to estimate T_e. Load deconvolution, however, asks for at least approximate mean crustal thickness and density—and mantle density—values, and with wavelet and windowed Fourier transform methods one can account for any spatial variability in these data sets. Therefore, we discuss here some contemporary models that provide such information. Finally, it would be poor form to discuss two different but powerful methods of spectral analysis without putting them head to head, so, at the very last, the battle is joined.

11.2 Data

As explained in Chap. 9, the effective elastic thickness of the lithosphere is estimated by comparing the gravity field and topography over a region using the admittance, coherency or coherence. Essentially, the topography forms the visible, surface load on the lithospheric plate and the gravity field—as a proxy for the Moho deflection—provides the flexural response. Therefore, the quantity represented by the 'topography' must include all surface loads on the lithosphere—including rock, water and ice—as all these materials will deflect the plate to varying amounts depending on their density and volume. So throughout this book the word 'topography' will refer to the surface mapped out by the interface between solid rock and air or water or ice. Over—or more accurately, under—the oceans or large lakes this interface is called the *bathymetry*, the depth of the oceans or lakes.

© Springer Nature Switzerland AG 2022
J. Kirby, *Spectral Methods for the Estimation of the Effective Elastic Thickness
of the Lithosphere*, Advances in Geophysical and Environmental Mechanics
and Mathematics, https://doi.org/10.1007/978-3-031-10861-7_11

Gravity and topography data are collected as either point observations or as time or area means. Either way, they are assigned to a particular latitude and longitude. The gravity field is generally measured directly using either gravimeters or gravity gradiometers; these can be placed on any solid surface, sailed in a ship, or flown in a plane or satellite (Torge 1989). The gravity field can be measured indirectly too, primarily using satellite altimetry, whereby the undulations of the mean ocean surface are measured with a radar altimeter and converted to a map of the geoid and then gravity anomaly (Fu and Cazenave 2001). Indirect measurements of gravity also include the tracking of artificial satellites using laser ranging or global navigation satellite systems (GNSS), since the satellites follow an equipotential surface (Sect. 8.2.2).

Topography has perhaps more measurement options available. Historically, trigonometric or spirit-levelled spot height surveys or barometric measurements provided most of the data on land, while bathymetry was measured using depth soundings or—more recently—sonar. Nowadays, sophisticated remote sensing technologies provide most of the data, such as airborne LiDAR (light detection and ranging) at sea or on land, or globally via satellite-borne synthetic aperture radar (SAR) or interferometric SAR (InSAR) observations.

There are now many models and data sets of the Earth's topography and gravity field freely available online. Most of these are provided as a global or near-global grid of values, where the grid cells contain a mean or median value of the topography or gravity field in that cell. Many topography models come in such a form (Sect. 11.2.2), as do the other data sets useful for T_e estimation (Sect. 11.2.4). In contrast, gravity field data are usually provided as a set of spherical harmonic coefficients from which the user can generate a space-domain grid to a desired resolution (Sect. 11.2.3). As these data sets—gravity and topography alike—are continually being updated, I will not recommend any particular one as such a recommendation would quickly become obsolete. I will, however, discuss several that have been widely used in the years preceding publication of this book.

11.2.1 Model Grid Spacing

A word on resolution and grid spacing first. In Sect. 6.5.2, we used the following reasoning to calculate a finest grid spacing that it is unnecessary to improve upon. If the minimum resolvable T_e from spectral methods is roughly 1 km (Sect. 11.8), then from Eq. 7.85 (or Eq. 6.15) the corresponding minimum flexural wavelength is ~36 km. According to the Nyquist criterion (Eqs. 2.43 and 6.17) features with this wavelength can be perfectly reconstructed from a grid of data if the grid spacing is no greater than half this wavelength, or ~18 km, which translates to approximately 10 arc-minutes (10′) in geodetic coordinates.

However, if the T_e-estimation method utilises the coherence (or SRC), then the data grid must be able to resolve the rollover wavelength for a given T_e, meaning one should use Eq. 9.32 to calculate wavelength. As Fig. 9.5 shows, coherence transition

Table 11.1 Smallest-resolvable T_e as a function of the grid spacing of the data model, for some commonly used grid spacings (on the Earth). The grid spacing is used to calculate a Nyquist wavelength using Eq. 2.46 (or Eq. 6.16), which is then used as the coherence transition ('rollover') wavelength. The corresponding smallest-resolvable T_e is calculated from the rollover wavelength using the rule of thumb, Eq. 11.1

Grid spacing (geodetic)	Grid spacing (surface equiv.)	Nyquist/rollover wavelength	Smallest T_e
1°	111.3 km	222.6 km	17.0 km
30′	55.7 km	111.3 km	6.8 km
15′	27.8 km	55.7 km	2.7 km
6′	11.1 km	22.3 km	0.8 km
5′	9.3 km	18.6 km	0.6 km
2′	3.7 km	7.4 km	0.2 km
1′	1.9 km	3.7 km	72 m
30″	928 m	1.9 km	29 m
15″	464 m	928 m	11 m
1″	31 m	62 m	0.3 m

wavelengths are smaller than flexural wavelengths. If we approximate Eq. 9.32 by a rule of thumb similar to Eq. 6.15, we can write

$$\lambda_t \approx 26.6 \, T_e^{3/4} \,, \tag{11.1}$$

for load ratio $f = 1$. So now, if we want to be able to resolve T_e values as low as 1 km, then Eq. 11.1 tells us that we need to be able to resolve features in the data grids with wavelengths of at least 26.6 km. The Nyquist criterion then tells us that the data grids must have a maximum spacing of 13.3 km, which translates to approximately 7.2′ in geodetic coordinates. In practice this would be set to 6′, giving ten grid nodes per degree.

Thus, any finer grid spacing corresponds to minimum elastic thicknesses that cannot meaningfully be resolved using the coherence (see Table 11.1), meaning that high-resolution data models provide no extra benefit and just take up space. That said, as technology advances it is not just the grid spacing of DEMs that gets better: the accuracy and coverage of data also improves. This means that the more recent models are likely to produce more accurate T_e estimates by virtue of their more reliable data sources rather than their resolution.

11.2.2   Topography Data

Digital elevation models (DEMs), also known as *digital terrain models* (DTMs), of global topography (and bathymetry) are now widely available, and their accuracy and resolution have improved greatly over the years, largely due to developments in satellite radar and airborne LiDAR technology (Hirt 2014). The first global DEM, released by the USA's National Oceanic and Atmospheric Administration (NOAA) in 1987, was ETOPO5 (NGDC 1988), which provided global mean elevations on land and at sea on a 5′ grid in latitude and longitude. Successor ETOPO models have been released over the years, culminating with ETOPO1, given on a 1′ geodetic grid (Amante and Eakins 2009). The first global model to be released on a 30 arc-second (30″) grid, in 1997, was the US Geological Survey's GTOPO30 model (Gesch et al. 1999), though it provided dry-land elevations only. Another continent-only global DEM followed soon after, GLOBE (Hastings and Dunbar 1999), produced by NOAA from a variety of sources, and also given on a 30″ grid. DEM resolution improved dramatically, however, with the observations collected by NASA's Shuttle Radar Topography Mission (SRTM), a set of InSAR data collected over land by a single space shuttle mission in 2000, but only between latitudes 60°N and 56°S (Rabus et al. 2003; Farr et al. 2007). Several global DEMs have been created from SRTM data alone, the most recent being version 4.1, provided on a 3″ grid (Jarvis et al. 2008). Another source of satellite data is the Advanced Spaceborne Thermal Emission and Reflection Radiometer (ASTER) sensor, onboard the Terra satellite (Tachikawa et al. 2011). Three versions of a land-only DEM have been created from these data, the most recent having a grid spacing of 1″, with coverage between latitudes 83°N and 83°S.

At sea, most bathymetric DEMs are constructed from satellite altimetry data due to their superior coverage when compared to shipborne sonar observations. However, while sonar is a direct measurement of the ocean depth, radar altimeters merely measure the height of the satellite above the ocean surface. This measurement is readily converted to the height of the sea surface above a reference ellipsoid such as GRS80, and then, if assumptions are made regarding oceanographic and atmospheric phenomena, the height of the geoid above the ellipsoid can be derived. Gravitational boundary value problems are then solved to obtain the marine gravity field (Heiskanen and Moritz 1967; Sandwell and Smith 1997). The conversion of marine gravity to bathymetry uses a (constrained) inverse theoretical admittance function, via the formula

$$H(\mathbf{k}) = Q^{-1}(|\mathbf{k}|)\, G(\mathbf{k})$$

(from Eq. 8.40), where G and H are the Fourier transforms of the (free-air) gravity anomaly and topography, respectively (Dixon et al. 1983; Smith and Sandwell 1994, 1997). However, the inverse admittance, Q^{-1}, approaches singularities at very long and very short wavelengths (Fig. 11.1), and is sensitive to elastic thickness in its medium-to-long wavelengths. Given the (then) lack of global T_e models, Smith and Sandwell (1997) chose to only predict the bathymetry from the marine gravity in a wavelength range of 15–160 km, a bandwidth in which, they said, Q^{-1} was independent of T_e variations—which is true to a certain extent for oceanic lithosphere, where T_e typically varies between 5 and 50 km (Fig. 11.1). So within this bandwidth they calculated $H(\mathbf{k})$ from $G(\mathbf{k})$ by approximating $Q^{-1}(|\mathbf{k}|)$ by a T_e-independent, Gaussian-based, analytic function. At shorter and longer wavelengths, they used shipborne observations of ocean depth.

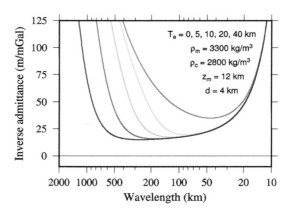

Fig. 11.1 Inverse free-air admittance (given by the reciprocal of Eq. 8.43) for surface loading of a single-layer crust with the parameter values shown in the figure. The red curve shows Airy isostasy ($T_e = 0$ km), while the other curves are (from lighter to darker tones) $T_e = 5$, 10, 20 and 40 km

Therefore, gravity-derived bathymetry contains an implicit assumption about the elastic thickness of the lithosphere, and such models should not really be used for T_e estimation using spectral methods. Unfortunately, there are very few global bathymetric models that are computed from shipborne observations only. One is the already-mentioned ETOPO5, another is the Centenary Edition of the General Bathymetric Chart of the Oceans (GEBCO) (IOC et al. 2003), which is supplied on a $1'$ grid. This GEBCO grid has recently been updated to the GEBCO 2020 product (GEBCO Compilation Group 2020), which is supplied on a $15''$ grid. However, GEBCO 2020 includes gravity-derived bathymetry data and so should be used with caution. The improvement in resolution offered by the newer product is readily seen in Fig. 11.2, though this fourfold decrease in grid spacing does not necessarily imply an improvement with regard to T_e estimates. While the newer model will probably contain more accurate depth observations—which should improve T_e estimates—Table 11.1 shows that its increase in resolution does not translate to an increase in accuracy of T_e estimates (which are limited by an error of roughly 1 km) and that the $1'$ grid spacing of the Centenary Edition is sufficient.

Both GEBCO grids also contain continental elevations. The Centenary Edition was supplemented with GLOBE data on land, while the GEBCO 2020 model uses land (and sea) data from a global DEM called SRTM15+ (Tozer et al. 2019). This latter model, provided on a $15''$ grid, is the latest in a series of models developed at the Scripps Institute of Oceanography that are based on SRTM and ASTER land elevations, gravity-derived bathymetry from satellite altimetry, shipborne bathymetric soundings and other sources.

Another global DEM—that is of particular use in T_e estimation—is the $1'$ Earth2014 model (Hirt and Rexer 2015) (Fig. 11.3). Although it uses now-superseded data sources, the model package includes four separate grids: the Earth's surface, bedrock, bedrock and ice, and rock-equivalent topography. For T_e-estimation purposes, the rock-equivalent topography model is the most useful, as will be discussed in Sect. 11.3. Another benefit of the Earth2014 model is that it is also provided as a

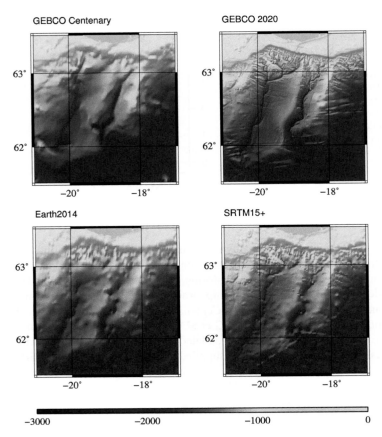

Fig. 11.2 Bathymetry south of Iceland from four global DEMs. The GEBCO Centenary model is supplied on a 1′ grid, GEBCO 2020 on a 15″ grid, Earth2014 on a 1′ grid and SRTM15+ on a 15″ grid. Mercator projections

set of spherical harmonic coefficients which can be expanded to any chosen degree and order (up to 10,800) and on any chosen grid size.

Another harmonic model (to degree and order 2190) is the DTM2006.0 global topography model that was released alongside the EGM2008 global geopotential harmonic model (Pavlis et al. 2007, 2012) (see Sect. 11.2.3). However, both these models used gravity-derived bathymetry in the oceans, so results should be treated with caution if they are used to estimate T_e in oceanic lithosphere.

Many models of the global topography provide only the elevation/depth of the solid surface: the ocean floor, continental rocky surface (including glaciers), and the surface of the major ice caps of Greenland and Antarctica (as shown in Fig. 11.3). As will be discussed in Sect. 11.3, flexural analyses over several kilometres of ice will generate errors if the thickness and density of the ice sheets is not accounted for. Fortunately there now exist several models of ice thickness, the most recent

Fig. 11.3 The Earth's topography, bathymetry and ice sheet surface from the Earth2014 model of Hirt and Rexer (2015). Mollweide projection; units in metres

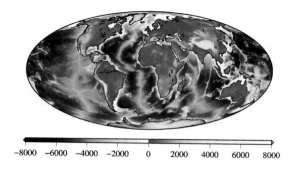

-8000 -6000 -4000 -2000 0 2000 4000 6000 8000

being the BedMachine models of Greenland (Morlighem et al. 2017) and Antarctica (Morlighem et al. 2020). The grids of the Earth2014 model (Hirt and Rexer 2015) also provide ice cap thicknesses of Greenland and Antarctica as part of a global DEM, though they use older data—Bedmap2 in Antarctica (Fretwell et al. 2013), and a Greenland model by Bamber et al. (2013).

11.2.3 Gravity Data

Modern global gravity models incorporate gravity data from a wide variety of sources: surveys on land using a gravity meter, shipborne and airborne gravity meters, airborne gravity gradiometers, satellite altimetry, satellite tracking, satellite gravity gradiometry and others. Most often these data are combined and provided as a set of *spherical harmonic coefficients*, but a few are supplied on space-domain grids. Spherical harmonics are used because they are solutions of Laplace's equation (Eq. 8.10) in spherical coordinates, with spherical harmonic coefficients essentially being a spherical version of Fourier coefficients. That is, spherical harmonic models provide the spectrum of data coordinated on the sphere (or ellipsoid, to an approximation). We shall not derive the expression for the spherical harmonic representation of the gravity anomaly—which can be found in most physical geodesy textbooks, such as Heiskanen and Moritz (1967), for example—but quote it here, as

$$\Delta g(r, \theta, \lambda) = \frac{\mathcal{G}M}{r^2} \sum_{n=2}^{n_{\max}} (n-1) \left(\frac{a}{r}\right)^n Y_n(\theta, \lambda) , \qquad (11.2)$$

where the Y_n are the *surface spherical harmonics* given by

$$Y_n(\theta, \lambda) = \sum_{m=0}^{n} \bar{P}_n^m(\cos\theta) \left(\delta\bar{C}_{nm}\cos m\lambda + \bar{S}_{nm}\sin m\lambda\right) .$$

The coordinate r is the spherical radial distance, θ is the geocentric colatitude and λ is the geodetic longitude; n is spherical harmonic *degree*, while m is the *order*; $\mathcal{G}$ is the Newtonian gravitational constant, M is the mass of the Earth, and a is the semi-major axis length. The C_{nm} and S_{nm} are the two spherical harmonic coefficients at each degree and order; the bar over them they are fully-normalised. The δ in front of the C_{nm} coefficient means that the C_{nm} coefficient from a reference ellipsoid model has been subtracted, thus providing gravity anomalies rather than absolute gravity values. Finally, the P_n^m are *associated Legendre functions* at each degree and order; the bar over them means they too are fully-normalised. These orthonormal functions (see the appendix to Chap. 3) are the spherical equivalent of the sines and cosines of the Fourier transform.

The maximum degree of expansion, $n_{\max}$, provides the resolution of the model, dictating the wavelength, $\lambda_{\min}$, of the smallest features that the model can portray. One can calculate this wavelength using the equation

$$\lambda_{\min} = \frac{2\pi a}{\sqrt{n_{\max}(n_{\max}+1)}} , \tag{11.3}$$

where a is the ellipsoid's semi-major axis length. For large values of harmonic degree, Eq. 11.3 can be written in two, convenient forms:

$$\lambda_{\min} \approx \frac{360°}{n_{\max}} \quad \text{and} \quad \lambda_{\min} \approx \frac{40{,}000 \text{ km}}{n_{\max}} ,$$

where the first equation gives the minimum wavelength in degrees of arc, while the second gives it in kilometres. Thus, if we acknowledge that we cannot resolve any T_e values less than 1 km, Eq. 11.1 tells us that the minimum-present wavelength in the model needs only be ~26.6 km. Then, Eq. 11.3 informs us that a spherical harmonic expansion of gravity (and topography) coefficients needs to be performed up to degree and order 1500, approximately. Such a harmonic model can be perfectly represented on a 7.2′ grid, according to the Nyquist criterion and Eq. 11.3, which, in practice, would be set to 6′. Table 11.2 gives the smallest-resolvable elastic thickness for some commonly-used maximum harmonic degrees.

One widely used harmonic model is EGM2008 (Pavlis et al. 2012), shown in Fig. 11.4, whose harmonic coefficients are given up to degree and order 2190 (~18 km minimum wavelength). Harmonic geopotential models are generally constructed from satellite and/or terrestrial (land, sea, airborne and altimetric) gravity observations, which are averaged into area means (or 'bins') before computation of the harmonic coefficients; in the case of EGM2008, the gravity data were averaged into 5′ bins. But due to poor coverage (on land), some bins contain few, or even no terrestrial observations. In such areas gravity data may be 'invented' using an assumed density and a high-resolution DEM, the reasoning going that at such short wavelengths the dominant gravity effect is from the topography and not deeper sources. Two options are available to simulate gravity data from topography. The first generates gravity anomalies implied by a *residual terrain model* (RTM) (Forsberg 1984), calculated in

Table 11.2 Smallest-resolvable T_e as a function of the maximum harmonic degree, n_{max}, in an expansion of geopotential model coefficients, for some commonly-used degrees. Also shown is the maximum allowable geodetic grid spacing of the data model so that the Nyquist criterion is satisfied, and the minimum wavelength present in the expansion, λ_{min}. The T_e values are obtained from the harmonic degrees using Eqs. 11.1 and 11.3

Harmonic degree	Maximum grid spacing	Minimum wavelength	Smallest T_e
180	60′	222 km	17 km
200	54′	200 km	15 km
250	43′	160 km	11 km
300	36′	133 km	8.6 km
360	30′	111 km	6.7 km
720	15′	55.6 km	2.7 km
1500	7.2′	26.7 km	1.0 km
2190	4.9′	18.3 km	607 m
10,800	1′	3.7 km	72 m

Fig. 11.4 The Earth's free-air gravity anomaly from the EGM2008 model of Pavlis et al. (2012). Mollweide projection; units in mGal

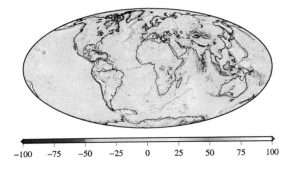

a similar manner to the terrain correction (Sect. 8.2.5). The second method simulates gravity anomalies from the topography and its isostatic (usually Airy) compensation (Pavlis and Rapp 1990); obviously any gravity model containing such information should never be used to estimate T_e. Fortunately, EGM2008 was deliberately constructed so as not to contain any information from an isostatic model, and the data infill was undertaken using an RTM based on the DTM2006.0 high-resolution topographic model (Pavlis et al. 2007).[1] Nevertheless, while the RTM gravity field is most often a good representation of the actual gravity field at short wavelengths, one must bear in mind that it is a model, and not real gravity data.

Most recent harmonic geopotential models make use of the three dedicated satellite gravity missions launched between 2000 and 2009, CHAMP, GRACE and GOCE (Pail 2014). These satellites were tracked (by other satellites or ground stations)

---

[1] Although EGM2008 incorporates an RTM implied by DTM2006.0—which uses gravity-derived bathymetry at sea (Sect. 11.2.2)—the RTM gravity field was not applied in the oceans.

as they journeyed along the equipotential surface at their orbit altitude, providing accurate information about the long wavelengths of the gravity field. To constrain its low-degree coefficients, EGM2008 used a harmonic model based on GRACE data to degree and order 180, and merged this with terrestrial and RTM data, as described above. Of all three, though, the GOCE satellite has provided the better data. First, it orbited at a lower altitude than CHAMP and GRACE, thus detecting higher-frequency harmonics of the gravity field than the other satellites.[2] Second, it housed an onboard gravity gradiometer that measured second-order derivatives of the gravity potential in all three spatial directions; such derivatives contain more high-frequency information than the potential itself. One such model constructed from GOCE data is TIM_R6 (Brockmann et al. 2019), which contains harmonic coefficients up to degree and order 300. However, since GOCE was in an orbit that could not observe the poles (latitudes higher than $\pm 83°$), a subsequent model, TIM_R6e (Zingerle et al. 2019), included terrestrial gravity data in these regions. Note that, from Table 11.2, these two models would be unable to reliably resolve T_e values less than approximately 9 km.

Finally, some gravity models are provided on space-domain grids rather than as spherical harmonic coefficients. One example is the GGMplus model (Hirt et al. 2013), provided on a 7.2″ grid (approximately 222 m) though only on land between $\pm 60°$ latitude. In this model, harmonics with wavelengths from 100 to 10,000 km came from GOCE and GRACE data, wavelengths from 10 to 100 km came from EGM2008, while wavelengths from 250 m to 10 km were synthesised from an RTM of SRTM data. Importantly, GGMplus does not contain any information from isostatic models.

11.2.4 Crustal Structure Data

In addition to gravity and topography data, T_e estimation using the coherence (SRC) or admittance also requires information about the densities and thicknesses of the various layers within the crust and the density of the lithospheric mantle, as described in Chap. 9. If the coherence is used, these data sets do not need to possess a very high resolution or accuracy, owing to the relative insensitivity of the coherence to variations in crustal thickness and density (Fig. 9.7). If the admittance is used though, then the depth and density information should be somewhat more accurate (Fig. 9.9).

In the early days of spectral T_e estimation, where a single estimate of the elastic thickness was made for an entire region, it was sufficient to know only the mean Moho depth and crust and mantle densities over the whole region. This would obviously present problems when the study area included provinces with very different values of these lithospheric parameters. However, given the current availability of maps

---

[2] Higher-frequency harmonics of the gravitational acceleration decay faster with increasing altitude than do lower-frequency harmonics. This is shown by the e^{-kz} term in Parker's formula, Eq. 8.23, and the r^{-n} term in Eq. 11.2.

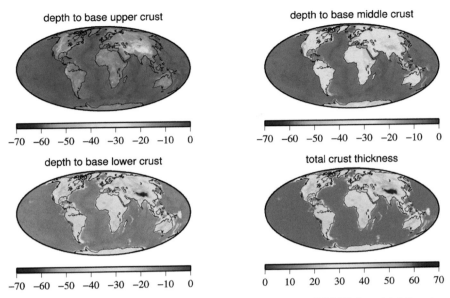

depth to base upper crust

depth to base middle crust

−70 −60 −50 −40 −30 −20 −10   0

−70 −60 −50 −40 −30 −20 −10   0

depth to base lower crust

total crust thickness

−70 −60 −50 −40 −30 −20 −10   0

0   10   20   30   40   50   60   70

Fig. 11.5 The depths to the bases of the three crustal layers of the CRUST1.0 model (kilometres below mean sea level), and total crust thickness (km). Mollweide projections

of crustal layer thicknesses and crust and mantle densities, when using wavelet or moving-window multitaper methods to construct maps of spatial T_e variations one can compute T_e at a grid node using the values of the lithospheric parameters at that grid node, rather than using area means.

While some regions of the Earth have high-resolution crust and mantle density and depth data, derived from detailed seismic surveys for instance, there are cruder global models available that are nevertheless of a sufficient resolution and accuracy for T_e estimation. One of the first was the CRUST5.1 model (Mooney et al. 1998), given on a $5° \times 5°$ grid. Successor models are CRUST2.0 (Bassin et al. 2000) and CRUST1.0 (Laske et al. 2013), provided on $2° \times 2°$ and $1° \times 1°$ grids, respectively. Figures 11.5 and 11.6 show some of the data in the CRUST1.0 model, which contains density and thickness information for the upper, middle and lower crust, the lithospheric mantle, and three layers of sediments.

The depth to the base of the crust, the Moho, may also be estimated from gravity data, in a method first proposed by Vening Meinesz (1931), and later developed by Parker (1972) and Oldenburg (1974) in planar coordinates, and by Moritz (1990) and Sjöberg (2009) in spherical coordinates. The approach rests upon the assumption that the density contrast at the Moho alone is the source of the variations in the Bouguer anomaly. The principle of the method can be shown by rearranging the linear approximation of Parker's formula, Eq. 8.25, to give

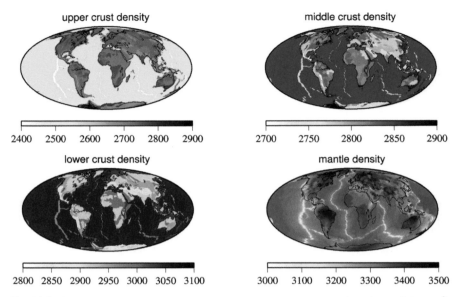

upper crust density

middle crust density

lower crust density

mantle density

Fig. 11.6 The densities of the three crustal layers and mantle of the CRUST1.0 model (kg m^{-3}). Mollweide projections

$$W(\mathbf{k}) \;=\; \frac{G(\mathbf{k})\, e^{k z_m}}{2\pi \mathcal{G} \Delta \rho_{mc}} \, ,$$

where $k \equiv |\mathbf{k}|$, and $G(\mathbf{k})$ and $W(\mathbf{k})$ are the Fourier transforms of the Bouguer anomaly and Moho relief, respectively. Note that one still needs to know, or guess at, a value for the mean Moho depth in the region (z_m), and assume a constant density contrast between mantle and crust ($\Delta \rho_{mc}$). The method also makes the implicit assumption that the topography is in isostatic equilibrium, there being no support from plate rigidity or mantle convection. Because of these assumptions, and the fact that a gravity-derived Moho does not represent an independent data source, such Moho models should be used with caution when estimating T_e from gravity and topography data. A brief review of other Moho estimation methods using gravity data can be found in Aitken et al. (2013).

As noted above, though, the accuracy and horizontal resolution of crust models do not need to be very high as their influence upon derived T_e estimates is relatively minor, as will be shown in Sect. 11.4.

11.2.5   Sediment Data

A last data source that can be useful is information about sediments. Given that sediment density is generally significantly less than that of continental or oceanic

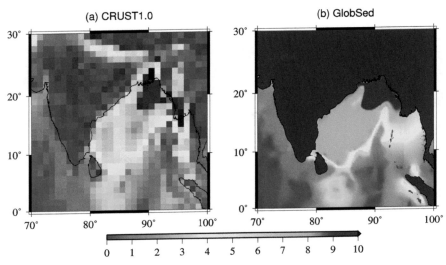

Fig. 11.7 Sediment thicknesses (km) in the Bay of Bengal and surrounds from **a** CRUST1.0, and **b** GlobSed. Mercator projections

crust—with a range of 1600–2600 kg m^{-3}, approximately—and that some sedimentary basins can reach several kilometres in thickness, it can be prudent to at least note the existence of large basins in one's study area, if one does not actually model their gravitational signature or flexural response (see Sect. 11.10). The CRUST2.0 (Bassin et al. 2000) and CRUST1.0 (Laske et al. 2013) models provide global maps of densities and thicknesses of three layers of sediments, though only on 2° × 2° and 1° × 1° grids, respectively. A higher-resolution model is GlobSed (Straume et al. 2019), provided on a 5′ × 5′ grid. However, this model only covers the oceans, and only provides the thickness of the whole sediment layer and not its density. Figure 11.7 shows an example of these two data sets over the Bay of Bengal.

11.3 Equivalent Topography

When one wishes to make a single estimate of T_e over a study area that includes both continent and ocean, the question of how to interpret the observed admittance or coherence arises: do we use continental or oceanic loading equations in the inversion models? The dilemma is highlighted in Fig. 11.8a, where the solid blue line shows the Bouguer admittance estimated over Australia and a considerable amount of the surrounding oceans.[3] Even though more than a quarter of the area is dry land, the long-

---

[3] The admittance follows the Airy isostasy prediction (dashed blue curve) down to about 800 km wavelength, reflecting the fact that such large loads are predominantly compensated by hydrostatic equilibrium (Sect. 1.1).

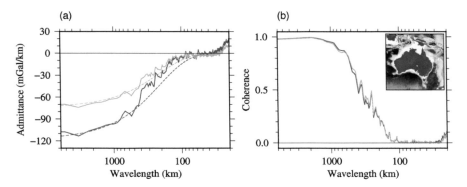

Fig. 11.8 Observed Bouguer admittance (**a**) and coherence (**b**) over Australia and surrounding oceans using the topography (blue lines) and equivalent topography (red lines), all computed using the multitaper method with NW = K = 3. The dashed curves in (**a**) show the theoretical Airy (T_e = 0) admittance for oceans (blue curve) and continents (red curve); other model parameters are z_m = 33 km, ρ_c = 2800 kg m^{-3}, ρ_m = 3300 kg m^{-3}. The inset in (**b**) shows the study area

wavelength admittance tends to the oceanic value of $-2\pi\mathcal{G}\Delta\rho_{cw} \approx -69$ mGal/km (for ρ_c = 2670 kg m^{-3} and ρ_w = 1030 kg m^{-3}, in Eq. 8.44). That is, the oceanic topography dominates that of the continents, not surprising since approximately 42% of the area has depths greater than 3000 m below sea level. The question, then, is how to invert the observed admittance: what value do we choose for the overlying fluid density in the deconvolution coefficients (Eqs. 9.5–9.8); ρ_f = 0 or ρ_f = ρ_w?

11.3.1   Calculation of the Equivalent Topography

One way to resolve the problem is to rescale the oceanic bathymetry to the depth it would have were all the seawater to be replaced by an equivalent mass of rock. Then, one can use continental loading equations in the inversion model. Consider Fig. 11.9. If a unit cross-sectional area of the ocean floor has a depth of h, then the column of water above it has a mass of $\rho_w h$ per unit area, where ρ_w is the density of seawater. An equivalent mass of rock of density ρ_c will thus have a mass of $\rho_c h'$ per unit area, where

$$\rho_c h' = \rho_w h ,$$

and where h' is the thickness of the column of rock. If the column of rock now replaces the column of seawater, then the new 'ocean' depth is

$$h_e = h - h' ,$$

giving

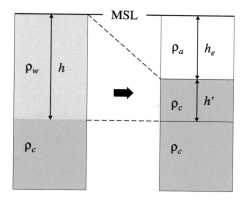

Fig. 11.9 Equivalent topography. At left is a column of seawater of bathymetric depth h and density ρ_w, overlying oceanic crust of density ρ_c. The right-hand figure shows the thickness, h', of a column of rock of the same mass as the water column. The new 'ocean' depth, h_e, from mean sea level (MSL) is the equivalent topography, though the oceanic crust is now overlain by air (of density $\rho_a = 0$)

$$h_e = \left(1 - \frac{\rho_w}{\rho_c}\right) h , \quad h < 0 . \tag{11.4}$$

The quantity h_e is the *equivalent topography*. Commonly, ρ_c is chosen to have the same value as the upper crust (or whole crust in a single-layer crust model, though see Sect. 11.3.3).

The same treatment can be applied to ice sheets. If an ice sheet of thickness t_i and density ρ_i lies on top of bedrock of density ρ_0, then an equivalent mass of rock will have thickness t_i', where

$$\rho_0 t_i' = \rho_i t_i .$$

If the surface of the ice sheet is at elevation h, then when the ice is replaced by rock, the new surface elevation is the equivalent topography,

$$h_e = h - (t - t_i') ,$$

or

$$h_e = h - \left(1 - \frac{\rho_i}{\rho_0}\right) t_i , \quad t_i > 0 . \tag{11.5}$$

Figure 11.10 shows a topographic profile through the North Atlantic Ocean, crossing the Greenland ice cap and Iceland. The equivalent topography as calculated by Eqs. 11.4 and 11.5 is shown by the red line.

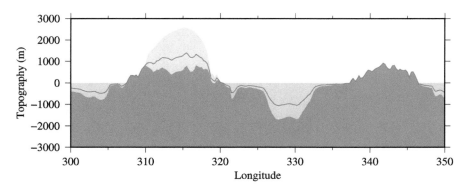

Fig. 11.10 Topographic profile through Greenland, Iceland and the North Atlantic Ocean, at 65°N latitude. Bedrock is shown in brown shading, ocean in blue and ice in cyan. The red line is the equivalent topography. Data from Earth2014 (Hirt and Rexer 2015) averaged onto a 15′ grid

11.3.2 Effect on the Admittance and Coherency

If all water and ice bodies in a study area are replaced by an equivalent mass of rock, then one may use the land-loading ($\rho_f = 0$) equations when inverting the admittance or coherence. The red line in Fig. 11.8a shows the Bouguer admittance over the study area using the equivalent topography transform, $H_e(\mathbf{k})$, in Eq. 9.11. Now, the long-wavelength admittance tends to the continental value of $-2\pi \mathcal{G}\rho_c \approx -112$ mGal/km (for $\rho_c = 2670$ kg m^{-3} in Eq. 8.44), suggesting that land-loading equations can be used in its inversion.

Note that the coherence (SRC), in contrast, is largely unaffected by the change from regular to equivalent topography (Fig. 11.8b). We can see this by considering the equations for the observed admittance and coherence computed at sea. If H is the Fourier transform of the regular topography, and H_e is the Fourier transform of the equivalent topography, then the Fourier transform of Eq. 11.4 can be written as

$$H_e = \eta H ,$$

where

$$\eta = \frac{\Delta\rho_{cw}}{\rho_c} , \tag{11.6}$$

for a single-layer crust of density ρ_c. Using Eq. 9.11, the observed admittance at sea using equivalent topography is

$$Q_{\mathrm{eq}}^{\mathrm{sea}} = \frac{\langle G H_e^* \rangle}{\langle H_e H_e^* \rangle} = \frac{\langle G \eta H^* \rangle}{\langle \eta H \eta H^* \rangle} = \frac{1}{\eta} \frac{\langle G H^* \rangle}{\langle H H^* \rangle} = \eta^{-1} Q_{\mathrm{reg}}^{\mathrm{sea}} , \tag{11.7}$$

where Q_{reg}^{sea} is the observed admittance at sea using regular topography. If we apply the same treatment to the observed coherency (Eq. 9.9), we find

$$\Gamma_{eq}^{sea} = \frac{\langle GH_e^* \rangle}{\langle GG^* \rangle^{1/2} \langle H_e H_e^* \rangle^{1/2}} = \frac{\langle G\eta H^* \rangle}{\langle GG^* \rangle^{1/2} \langle \eta H \eta H^* \rangle^{1/2}} = \Gamma_{reg}^{sea}.$$

That is, the observed coherency computed using equivalent topography is equal to that computed using regular topography, shown in Fig. 11.8b for the SRC.

Equivalent topography is used at sea or over ice caps so that land-loading flexural equations can be used to invert the admittance over the whole study area. However, this is only strictly valid when the loading is surface-only, as will now be shown. Inversion of the admittance (or SRC) is performed by minimising the chi-squared misfit between observed and predicted admittances, weighted by the observed errors, as in Eq. 9.38:

$$\chi^2 = \sum_k \Delta Q^2 ,$$

where

$$\Delta Q = \frac{Q_{obs} - Q_{pr}}{\sigma_{Q_{obs}}} .$$

Ideally, we want the chi-squared statistic to be independent of the choice of equivalent or regular topography. This is satisfied, for example, in a surface loading environment at sea, where the ΔQ term using equivalent topography is

$$\Delta Q_{eq}^{sea,T} = \frac{Q_{eq}^{sea} - Q_T^{land}}{\sigma_{eq}^{sea}} .$$

Using Eq. 11.7, and assuming that $\sigma_{eq}^{sea} = \eta^{-1} \sigma_{reg}^{sea}$ (where η is given by Eq. 11.6), the ΔQ term becomes

$$\Delta Q_{eq}^{sea,T} = \frac{\eta^{-1} Q_{reg}^{sea} + 2\pi \mathcal{G} \rho_c \, e^{-kz_m} \xi^{-1}}{\eta^{-1} \sigma_{reg}^{sea}} ,$$

where we have used Eq. 8.43 for the theoretical, surface-loading, Bouguer admittance, or

$$Q_{pr} = Q_T = -2\pi \mathcal{G} \Delta \rho_{cf} \, e^{-kz_m} \xi^{-1} ,$$

where

$$\xi = 1 + \frac{Dk^4}{\Delta \rho_{mc} g} ,$$

from Eq. 9.28. If we use the fact that $\eta \rho_c = \Delta \rho_{cw}$ from Eq. 11.6, then we find that

$$\Delta Q_{\mathrm{eq}}^{\mathrm{sea},T} \;=\; \frac{Q_{\mathrm{reg}}^{\mathrm{sea}} \;+\; 2\pi \mathcal{G} \Delta\rho_{cw}\, e^{-kz_m}\xi^{-1}}{\sigma_{\mathrm{reg}}^{\mathrm{sea}}} \;=\; \Delta Q_{\mathrm{reg}}^{\mathrm{sea},T} \;,$$

the final equality being made because ξ has the same value in continental and oceanic environments. Thus, at sea under surface loading, the chi-squared statistic is invariant under the replacement of regular by equivalent topography.

The same cannot be said when the loading is exclusively internal. If we want to invert the equivalent-topography admittance at sea using land-loading equations, the appropriate ΔQ term is now

$$\Delta Q_{\mathrm{eq}}^{\mathrm{sea},B} \;=\; \frac{Q_{\mathrm{eq}}^{\mathrm{sea}} \;-\; Q_{B}^{\mathrm{land}}}{\sigma_{\mathrm{eq}}^{\mathrm{sea}}} \;.$$

As before, we have that $Q_{\mathrm{eq}}^{\mathrm{sea}} = \eta^{-1} Q_{\mathrm{reg}}^{\mathrm{sea}}$ and $\sigma_{\mathrm{eq}}^{\mathrm{sea}} = \eta^{-1}\sigma_{\mathrm{reg}}^{\mathrm{sea}}$. The theoretical, internal-loading, Bouguer admittance is given by Eq. 8.46, or

$$Q_B \;=\; -2\pi\mathcal{G}\Delta\rho_{cf}\, e^{-kz_m}\phi \;,$$

where

$$\phi \;=\; 1 + \frac{Dk^4}{\Delta\rho_{cf}\,g} \;,$$

from Eq. 9.28. Importantly, ϕ does not have the same value in continental and oceanic environments, unlike ξ, because of the presence of the term in ρ_f. Thus, the ΔQ term becomes

$$\Delta Q_{\mathrm{eq}}^{\mathrm{sea},B} \;=\; \frac{\eta^{-1} Q_{\mathrm{reg}}^{\mathrm{sea}} \;+\; 2\pi\mathcal{G}\rho_c\, e^{-kz_m}\phi_{\mathrm{land}}}{\eta^{-1}\sigma_{\mathrm{reg}}^{\mathrm{sea}}} \;,$$

or, because $\eta\rho_c = \Delta\rho_{cw}$,

$$\Delta Q_{\mathrm{eq}}^{\mathrm{sea},B} \;=\; \frac{Q_{\mathrm{reg}}^{\mathrm{sea}} \;+\; 2\pi\mathcal{G}\Delta\rho_{cw}\, e^{-kz_m}\phi_{\mathrm{land}}}{\sigma_{\mathrm{reg}}^{\mathrm{sea}}} \;\neq\; \Delta Q_{\mathrm{reg}}^{\mathrm{sea},B} \;.$$

The last inequality arises because we cannot turn the ϕ_{land} into a ϕ_{sea}, and a true $\Delta Q_{\mathrm{reg}}^{\mathrm{sea},B}$ would contain the term $2\pi\mathcal{G}\Delta\rho_{cw}\, e^{-kz_m}\phi_{\mathrm{sea}}$, where

$$\phi_{\mathrm{land}} \;=\; 1 + \frac{Dk^4}{\rho_c g} \;, \qquad \phi_{\mathrm{sea}} \;=\; 1 + \frac{Dk^4}{\Delta\rho_{cw} g} \;.$$

The error incurred in using land-loading equations with equivalent topography at sea is illustrated in Fig. 11.11, which shows the observed admittances from Fig. 11.8a, but now fitted (at wavelengths above 400 km) with combined-loading, theoretical admittance curves. The equivalent-topography admittance (solid red line) is best-fitted by a land-loading (dashed red) curve of $T_e = 40$ km. However, when the regular-

Fig. 11.11 As Fig. 11.8a, except the dashed curves are now theoretical, combined-loading, Bouguer admittances (Eq. 9.33, with $f = 1$). The red dashed curve is continental with $T_e = 40$ km; the blue dashed curve is oceanic with $T_e = 33$ km; the red dotted curve is continental with $T_e = 33$ km; other model parameters are given in Fig. 11.8

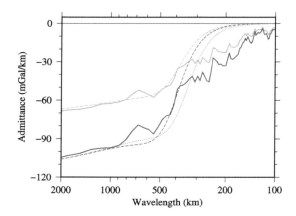

topography admittance (solid blue line) is fitted with a sea-loading (dashed blue) curve, the best-fitting T_e is 33 km, showing that use of equivalent topography can sometimes provide overestimates of T_e when inverting the admittance. This is also shown by the red dotted curve—from a land-loading model with $T_e = 33$ km—which does not fit the observed equivalent-topography admittance as well as the $T_e = 40$ km curve.

11.3.3 *Equivalent Topography and the Bouguer Correction*

The simple Bouguer correction can be calculated directly from equivalent topography. This nice shortcut will obviously save a bit of time if one's study area includes oceanic and ice-capped regions, the Bouguer correction not having to be computed in a separate step to the equivalent topography.

Consider the case of the oceans first. In Sect. 8.2.5, we saw that the simple Bouguer correction in the oceans is given by the equation at scenario D in Table 8.2, or

$$\delta g_B = -2\pi \mathcal{G} (\rho_0 - \rho_w) h , \tag{11.8}$$

for $h < 0$. That is, we remove the gravitational attraction of seawater, and add back the gravitational attraction of rock. Furthermore, we have just seen that the equivalent topography in the oceans is given by Eq. 11.4, or

$$h_e = \frac{\rho_0 - \rho_w}{\rho_0} h , \tag{11.9}$$

for $h < 0$ (where we have used the Bouguer reduction density, ρ_0, in place of the crustal density, ρ_c). That is, the seawater has been replaced by an equivalent mass of rock. Now, recognising the conceptual similarity between the Bouguer correction and the equivalent topography, if we substitute Eq. 11.9 into Eq. 11.8 we obtain

$$\delta g_B = -2\pi \mathcal{G} \rho_0 h_e . \tag{11.10}$$

That is, the Bouguer correction at sea corresponds to the addition of the gravitational attraction of rock of density ρ_0, but smaller thickness h_e.

Now consider the case of an ice cap with its base above mean sea level. The simple Bouguer correction here is given by the equation at scenario E in Table 8.2, or

$$\delta g_B = -2\pi \mathcal{G} \left[\rho_i t_i + \rho_0 (h - t_i) \right] , \tag{11.11}$$

for $h > t_i > 0$. We have also just seen that the equivalent topography of such an ice cap is given by Eq. 11.5, or

$$h_e = h - \left(\frac{\rho_0 - \rho_i}{\rho_0} \right) t_i , \tag{11.12}$$

for $h > t_i > 0$. Now, rearranging Eq. 11.12 we obtain

$$\rho_0 h_e = \rho_0 (h - t_i) + \rho_i t_i .$$

Substituting this into Eq. 11.11 then gives

$$\delta g_B = -2\pi \mathcal{G} \rho_0 h_e ,$$

which is identical to Eq. 11.10.

Indeed, we can perform similar algebraic endeavours with all the environmental scenarios in Table 8.2, and will always obtain Eq. 11.10. This means that if one is undertaking a T_e study using equivalent topography, one can compute the simple Bouguer correction directly from the equivalent topography and then apply that to the free-air anomaly in order to obtain the simple Bouguer anomaly. Note, though, that the choice of upper crustal density in the equivalent topography calculation must equal the Bouguer reduction density (though one is free, of course, to choose any realistic value, and is not limited to $\rho_0 = 2670\,\mathrm{kg\,m^{-3}}$). Table 11.3 gives equivalent topography formulae corresponding to the scenarios in Table 8.2 and Fig. 8.4.

11.4 Depth and Density Tests

In this section, we test the effect upon T_e estimates of variations in the Bouguer reduction density, and in the densities and depths to interfaces of a crust model.

Table 11.3 Equivalent topography formulae for the environmental scenarios shown in Fig. 8.4 (see also Table 8.2). The densities are: ρ_0 is the density of the upper crust (2670 kg m^{-3}), ρ_w is the density of seawater (1030 kg m^{-3}), ρ_l is the density of fresh lake water (1000 kg m^{-3}) and ρ_i is the density of ice (917 kg m^{-3}). The depths and thicknesses are: h is the topographic surface (i.e. the distance from mean sea level (MSL) to either the surface of the rocky topography, the sea floor, the lake surface or the ice sheet surface), d_l is the depth of the lake and t_i is the thickness of the ice sheet/cap

Environment	Equivalent topography	Conditions
A: Land	$h_e = h$	$h > 0$
B: Lake (bottom above MSL)	$h_e = h - \frac{\rho_0 - \rho_l}{\rho_0} d_l$	$h > d_l > 0$
C: Lake (bottom below MSL)	$h_e = h - \frac{\rho_0 - \rho_l}{\rho_0} d_l$	$d_l > h > 0$
D: Ocean	$h_e = \frac{\rho_0 - \rho_w}{\rho_0} h$	$h < 0$
E: Ice (bottom above MSL)	$h_e = h - \frac{\rho_0 - \rho_i}{\rho_0} t_i$	$h > t_i > 0$
F: Ice (bottom below MSL)	$h_e = h - \frac{\rho_0 - \rho_i}{\rho_0} t_i$	$t_i > h > 0$

11.4.1 Bouguer Reduction Density and Terrain Corrections

Our first test of density assumptions lies with the choice of reduction density when forming the simple Bouguer anomaly (see Sect. 8.2.5). Conventionally, a value of 2670 kg m^{-3} is chosen, though this is not set in stone. As the simple Bouguer correction depends upon both the height and density of the topography, we choose a mountainous region as the test area, specifically the Rocky Mountains of the western United States where the topography varies between approximately 3600 m above sea level to 5700 m below sea level (on the 10 km grid used here). Figure 11.12 shows two T_e maps over the area,[4] one using a reduction density of 2500 kg m^{-3}, the other using 2800 kg m^{-3}. There are noticeable differences between the two maps, often revealed in the presence of 'bullseyes' such as that near 121°W, 39°N in Fig. 11.12b, which is not present in Fig. 11.12a. Thus, the choice of Bouguer reduction density can have an effect upon T_e estimates.

We can also ascertain whether or not the application of terrain corrections to the simple Bouguer anomaly to form a complete Bouguer anomaly (see Sect. 8.2.5) makes much difference to T_e estimates. Using a mountainous region ensures high values of the terrain correction since it is essentially proportional to the gradient of the topography rather than its absolute value; though it should be appreciated that the rather coarse grid (10 km) will reduce these gradients, resulting in smaller terrain corrections. Nevertheless, comparison of Fig. 11.13a and b shows that, on land, T_e

---

[4] Note, T_e was computed using data over the whole of the North American continent, not just the area shown.

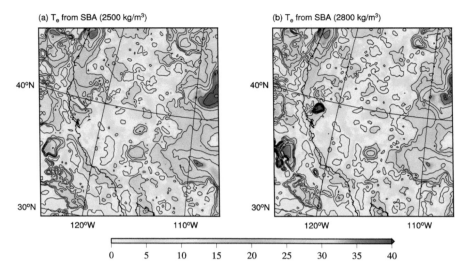

Fig. 11.12 T_e over the western United States from the simple Bouguer anomaly using two different reduction densities: **a** 2500 kg m^{-3}; **b** 2800 kg m^{-3}. Fan wavelet method with $|\mathbf{k}_0| = 3.081$; gravity and topography data from EGM2008; load deconvolution with the Bouguer SRC; single-layer crust with $z_m = 35$ km, $\rho_c = 2800$ kg m^{-3}, $\rho_m = 3300$ kg m^{-3}; contours every 5 km; transverse Mercator projection

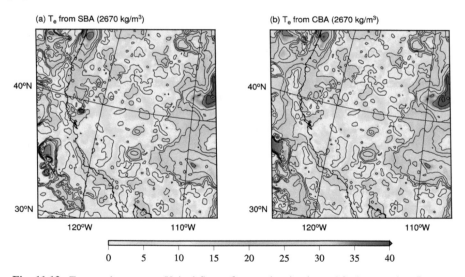

Fig. 11.13 T_e over the western United States from **a** the simple, and **b** the complete Bouguer anomaly, both using a reduction density of 2670 kg m^{-3}. Other details as Fig. 11.12

is not significantly affected, though the same is not true over the oceans. Continental terrain corrections are always greater than or equal to zero, while those computed at sea can be positive or negative and generally have larger magnitudes. Thus, in general, the use of complete Bouguer anomalies is more important at sea than on land, where simple Bouguer anomalies may suffice. Note though that it is essential to compute terrain corrections using the same reduction density as was used to compute the simple Bouguer correction.

11.4.2 Crustal Structure

The sensitivity of T_e estimates to assumptions about crustal structure is also readily tested. Figure 11.14a shows T_e estimated using a single-layer crust model with a uniform depth to Moho of 30 km and a constant crust density of 2835 kg m^{-3} (these values are the means taken from the CRUST1.0 model over the region). If the Moho depth of the inversion model is increased by 5 km but the crustal density held fixed, then the T_e map shown in Fig. 11.14b is obtained. Conversely, if the crust density is increased by 100 kg m^{-3} but the Moho depth held fixed, then the T_e map shown in Fig. 11.14c is obtained. It is clear that the Moho depth change has a much greater effect than the density change, with the thicker crust (deeper internal load) resulting in a higher T_e.

Figure 11.14d shows the T_e obtained when the crust of the inversion model has three layers, and the thicknesses and densities of each layer are allowed to vary horizontally, using the values given by the CRUST1.0 model (Sect. 11.2.4). Much of Fig. 11.14d is similar to Fig. 11.14a, except in regions where the depth to Moho is very different from the mean value of 30 km, that is, over the far eastern extents and over oceanic lithosphere, where the depth to Moho is in the range 10–16 km. It is therefore advisable to use a crust model when estimating T_e, at the very least one that reflects actual variations in Moho depth.

11.5 Estimation of the Load Ratio

As described in Sects. 9.4.2 and 9.4.3, the load ratio, $f(k)$ (Sect. 9.3.2), may be estimated as a 'by-product' of load deconvolution. This is because load deconvolution needs to reconstruct the initial loads in order to evaluate the predicted coherence. The load ratio is a wavenumber-dependent function, however, so rather than quoting its value at every wavenumber, most studies provide the value of f at the predicted coherence transition (rollover) wavenumber (k_t; Sect. 9.3.3). The solid magenta line in Fig. 11.15 shows the load ratio (plotted as the internal load fraction, F, using Eq. 9.22) estimated from a synthetic model with uniform $T_e = 50$ km and $f = 1$ (see Sect. 10.3). The magenta dot on that line shows the recovered value of the internal load fraction at the predicted SRC rollover wavelength as $F(k_t) = 0.55$, or $f(k_t) =$

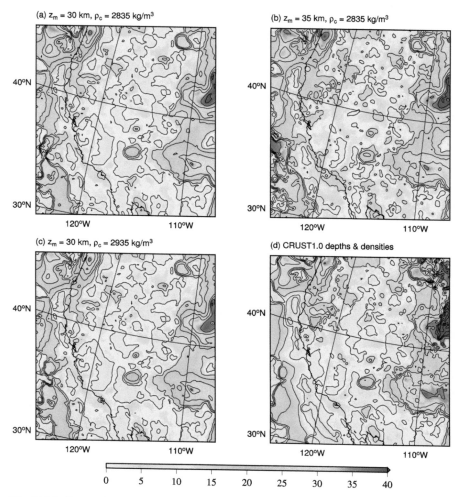

(a) $z_m = 30$ km, $\rho_c = 2835$ kg/m³

(b) $z_m = 35$ km, $\rho_c = 2835$ kg/m³

(c) $z_m = 30$ km, $\rho_c = 2935$ kg/m³

(d) CRUST1.0 depths & densities

Fig. 11.14 T_e over the western United States from the complete Bouguer anomaly using different crust models. **a** Single-layer crust with indicated depth to Moho and density. **b** Single-layer crust with depth to Moho increased by 5 km and density the same. **c** Single-layer crust with crustal density increased by 100 kg m⁻³, and depth to Moho the same. **d** Three-layer crust as given by CRUST1.0 model. In **a**–**c** mantle density is 3271 kg m⁻³. Fan wavelet method with $|\mathbf{k}_0| = 3.081$; gravity and topography data from EGM2008; load deconvolution with the Bouguer SRC; contours every 5 km; transverse Mercator projection

Fig. 11.15 Internal load fractions ($F(k)$, solid lines) recovered from two synthetic plate models with uniform T_e = 50 km and $f = 1$ ($F = 0.5$), shown with their respective predicted SRCs (dashed lines). The models have initial loads with equal (magenta lines), and differing (green lines) fractal dimensions. The circles show the values of F at the SRC rollover wavelength. Multitaper method in global mode with NW = 1, $K = 1$

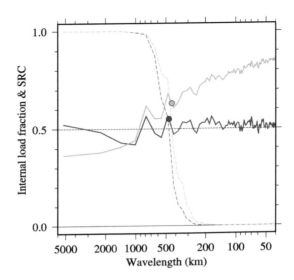

1.22. Since the synthetic model was generated using $f = 1$, the recovered value is reasonable.

The load ratio recovered from real data often exhibits a trend that sees it increase with increasing wavenumber. Swain and Kirby (2003) found that this trend could be reproduced in synthetic data by assigning a higher fractal dimension to the initial internal load than that assigned to the initial surface load (see Sect. 10.2). The solid green line in Fig. 11.15 shows the load fraction estimated from a synthetic model with an initial surface load of fractal dimension 2.5 and an initial internal load of fractal dimension 2.99 (the synthetic model that gave the magenta line has loads with equal fractal dimensions of 2.5). The recovered value of $f(k_t)$ is 1.7 ($F(k_t) = 0.63$), higher than the case with equal fractal dimensions.

It is possible that this trend in the load ratio causes a correlation between recovered values of T_e and $f(k_t)$ that has been observed in some studies (Tassara et al. 2007). Such a correlation is shown in Fig. 11.16, where the synthetic gravity and topography were constructed using the elliptical T_e model shown in Fig. 10.10a, with initial loads of unequal fractal dimension and $f = 1$. It is clear that the map of internal load fraction exhibits the elliptical structure of the model T_e. Why this should be so is revealed in Fig. 11.16c, which shows a cross-section in the xs-plane through $F(k, \mathbf{x})$—the 3D function containing $F(k)$ at each spatial location, calculated at step 10 in Sect. 9.4.3. Also shown is the locus of the predicted SRC rollover (the green line), and it is the value of $F(k, \mathbf{x})$ on this green line that gives $F(k_t, \mathbf{x})$, plotted in Fig. 11.16b. An upward trend in the load ratio with increasing wavenumber (decreasing wavelength) can be seen at all spatial locations, but the trend has different patterns depending upon the model (and recovered) T_e values. So as the predicted SRC rollover migrates to longer wavelengths over the high-T_e ellipse, it intersects regions of $F(k, \mathbf{x})$ with differing trends, compared to the low-T_e regions, and this imprints the pattern of

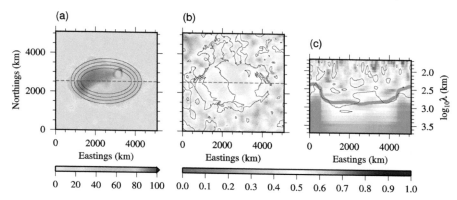

Fig. 11.16 Maps of **a** T_e and **b** internal load fraction at the SRC rollover wavelength, recovered from a synthetic model with an elliptical-shaped T_e distribution and $f = 1$; the contours in (**a**) show the model values at $T_e = 25$, 50 75 and 100 km. **c** Slice in the xs-plane—along the dashed line in (**b**)—through $F(k, \mathbf{x})$; the green line shows the predicted SRC rollover wavelength. The $F = 0.5$ ($f = 1$) contour is shown in black in (**b**) and (**c**). Fan wavelet method with $|\mathbf{k}_0| = 5.336$

the local T_e variations into the resulting map of $F(k_t)$. It is not fully understood at present why the recovered load ratio should be influenced by the regional T_e value. But since the load ratio can potentially provide useful geological information (e.g. Fig. 11.17), this topic should be investigated further.

11.6 Deconvolution with the Admittance

Although Forsyth (1985) developed it as such, the load deconvolution method does not exclusively apply to the Bouguer coherence or SRC, and one can readily use the (Bouguer or free-air) admittance, as Pérez-Gussinyé and Watts (2005) did when they estimated T_e over Europe, comparing it with T_e from the Bouguer coherence. Instead of computing the predicted SRC (Eqs. 9.36 or 9.37), one simply uses the predicted admittance, as given by Eqs. 9.18 or 9.20. In practice, if one is using the multitaper method, at step 15 in Sect. 9.4.2 one evaluates the predicted, radially-averaged admittance for the window with

$$Q_{pr,R}(k) = \frac{\mathsf{Re}\big[\bar{S}_{gh}(k)\big]}{\bar{S}_{hh}(k)} .$$

Alternatively, if one is using the wavelet method, at step 11 in Sect. 9.4.3 one evaluates the predicted, local wavelet admittance at location $\mathbf{x}_i$ with

$$Q_{pr,R}(k_e, \mathbf{x}_i) = \frac{\mu_T(k_e)\kappa_T(k_e) + \mu_B(k_e)\kappa_B(k_e)f^2(k_e, \mathbf{x}_i)}{\kappa_T^2(k_e) + \kappa_B^2(k_e)f^2(k_e, \mathbf{x}_i)} .$$

Fig. 11.17 Internal load fraction at the SRC rollover wavenumber, $F(k_t)$, over the western United States, corresponding to the T_e map in Fig. 11.14d. Fan wavelet method with $|\mathbf{k}_0| = 3.081$; gravity and topography data from EGM2008; load deconvolution with the Bouguer SRC; transverse Mercator projection

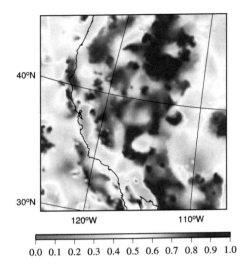

One then evaluates the chi-squared statistic (Eq. 9.38) between observed and predicted admittances, using the errors on the observed admittance as weights

$$\chi^2(T_e) = \sum_k \left(\frac{Q_{obs,R}(k) - Q_{pr,R}(k; T_e)}{\sigma_{Q_{obs,R}}(k)} \right)^2 . \qquad (11.13)$$

Figure 11.18a shows T_e estimated over the western United States from the free-air admittance using load deconvolution; it can be compared with Fig. 11.14d which was estimated from the Bouguer SRC. Indeed, a comparison shows that the admittance-derived estimate has much more variability than that derived from the SRC, containing 'bulls-eyes' and regions of exceptionally steep T_e gradient. Rather than being real features, these are artefacts, reflecting the observations in Sects. 8.6 and 9.3.3 that the admittance is very sensitive to variations in crust and mantle parameters, more so than is the coherence. This sensitivity was also noted by Kirby and Swain (2008) when using synthetic data, and is somewhat ironic considering that the observed admittance is noted for being able to deal with noise better than the coherence or SRC (Sect. 5.3.2).

11.7   Uniform-f Inversion

Figure 11.18 also shows T_e estimated over the region from the free-air admittance using the uniform-f method (Sect. 9.4.1.1). While the general trends are similar, there is more detail in the load deconvolution map (Fig. 11.18a). Recall that load deconvolution allows for the reconstruction of the initial loads—given a certain T_e—

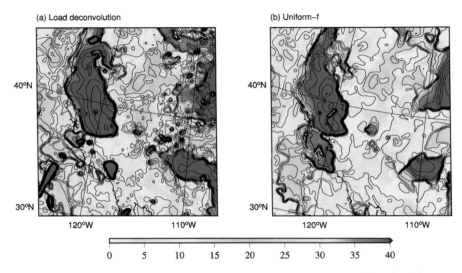

Fig. 11.18 T_e over the western United States from the free-air admittance using **a** load deconvolution, and **b** the uniform-f method. Other details as for Fig. 11.14d

from the observed gravity and topography, which imparts a wavenumber-dependence to the load ratio, $f(k)$; it then adjusts T_e until the (predicted) coherence between the reconstructed loads matches the observed coherence between the actual gravity and topography. In contrast, uniform-f inversion seeks only to fit the theoretical admittance to the observed admittance[5] by varying two, constant parameters: T_e and (a wavenumber-independent) f. The chi-squared statistic (Eq. 11.13) between observed and theoretical admittances is thus given by

$$\chi^2(T_e, f) = \sum_k \left(\frac{Q_{obs,R}(k) - Q_{th}(k; T_e, f)}{\sigma_{Q_{obs,R}}(k)} \right)^2, \qquad (11.14)$$

where the theoretical admittance is given by Eq. 9.24, reproduced here as

$$Q_{th}(k) = \frac{\mu_T(k)\kappa_T(k) + \mu_B(k)\kappa_B(k)f^2}{\kappa_T^2(k) + \kappa_B^2(k)f^2}.$$

So the detail observed in Fig. 11.18a is likely due to the fact that load deconvolution allows for a wavenumber-dependent load ratio, effectively providing an extra degree of freedom when solving for T_e that is not present in the wavenumber-independent load ratio of the uniform-f method, rendering that smoother, in general (Fig. 11.18b).

---

[5] Note that the coherence can be used with the uniform-f method, though this has not been common practice.

11.8 T_e Errors

Errors on T_e estimated from load deconvolution are obtained from the chi-squared statistic, Eq. 9.38, as discussed in Sect. 9.4.1.6. They can be presented as the lower and upper confidence limits (T_{lo} and T_{hi}, respectively), or as the mean of the differences of these limits with the best-fitting T_e value, thus:

$$\varepsilon_{T_e} = \frac{(T_{hi} - T_e) + (T_e - T_{lo})}{2} = \frac{T_{hi} - T_{lo}}{2} = \frac{\Delta T_e}{2},$$

where ΔT_e is the confidence interval (see Fig. 9.13). An example of this mean error corresponding to the T_e map in Fig. 11.14d is shown in Fig. 11.19.

Occasionally, though, one or both of the confidence limits is undefined. If this occurs the affected limit is usually the upper limit, as shown in Fig. 11.20. Here, the chi-squared curve rises reasonably steeply at values of T_e below the best-fit value, providing a well-defined estimate of T_{lo}, and indicating that the predicted SRC does not provide a good fit to the observed SRC when $T_e < T_{lo}$. However, at values of T_e greater than the best-fit value, the flatter chi-squared curve indicates that a very wide range of T_e values gives predicted SRCs that all match the observed SRC fairly well. In this case the increase in chi-squared is not sufficient to exceed the $\Delta \chi^2$ value corresponding to the chosen 95% confidence level, and T_{hi} is undefined. An upper confidence limit could be identified, though, if one is prepared to choose a smaller confidence level, say 68% where $\Delta \chi^2 = 1$ (see Tables 9.1 or 11.4).

When the uniform-f method is used there are two parameters for which the errors must be estimated: T_e and f. Unlike load deconvolution, which does not provide a mechanism for the estimation of errors on the load ratio, errors on f may indeed be

Fig. 11.19 Mean error (95% confidence) on T_e over the western United States, corresponding to the T_e map in Fig. 11.14d, using jackknife error estimates on the SRC. Fan wavelet method with $|\mathbf{k}_0| = 3.081$; gravity and topography data from EGM2008; load deconvolution with the Bouguer SRC; contours every 2 km; transverse Mercator projection

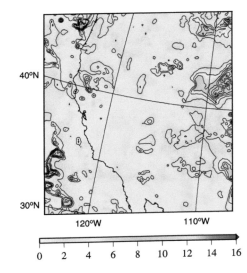

Fig. 11.20 The chi-squared statistic (red curve) for a range of T_e values. The best-fit T_e is 105.2 km. The lower 95% confidence limit (T_{lo}) is 48.2 km, while the upper 95% confidence limit (T_{hi}) is undefined. $\Delta\chi^2 = 4$. See also Fig. 9.13

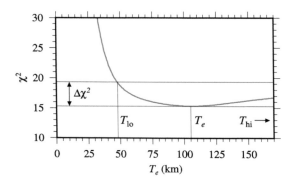

Table 11.4 $\Delta\chi^2$ values corresponding to the indicated confidence level (CL) for one and two inversion parameters (Press et al. 1992)

CL:	68.27%	90%	95.45%	99%	99.73%	99.99%
1 parameter:	1.00	2.71	4.00	6.63	9.00	15.10
2 parameters:	2.30	4.61	6.18	9.21	11.80	18.40

Fig. 11.21 The chi-squared misfit surface from uniform-f inversion of the free-air admittance. The red cross shows the location of the best-fitting parameter values (at $T_e = 128.6$ km, $F = 0.47$). The thick black contour shows the 95% confidence limits ($\Delta\chi^2 = 6.18$), giving $T_{lo} = 101.2$ km, $T_{hi} = 152.5$ km, $F_{lo} = 0.39$, $F_{hi} = 0.52$. Colours and contours show $\Delta\chi^2$ for each T_e and F in the grid, relative to the global minimum χ^2 value

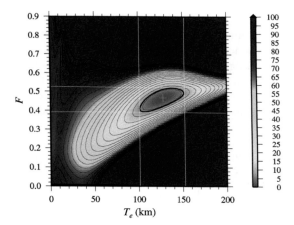

estimated in the uniform-f method. Here, a 2D plot of $\Delta\chi^2$ values is created using Eq. 11.14 for T_e and F values over a large range, as in Fig. 11.21. Note that it is simpler to use F, rather than f, as the former is bounded. For a given confidence level, a contour of its corresponding $\Delta\chi^2$ value (Table 11.4) provides the confidence limits on both T_e and F, as shown in Fig. 11.21.

11.9 Noise Detection

The issue of noise (as defined in Sect. 9.6.1) and how it affects T_e estimates made using load deconvolution with the coherence (or SRC) has been a topic of much debate recently (see the summaries by Kirby (2014) and Watts (2021)). A model describing such noise—unexpressed loading—was proposed by McKenzie (2003), but as discussed in Sect. 9.6.2 it seems as though such a model has very limited correspondence with reality. Nevertheless, the question of whether or not noise—or rather, signals that are not modelled by load deconvolution—is present in a study area cannot be dismissed out of hand.

Ideally, we would like to have some signal processing technique that would remove the noise from the data before we set about estimating T_e. Unfortunately, this does not yet exist. Our second choice would then be some method that detects noise, even though it may not be able to remove it. And, fortunately, such a method does exist and has been described in Sect. 5.3.3: it is simply the imaginary part of the observed coherency, or more specifically, its normalised square (NSIC, Eq. 5.19). As shown in Fig. 5.7, when band-limited noise was added to one of two (non-flexural) synthetic signals, their SRC decreased and their NSIC increased in that waveband.

The method also has the advantage that it can distinguish between 'natural' and noise-induced decreases in coherence, with the NSIC registering an increase with the latter but not the former (Fig. 5.7d). This feature is important in flexural studies, because the load deconvolution model predicts low coherence—in certain wavebands—between both Bouguer and free-air anomalies and the topography.

We have not discussed the free-air coherence in this book yet. This is because the Bouguer coherence (SRC)—with its sharp, diagnostic rollover from 1 to 0—is much more suited to T_e estimation. In contrast, free-air coherence curves are much more variable, meaning that observed free-air coherence data could be fit equally well by curves with different combinations of parameters such as T_e, f and z_m. If we use our familiar example of a single-layer crust, then the procedure in Sect. 9.3.3 gives us the equation for the free-air coherency (i.e. with $\beta_g = 1$) as

$$\Gamma_{th}(k) = \frac{\left(\xi e^{-kd} - e^{-kz_m}\right)\xi - \left(\phi e^{-kz_m} - e^{-kd}\right)f^2r^2}{\left[\left(\xi e^{-kd} - e^{-kz_m}\right)^2 + \left(\phi e^{-kz_m} - e^{-kd}\right)^2 f^2r^2\right]^{1/2}\left[\xi^2 + f^2r^2\right]^{1/2}},$$
$$(11.15)$$

where ξ and ϕ are given by Eq. 9.28, r is given by Eq. 9.30, z_m is the Moho depth, and one would set ocean depth, d, to zero for continental environments. The variation of the free-air coherence (SRC) with T_e and f is shown in Fig. 11.22. It can be seen that the free-air coherence is uniformly high when loading is predominantly surface-only, no matter how rigid the lithosphere. As the contribution from the internal load increases, though, the free-air coherence decreases, the more so for stronger

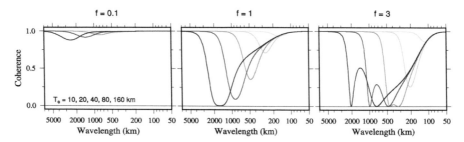

Fig. 11.22 Theoretical free-air coherence curves from the square of Eq. 11.15 for a single-layer continental crust, showing the variation with T_e for three different load ratio (f) values. The T_e values are 10, 20, 40, 80 and 160 km, and lighter curves show smaller values of T_e, darker curves show higher values, also indicated in the left-hand panel. Other plate parameters are $\rho_c = 2800\,\mathrm{kg\,m^{-3}}$, $\rho_m = 3300\,\mathrm{kg\,m^{-3}}$ and $z_m = 35\,\mathrm{km}$

lithosphere. Hence low free-air coherence is symptomatic of high T_e and dominant internal loads.

The free-air coherence has been erroneously proposed as an alternative method of noise detection (McKenzie 2003) because, the proposal went, it should have a value of 1 if all internal loads are expressed in the topography (see Sect. 9.6). Unlike the Bouguer anomaly, the free-air anomaly retains the gravitational signature of the topographic masses, so if the free-air coherence is low, then this must be due to the presence of unexpressed internal loads—or so McKenzie (2003) argued. However, as is clear from Fig. 11.22, there are many causes of low free-air coherence (high T_e, dominant internal loads, etc. as noted above), all predicted and explained by Forsyth (1985)'s load deconvolution model, making its use as a diagnostic for noise questionable.[6]

In Sect. 5.3.3, the NSIC noise detection method was applied to two random, fractal signals generated with a synthetic coherence described in the Appendix to Chap. 5 (Fig. 5.7). We now repeat that experiment, but use synthetic flexural signals instead (generated as described in Sect. 10.4, with $T_e = 40\,\mathrm{km}$ and $f = 1.5$). Figure 11.23a shows the observed global wavelet SRC between the synthetic free-air anomaly and topography, which agrees very well with the theoretical prediction for the model parameters. And even though the SRC has low values around 500 km wavelength, the NSIC (Fig. 11.23c) is approximately zero here, showing that low free-air SRC does not necessarily imply high free-air NSIC. When band-limited noise was added to the free-air anomaly, the free-air SRC was reduced in the waveband (with some leakage into adjacent harmonics), as expected (Fig. 11.23b). The corresponding NSIC (Fig. 11.23d), though, now shows high values within the waveband of the noise, revealing the presence of uncorrelated harmonics not predicted by the load deconvolution flex-

---

[6] The argument is an example of petitio principii (begging the question, or assuming the conclusion), because the agenda in McKenzie (2003) was to show that continental T_e could never exceed 25 km and that internal (expressed) loading was non-existent ($f \to 0$): T_e can only be reliably estimated when the free-air coherence is high, but the free-air coherence is only high when T_e and f are low, therefore any study that finds that T_e and f are high must be unreliable.

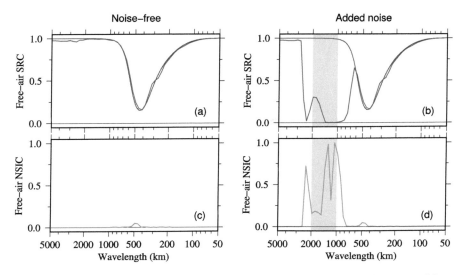

Fig. 11.23 **a** Observed SRC (red line) between the free-air anomaly and topography generated from a synthetic plate of $T_e = 40$ km and $f = 1.5$; the theoretical free-air SRC for these plate parameters is shown by the blue line. **b** When band-limited noise (between 1000–2000 km wavelength, shown by the grey-shaded region) was added to the free-air anomaly, the SRC given by the red line was observed. The solid blue line is identical to that from panel (**a**). **c** The normalised squared imaginary free-air coherency (NSIC) corresponding to the noise-free model in (**a**). **d** The NSIC corresponding to the added-noise model in (**b**). Fan wavelet method in global mode with $|\mathbf{k}_0| = 7.547$

ural model. Hence, a comparison of the free-air SRC and NSIC shows the difference between 'natural' and noise-induced reductions in the SRC.

Figure 11.24a shows that the observed Bouguer SRC for the model in Fig. 11.23b is greatly affected by the added noise (see also Figs. 5.6a and 9.15a, and the accompanying text). Indeed, an inversion of this SRC might very well pick the longer-wavelength rollover as being diagnostic of the local T_e, greatly overestimating the reality (dashed blue line). Importantly, however, if the noise has a waveband far from the Bouguer SRC rollover, T_e estimated from the Bouguer SRC will not be biased. Figure 11.24b shows the Bouguer SRC when the noise is restricted to a waveband below the Bouguer SRC rollover, which is not affected: T_e may be reliably estimated in such a case.

So much for synthetic data. Figure 11.25 shows a real-world example in North America, similar to the study by Kirby and Swain (2009) who masked out T_e values (Fig. 11.25b) over regions that were potentially contaminated by uncorrelated, model noise. The Mid-continental Rift (MCR) is such an area. Here, inversion of the Bouguer SRC yields T_e estimates of around 100 km or more (Fig. 11.25a), but an analysis of the actual SRC (observed and predicted) together with the free-air NSIC reveals that noise in the waveband ~300–900 km might be reducing the Bouguer SRC, creating an artificial rollover at ~900 km wavelength (Fig. 11.25d). Indeed, there is a potential remnant of the true, flexural rollover at ~300 km wavelength,

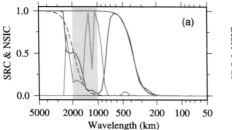 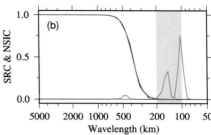

Fig. 11.24 **a** Observed Bouguer SRC (red line) corresponding to the free-air SRC in Fig. 11.23b, with the free-air NSIC from Fig. 11.23d reproduced (green line). The grey-shaded region shows the bandwidth of the added noise (1000–2000 km wavelength), while the solid, blue line shows the theoretical Bouguer SRC for $T_e = 40$ km and $f = 1.5$. The dashed blue line shows the SRC for $T_e = 300$ km and $f = 1$, values which might be obtained from an inversion of the longer-wavelength rollover. **b** When noise with a bandwidth of 100–200 km wavelength (shown by the grey-shaded region) is added to the Bouguer anomaly, the Bouguer SRC rollover is unaffected. The free-air NSIC is shown in green. Fan wavelet method in global mode with $|\mathbf{k}_0| = 7.547$

and a crude fit of a theoretical curve to this remnant yields a T_e of ∼30 km. Hence, a map like that in Fig. 11.25c can be created, which shows the mean value of the free-air NSIC in the waveband just below the Bouguer SRC rollover; if this mean value is greater than 0.5, then it is possible that noise has reduced the Bouguer SRC at the location, creating an artificial rollover which might have been erroneously interpreted by the inversion algorithm, thus giving a T_e overestimate. The message, though, is that it is worth checking such regions by generating plots such as Fig. 11.25d.

Sometimes plots such as Fig. 11.25e are observed. There is also noise at this location, but it is at longer wavelengths than the Bouguer SRC rollover and has not reduced it. There is also noise in the ∼160–400 km waveband, but importantly there is none in the band below the rollover (grey box), meaning that the best-fit T_e value of 123 km here is reliable (Kirby and Swain 2014).

11.10   Accounting for Sediments

Most of the surface of the Earth, continental and oceanic alike, is covered with a layer of sediments (Fig. 11.26). The majority (just over half of the Earth's surface area) of these sediments are in layers less than 500 m thick, and so will play only a minor role in flexure. However, some sedimentary basins have thicknesses greater than 5 km. So it is advisable, when estimating T_e, to account for the mass of the sediments, especially since their density is often quite less than the crustal densities assumed in flexural models.

We can get an idea of the effect of sediments on T_e estimates by choosing a two-layer crust model where the upper layer represents the sediments. The theoretical

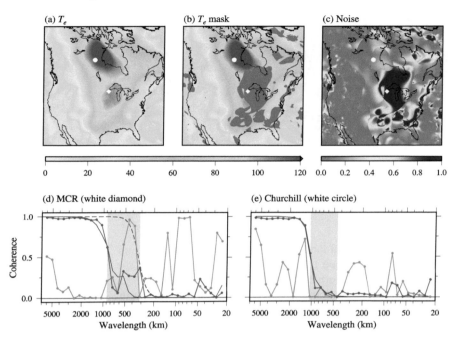

Fig. 11.25 a T_e over North America, from inversion of the Bouguer SRC using the fan wavelet method with $|\mathbf{k_0}| = 5.336$. **b** As (**a**), except with regions potentially biased by noise masked out in grey. **c** The noise field (free-air NSIC). **d** The Bouguer SRC (red) at a location near the Mid-continental Rift (MCR, white diamonds in (**a**)–(**c**)), showing also the free-air NSIC (green), and the predicted SRC corresponding to the best-fitting T_e of 103 km (solid blue); the dashed blue line is the theoretical Bouguer SRC for $T_e = 30$ km; the grey box shows the waveband potentially affected by noise. **e** As (**d**) but for a location in the Churchill province (white circles in (**a**)–(**c**)), where there is no noise in the grey box and the best-fitting T_e is 123 km

Fig. 11.26 Global sediment thickness (km) on a $1°$ grid from the CRUST1.0 model (Laske et al. 2013). Mollweide projection

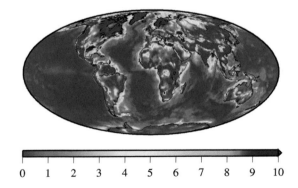

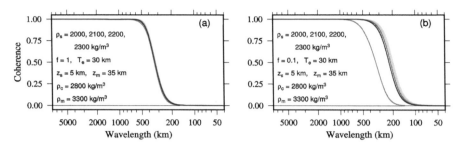

Fig. 11.27 Bouguer coherence curves for a two-layer continental crust with **a** $f = 1$, and **b** $f = 0.1$, showing the variation with layer-one (sediment) density (ρ_s); lighter curves show smaller values of the parameter, darker curves show higher values, indicated in the panels. The red curves show the coherence for a single-layer model

Bouguer coherence on land for such a model is plotted in Fig. 11.27 for $f = 1$ and $f = 0.1$. It can be seen that the effect of sediments upon the coherence is minor when surface and internal initial loads are of equal magnitude, but is significant when surface loading dominates, the coherence rollovers in the sedimented models being clustered together but markedly different from the unsedimented (single layer) coherence curve. In the oceans, however, the difference between sedimented and unsedimented coherence curves is less pronounced.

Another way to account for the effect of sediments is to remove them. A two-stage method gaining traction in the recent literature is to apply Bouguer and equivalent topography corrections to the gravity and topography data, respectively, based on sediment density and thickness (Chen et al. 2015; Shi et al. 2017). Dealing with the gravity first, one computes the gravitational attraction of the sediment layer (of thickness $t_s(\mathbf{x})$ and density $\rho_s(\mathbf{x}, z)$) and removes it from the Bouguer anomaly, but then adds back the gravitational attraction of a layer of rock (of the same thickness $t_s(\mathbf{x})$ but density ρ_0). The net effect, therefore, is to remove the gravitational attraction of a layer of thickness $t_s(\mathbf{x})$ and density $\Delta\rho_{0s}(\mathbf{x}, z)$. As the reader may recall from Sect. 8.2.5, this procedure mirrors the standard Bouguer correction in which water or ice is replaced by rock. This ensures that the Bouguer anomaly is free from the rock-sediment density contrast.

Several methods have been used to compute the gravitational attraction of the sediment layer, δg_s. The simplest is to utilise Parker's complete formula, Eq. 8.22, written here as

$$F\{\delta g_s(\mathbf{x})\} = -2\pi\mathcal{G}(\rho_0 - \bar{\rho}_s)\, e^{-kz_0} \sum_{n=1}^{\infty} \frac{k^{n-1}}{n!}\, F\{h^n(\mathbf{x}) - [h(\mathbf{x}) - t_s(\mathbf{x})]^n\} \ ,$$

where we have let $\rho(\mathbf{x})$ in Eq. 8.22 be a constant density contrast, $\rho_0 - \bar{\rho}_s$, where $\bar{\rho}_s$ is the mean density of the sediments, and where the upper surface, h_1, is the topographic/bathymetric surface h, and the lower surface, h_2, is the depth to the base of the sediment layer, $h - t_s$, and where z_0 is the mean vertical distance from the measurement point to the top of the sediment layer. The sediment-corrected Bouguer anomaly is then

$$\Delta g_{B,s} = \Delta g_B - \delta g_s \, ,$$

where Δg_B is the complete Bouguer anomaly.

More sophisticated methods account for not only the variation of sediment density with spatial location, but also depth. As sediments are deposited over time, the oldest and deepest are compacted by the weight of those above them, and their density increases. There are many ways to model this increase with depth. One can use an analytic function with a quadratic (Rao and Babu 1991) or exponential (Cowie and Karner 1990; Ji et al. 2021) increase with depth, which take into account sediment compaction and porosity. Or one can use numeric density-depth profiles based on well log and seismic data (Artemjev and Kaban 1994; Kaban and Mooney 2001; Stolk et al. 2013). Once a density model has been chosen, its gravitational attraction is determined from forward gravity modelling, in either the space or wavenumber domains. Space-domain methods are implementations of Eq. 8.6, where the sediment layer is broken up into rectangular cells, prisms or tesseroids (Rao and Babu 1991; Kaban et al. 2016). Wavenumber-domain methods are variations of Parker's formula, above (Parker 1972; Cowie and Karner 1990), which itself is also an implementation of Eq. 8.6, of course. Remember, the gravitational attraction of the sediments must always be computed for a density contrast (rather than absolute density) relative to a reference density for the upper crust, ρ_0, at least for that part of the layer that is below mean sea level.

The equivalent topography correction is more straightforward to calculate, and follows the derivation to correct for a layer of ice in Sect. 11.3.1. For a sediment layer of density ρ_s and thickness t_s, its rock-equivalent topography is given by

$$h_e = h - \left(1 - \frac{\rho_s}{\rho_0}\right) t_s$$

(cf. Eq. 11.5). If one wishes to account for sediment compaction and a density increase with depth, then a method such as that employed by Ji et al. (2021) is possible, whereby the sediment layer is split into several sub-layers of different density, and an equivalent topography calculated for each. Such an approach also allows for Parker's method to be applied to each sub-layer, and the results summed.

Accounting for sediments in such a fashion when determining elastic thickness will generally lower the T_e estimates from what they would otherwise have been (Artemjev and Kaban 1994; Shi et al. 2017; Kaban et al. 2018; Y Lu et al. 2021). But accounting for sediments also reduces the amount of noise in the system, where 'noise' is defined in Sect. 9.6.1. In regions with subdued topography and substantial sedimentary cover, the surface load is dominated by density variations rather than topographic relief, as sediments tend to fill the valleys and topographic depressions, obscuring the basement relief. Even if there is some secondary flexural adjustment due to the mass of the sediments, the resultant topography will be flattened, as shown in Fig. 11.28. The coherence between Bouguer anomalies and topography would then

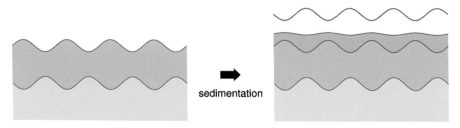

Fig. 11.28 When a post-flexure crust (left) is sedimented such that the sediments (orange shading, right) cover the basement, the Bouguer anomaly (red curve) is unchanged but the topography becomes subdued. In this case, inversion of the Bouguer coherence using load deconvolution might yield an overestimate of T_e

most likely give an overestimate of T_e, due to the presence of the subdued topography coupled with the mistaken assumption of a high and uniform (upper) crustal density. However, if the thickness and density of sediments are taken into account through the methods discussed above, then such overestimation should be avoided. Indeed, Kaban et al. (2018) not only found this to be so in the European lithosphere, but also found that accounting for sediments reduced the value of the free-air NSIC (Sect. 11.9) in regions with subdued topography and significant sedimentary cover.

Note that the treatment of sediments discussed here is not the same as sediment *backstripping* (Watts 2001). In its original form, backstripping took a heavily sedimented region, removed (or stripped back) the layer of sediment, and calculated the new position of the basement and Moho without the weight of the overlying sediments, i.e. after they had rebounded (Watts and Ryan 1976; Watts 1988). Naturally, the amount of rebound depends upon the local T_e value. In later work, the effect upon the gravity field was also calculated (Watts and Torné 1992; Stewart et al. 2000; Wyer and Watts 2006). As it is not (yet?) part of a spectral T_e estimation method, backstripping is not discussed further in this book.

11.11 Wavelet Versus Multitaper

Wavelet or multitaper? The big question, to which—perhaps unfortunately—there is no clear answer, as both have their merits and deficiencies. Both may be used in *local* mode, which creates maps of the variation of T_e over the study area, but both may also be used in *global* mode, whereby a single T_e estimate is made for a large study area. While the global mode has been presented throughout this book, this was mainly for educational purposes and the technique is now rather obsolete and archaic, mainly because it is now well-established that T_e is highly variable, even over relatively short distances. That is, selecting a large study area over which to compute a single estimate of T_e runs the risk of blending data from diverse tectonic provinces, as originally noted by Forsyth (1985). As Fig. 11.29 shows, when a single

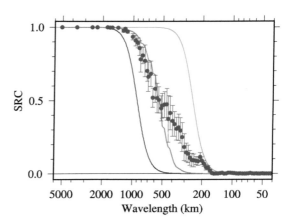

Fig. 11.29 The global Bouguer SRC of synthetic data generated from the elliptical T_e plate shown in Fig. 10.10, which varies from $T_e = 20$ to 120 km. The red circles and error bars are the observed SRC, the blue line is the best-fitting predicted SRC ($T_e = 58$ km), and the green lines are theoretical curves for $T_e = 120$ km (dark green) and 20 km (light green). Multitaper method with NW = 3, $K = 3$

T_e estimate is made over a region with a large T_e range, the observed SRC does not fit theoretical predictions, and the best-fitting T_e is not representative of the region.

One tactic to avoid blending data from different provinces is to make the study area smaller, or indeed to estimate T_e within many small windows covering a large study area as in the short-time Fourier transform (Sect. 3.5). In such a fashion, a map of T_e variations will be produced. A few early studies did this (Ebinger et al. 1989; Bechtel et al. 1990), but they used the periodogram to estimate the gravity and topography spectra, and as we saw in Sect. 3.2 the periodogram is a convolution of the true power spectrum with Fejér's kernel. Recall (Sect. 2.5), truncation of a signal by windowing it with a rectangular function gives rise to three artefacts: degraded resolution (smoothing of higher frequencies), spectral leakage (power that should be concentrated at one harmonic leaking into adjacent harmonics), and the Gibbs phenomenon (spurious oscillations appearing at discontinuities in the signal).

These artefacts—especially leakage—can be mitigated by instead tapering the data in the extracted window with Slepian functions (the discrete prolate spheroidal sequences, Sect. 3.3). However, the data over the study area still need to be extracted over a finite window, and, as several authors have noted (Audet and Mareschal 2004; Pérez-Gussinyé et al. 2004; Pérez-Gussinyé and Watts 2005; Audet et al. 2007; Pérez-Gussinyé et al. 2007), the size of the window greatly affects the auto- and cross-spectra and thus the resulting coherence and admittance. Figure 11.30 shows some examples from synthetic models. Here, gravity and topography data were generated from uniform-T_e plates over a 6000×6000 km^2 area, but to simulate the windowing performed when generating T_e maps, data were extracted over single windows of size 1000×1000 km^2 and 400×400 km^2, and then the observed and predicted SRC were computed from the extracted data. When T_e is small and the coherence rollover at short wavelengths, the 1000 km window is large enough to resolve the rollover wavelength (Fig. 11.30a). In contrast, when T_e is large and the rollover at long wavelengths, the 400 km window cannot resolve the rollover wavelength and the observed and predicted SRCs struggle to reach values above 0.1 (Fig. 11.30d). It

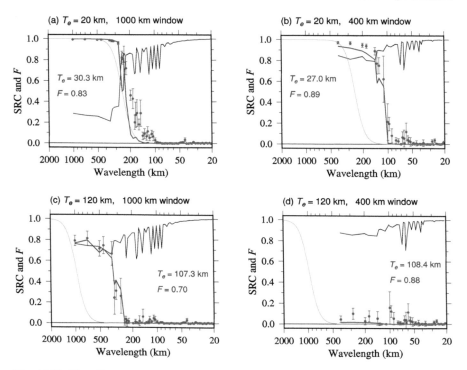

Fig. 11.30 The effect of windowing data using the multitaper method (with NW = 3, K = 3). Bouguer anomaly and topography data were generated from synthetic plates with T_e = 20 and 120 km, and F = 0.5 (f = 1), over a 6000×6000 km² area. Sub-sets were extracted over two windows of size 1000×1000 km², and 400×400 km², and the observed SRC (red circles with error bars), predicted SRC (blue lines), and load fraction ($F(k)$, black lines) computed. The best fitting T_e and F values are shown in each panel. The green curves show the theoretical Bouguer SRC for the corresponding synthetic model T_e and f

is also interesting to note in Fig. 11.30 that the load fraction, $F(k)$, is consistently overestimated, even when the 1000 km window is used to analyse a low-T_e plate. This has not been commented on in the literature and explains why the predicted SRC departs from the theoretical SRC (which is evaluated for F = 0.5) in three out of the four panels in that figure.

It is possible that the high F values are caused by the relatively small window sizes, because when the analysis is performed over a large area the recovered load fraction is faithful to that of the model (Fig. 11.15, where the size of the study area is 5000×5000 km²). But besides the dimensions of the analysis area, another major difference between the F results in Figs. 11.15 and 11.30 is that the synthetic gravity and topography data in the former are periodic while the windowed data in the latter are aperiodic (or more accurately, the former have continuous boundary conditions while the latter have discontinuous boundary conditions; see Sects. 10.3 and 10.4, and Fig. 10.4). With periodic data, the wrap-around effect in the cyclic convolution (Sect.

2.5.5) between gravity and topography[7] does not encounter any change in the signal or its derivative at the data edges; there is a smooth transition over the boundaries. In contrast, with aperiodic data the wrap-around encounters the discontinuity and the Gibbs phenomenon is generated (Sect. 2.5.4).[8] Hence it is plausible that one or both of these effects (small window and/or aperiodic data) causes an overestimation of the initial internal load during the load deconvolution procedure.

Despite the mitigation of leakage through Slepian tapering, the multitaper method's biggest issue in T_e map-making is still that large windows are needed to resolve the long rollover wavelengths of high-T_e provinces, but they potentially incorporate data from diverse tectonic provinces. This is seen in Fig. 11.31, which shows T_e estimated over North America using the multitaper and wavelet methods. In Fig. 11.31b the large moving window captures gravity and topography data from regions with a wide range of T_e values. And as was seen in Fig. 3.9a, a large window causes spatial leakage of harmonics into neighbouring regions. So in terms of T_e estimation these features will result in a smoothed T_e map lacking in detail. Small windows, in contrast, capture much less data and so have a lower risk of blending data from diverse tectonic provinces (Fig. 11.31a). The boundary between spectrum changes is also much better resolved when using small windows (Fig. 3.9c). However, as was noted in Sect. 3.5, larger windows give more accurate spectra than small windows due to their better sampling of the wavenumber domain, thus providing a better estimation of the SRC rollover wavelength and hence T_e. And small windows not only have a comparatively poor wavenumber resolution (Sect. 9.8), but they also can be too small to resolve long rollover wavelengths (e.g. Fig. 11.30d). A noticeable deficiency with the multitaper method, though, is the border around the edges of the study area, in which no T_e estimates are made (Figs. 11.31a, b). The width of this border increases with window size.

The wavelet transform, in contrast, avoids these issues to a certain extent because the data are not windowed. The convolution of the data with the wavelet occurs over the whole study area, and the Morlet wavelets themselves—being based on Gaussian functions—never decay to zero, thus utilising all available data without truncation. Nevertheless, the wavelet method is not immune from the aforementioned need to sample more data when the local T_e is high. In Figs. 11.31c, d the concentric red circles are contours of the Gaussian envelope of two Morlet wavelets, which we will call *transition wavelets*. A transition wavelet is one whose equivalent Fourier wavelength (λ_e) is equal to the coherence transition wavelength (λ_t) for that T_e value; that is, such wavelets are of the correct scale to resolve the rollover. In each figure, the smaller wavelet is centred on a grid node where $T_e = 20$ km, while the larger wavelet is centred on a grid node where $T_e = 100$ km. So from Eq. 9.32, when $T_e = 20$ km we have $\lambda_t = 251$ km, and when $T_e = 100$ km we have $\lambda_t = 840$ km (with f

---

[7] Remember, the admittance and coherency formulae contain the gravity and topography auto- and cross-spectra (e.g. $G(\mathbf{k})H^*(\mathbf{k})$). As multiplication in the wavenumber domain is the same as convolution in the space domain, whatever happens in one domain will automatically affect the other, even if one does not explicitly perform operations in that other domain.

[8] See also Fig. 2.13 because, remember, the discrete Fourier transform assumes that the space domain signal is infinitely long and periodic (Sect. 2.4.2).

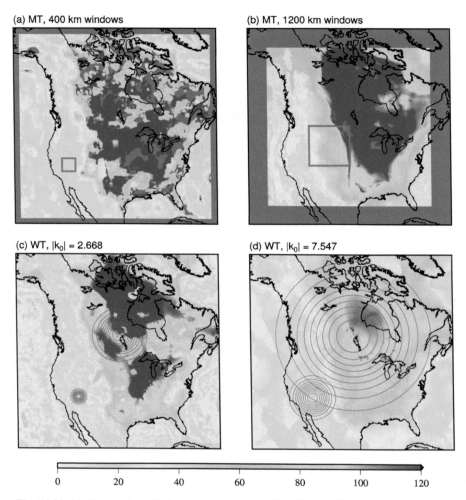

Fig. 11.31 North American T_e from the multitaper (MT; NW = 3, K = 3) and wavelet (WT) methods. The red boxes in (**a**) and (**b**) show the size of the $400 \times 400 \, \text{km}^2$ and $1200 \times 1200 \, \text{km}^2$ windows used in the multitaper method; the grey-shaded regions in (**a**) are not noise, but are areas where the observed SRC is too low to be inverted. The red circles in (**c**) and (**d**) show the contours (at 0.1 units) of the Gaussian envelope of two transition wavelets (see text) for $T_e = 20 \, \text{km}$ (smaller wavelet) and $100 \, \text{km}$ (larger wavelet)

= 1). Now, from Eq. 4.42, the scale (s) of a $|\mathbf{k}_0| = 2.668$ wavelet with an equivalent Fourier wavelength of 251 km is $s = 107$ km, while that at $\lambda_e = 840$ km is $s = 357$ km (Fig. 11.31c). Thus, in order to resolve a higher T_e value a larger-scale wavelet is needed, which draws on data over a greater spatial area. And when $|\mathbf{k}_0|$ is high, as in Fig. 11.31d, the corresponding wavelet scales are larger still, drawing on even more data. To resolve a T_e of 20 km, a $|\mathbf{k}_0| = 7.547$ wavelet has a scale of 302 km, while a $T_e = 100$ km region needs a wavelet with a scale of 1009 km to resolve it.

While the wavelet transform was specifically designed to estimate spectra of non-stationary signals, no spectral analysis method is immune from the trade-off between accuracy in space or wavenumber domains. Throughout Chap. 4—but especially in Sects. 4.8.2 and 4.11.3—we saw how the spatial and wavenumber resolution of wavelets is scale-dependent (unlike the resolution of the short-time Fourier transform, which is fixed). Large-scale wavelets have a good wavenumber resolution but a poor spatial resolution, and vice versa for small-scale wavelets. Furthermore (Sect. 4.3.3), we saw how wavelets with a high $|\mathbf{k}_0|$ value have better wavenumber resolution than low-$|\mathbf{k}_0|$ wavelets, but a worse spatial resolution. These properties suggest that T_e maps will show more detail in regions with a low elastic thickness (because the transition wavelets have a smaller scale, with good spatial resolution), but in regions where T_e is high the maps will be smoother, with less detail (because the transition wavelets have a larger scale, with comparatively poor spatial resolution). In addition, if one desires T_e maps with a high spatial resolution, it is better to use low-$|\mathbf{k}_0|$ wavelets as these provide a better spatial resolution than high-$|\mathbf{k}_0|$ wavelets. However, if absolute T_e values are more important, then the superior wavenumber resolution afforded by high-$|\mathbf{k}_0|$ wavelets means these should be chosen for the analysis because they can capture the coherence rollover wavelength better. A convenient summary of factors affecting the resolution of the two methods is provided in Table 11.5.

Detailed comparisons of the two methods have been made elsewhere, when used in local mode (Audet et al. 2007; Pérez-Gussinyé et al. 2009; Kirby and Swain 2011; Jiménez-Díaz et al. 2014; Chen et al. 2014; Kirby and Swain 2014), and global mode (Kirby and Swain (2014), and Figs. 10.7 and 10.8 in this book). Many of these studies also contain resolution tests of the two methods (Pérez-Gussinyé et al. 2004, 2007; Kirby and Swain 2008). Such tests typically create synthetic gravity and topography models from a plate with a spatially-variable T_e distribution using the methods discussed in Chap. 10. An example of a model plate is shown in Fig. 11.32, with T_e recovered from wavelet and multitaper methods for a single synthetic model shown in Fig. 11.33. First, much of the variability in the T_e maps comes from random correlations between the synthetic initial loads (Sects. 10.2 and 10.4). However, it is clear that there is generally more detail in the low- rather than high-T_e regions, and that the maps generated using low values of $|\mathbf{k}_0|$ (wavelet method) and small windows (multitaper method) show more spatial variability, though the 400 km and even 800 km windows of the multitaper results are not able to resolve the SRCs at some locations within high-T_e regions.

Table 11.5 Summary of how the various parameters of the wavelet and multitaper methods affect their resolution properties in the space and wavenumber domains. The parameters discussed are the time-bandwidth product (NW), number of tapers (K) and window size for the multitaper method, and central wavenumber ($|\mathbf{k}_0|$) for the wavelet method. These results can be verified by reproducing Figs. 3.9 and 4.6 for various parameter values

Multitaper	Wavelet
Data are truncated, but Slepian tapers minimise spectral leakage.	No need for windowing; data not truncated
Three parameters to adjust: NW, K, window size	Only one parameter to adjust: $\|\mathbf{k}_0\|$
Small NW: good wavenumber resolution Large NW: poor wavenumber resolution NW does not affect spatial resolution	Small $\|\mathbf{k}_0\|$: good spatial resolution, poor wavenumber resolution Large $\|\mathbf{k}_0\|$: poor spatial resolution, good wavenumber resolution
Low K: higher spectral variance High K: lower spectral variance; degraded wavenumber resolution (esp. at smaller window sizes) and slightly degraded spatial resolution (esp. at larger window sizes)	
Small window: poor wavenumber resolution, good spatial resolution Large window: good wavenumber resolution, poor spatial resolution	
Wavenumber-independent resolution in space and wavenumber domains	Long wavelengths: poor spatial resolution, good wavenumber resolution Short wavelengths: good spatial resolution, poor wavenumber resolution

Fig. 11.32 A model T_e distribution comprising an ellipse of magnitude range 20–110 km superimposed with sinusoidal oscillations of 10 km peak-to-trough amplitude and 500 km wavelength. T_e contours are from 10 to 110 km, every 10 km

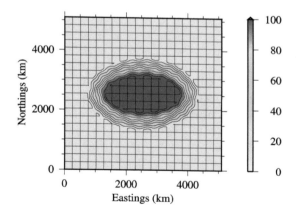

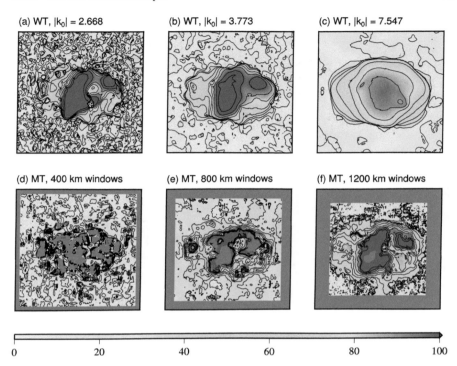

Fig. 11.33 T_e recovered from a single synthetic model using (**a**)–(**c**) the fan wavelet method with the indicated $|\mathbf{k}_0|$ values, and (**d**)–(**f**) using the multitaper method (NW = 3, $K = 3$) with the indicated window sizes. The grey-shaded regions in (**d**)–(**f**) show where the observed SRC is too low to be inverted. T_e contours are from 10 to 110 km, every 10 km

In order to remove the distracting effects of random initial load correlations it is usual practice to average the T_e maps recovered from many synthetic models, shown in Fig. 11.34. However, these images are too crude to show the fine detail in the spatial variability, so Figs. 11.35 and 11.36 show extractions over the low- and high-T_e areas, respectively. If one had to choose a best reproduction of the model T_e, one would probably choose the wavelet method with $|\mathbf{k}_0| = 2.668$ for the low-T_e regions (Fig. 11.35b), and the multitaper method with $1200 \times 1200 \, \text{km}^2$ windows for the high-T_e region (Fig. 11.36g). Note though that other values for NW and K are likely to give different results (NW = $K = 3$ were used here), with Pérez-Gussinyé et al. (2007) finding that the use of small windows increases the already high variance that one usually gets when using a low number of tapers.

When it comes to computation execution time, the wavelet method is a clear winner, at least for the recovered models shown in Fig. 11.34. Table 11.6 shows the length of wall-clock time taken to compute each set of 100 models, relative to the 30 min for the wavelet method with $|\mathbf{k}_0| = 2.668$ (which was the fastest). Actual processing (CPU) time depends on many factors of course, but here each set of 100

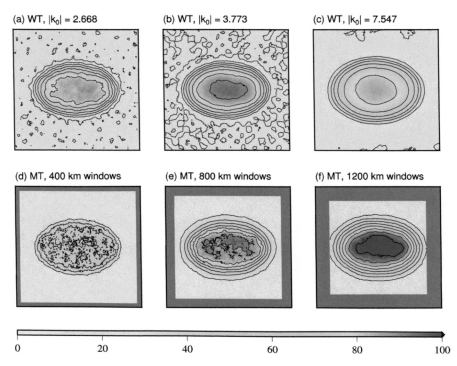

(a) WT, $|k_0| = 2.668$ (b) WT, $|k_0| = 3.773$ (c) WT, $|k_0| = 7.547$

(d) MT, 400 km windows   (e) MT, 800 km windows   (f) MT, 1200 km windows

| 0 | 20 | 40 | 60 | 80 | 100 |

Fig. 11.34 As Fig. 11.33, except showing the average T_e over 100 model results

models was run on a single processor with no other jobs running. In general, the fan wavelet method is much quicker than moving-window multitaper methods, even for larger values of $|k_0|$ which require more Morlet wavelets in the fan than smaller values of $|k_0|$. With moving-window methods, smaller windows provide a faster run-time than do larger windows, but even the smallest acceptable window size (400 km) still takes much longer to execute than the slowest wavelet method.

11.12   Other Considerations

Although this book is primarily concerned with T_e estimation using load deconvolution—in which two initial loads are the only forces acting on the lithosphere—it is worthwhile mentioning, albeit briefly, some recent research that has considered other forces and phenomena that may impact T_e recovery.

In some regions of the Earth, the topography is supported dynamically by convective upwellings of hot mantle material, as well as by isostasy or flexure, so the theory presented in this book cannot be used to accurately estimate T_e there. But such *dynamic topography* and its corresponding gravity field can nevertheless be

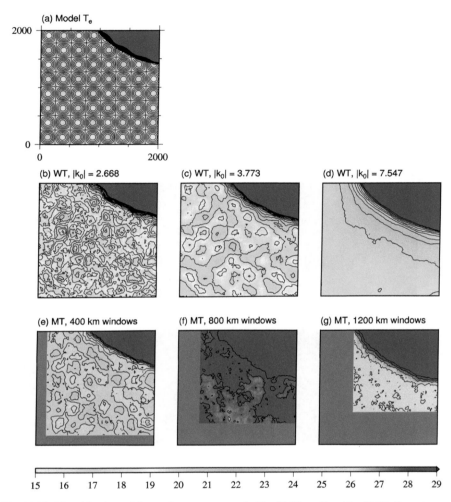

Fig. 11.35 Magnifications of the south-west corners of **a** Fig. 11.32, and **b–g** Fig. 11.34. T_e contours are from 15–29 km, every 1 km

described by models which are dependent upon T_e (McKenzie 2010). While it is not possible to readily incorporate these models into the load deconvolution formalism, a combined convection-flexure free-air admittance function can be developed and used to invert the observed admittance for T_e, though only by uniform-f inversion. This was attempted by Kirby and Swain (2014), who found that the elastic thickness of the Canadian shield remained high when convection (and glacial isostatic adjustment) was considered.

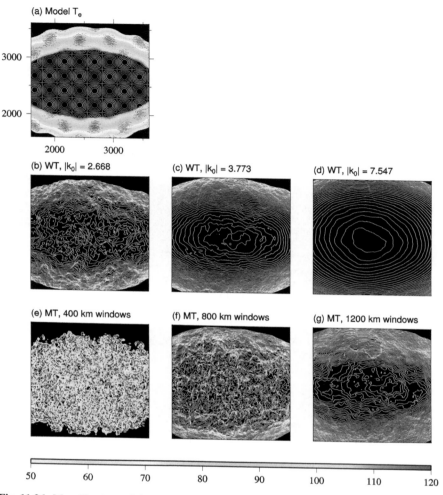

Fig. 11.36 Magnifications of the central regions of **a** Fig. 11.32, and **b–g** Fig. 11.34. T_e contours are from 50 to 120 km, every 1 km

Table 11.6 Relative wall-clock processing times when computing the 100 T_e models used to form the averages in Fig. 11.34, using the wavelet method with $|\mathbf{k}_0| = 2.668$, 3.773 and 7.547, and the multitaper method with windows of size $400 \times 400\,\mathrm{km}^2$, $800 \times 800\,\mathrm{km}^2$ and $1200 \times 1200\,\mathrm{km}^2$. The time taken to compute the 100 models using the wavelet method with $|\mathbf{k}_0| = 2.668$ was 30 min

2.668	3.773	7.547	400	800	1200
1	1.13×	1.7×	58×	284×	645×

Subduction zones constitute another environment where load deconvolution and the use of the coherence or admittance might not provide correct T_e estimates. Indeed, even the use of a thin elastic plate may be inappropriate in subduction zones, where the curvature experienced by the subducted plate generates stresses that exceed the yield strength of its rock, weakening the plate (Sect. 1.3.1). The upper part of the plate fractures while its lower part undergoes ductile deformation, both serving to reduce the thickness of the elastic part of the plate. Additionally, the faulting in the upper part potentially admits seawater, causing hydration and serpentinization of olivine in the uppermost mantle, further weakening the subducting plate. In attempting to overcome the failure of elastic models to explain the stresses in such regimes, authors have invoked such diverse rheological models as elastic-plastic, elastic-perfectly plastic, viscous and viscoelastic. Encouragingly for flexural studies using load deconvolution, though, it has been found that the curvature-induced strength reduction can be modelled using a thin, elastic plate with spatially variable flexural rigidity.

However, while studies using forward models of the gravity and bathymetry predicted from (1D) beam flexure have recovered low T_e estimates over ocean trenches, 2D admittance/coherence studies of the kind discussed in this book invariably find T_e over subduction zones to be higher than the surroundings. Possible reasons for this include the fact that such 2D studies use a continuous, rather than broken, plate model in the inversion, although this usage could under- rather than overestimate T_e (Karner and Watts 1983; McNutt et al. 1988). Another consideration is that an oceanic trench is highly anisotropic, and any weakening induced by plate curvature should only show in the trench-perpendicular direction. An isotropic analysis will of course measure an azimuthal average of the local T_e, though one might still expect this average to be lower than the regional value. Some authors, though, have succeeded in reducing the T_e estimates over trenches by modelling and removing the gravity and topography effect of the downgoing slab before computation of the admittance or coherence (Pérez-Gussinyé et al. 2008; Bai et al. 2018). Alternatively, it has been suggested that T_e might actually be higher than anticipated in some subduction zones due to mechanisms such as strain hardening (Z Lu et al. 2021).

Further information on T_e estimation in subduction zones may be found in Judge and McNutt (1991), Watts (2001), Bry and White (2007), Pérez-Gussinyé et al. (2008), Contreras-Reyes and Osses (2010), Arredondo and Billen (2012), Zhang et al. (2014), Garcia et al. (2015), Hunter and Watts (2016), Bai et al. (2018), Ji et al. (2020) and Z Lu et al. (2021).

11.13 Summary

Chapter 11 provided examples and scenarios of T_e estimation using actual Earth data. It therefore began with a discussion of various gravity, topography, sediment and crustal structure data sets, and provided an explanation of the data resolution needed to accurately estimate the effective elastic thickness. A discussion of the utility of the rock-equivalent topography followed. The chapter also looked at how the choice of

Bouguer reduction density and gravimetric terrain corrections affects T_e estimates, plus the influence thereon of crustal structure models and sediments. Topics such as deconvolution with the admittance, uniform-f inversion, the interpretation of the load ratio, and T_e errors were explored. The chapter also expanded upon the method to detect model noise, which began in Chaps. 5 and 9. Finally, we concluded with a face-off between wavelet and multitaper methods—with no clear winner—and a brief mention of environments where combined loading and load deconvolution may not be the only T_e-dependent model that applies.

11.14   Further Reading

Due to the somewhat diverse nature of topics in this chapter, I have attempted to provide pertinent citations within each section, rather than here in the 'further reading' section.

References

Aitken ARA, Salmon ML, Kennett BLN (2013) Australia's Moho: a test of the usefulness of gravity modelling for the determination of Moho depth. Tectonophys 609:468–479

Amante C, Eakins BW (2009) ETOPO1 1 Arc-minute global relief model: procedures, data sources and analysis. National Oceanic and Atmospheric Administration, Boulder, Colorado. https://www.ngdc.noaa.gov/mgg/global/relief/ETOPO1/docs/ETOPO1.pdf. Accessed 16 June 2020

Arredondo KM, Billen MI (2012) Rapid weakening of subducting plates from trench-parallel estimates of flexural rigidity. Phys Earth Planet Inter 196–197:1–13

Artemjev ME, Kaban MK (1994) Density inhomogeneities, isostasy and flexural rigidity of the lithosphere in the Transcaspian region. Tectonophys 240:281–297

Audet P, Mareschal J-C (2004) Variations in elastic thickness in the Canadian Shield. Earth Planet Sci Lett 226:17–31

Audet P, Jellinek M, Uno H (2007) Mechanical controls on the deformation of continents at convergent margins. Earth Planet Sci Lett 264:151–166

Bai Y, Dong D, Kirby JF, Williams SE, Wang Z (2018) The effect of dynamic topography and gravity on lithospheric effective elastic thickness estimation: a case study. Geophys J Int 214:623–634

Bamber JL, Griggs JA, Hurkmans RTWL, Dowdeswell JA, Gogineni SP, Howat I, Mouginot J, Paden J, Palmer S, Rignot E, Steinhage D (2013) A new bed elevation dataset for Greenland. Cryosphere 7:499–510

Bassin C, Laske G, Masters G (2000) The current limits of resolution for surface wave tomography in North America. EOS Trans Am Geophys Union 81:F897

Bechtel TD, Forsyth DW, Sharpton VL, Grieve RAF (1990) Variations in effective elastic thickness of the North American lithosphere. Nature 343:636–638

Brockmann JM, Schubert T, Mayer-Gürr T, Schuh W-D (2019) The Earth's gravity field as seen by the GOCE satellite—an improved sixth release derived with the time-wise approach (GO_CONS_GCF_2_TIM_R6). GFZ Data Services. https://doi.org/10.5880/ICGEM.2019.003. Accessed 20 Jan 2021

Bry M, White N (2007) Reappraising elastic thickness variation at oceanic trenches. J Geophys Res 112:B08414. https://doi.org/10.1029/2005JB004190

Chen B, Liu J, Chen C, Du J, Sun Y (2014) Elastic thickness of the Himalayan-Tibetan orogen estimated from the fan wavelet coherence method, and its implications for lithospheric structure. Earth Planet Sci Lett 409:1–14

Chen B, Kaban MK, El Khrepy S, Al-Arifi N (2015) Effective elastic thickness of the Arabian plate: weak shield versus strong platform. Geophys Res Lett 42:3298–3304

Contreras-Reyes E, Osses A (2010) Lithospheric flexure modelling seaward of the Chile trench: implications for oceanic plate weakening in the Trench Outer Rise region. Geophys J Int 182:97–112

Cowie PA, Karner GD (1990) Gravity effect of sediment compaction: examples from the North Sea and Rhine Graben. Earth Planet Sci Lett 99:141–153

Dixon TH, Naraghi M, McNutt MK, Smith SM (1983) Bathymetric prediction from SEASAT altimeter data. J Geophys Res 88(C3):1563–1571

Ebinger CJ, Bechtel TD, Forsyth DW, Bowin CO (1989) Effective elastic plate thickness beneath the East African and Afar plateaus and dynamic compensation of the uplifts. J Geophys Res 94(B3):2883–2901

Farr TG, Rosen PA, Caro E, Crippen R, Duren R, Hensley S, Kobrick M, Paller M, Rodriguez E, Roth L, Seal D, Shaffer S, Shimada J, Umland J, Werner M, Oskin M, Burbank D, Alsdorf D (2007) The shuttle radar topography mission. Rev Geophys 45:RG2004. https://doi.org/10.1029/2005RG000183

Forsberg R (1984) A study of terrain reductions, density anomalies and geophysical inversion methods in gravity field modelling. Ohio State University Report 355, Department of Geodetic Science and Surveying, Ohio State University, Columbus

Forsyth DW (1985) Subsurface loading and estimates of the flexural rigidity of continental lithosphere. J Geophys Res 90(B14):12,623–12,632

Fretwell P, Pritchard HD, Vaughan DG, Bamber JL, Barrand NE, Bell R, Bianchi C, Bingham RG, Blankenship DD, Casassa G, Catania G, Callens D, Conway H, Cook AJ, Corr HFJ, Damaske D, Damm V, Ferraccioli F, Forsberg R, Fujita S, Gim Y, Gogineni P, Griggs JA, Hindmarsh RCA, Holmlund P, Holt JW, Jacobel RW, Jenkins A, Jokat W, Jordan T, King EC, Kohler J, Krabill W, Riger-Kusk M, Langley KA, Leitchenkov G, Leuschen C, Luyendyk BP, Matsuoka K, Mouginot J, Nitsche FO, Nogi Y, Nost OA, Popov SV, Rignot E, Rippin DM, Rivera A, Roberts J, Ross N, Siegert MJ, Smith AM, Steinhage D, Studinger M, Sun B, Tinto BK, Welch BC, Wilson D, Young DA, Xiangbin C, Zirizzotti A (2013) Bedmap2: improved ice bed, surface and thickness datasets for Antarctica. Cryosphere 7:375–393

Fu L-L, Cazenave A (eds) (2001) Satellite altimetry and earth sciences: a handbook of techniques and applications. In: International Geophysics Series, vol 69. Academic Press, San Diego

Garcia ES, Sandwell DT, Luttrell KM (2015) An iterative spectral solution method for thin elastic plate flexure with variable rigidity. Geophys J Int 200:1010–1026

GEBCO Compilation Group (2020) GEBCO 2020 Grid. National Oceanography Centre, Southampton, UK. https://doi.org/10.5285/a29c5465-b138-234d-e053-6c86abc040b9. Accessed 15 Jan 2021

Gesch DB, Verdin KL, Greenlee SK (1999) New land surface digital elevation model covers the Earth. Eos 80:69–70

Hastings DA, Dunbar PK (1999) Global land one-kilometer base elevation (GLOBE) digital elevation model. National Oceanic and Atmospheric Administration, Boulder, Colorado. https://www.ngdc.noaa.gov/mgg/topo/report/globedocumentationmanual.pdf. Accessed 16 June 2020

Heiskanen WA, Moritz H (1967) Physical geodesy. WH Freeman, San Francisco

Hirt C (2014) Digital terrain models. In: Grafarend E (ed) Encyclopedia of geodesy. Springer, Cham

Hirt C, Rexer M (2015) Earth 2014: 1 arc-min shape, topography, bedrock and ice-sheet models—available as gridded data and degree-10,800 spherical harmonics. Int J Appl Earth Obs Geoinf 39:103–112

Hirt C, Claessens S, Fecher T, Kuhn M, Pail R, Rexer M (2013) New ultrahigh-resolution picture of Earth's gravity field. Geophys Res Lett 40:4279–4283

Hunter J, Watts AB (2016) Gravity anomalies, flexure and mantle rheology seaward of circum-Pacific trenches. Geophys J Int 207:288–316

IOC, IHO, BODC (2003) Centenary edition of the GEBCO digital atlas. Intergovernmental oceanographic commission and the international hydrographic organization as part of the general bathymetric chart of the oceans. British Oceanographic Data Centre, Liverpool

Jarvis A, Reuter HI, Nelson A, Guevara E (2008) Hole-filled seamless SRTM data V4. International Centre for Tropical Agriculture (CIAT). http://srtm.csi.cgiar.org. Accessed 25 June 2020

Ji F, Zhang Q, Zhou X, Bai Y, Li Y (2020) Effective elastic thickness for Zealandia and its implications for lithospheric deformation. Gondwana Res 86:46–59

Ji F, Zhang Q, Xu M, Zhou X, Guan Q (2021) Estimating the effective elastic thickness of the Arctic lithosphere using the wavelet coherence method: tectonic implications. Phys Earth Planet Inter 318:106770

Jiménez-Díaz A, Ruiz J, Pérez-Gussinyé M, Kirby JF, Álvarez-Gómez JA, Tejero R, Capote R (2014) Spatial variations of effective elastic thickness of the lithosphere in Central America and surrounding regions. Earth Planet Sci Lett 391:55–66

Judge AV, McNutt MK (1991) The relationship between plate curvature and elastic plate thickness: a study of the Peru-Chile trench. J Geophys Res 96(B10):16,625–16,639

Kaban MK, Mooney WD (2001) Density structure of the lithosphere in the southwestern United States and its tectonic significance. J Geophys Res 106(B1):721–739

Kaban MK, El Khrepy S, Al-Arifi N (2016) Isostatic model and isostatic gravity anomalies of the Arabian plate and surroundings. Pure Appl Geophys 173:1211–1221

Kaban MK, Chen B, Tesauro M, Petrunin AG, El Khrepy S, Al-Arifi N (2018) Reconsidering effective elastic thickness estimates by incorporating the effect of sediments: a case study for Europe. Geophys Res Lett 45:9523–9532

Karner GD, Watts AB (1983) Gravity anomalies and flexure of the lithosphere at mountain ranges. J Geophys Res 88(B12):10,449–10,477

Kirby JF (2014) Estimation of the effective elastic thickness of the lithosphere using inverse spectral methods: the state of the art. Tectonophys 631:87–116

Kirby JF, Swain CJ (2008) An accuracy assessment of the fan wavelet coherence method for elastic thickness estimation. Geochem Geophys Geosyst 9:Q03022 https://doi.org/10.1029/2007GC001773, (Correction, Geochem Geophys Geosyst 9:Q05021). https://doi.org/10.1029/2008GC002071

Kirby JF, Swain CJ (2009) A reassessment of spectral T_e estimation in continental interiors: the case of North America. J Geophys Res 114(B8):B08401. https://doi.org/10.1029/2009JB006356

Kirby JF, Swain CJ (2011) Improving the spatial resolution of effective elastic thickness estimation with the fan wavelet transform. Comput Geosci 37:1345–1354

Kirby JF, Swain CJ (2014) The long wavelength admittance and effective elastic thickness of the Canadian shield. J Geophys Res Solid Earth 119:5187–5214

Laske G, Masters G, Ma Z, Pasyanos M (2013) Update on CRUST1.0 – a 1-degree global model of Earth's crust. Geophys Res Abstr 15:EGU2013-2658

Lu Y, Lu Z, Li C-F, Zhu S, Audet P (2021) Lithosphere weakening during Arctic Ocean opening: evidence from effective elastic thickness. Geophys Res Lett 48:e2021GL094090. https://doi.org/10.1029/2021GL094090

Lu Z, Audet P, Li C-F, Zhu S, Wu Z (2021) What controls the effective elastic thickness of the lithosphere in the Pacific Ocean? J Geophys Res Solid Earth 126:e2020JB021074. https://doi.org/10.1029/2020JB021074

McKenzie D (2003) Estimating T_e in the presence of internal loads. J Geophys Res 108(B9):2438. https://doi.org/10.1029/2002JB001766

McKenzie D (2010) The influence of dynamically supported topography on estimates of T_e. Earth Planet Sci Lett 295:127–138

McNutt MK, Diament M, Kogan MG (1988) Variations of elastic plate thickness at continental thrust belts. J Geophys Res 93(B8):8825–8838

Mooney WD, Laske G, Masters TG (1998) CRUST 5.1: a global crustal model at 5° × 5°. J Geophys Res 103(B1):727–747

Moritz H (1990) The inverse Vening Meinesz problem in geodesy. Geophys J Int 102:733–738

Morlighem M, Williams CN, Rignot E, An L, Arndt JE, Bamber JL, Catania G, Chauché N, Dowdeswell JA, Dorschel B, Fenty I, Hogan K, Howat I, Hubbard A, Jakobsson M, Jordan TM, Kjeldsen KK, Millan R, Mayer L, Mouginot J, Noël BPY, O'Cofaigh C, Palmer S, Rysgaard S, Seroussi H, Siegert MJ, Slabon P, Straneo F, van den Broeke MR, Weinrebe W, Wood M, Zinglersen KB (2017) BedMachine v3: complete bed topography and ocean bathymetry mapping of Greenland from multibeam echo sounding combined with mass conservation. Geophys Res Lett 44:11,051–11,061

Morlighem M, Rignot E, Binder T, Blankenship D, Drews R, Eagles G, Eisen O, Ferraccioli F, Forsberg R, Fretwell P, Goel V, Greenbaum JS, Gudmundsson H, Guo J, Helm V, Hofstede C, Howat I, Humbert A, Jokat W, Karlsson NB, Lee WS, Matsuoka K, Millan R, Mouginot J, Paden J, Pattyn F, Roberts J, Rosier S, Ruppel A, Seroussi H, Smith EC, Steinhage D, Sun B, van den Broeke MR, van Ommen TD, van Wessem M, Young DA (2020) Deep glacial troughs and stabilizing ridges unveiled beneath the margins of the Antarctic ice sheet. Nature Geosci 13:132–137

NGDC (1988) Digital relief of the surface of the Earth; NGDC data announcement 88-MGG-02. National Geophysical Data Center, National Oceanic and Atmospheric Administration, Boulder, Colorado. https://www.ngdc.noaa.gov/mgg/global/etopo5.HTML. Accessed 13 Jan 2021

Oldenburg DW (1974) The inversion and interpretation of gravity anomalies. Geophysics 39:526–536

Pail R (2014) CHAMP-, GRACE-, GOCE-satellite projects. In: Grafarend E (ed) Encyclopedia of geodesy. Springer, Cham

Parker RL (1972) The rapid calculation of potential anomalies. Geophys J R Astron Soc 31:447–455

Pavlis NK, Rapp RH (1990) The development of an isostatic gravitational model to degree 360 and its use in global gravity modelling. Geophys J Int 100:369–378

Pavlis NK, Factor JK, Holmes SA (2007) Terrain-related gravimetric quantities computed for the next EGM. In: Kılıçoğlu A, Forsberg R (eds) Gravity field of the Earth: 1st International symposium of the international gravity field service, Istanbul, Turkey. Harita Dergisi (Journal of Mapping), Special Issue 18. General Command of Mapping, Ankara, pp 318–323

Pavlis NK, Holmes SA, Kenyon SC, Factor JK (2012) The development and evaluation of the Earth Gravitational Model 2008 (EGM2008). J Geophys Res 117:B04406 https://doi.org/10.1029/2011JB008916, (Correction, J Geophys Res Solid Earth 118:2633). https://doi.org/10.1002/jgrb.50167

Pérez-Gussinyé M, Lowry AR, Watts AB, Velicogna I (2004) On the recovery of effective elastic thickness using spectral methods: examples from synthetic data and from the Fennoscandian Shield. J Geophys Res 109(B10):B10409. https://doi.org/10.1029/2003JB002788

Pérez-Gussinyé M, Watts AB (2005) The long-term strength of Europe and its implications for plate-forming processes. Nature 436:381–384

Pérez-Gussinyé M, Lowry AR, Watts AB (2007) Effective elastic thickness of South America and its implications for intracontinental deformation. Geochem Geophys Geosyst 8:Q05009. https://doi.org/10.1029/2006GC001511

Pérez-Gussinyé M, Lowry AR, Phipps-Morgan J, Tassara A (2008) Effective elastic thickness variations along the Andean margin and their relationship to subduction geometry. Geochem Geophys Geosyst 9:Q02003. https://doi.org/10.1029/2007GC001786

Pérez-Gussinyé M, Swain CJ, Kirby JF, Lowry AR (2009) Spatial variations of the effective elastic thickness, T_e, using multitaper spectral estimation and wavelet methods: examples from synthetic data and application to South America. Geochem Geophys Geosyst 10:Q04005. https://doi.org/10.1029/2008GC002229

Press WH, Teukolsky SA, Vetterling WT, Flannery BP (1992) Numerical recipes in Fortran 77, 2nd edn. Cambridge University Press, Cambridge

Rabus B, Eineder M, Roth A, Bamler R (2003) The shuttle radar topography mission—a new class of digital elevation models acquired by spaceborne radar. ISPRS J Photogramm Remote Sens 57:241–262

Rao DB, Babu NR (1991) A Fortran-77 computer program for three-dimensional analysis of gravity anomalies with variable density contrast. Comput Geosci 17:655–667

Sandwell DT, Smith WHF (1997) Marine gravity anomaly from Geosat and ERS-1 satellite altimetry. J Geophys Res 102(B5):10,039–10,054

Shi X, Kirby J, Yu C, Jiménez-Díaz A, Zhao J (2017) Spatial variations in the effective elastic thickness of the lithosphere in Southeast Asia. Gondwana Res 42:49–62

Sjöberg LE (2009) Solving Vening Meinesz-Moritz inverse problem in isostasy. Geophys J Int 179:1527–1536

Smith WHF, Sandwell DT (1994) Bathymetric prediction from dense satellite altimetry and sparse shipboard bathymetry. J Geophys Res 99(B11):21,803–21,824

Smith WHF, Sandwell DT (1997) Global seafloor topography from satellite altimetry and ship depth soundings. Science 277:1957–1962

Stewart J, Watts AB, Bagguley JG (2000) Three-dimensional subsidence analysis and gravity modelling of the continental margin offshore Namibia. Geophys J Int 141:724–746

Stolk W, Kaban MK, Beekman F, Tesauro M, Mooney WD, Cloetingh S (2013) High resolution regional crustal models from irregularly distributed data: application to Asia and adjacent areas. Tectonophys 602:55–68

Straume EO, Gaina C, Medvedev S, Hochmuth K, Gohl K, Whittaker JM, Fattah RA, Doornenbal JC, Hopper JR (2019) GlobSed: updated total sediment thickness in the world's oceans. Geochem Geophys Geosyst 20:1756–1772

Swain CJ, Kirby JF (2003) The effect of 'noise' on estimates of the elastic thickness of the continental lithosphere by the coherence method. Geophys Res Lett 30:1574. https://doi.org/10.1029/2003GL017070

Tachikawa T, Kaku M, Iwasaki A, Gesch D, Oimoen M, Zhang Z, Danielson J, Krieger T, Curtis B, Haase J, Abrams M, Crippen R, Carabajal C (2011) ASTER global digital elevation model version 2—summary of validation results. https://www.jspacesystems.or.jp/ersdac/GDEM/ver2Validation/Summary_GDEM2_validation_report_final.pdf. Accessed 13 Jan 2021

Tassara A, Swain C, Hackney R, Kirby J (2007) Elastic thickness structure of South America estimated using wavelets and satellite-derived gravity data. Earth Planet Sci Lett 253:17–36

Torge W (1989) Gravimetry. de Gruyter, Berlin

Tozer B, Sandwell DT, Smith WHF, Olson C, Beale JR, Wessel P (2019) Global bathymetry and topography at 15 arc sec: SRTM15+. Earth Space Sci 6:1847–1864

Vening Meinesz FA (1931) Une nouvelle méthode pour la réduction isostatique régionale de l'intensité de la pésanteur. Bull Géod 29:33–51

Watts AB (1988) Gravity anomalies, crustal structure and flexure of the lithosphere at the Baltimore Canyon Trough. Earth Planet Sci Lett 89:221–238

Watts AB (2001) Isostasy and flexure of the lithosphere. Cambridge University Press, Cambridge

Watts AB (2021) Isostasy. In: Gupta HK (ed) Encyclopedia of solid earth geophysics, 2nd edn. Springer, Cham, pp 831–847

Watts AB, Ryan WBF (1976) Flexure of the lithosphere and continental margin basins. Tectonophys 36:25–44

Watts AB, Torné M (1992) Crustal structure and the mechanical properties of extended continental lithosphere in the Valencia trough (western Mediterranean). J Geol Soc Lond 149:813–827

Wyer P, Watts AB (2006) Gravity anomalies and segmentation at the East Coast, USA continental margin. Geophys J Int 166:1015–1038

Zhang F, Lin J, Zhan W (2014) Variations in oceanic plate bending along the Mariana trench. Earth Planet Sci Lett 401:206–214

Zingerle P, Brockmann JM, Pail R, Gruber T, Willberg M (2019) The polar extended gravity field model TIM_R6e. GFZ Data Services. https://doi.org/10.5880/ICGEM.2019.005. Accessed 14 Oct 2020

Index

© Springer Nature Switzerland AG 2022
J. Kirby, *Spectral Methods for the Estimation of the Effective Elastic Thickness
of the Lithosphere*, Advances in Geophysical and Environmental Mechanics
and Mathematics, https://doi.org/10.1007/978-3-031-10861-7